Nanotechnology and Drug Delivery Innovations in Lung Cancer Treatment

This book examines the role of nanotechnology in transforming the diagnosis, treatment, and management of lung cancer. The initial chapters introduce lung cancer through its epidemiology, pathophysiology, and immunological mechanisms. The subsequent chapters emphasize the role of novel nanoparticles in early detection and imaging of lung cancer. The book then explores the potential of different nanomaterials, such as metal and metal oxide nanoparticles, nano micelles, carbon nanotubes, liposomes, polymeric nanoparticles, solid lipid nanoparticles, dendrimers, nanofibers, and quantum dots, in lung cancer treatment. Each chapter focuses on the unique properties from targeted drug delivery to improved bioavailability and reduced side effects. Exploring theragnostic nanoparticles, the book highlights their dual role in diagnosis and therapy, alongside nano-immunotherapy strategies for improved cancer treatment.

The book concludes with a critical review of ongoing clinical trials, offering insights into the future of nanotechnology-based lung cancer therapies and emerging trends in the field.

Key Features:

- Details the epidemiology and pathophysiology of lung cancer, covering disease prevalence, risk factors, and cellular mechanisms
- Examines the diagnostic capabilities of nanotechnology, emphasizing the use of nanoparticles in early detection and imaging
- Explores the therapeutic potential of various nanomaterials, such as liposomes, dendrimers, and quantum dots, for enhanced drug delivery and treatment
- Highlights theranostic nanoparticles, combining diagnostic and therapeutic functions for personalized cancer care
- Reviews clinical trials and emerging trends, providing insights into the future of nanotechnology-based lung cancer therapies

The book is intended for researchers of oncology, nanomedicine, pharmaceutical sciences, and biomedical engineering.

Nanotechnology and Drug Delivery Innovations in Lung Cancer Treatment

Edited by Parteek Prasher, Mousmee Sharma,
Indu Pal Kaur, Ronan MacLoughlin, Raimar Loebenberg,
Neal M. Davies, and Kamal Dua

CRC Press
Taylor & Francis Group
Boca Raton London New York

CRC Press is an imprint of the
Taylor & Francis Group, an **informa** business

Designed cover image: Shuttershock

First edition published 2026
by CRC Press
2385 NW Executive Center Drive, Suite 320, Boca Raton FL 33431

and by CRC Press
4 Park Square, Milton Park, Abingdon, Oxon, OX14 4RN

CRC Press is an imprint of Taylor & Francis Group, LLC

ISBN: 978-1-032-97861-1 (hbk)
ISBN: 978-1-032-97862-8 (pbk)
ISBN: 978-1-003-59583-0 (ebk)

DOI: 10.1201/9781003595830

Typeset in Times
by Apex CoVantage, LLC

Contents

Preface

Lung cancer continues to represent one of the most formidable challenges in global healthcare, accounting for the highest cancer-related mortality rates worldwide. Despite remarkable advances in surgery, chemotherapy, radiotherapy, immunotherapy, and targeted therapies, the prognosis for many patients remains poor due to late-stage diagnosis, limited treatment options, and the frequent emergence of drug resistance. These limitations underscore the pressing need for novel therapeutic strategies that not only extend survival but also enhance the quality of life for patients.

Nanotechnology has emerged as a transformative force in medicine, particularly in oncology. Its ability to enable targeted drug delivery, enhance therapeutic efficacy, overcome multidrug resistance, and support early detection positions it as a cornerstone of next-generation lung cancer management. This volume seeks to comprehensively explore the application of nanotechnology in lung cancer treatment, bridging fundamental research with translational and clinical perspectives.

The chapters included in this book traverse a carefully curated pathway: beginning with a foundation in the epidemiology, pathophysiology, and immunology of lung cancer; moving through detailed explorations of a wide range of nanostructures, including liposomes, micelles, dendrimers, quantum dots, solid lipid nanoparticles, and carbon nanotubes; and culminating in clinical trials, theranostics strategies, and future directions. Together, these chapters provide readers with an integrated resource that highlights both the opportunities and the challenges inherent in bringing nanotechnology-based drug delivery systems from the laboratory bench to the patient bedside.

This book is intended for a diverse audience, including graduate students, early-career researchers, clinicians, and industry professionals engaged in drug delivery, pharmaceutical sciences, biotechnology, nanomedicine, and oncology. It is our hope that the knowledge compiled herein will serve as both a reference and an inspiration, stimulating further research and innovation in the fight against lung cancer.

We, the editors, extend our sincere gratitude to all contributing authors for their scholarly work and to our academic and clinical colleagues whose pioneering efforts continue to illuminate new possibilities in lung cancer treatment. We also acknowledge the patients and families whose experiences motivate this research. It is with a vision for improved therapeutic outcomes and a future where lung cancer is no longer a leading cause of global mortality that we present this work.

The Editors

Parteek Prasher, Associate Professor, Department of Chemistry, University of Petroleum & Energy Studies, holds research experience of eight years in medicinal chemistry and biomaterials, while working in the field of anti-inflammatory and anticancer chemotherapy. The research team led by Dr. Prasher focuses on the development of rationally designed molecules for targeting various disorders, including inflammation, cancer, and antimicrobial MDR, which also includes the identification of novel pharmacophores, chemical moieties, and APIs for developing the potential chemotherapeutics. With several national/international collaborations, Dr. Prasher has published more than 100 peer-reviewed publications in journals of repute. Dr. Prasher has been a recipient of the prestigious Junior/Senior Research Fellowship by the Council of Scientific and Industrial Research, Government of India (2012–2015). He received the Lectureship award from the Government of India (2012). He is also a recipient of the Teachers Associateship for Research Fellowship (TARE) and research grant from the Science and Engineering Research Board of the Department of Science and Technology, Government of India for three years (December 2021 to December 2024). Dr. Prasher serves as the Associate Editor of the journal *Frontiers in Drug Delivery.*

Mousmee Sharma, Assistant Professor, Department of Chemistry, Uttaranchal University, has research experience of more than five years in physical chemistry and biomaterials, while working in the field of biomembranes. The recent research work of Dr. Mousmee is focused on the development of mucoadhesive materials for drug delivery across the physiological mucus barrier. Dr. Mousmee has published 50 research/review articles and 20 book chapters in journals of high repute. A gold medallist in her post-graduation, Dr. Mousmee recently received the "Recognized Reviewer" status from Elsevier and "Brand Ambassador" for the "Bentham Science Publishers" for the year 2021.

Indu Pal Kaur is presently Professor and Chairperson at the University Institute of Pharmaceutical Sciences (UIPS), Panjab University, Chandigarh, India. She was Director, Sophisticated Analytical Instrumentation Facility, Panjab University, from 2012 to 2015 and Dean—Faculty of Pharmaceutical Sciences for the period 2017–18. She has been Secretary, Indian Pharmaceutical Association, Punjab Branch, since 2017. She has been named amongst the top 2% of the most cited scientists across the globe as per Stanford University, USA (2020). Her research forte is enhancing the bioperformance of drugs, plant extracts/phytochemicals, and small/large biomolecules, namely siRNA and probiotics, using active-tailored delivery systems like solid lipid nanoparticles (SLNPs), nano vesicles, nano emulgel, beads, and lipid nanostructured herbal scaffolds and dressings. Apart from this, her group has worked extensively on novel platform technologies (SLNPs, nanovesicles, and Cyclodextrine complexes) to overcome challenges in ocular delivery of molecules like atorvastatin, bimatoprost, acetazolamide, ketoconazole, fluconazole, amphotericin B, timolol, and natamycin. She has been granted nine Indian patents and one US patent and has filed 19 patents in the past ten years. She has produced one post-doctorate, 25 PhD, and 55 post-graduate students and has received funding of approximately INR 100 million from government agencies and industrial consultancies. She is a US-Fulbright Fellow (2017–18) and was awarded Women Scientist Award 2018 by the Organisation of Pharmaceutical Producers of India (OPPI). She has delivered 50 invited talks and published more than 145 high-impact international publications. Her research work was funded twice (2017 and 2020) by DST and Science and Technology Facilities Council, ISIS, UK, for SANS characterization at the ISIS facility—Rutherford Appleton Laboratory, UK. She has received the BRIC Technology Exposition Award, consecutively for two years (2019, 2020), and the Researcher of the Year award 2019 from Sunpure Research Incubation Center, both for her work on solid lipid nanoparticles of curcumin. Her research work has been honoured by the Edward Kennedy Memorial Award

(2019) and the Mekaster Young Scientist Award (2018). Her work emphasizes industrial and clinical translation, reflected in her four technologies transfers to three Indian industries.

Ronan MacLoughlin is currently Associate Director of R&D, Science and Emerging Technologies at Aerogen Limited. Dr. MacLoughlin has more than 20 years' experience in respiratory drug delivery with several nebulizer and accessory product launches over that time. He has responsibility for new product development, establishing and exploiting the science underpinning respiratory drug delivery and identifying new and potentially disruptive emerging technologies. To this end, he has developed multiple technologies and products with several patents granted and pending that cover a range of drug, device, drug–device combination products, patient interventions, and patient interfaces. Dr. MacLoughlin currently serves as Chair of the Industry Representative Group in CURAM, the Science Foundation Ireland Center for the development of the next generation of smart medical devices, and sits on the board (previously chair) of the Medical and Engineering Technologies Center (MET), the Enterprise Ireland Technology Gateway. Additionally, he is currently chair of the Paediatric and Cystic Fibrosis working group within the International Society for Aerosols in Medicine (ISAM). Dr. MacLoughlin holds the position of adjunct Associate Professor in Trinity College Dublin (School of Pharmacy and Pharmaceutical Sciences) and Senior Honorary Lecturer in the Royal College of Surgeons, Ireland (School of Pharmacy & Biomolecular Sciences).

Raimar Loebenberg holds a BSc in pharmacy from the Johannes Gutenberg-University in Mainz, Germany. He received his PhD in pharmaceutics from the Johann Wolfgang Goethe University in Frankfurt in 1996. He joined the University of Alberta in 2000. His research interests are in biopharmaceutics, to predict the oral performance of drugs and botanicals and inhalable nanoparticles to treat lung diseases like cancer or tuberculosis. He is the founder and director of the Drug Development and Innovation Centre at the University of Alberta. He was president of the Canadian Society for Pharmaceutical Sciences in 2014 and 2015. He is a member of the United States Pharmacopeia Dietary Supplement Expert Committee. He is Vice Chair of the Specialty Committee of Traditional Chinese Medicine in Pharmaceutics of the World Foundation of Chinese Medicine Science. He is a member of the Health Canada Scientific Advisory Committee on Pharmaceutical Sciences and Clinical Pharmacology. He is a fellow of the Canadian Society for Pharmaceutical Sciences.

Neal M. Davies received his undergraduate degree in pharmacy from the University of Alberta in 1991. In 1996, he completed his PhD in pharmaceutical sciences, specializing in pharmacokinetics, at the University of Alberta. From 1995 to 1998, he undertook post-doctoral training in pharmacology and toxicology at the University of Calgary. He then joined the University of Sydney as a lecturer in 1998. In 2002, he returned to North America to join Washington State University's (WSU's) College of Pharmacy as an academic staff member, where he has held the positions of Director of the Pharmaceutical Sciences Graduate Program, Director of Professional and Undergraduate Research, and Director of the Summer Undergraduate Research Fellowship program. He has been active on numerous university-wide committees, including service as a member of WSU's Faculty Senate and as Chair of the Institutional Animal Care and Use Committee. From 2011 to 2016 he was Dean of the Faculty of Pharmacy at the University of Manitoba. He has been involved in several administrative and operational initiatives, including strategic planning, faculty accreditation review processes, and development of a first professional degree in pharmacy (PharmD). He has played key roles in faculty and graduate student recruitment and in the success and development of the Pharmaceutical Analysis Laboratory (PAL). Throughout his career, Dr. Davies has maintained the highest standards of teaching and research and his achievements in both areas have been recognized through numerous awards and honours, including awards from the United States Pharmacopeia and the Canadian Society of Pharmaceutical Sciences. He has taught at the undergraduate and graduate levels and has an ongoing and significant commitment towards graduate supervision and training.

He has been at the forefront of the use of technology in the classroom, has re-engineered curricular content, and has maintained high standards in student mentoring. He was inducted into the Rho Chi honours society for his outstanding teaching. Dr. Davies also maintains an active and highly collaborative externally funded research program, focused on pharmacokinetics and drug delivery as well as basic and clinical pharmaceutical sciences, which has attracted funding from the National Institutes of Health and the industry. He holds several patent applications and has disseminated his work broadly through numerous journal publications, many books and book chapters, abstracts, and invited conference presentations. He has approximately 20,000 citations and an H-index of ~70. Dr. Davies has been a registered pharmacist with the Alberta College of Pharmacists since 1991. He is active in numerous professional organizations/associations, including the Canadian Society of Pharmaceutical Sciences and the American Association of Pharmaceutical Scientists, and serves as a reviewer on the editorial board of several pharmaceutical and pharmacological science journals. Over the past decade Dr. Davies has been heavily involved in pharmacy education research, including curriculum mapping, interprofessional education, online teaching, and implementing expanded scopes of practice programs. On September 1, 2016, Dr. Davies commenced his term as Dean and Professor in the Faculty of Pharmacy & Pharmaceutical Sciences at the University of Alberta. He has also served as President of the Canadian Society for Pharmaceutical Sciences and the Canadian Pharmaceutical Sciences Foundation and is a fellow of the Canadian Academy of Health Sciences and Canadian Society for Pharmaceutical Sciences.

Kamal Dua is Associate Professor in the Discipline of Pharmacy at the University of Technology Sydney (UTS). Dr. Dua has been recognized and named as one of Australia's Top 250 researchers in 2025 and 2024 in Research Magazine by "The Australian" in the research discipline of Health and Medical sciences, and #1/research leader in the area of Pharmacology and Pharmacy (2025) and Toxicology (2024), respectively. With over 17 years of research experience, he has specialized in drug delivery targeting inflammatory diseases. Additionally, Dr. Dua also leads drug delivery research at the Woolcock Institute of Medical Research. In this role, he focuses on advancing targets identified in research projects to develop innovative formulations, taking the first steps toward clinical translation. Dr. Dua's research encompasses two complementary areas: drug delivery and immunology. His work explores how these disciplines can mutually benefit, contributing to the goal of promoting longer and healthier lives for the community. His commitment to synergy is reflected in his extensive publication record in reputable journals. Dr. Dua's research interests are centered on harnessing the pharmaceutical potential of modulating critical regulators, such as interleukins and microRNAs. He also specializes in developing new and effective drug delivery formulations for managing inflammation in chronic airway diseases and cancer.

1 Introduction to Lung Cancer

Rahamat Unissa Syed[a*], *S. Afsar*[b], *Jowaher Alanazi*[c], *and Muteb Alanazi*[d]

[a] Department of Pharmaceutics, College of Pharmacy, University of Hail, Hail, Saudi Arabia

[b] Department of Virology, Sri Venkateswara University, Tirupathi, Andhra Pradesh, India

[c] Department of Pharmacology and Toxicology, College of Pharmacy, University of Ha'il, Ha'il, Saudi Arabia

[d] Department of Clinical Pharmacy, College of Pharmacy, University of Ha'il, Ha'il, Saudi Arabia

*Corresponding author: Rahamat Unissa Syed.

1.1 INTRODUCTION TO LUNG CANCER

1.1.1 Overview of Lung Cancer

The two main types of lung cancer are small cell lung cancer (SCLC) and non–small cell lung cancer (NSCLC); these constitute the malignant diseases originating inside the lungs. NSCLC occurs in 85% of lung cancer patients; however, SCLC remains rare but develops rapidly with poor prognosis [1]. Lung cancer constitutes the leading cause of cancer mortality worldwide as it leads to approximately 1.8 million deaths yearly, thus demonstrating its intense impact on public health. The prevalence of the condition shows regional differences because countries which maintain high pollution levels and smoking rates experience elevated incidence figures [2–4]. Lung cancer develops primarily from smoking behaviour since this activity leads to this disease in approximately 85% of patients. The development of the illness is affected by heredity, which works together with family medical background and environmental factors such as radon pollution, asbestos, and air pollutants. Persons who carry inherited genetic abnormalities are prone to developing lung cancer even if they abstain from smoking [5, 6]. The abnormal DNA in lung cells creates pathophysiological conditions leading to uncontrolled cell growth and immune system avoidance plus metastatic spreading. The classification of lung cancer occurs through tumour behaviour evaluation together with histological analysis. Three subtypes of NSCLC exist: squamous cell carcinoma, adenocarcinoma, and large cell carcinoma, which exhibit different biological traits that result in various treatment responses [7–10]. SCLC spreads quickly to different parts of the body due to its lower differentiation state. Researchers have identified distinct genetic changes that stand out during NSCLC evaluations, especially mutations of epidermal growth factor receptor (EGFR) and the occurrence of anaplastic lymphoma kinase (ALK) rearrangements. Early-stage detection of the disease meets significant challenges since the condition rarely presents symptoms during its initial development [11–14]. Patients survive at low rates mainly because their diagnosis occurs too late into their condition but advancements in biomarker-based diagnoses and customized therapy and imaging techniques have shown positive outcomes. Resolving this dangerous medical condition requires deep knowledge about lung cancer risk elements and classification schemes and disease progression for designing personalized therapies and developing better diagnostic methods alongside prevention strategies [15–17].

DOI: 10.1201/9781003595830-1

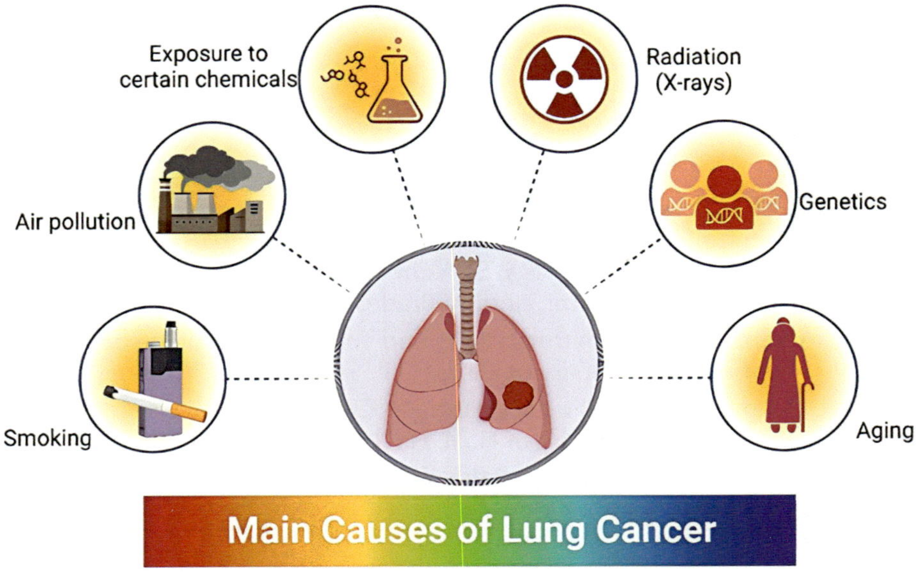

FIGURE 1.1 The leading contributors to lung cancer

1.2 ETIOLOGY AND PATHOGENESIS

The processes that lead to lung cancer come from three main influences: genetic causes, environmental elements, and molecular factors. Genetic abnormalities that enable cells to proliferate without control and survive and scatter beyond their original location include mutations that affect oncogenes EGFR, KRAS, and ALK [18–21]. Such mutations function as oncogenic drivers to convert normal lung cells into cancerous ones. People continue to develop lung cancer due to smoking because carcinogenic substances including nitrosamines and polycyclic aromatic hydrocarbons present in tobacco break DNA, thus initiating the carcinogenic process [22]. Lung cancer represents an untreatable condition that becomes more likely due to occupational radon exposure and occupational asbestos exposure and occupational exposure to industrial toxins as well as air pollution exposure. Lung carcinogenesis receives significant impact from inflammatory pathways because persistent inflammatory responses typically originating from long-term smoking or environmental exposure create gene abnormalities combined with changes to the lung tissue microenvironment [23–25]. The inflammatory process produces three harmful substances, which are growth factors and pro-inflammatory cytokines and reactive oxygen species that promote tumour growth acceleration. The tumour microenvironment significantly impacts disease development since it contains fibroblasts together with immune cells and cancer cells along with extracellular matrix [26]. Both fibroblasts together with elements in the extracellular matrix create conditions which enable tumour growth and metastasis progression while cancer cells work to avoid detection by immune cells. The tumour becomes better suited to grow through these tissue interactions as well as spreads into adjacent areas while building resistance against treatment [27–29]. Various factors starting with genetic changes and extending to environmental exposures and inflammatory reactions and the dynamic interactions within the tumour environment control both progression and development and treatment resistance of lung cancer [30].

1.3 MOLECULAR AND CELLULAR MECHANISMS

Complex cellular and molecular processes within the lung drive tumour growth along with the spread of metastasis and survival of malignant cells. Three fundamental characteristics of lung cancer include resistance to cell death, evading growth suppressors, and persistent signalling for cellular proliferation [31–34]. Tumour cells develop continuous cellular signalling mechanisms to trigger division through pathway activation, which includes receptors for EGFR and KRAS together with growth factor receptors. Lung cancer develops the ability to bypass p53 pathway-based tumour-suppressor systems thereby permitting its survival through uncontrolled cell expansion [35]. Among all fundamental lung cancer traits stands its ability to defend against apoptosis because tumour cells acquire death signal avoidance capabilities to preserve themselves while their chromosomes remain variable. The three classes of non-coding RNAs including miRNAs, lncRNAs, and circRNAs maintain the progression of lung cancer by managing important genes essential for cellular processes like proliferation and apoptosis and metastasis [36–38]. The cellular impact of cancer cells and their metastasis can be modified through chromatin and mRNA interactions carried out by the lesser known lncRNAs and circRNAs. Drug resistance together with lung cancer expansion depends heavily on modifications taking place at the epigenetic level including DNA methylation and histone modification [39]. The changes in histone structure affect chromatin structure while abnormal DNA methylation causes tumour suppressor genes to become inactivated. The medicinal impact of epigenetic changes makes them essential targets for new therapeutic development because they lead to drug resistance. Lung cancer development occurs mostly because high levels of reactive oxygen species (ROS) create DNA damage, leading to gene mutations that promote genomic instability [40–42]. ROS creates two harmful effects on cancer development through NF-κB pathway activation and other pro-tumorigenic signalling pathways, which leads to increased tumour development as well as inflammation and therapeutic resistance. The treatment of lung cancer requires specific interventions which aim at oxidative stress pathways together with ncRNAs and epigenetic changes [43].

1.4 HISTOLOGICAL AND MOLECULAR SUBTYPES

NSCLC and SCLC represent the principal forms of lung cancer whereas these varieties exhibit different histological patterns and molecular profiles. NSCLC represents 85% of total lung cancer incidents; therefore, it surpasses all other forms of this disease. NSCLC consists of three primary subtypes: squamous cell carcinoma, big cell carcinoma, and adenocarcinoma [44, 45]. Most occurrences of lung cancer start in the peripheral tissue where adenocarcinoma emerges as the leading subtype primarily affecting individuals who do not smoke. ALK, EGFR, and KRAS are important genes that play a central role in tumour growth along with treatment resistance and show defining characteristics as mutations in these genes [46]. When smoking causes squamous cell carcinoma, it develops in central airways while showing prominent keratinized patterns under histological examination. The probability of this subtype recurring is high but radiotherapy shows better effectiveness in treating it. Large cell carcinoma presents an unfavourable treatment situation because this improperly differentiated cancer type shows no identifiable features of NSCLC subtypes and usually emerges during advanced stages of disease [47–50]. Lung cancer consists of 15% SCLC cases because this subtype spreads aggressively while growing rapidly and producing quick metastasis. Neuroendocrine attributes combined with extensive tumour mass at initial diagnosis make SCLC treatment difficult because of its association with smoking behaviour [51]. The treatment outcomes remain poor for SCLC because this cancer type commonly develops resistance upon initial response to chemotherapy and radiation while leading to treatment relapse. Additional challenges exist for clinical management of two rare types of lung cancer, which include pulmonary sarcomatoid carcinoma and carcinoid tumours alongside the main subtypes [52]. The development pace of carcinoid

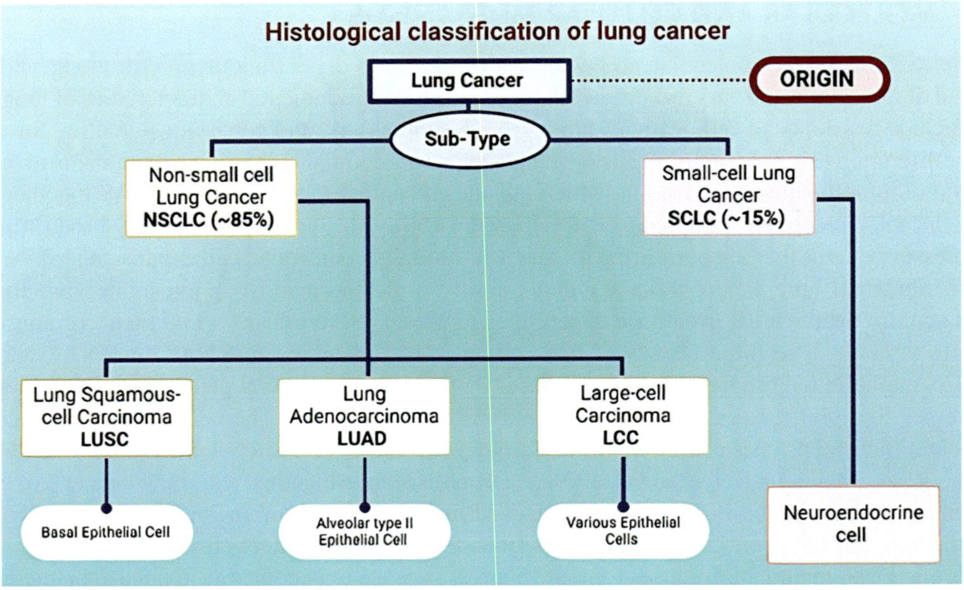

FIGURE 1.2 Lung cancer histological categories shown through two primary groups, whereby SCLC makes up 15% of cases while NSCLC represents approximately 85%. LUSC together with LUAD, and LCC represent three distinct subtypes of NSCLC that originate from different epithelial cells. Scientists confirm that neuroendocrine cells form the origin of SCLC.

tumours is slow although they resist spread; yet the resistant standard treatments are common for sarcomatoid carcinoma because of its aggressive behaviour [53]. The increasing importance of targeted therapy and immunologic approaches requires clinicians to use molecular and histological characteristics when developing treatment strategies for common and rare types of lung cancer. Tailored treatments require this knowledge because it lets healthcare providers improve lung cancer patients' disease course and life expectancy [54].

1.5 DIAGNOSIS AND STAGING

The procedure of diagnosing and staging lung cancer helps doctors make prognosis assessments as well as select appropriate treatments. LDCT serves as the optimal imaging examination for testing patients with chronic smoking history who face the highest risk. Identifying lung cancer at an early stage serves as a critical key to higher survival possibilities [55]. Biomarkers continue to become more crucial for conducting non-invasive early diagnosis in medical fields. The active markers consist of protein markers alongside microRNA and circulating tumour DNA (ctDNA). Lung cancer treatment plans and disease assessment depend on a standardized framework called TNM staging, which examines tumour size (T) together with nodal (N) stage and metastasis (M) [56]. Advanced lung cancer stages III and IV require systemic treatments including chemotherapy and targeted therapy and immunotherapy; yet surgery and localized therapies are effective methods mostly used to treat early-stage lung cancers Stages I and II. The medical field has recognized liquid biopsy as a revolutionary procedure which replaces conventional tissue biopsies for both lung cancer diagnosis and monitoring purposes [57]. The method allows researchers to find ctDNA as well as exosomal RNA and diverse molecular markers within blood specimens. Based on real-time visualization of patients' tumour behaviour as well as their response to treatment, the technique enhances targeted treatment strategies [58]. Next-generation sequencing (NGS)- and artificial intelligence (AI)-assisted imaging improve diagnosis accuracy while enhancing operational efficiency in lung cancer screening in conjunction with novel breath-based volatile organic compound (VOC) analysis. The detailed genetic analysis of NGS allows healthcare providers to personalize medication selection in addition to radiomics software, which extracts hidden imaging signals of cancer [59, 60]. The medical field sees a transformation in lung cancer management because these innovations help doctors identify cancer early while creating precise treatment designs which will enhance patient results. The progression of scientific discovery will enable clinical practitioners to integrate these modern diagnostic technologies, which enhances lung cancer treatment methods while improving diagnostic precision [61].

1.6 CURRENT TREATMENT STRATEGIES

Health status and disease subtype and staging determine which treatments will be used to manage lung cancer, including surgery and chemotherapy along with radiation and multi-treatment plans. Surgery remains the optimal treatment method when lung cancer starts at an early stage as NSCLC [62]. Surgical treatments such as lobectomy and pneumonectomy together with video-assisted thoracoscopic surgery (VATS) provide different recovery benefits along with varying degrees of surgical complexity. The determination of surgical candidates happens through tumour measurement and lymph node examination and lung function tests and VATS offers an alternative method to minimize post-operative problems [63]. The main treatment for advanced lung cancer involves chemotherapy, where cisplatin together with pemetrexed or paclitaxel constitutes the precious metal standard along with carboplatin. New treatment applications of immuno-checkpoint inhibitors along with targeted agents have reduced side effects while improving treatment success rates [64]. EBRT represents the traditional therapy, whereas SBRT provides precise delivery of radiation for early-stage inoperable tumours leading to effective treatment with minimized tissue damage both during curative and palliative care. The combination of immunotherapy along with molecular targeting drugs has expanded survival rates so that local advanced and metastatic lung cancer now often receives multidisciplinary treatments including surgery together with chemotherapy and radiation therapy [65]. The treatment of advanced NSCLC experienced an important transformation because of immune checkpoint drugs and chemotherapy that attack the PD-1/PD-L1 signalling systems. Genetic profiles of patients should be utilized as part of tailor-made treatment plans so as to provide better therapeutic efficacy; these plans in particular, need to identify the mutations of EGFR, ALK, and KRAS. The transformation throughout lung cancer therapeutic developments using precise medicine alongside new treatment approaches has been a successful sign for better patient prognostics and extended survival time [66, 67].

1.7 TARGETED THERAPY AND PRECISION MEDICINE

The development of lung cancer treatment resorts to precision medicine practices to find genetic tumour activators that yield both risk-reduced and performance-enhanced therapeutic approaches. Two principal mutations which occur in NSCLC cells turn on both EGFR and ALK and Kirsten rat sarcoma virus oncogene (KRAS) [68]. Lung cancer patients with EGFR-mutant cells respond much better to osimertinib treatment and survive longer while ALK-positive patients achieve higher success rates through alectinib treatment. The pharmaceutical agent sotorasib is available for treating KRAS G12C-mutant lung cancer patients by specifically targeting KRAS mutations despite their initial classification as incurable [69, 70]. Bevacizumab serves as an important precision medicine strategy along with angiogenesis inhibitors. These medications block vascular endothelial growth factor (VEGF) signalling to produce reduced blood supply for tumours and delays their growth. Targeted treatment resistance happens to patients during their therapeutic journey despite early positive results presenting a considerable challenge to clinicians [71, 72]. Novel treatment strategies must be developed because patients develop resistance through phenotypic changes and bypass signalling mechanisms together with secondary genetic variations. The development of next-generation TKI drugs represents an approach to manage resistance mutations in patients with ALK-rearranged lung cancer through the use of lorlatinib [73, 74]. Combined treatment of immune checkpoint inhibitors with both anti-angiogenic drugs and targeted inhibitors creates more effective treatment outcomes that fight drug resistance. The advancement of genetic profiling technology together with liquid biopsy tests allows healthcare providers to enact new treatments during resistance mutation-free timeframes. The expansion of precision medicine occurs when new inhibitors and molecular targets emerge because of research advances, which gives lung cancer patients better personalized treatments options [75, 76].

TABLE 1.1
A Breakdown of Primary Cellular Models and Lung Cancer Processes

Cellular/Model	Mechanism	Signalling Pathway	Therapeutic Potential	Clinical Implementation	Reference
NSCLC (adenocarcinoma model)	Driven by genetic alterations (e.g. EGFR, KRAS, ALK mutations) leading to uncontrolled cell proliferation and evasion of apoptosis	EGFR, KRAS, PI3K/AKT, MAPK	Targeted therapies using tyrosine kinase inhibitors (TKIs) (e.g. osimertinib for EGFR-mutant cases); potential for combination with immunotherapy	Widely adopted targeted therapy regimens based on molecular profiling in clinical settings	[77]
SCLC model	Characterized by rapid proliferation, neuroendocrine differentiation, and early metastasis driven by complex genetic and epigenetic changes	Notch, MYC, and neuroendocrine pathways	Standard chemotherapy and radiotherapy with emerging immunotherapy strategies; potential for novel agents addressing high tumour burden and resistance	Platinum-based chemotherapy and radiation are standard; immunotherapeutic approaches under investigation	[78]
Liquid biopsy biomarkers	Involve shedding of circulating tumour cells (CTCs), exosomal miRNAs, and ctDNA from tumours into the bloodstream, reflecting tumour dynamics and mutations	Reflects underlying mutation pathways (e.g. EGFR, ALK)	Early detection, monitoring therapeutic response, and identifying minimal residual disease; supports personalized treatment adjustments	Increasing integration into precision oncology workflows for non-invasive, real-time tumour profiling	[79, 80]
Tumour microenvironment model	Complex interplay between cancer cells, immune cells, fibroblasts, and extracellular matrix components, leading to immune evasion and metastasis	NF-κB, PD-1/ PD-L1, TGF-β	Immunotherapies (e.g. checkpoint inhibitors) and anti-angiogenic agents to disrupt supportive niches; potential for combination strategies to overcome resistance	Immune checkpoint inhibitors (pembrolizumab, nivolumab) are in clinical use; novel combination approaches are in clinical trials	[81, 82]

TABLE 1.1 *(Continued)*

A Breakdown of Primary Cellular Models and Lung Cancer Processes

Cellular/Model	Mechanism	Signalling Pathway	Therapeutic Potential	Clinical Implementation	Reference
Non-coding RNA (miRNAs, lncRNAs, circRNAs)	Modulate gene expression by targeting oncogenes and tumour suppressors, influencing cell proliferation, apoptosis, and metastatic behaviour	Indirect modulation of pathways (p53, Wnt, MAPK)	Emerging as diagnostic biomarkers and potential therapeutic targets; ability to modulate multiple oncogenic pathways simultaneously	Under clinical investigation; several miRNA-based therapies and diagnostics are in early-phase clinical trials	[83]
Stem cell therapy models	Utilize mesenchymal stem cells (MSCs), induced pluripotent stem cells (iPSCs), and lung progenitor cells to repair damaged lung tissue and modulate inflammation	Involvement of growth factors and cytokine signalling (e.g. VEGF, TGF-β)	Promising approach for lung regeneration, repair after injury (e.g. post-cancer treatment, COPD); potential to restore pulmonary function and reduce inflammation	Early-phase clinical trials investigating safety and efficacy in regenerative lung medicine	[84]
AI-driven computational models	Leverage machine learning and big data analytics to analyse imaging, genomic, and clinical datasets for pattern recognition and outcome prediction	N/A (computational approach)	Enhance early diagnosis, prognostic predictions, and personalized treatment planning; accelerate drug discovery by identifying novel targets through data integration	Increasingly integrated into diagnostic imaging, genomic profiling, and clinical decision-support systems	[85]
Precision medicine models	Involve comprehensive genomic profiling to identify actionable mutations and tailor treatment strategies based on individual molecular signatures	Variable (e.g. EGFR, ALK, KRAS, etc.)	Enable the selection of targeted therapies and combination regimens tailored to patient-specific genetic profiles; potential for improved efficacy and reduced toxicity	Routinely employed in personalized treatment regimens for lung cancer with ongoing advancements in molecular diagnostics and targeted therapy development	[86]

1.8 IMMUNOTHERAPY IN LUNG CANCER

Evidence shows that immuno-checkpoint inhibitors have emerged as optimal immunotherapy agents that utilize immune system mechanisms to combat and eliminate tumour cells in lung cancer patients. Cancer cells evade immune responses through critical immunological checkpoints which comprise PD-1 protein and its PD-L1 ligand [87, 88]. Nivolumab and pembrolizumab medications function by eliminating pathways, which makes T-cells more active and longer lasting for treating NSCLC patients. People who exhibit high levels of PD-L1 can benefit from ICIs, which show effective tumour response whether given independently or alongside chemotherapy [89]. Prediction markers and combination therapies need development because acquired resistance and primary resistance prove to be ongoing issues. The medical community continues to study chimeric antigen receptor (CAR) T-cell therapy for treating lung cancer because it shows promising results against hematologic malignancies [90]. The treatment of solid tumours including lung cancer by CAR T-cells remains limited because of tumour heterogeneity along with restricted microenvironment and serious immune-related toxicity potential during antigen-specific T-cell training. The effectiveness of CAR T-cells receives enhancement attempts from researchers through combination therapies involving checkpoint inhibitors and co-stimulatory signalling and cytokine medications and cell penetration methods [91, 92]. Tumour-infiltrating lymphocytes (TILs) represent a fundamental part of lung cancer immunotherapy because these cells directly produce the elements inside the tumour microenvironment. A tumour environment suppressing the immune system that includes regulatory T cells and myeloid-derived suppressor cells worsens resistance yet strong CD8+ cytotoxic T cell presence and extensive TIL numbers improve both treatment response and patient outcomes [93]. Scientists are investigating different methods for enhancing TIL activity because it improves patient responses towards immunotherapy treatment. Research shows that combining TIL-based procedures and CAR T-cell treatment with ICIs and unique combination strategies have the potential to defeat resistance mechanisms and extend the survival time for lung cancer patients [94].

1.9 LUNG CANCER BIOMARKERS AND THEIR CLINICAL RELEVANCE

The advancement of precision oncology depends heavily on biomarkers of lung cancer, which deliver non-invasive assessment tools for theragnostic process and clinical disease monitoring and cancer detection. Among the promising biomarkers which provide real-time tumour monitoring are circulating tumour cells (CTCs) and exosomal microRNAs (miRNAs) [95]. CTCs play a crucial role in prognosis because higher counts worsen patient outcomes while these cells release from primary or metastatic conditions and circulate through blood. Medical research continues to explore exosomal miRNAs contained in extracellular vesicles because these small non-coding RNAs influence gene expression during tumour development and aid immunological evasion and medication resistance [96, 97]. Lung cancer diagnosis and treatment possibilities have completely altered with liquid biopsy because it provides minimally invasive extracellular vesicle and CTC and ctDNA testing from blood samples. Standard tissue biopsy procedures get replaced by liquid biopsy to enable ongoing tumour monitoring while detecting early return of disease and actionable mutations such as EGFR, ALK, and KRAS changes which guide selection of appropriate targeted medicines [98]. The therapeutic value of liquid biopsy is extended for both treatment assessment and recurrence prediction because it detects tiny residual disease within the body. The strong presence of tumour heterogeneity remains an important barrier against biomarker-based therapy methods. More advanced multi-biomarker approaches in medicine customization become essential because intra-tumoral genetic variations lead to resistant clone development that needs better treatment plan management [99]. The current limitations of tumour heterogeneity research will eventually enable the combination of liquid biopsy with single-cell sequencing and AI-based biomarker analysis to improve precision medicine strategies. The incorporation of biomarkers enables hospitals to create better lung cancer treatments because they develop detection methods that are quicker while monitoring patients continuously with individualized drugs that boost survival rates and quality of life [100, 101].

1.10 LUNG CANCER AND COMORBIDITIES

Multiple health conditions that occur together with lung cancer influence both disease progression and therapeutic outcomes as well as patient survival rates. The parallel risk elements of smoking and chronic inflammation between these conditions make chronic obstructive pulmonary disease (COPD) the most dangerous comorbidity [102]. The developing lung cancer benefits from epithelial-mesenchymal transition and oxidative stress and structural lung damage which result from persistent COPD-related airway inflammation. The tumour environment becomes complicated because COPD-induced remodelling of airways allows cancer development while hindering treatment strategies [5]. Patients with lung cancer experience greater cardiovascular disorders because their endothelial function and systemic inflammation become worse along with pulmonary difficulties. This health risk gets more dangerous when such common risk elements as smoking and high blood pressure and dyslipidaemia exist together [103]. The treatment of lung cancer through chemotherapy and radiation therapy leads to cardiotoxicity that intensifies heart function deterioration. Medical conditions involving inflammation along with clotting problems both elevate the danger of serious blood clots that badly affect life expectancy and patient quality of life. Lung tumour development becomes much harder due to metabolic conditions including diabetes along with dyslipidaemia and obesity which disrupt essential metabolic operations [104]. Insulin resistance together with high blood sugar drive tumour-promoting cell signalling pathways inside the body but obesity leads to persistent low-grade inflammation that increases tumour growth via cytokine activity. The growth and metastasis of cancer cells are associated with the dyslipidaemia-caused dysregulation of lipid metabolism [105, 106]. Metabolic abnormalities that affect medication absorption together with therapeutic response make both disease progression faster and treatment less effective because insulin resistance causes metabolic changes. The relationship between various medical conditions and lung cancer calls for complete multidisciplinary patient treatment strategies thus becoming essential to care delivery [107]. The treatment outcomes and patient quality of life for those with lung cancer need improvement through enhanced cardiovascular tests combined with anti-inflammatory drug treatments and metabolic therapy plans. Research exploring lung cancer along with its linked diseases supports the development of specialized therapy programs, which include heart disease protection medications alongside inflammatory medications and metabolic control medicines for better disease outcome [108, 109].

1.11 ADVANCES IN REGENERATIVE AND PERSONALIZED MEDICINE

Medical advances in both regenerative and personalized medicine have led to a revolution in lung disease therapy that provides the chance for enhanced treatments alongside improved patient recovery rates. The application of stem cell regenerative medicine demonstrates high potential for treating damaged lungs through the utilization of MSCs, iPSCs and lung progenitor cells [110]. Stem cells are utilized for lung tissue repair after COPD and idiopathic pulmonary fibrosis as well as in patients who have suffered lung injury due to cancer therapy. The therapeutic capability of iPSCs includes functional lung cell manufacture, whereas MSCs promote tissue healing through their secretion of anti-inflammatory biochemicals that enhance tissue regeneration [111, 112]. Researchers study lung progenitor cells because of their ability to fix alveolar structural damage in lungs while improving gas transport functions. The creation of organoids using both extracellular matrix-based models and bioengineered lung scaffolds takes place at once to support potential lung healing and to allow advanced tissue engineering and 3D bioprinting applications [113]. Lung tissue production in vitro laboratories has become possible through recent developments for future use in medical fields including personalized treatment and drug test methods and lung transplant possibilities. Scientists would gain better disease change comprehension and treatment accuracy by printing detailed lung structures and building models which mimic human lung conditions [114]. Future treatment of lung cancer will undergo complete transformation through personalized medicine when medical professionals base care plans on genetic and molecular patient profiles. Medical developers can produce specific treatments for tumours by using genomic profiling methods to determine specific mutations including EGFR, ALK, and KRAS in patients' cancer [115]. AI technology enables precision medicine to provide better treatment plans through its evaluation of vast datasets that enables assessment of best therapeutic choices and outcome monitoring. The individualized treatment methods help doctors select the most appropriate drugs for patients, which leads to more effective lung cancer therapy while reducing patients' adverse effects [116, 117]. Changes in lung disease treatment will arise from three major developments: stem cell therapy in combination with bioprinting and personalized medicine. These approaches enable specific patient-centred solutions and quality medical procedures and tissue regeneration [118].

1.12 PALLIATIVE CARE AND QUALITY OF LIFE

Patients who have advanced lung cancer obtain significant advantages from palliative care treatment methods. The core element of advanced lung cancer treatment consists of controlling symptoms such as pain and dyspnoea and fatigue along with anorexia. Non-pharmacological treatments such as oxygen therapy and physical therapy as well as nutritional assistance combine with pharmaceutical drugs such as opioids for pain relief and corticosteroids for inflammation to decrease discomfort [119]. Each symptom-relieving approach targets individual patient needs in order to achieve both comfort and dignity. The mental well-being and supportive needs within lung cancer therapy maintain an equivalence with treatment of bodily symptoms. Managing the terminal nature of their disease alongside anxiety and despondence and coping with its emotional burdens represent major emotional challenges faced by patients alongside their relatives [120]. Medical services through counselling and psychotherapy enable patients to manage psychological barriers. Patients achieve better emotional health results together with improved coping ability when healthcare professionals join with families to establish open dialogue for addressing psychological requirements. Palliative therapy needs hospice and supportive care services at all times but particularly when lung cancer reaches its advanced stage [121, 122]. The care approaches provide patients with a complete level of care and unite curative treatments with palliative interventions to preserve their comfort and dignity along their journey. Hospice care as an end-of-life support system focuses on essential aspects including pain control and providing comfort to families and offering compassionate treatment in the last life stages. This complete treatment method prioritizes attending to the physical, psychological, and social needs of patients and their families to improve life quality when facing challenging times [123].

1.13 FUTURE DIRECTIONS AND CHALLENGES

Research and treatment of lung cancer will prioritize the advancement of therapeutic strategies as well as the application of artificial intelligence technologies while running international disease prevention programs. The discovery of new therapeutic targets creates prospects to develop specific treatments with better efficiency. Researchers focus on cancer stem cells and selected genetic faults such as EGFR and ALK and KRAS and the tumour environment. Scientists dedicate their studies to these regions because they aim to create personalized therapeutic approaches which decrease adverse effects and improve survival statistics. The medical industry has initiated scientific studies of three new treatment approaches: body immune system enhancement through immunotherapies, targeted treatments for molecular irregularities, and the combination of various therapeutic modalities. The literature on lung cancer shows the most promising research involves finding treatments which combat resistance mechanisms together with heterogeneity issues. AI technologies revolutionize both medical procedures and lung cancer research operations. AI revolutionizes prognosis prediction and early diagnosis and new medicinal medicine development because AI handles massive information effectively. Researchers can now use machine learning algorithms to analyse massive datasets to accelerate drug development, identify new therapeutic biomarkers, and refine medical imaging. AI-powered precision medicines enable developers to create bespoke medications based on individual genetic and molecular data, increasing therapeutic efficacy. The fight to prevent lung cancer remains essential worldwide because new treatments and technological developments persist. The frequency of lung cancer cases lowers through smoking cessation programs because they react directly with the underlying lung cancer origin. Lung cancer risks decrease through active preventive measures, including disease-related immunization campaigns and high-risk screening initiatives. Long-term prevention relies heavily on broad-based public health activities that both teach people about wellness practices and motivate people to select healthier ways of life. Healthcare organizations must collaborate with government bodies and research facilities to expand medical access and create unified initiatives and deliver vital prevention services to all global vulnerable communities. Although research on lung cancer therapy keeps advancing, there are ongoing difficulties to resolve regarding late diagnosis and drug resistance alongside unequal access to new drugs. Collaboration among research institutions enables innovative treatment targets, computer-assisted diagnosis integration, and preventive programs to provide hope for lung cancer patient health improvement worldwide.

REFERENCES

1. Molina, J.R., et al., Non-small cell lung cancer: epidemiology, risk factors, treatment, and survivorship. *Mayo Clin Proc*, 2008. **83**(5): p. 584–594.
2. Zhou, J., et al., Global burden of lung cancer in 2022 and projections to 2050: incidence and mortality estimates from GLOBOCAN. *Cancer Epidemiol*, 2024. **93**: p. 102693.
3. Alharbi, K.S., et al., PI3K/Akt/mTOR pathways inhibitors with potential prospects in non-small-cell lung cancer. *J Environ Pathol Toxicol Oncol*, 2022. **41**(4): p. 85–102.
4. Bhat, A.A., et al., MALAT1: a key regulator in lung cancer pathogenesis and therapeutic targeting. *Pathol Res Pract*, 2024. **253**: p. 154991.
5. Parris, B.A., et al., Chronic obstructive pulmonary disease (COPD) and lung cancer: common pathways for pathogenesis. *J Thorac Dis*, 2019. **11**(Suppl 17): p. S2155–S2172.
6. Majewski, S. and W.J. Piotrowski, Air pollution: an overlooked risk factor for idiopathic pulmonary fibrosis. *J Clin Med*, 2020. **10**(1).
7. Massion, P.P. and D.P. Carbone, The molecular basis of lung cancer: molecular abnormalities and therapeutic implications. *Respir Res*, 2003. **4**(1): p. 12.
8. Dorantes-Heredia, R., J.M. Ruiz-Morales, and F. Cano-García, Histopathological transformation to small-cell lung carcinoma in non-small cell lung carcinoma tumors. *Transl Lung Cancer Res*, 2016. **5**(4): p. 401–412.
9. Bhat, A.A., et al., Polysaccharide-based nanomedicines targeting lung cancer. *Pharmaceutics*, 2022. **14**(12).
10. Chan, Y., et al., Sugar-based nanoparticles for respiratory diseases: a new paradigm in the nanoworld. *Future Med Chem*, 2020. **12**(21): p. 1887–1890.
11. Semenova, E.A., R. Nagel, and A. Berns, Origins, genetic landscape, and emerging therapies of small cell lung cancer. *Genes Dev*, 2015. **29**(14): p. 1447–1462.
12. Yoneda, K., et al., Treatment of non-small cell lung cancer with EGFR-mutations. *J UOEH*, 2019. **41**(2): p. 153–163.
13. Chellappan, D.K., et al., Targeting the mitochondria in chronic respiratory diseases. *Mitochondrion*, 2022. **67**: p. 15–37.
14. Dahiya, R., et al., Harnessing pyroptosis for lung cancer therapy: the impact of NLRP3 inflammasome activation. *Pathol Res Pract*, 2024. **260**: p. 155444.
15. Kumar, P., S. Gupta, and B.C. Das, Saliva as a potential non-invasive liquid biopsy for early and easy diagnosis/prognosis of head and neck cancer. *Transl Oncol*, 2024. **40**: p. 101827.
16. Gupta, G., et al., Tumor suppressor role of miR-503. *Panminerva Med*, 2018. **60**(1): p. 17–24.
17. Hakami, M.A., et al., PVT1 lncRNA in lung cancer: a key player in tumorigenesis and therapeutic opportunities. *Pathol Res Pract*, 2024. **253**: p. 155019.

18. Ramazi, S., et al., Epigenetic regulation in lung cancer. *MedComm (2020)*, 2023. **4**(6): p. e401.
19. Hecht, S.S., Lung carcinogenesis by tobacco smoke. *Int J Cancer*, 2012. **131**(12): p. 2724–2732.
20. Gupta, G., et al., A clinical update on metformin and lung cancer in diabetic patients. *Panminerva Med*, 2018. **60**(2): p. 70–75.
21. Hussain, M.S., et al., Long non-coding RNAs in lung cancer: unraveling the molecular modulators of MAPK signaling. *Pathol Res Pract*, 2023. **249**: p. 154738.
22. Holme, J.A., et al., Lung cancer associated with combustion particles and fine particulate matter (PM2.5): the roles of polycyclic aromatic hydrocarbons (PAHs) and the aryl hydrocarbon receptor (AhR). *Biochem Pharmacol*, 2023. **216**: p. 115801.
23. Shankar, A., et al., Environmental and occupational determinants of lung cancer. *Transl Lung Cancer Res*, 2019. **8**(Suppl 1): p. S31–S49.
24. Hussain, M.S., et al., Unraveling NEAT1's complex role in lung cancer biology: a comprehensive review. *EXCLI J*, 2024. **23**: p. 34–52.
25. Jha, S.K., et al., Cellular senescence in lung cancer: molecular mechanisms and therapeutic interventions. *Ageing Res Rev*, 2024. **97**: p. 102315.
26. Tripathi, S., Y. Sharma, and D. Kumar, Unveiling the link between chronic inflammation and cancer. *Metabolism Open*, 2025. **25**: p. 100347.
27. Caligiuri, G. and D.A. Tuveson, Activated fibroblasts in cancer: perspectives and challenges. *Cancer Cell*, 2023. **41**(3): p. 434–449.
28. Hussain, M.S., et al., From carcinogenesis to therapeutic avenues: lncRNAs and mTOR crosstalk in lung cancer. *Pathol Res Pract*, 2024. **253**: p. 155015.
29. Rohilla, S., et al., Recent developments and challenges in molecular-targeted therapy of non-small-cell lung cancer. *J Environ Pathol Toxicol Oncol*, 2023. **42**(1): p. 27–50.
30. Greten, F.R. and S.I. Grivennikov, Inflammation and cancer: triggers, mechanisms, and consequences. *Immunity*, 2019. **51**(1): p. 27–41.
31. Larsen, J.E. and J.D. Minna, Molecular biology of lung cancer: clinical implications. *Clin Chest Med*, 2011. **32**(4): p. 703–740.
32. Xie, S., et al., The metastasizing mechanisms of lung cancer: recent advances and therapeutic challenges. *Biomed Pharmacother*, 2021. **138**: p. 111450.
33. Hussain, M.S., et al., Groundbreaking mRNA lung cancer vaccine trials: a new dawn in cancer treatment. *Curr Cancer Drug Targets*, 2025. **25**.
34. Sadique Hussain, M., et al., Exosomal ncRNAs in liquid biopsies for lung cancer. *Clin Chim Acta*, 2025. **565**: p. 119983.
35. Ferro, A., et al., The study of primary and acquired resistance to first-line osimertinib to improve the outcome of EGFR-mutated advanced Non-small cell lung cancer patients: the challenge is open for new therapeutic strategies. *Crit Rev Oncol Hematol*, 2024. **196**: p. 104295.
36. Chu, X., et al., Cancer stem cells: advances in knowledge and implications for cancer therapy. *Signal Transduct Target Ther*, 2024. **9**(1): p. 170.
37. Sharma, A., et al., Advances in lung cancer treatment using nanomedicines. *ACS Omega*, 2023. **8**(1): p. 10–41.
38. Shrivastava, G., et al., Targeting LIN28: a new hope in prostate cancer theranostics. *Future Oncol*, 2021. **17**(29): p. 3873–3880.
39. Ge, X., Z. Shen, and Y. Yin, Comprehensive review of LncRNA-mediated therapeutic resistance in non-small cell lung cancer. *Cancer Cell Int*, 2024. **24**(1): p. 369.
40. Yang, Y., M. Zhang, and Y. Wang, The roles of histone modifications in tumorigenesis and associated inhibitors in cancer therapy. *J Natl Cancer Cent*, 2022. **2**(4): p. 277–290.
41. Subramaniyan, V., et al., A review on epidermal growth factor receptor's role in breast and non-small cell lung cancer. *Chem Biol Interact*, 2022. **351**: p. 109735.
42. Thapa, R., et al., New horizons in lung cancer management through ATR/CHK1 pathway modulation. *Future Med Chem*, 2023. **15**(19): p. 1807–1818.
43. Yu, Y., et al., Roles of reactive oxygen species in inflammation and cancer. *MedComm (2020)*, 2024. **5**(4): p. e519.
44. Padinharayil, H., et al., Non-small cell lung carcinoma (NSCLC): implications on molecular pathology and advances in early diagnostics and therapeutics. *Genes Dis*, 2023. **10**(3): p. 960–989.
45. PDQ Adult Treatment Editorial Board, Non-small cell lung cancer treatment (PDQ®): health professional version, in *PDQ cancer information summaries*. 2002, National Cancer Institute, Bethesda, MD.

46. da Cunha Santos, G., F.A. Shepherd, and M.S. Tsao, EGFR mutations and lung cancer. *Annu Rev Pathol*, 2011. **6**: p. 49–69.
47. Park, H.J., et al., Keratinization of lung squamous cell carcinoma is associated with poor clinical outcome. *Tuberc Respir Dis (Seoul)*, 2017. **80**(2): p. 179–186.
48. Souto, E.B., et al., Non-melanoma skin cancers: physio-pathology and role of lipid delivery systems in new chemotherapeutic treatments. *Neoplasia*, 2022. **30**: p. 100810.
49. Thapa, R., et al., Epithelial-mesenchymal transition to mitigate age-related progression in lung cancer. *Ageing Res Rev*, 2024. **102**: p. 102576.
50. Thapa, R., et al., Influence of gut and lung dysbiosis on lung cancer progression and their modulation as promising therapeutic targets: a comprehensive review. *MedComm (2020)*, 2024. **5**(12): p. e70018.
51. Rudin, C.M., et al., Small-cell lung cancer. *Nat Rev Dis Primers*, 2021. **7**(1): p. 3.
52. Passaro, A., et al., Overcoming therapy resistance in EGFR-mutant lung cancer. *Nat Cancer*, 2021. **2**(4): p. 377–391.
53. Perez, K. and J. Chan, Medical management of gastrointestinal neuroendocrine tumors. *Curr Opin Endocrinol Diabetes Obes*, 2022. **29**(2): p. 219–224.
54. Salmani-Javan, E., M. Farhoudi Sefidan Jadid, and N. Zarghami, Recent advances in molecular targeted therapy of lung cancer: possible application in translation medicine. *Iran J Basic Med Sci*, 2024. **27**(2): p. 122–133.
55. Amicizia, D., et al., Systematic review of lung cancer screening: advancements and strategies for implementation. *Healthcare (Basel)*, 2023. **11**(14).
56. Oh, H.J., et al., Clinical applications of circulating biomarkers in non-small cell lung cancer. *Front Cell Dev Biol*, 2024. **12**: p. 1449232.
57. Petrella, F., et al., Stage III non-small-cell lung cancer: an overview of treatment options. *Curr Oncol*, 2023. **30**(3): p. 3160–3175.
58. Caputo, V., et al., Diagnostic value of liquid biopsy in the era of precision medicine: 10 years of clinical evidence in cancer. *Explor Target Antitumor Ther*, 2023. **4**(1): p. 102–138.
59. Cainap, C., et al., Next generation sequencing technology in lung cancer diagnosis. *Biology (Basel)*, 2021. **10**(9).
60. Marshall, H.M. and K.M. Fong, Lung cancer screening—time for an update? *Lung Cancer*, 2024. **196**: p. 107956.
61. Zhang, H., et al., The application of artificial intelligence in lung cancer: a narrative review. *Transl Cancer Res*, 2021. **10**(5): p. 2478–2487.
62. Lemjabbar-Alaoui, H., et al., Lung cancer: biology and treatment options. *Biochim Biophys Acta*, 2015. **1856**(2): p. 189–210.
63. Falcoz, P.E., et al., Video-assisted thoracoscopic surgery versus open lobectomy for primary non-small-cell lung cancer: a propensity-matched analysis of outcome from the European society of thoracic surgeon database. *Eur J Cardiothorac Surg*, 2016. **49**(2): p. 602–609.
64. Pereira, D., et al., Metal-based approaches to fight cervical cancer. *Drug Discov Today*, 2024. **29**(8): p. 104073.
65. Koka, K., et al., Technological advancements in external beam radiation therapy (EBRT): an indispensable tool for cancer treatment. *Cancer Manag Res*, 2022. **14**: p. 1421–1429.
66. Parvez, A., et al., PD-1 and PD-L1: architects of immune symphony and immunotherapy breakthroughs in cancer treatment. *Front Immunol*, 2023. **14**: p. 1296341.
67. Lin, X., et al., Regulatory mechanisms of PD-1/PD-L1 in cancers. *Mol Cancer*, 2024. **23**(1): p. 108.
68. Krzyszczyk, P., et al., The growing role of precision and personalized medicine for cancer treatment. *Technology (Singap World Sci)*, 2018. **6**(3–4): p. 79–100.
69. Osoegawa, A., et al., Osimertinib as first-line treatment for recurrent lung cancer patients with EGFR mutation. *J Thorac Dis*, 2023. **15**(10): p. 5566–5573.
70. Remon, J., et al., Perioperative treatment strategies in EGFR-mutant early-stage NSCLC: current evidence and future challenges. *J Thorac Oncol*, 2024. **19**(2): p. 199–215.
71. Garcia, J., et al., Bevacizumab (Avastin®) in cancer treatment: a review of 15 years of clinical experience and future outlook. *Cancer Treat Rev*, 2020. **86**: p. 102017.
72. Haibe, Y., et al., Resistance mechanisms to anti-angiogenic therapies in cancer. *Front Oncol*, 2020. **10**: p. 221.
73. Johnson, M., et al., Treatment strategies and outcomes for patients with EGFR-mutant non-small cell lung cancer resistant to EGFR tyrosine kinase inhibitors: focus on novel therapies. *Lung Cancer*, 2022. **170**: p. 41–51.

74. Westover, D., et al., Mechanisms of acquired resistance to first- and second-generation EGFR tyrosine kinase inhibitors. *Ann Oncol*, 2018. **29**: p. i10–i19.
75. Song, Y., et al., Anti-angiogenic agents in combination with immune checkpoint inhibitors: a promising strategy for cancer treatment. *Front Immunol*, 2020. **11**: p. 1956.
76. Ciciola, P., et al., Combining immune checkpoint inhibitors with anti-angiogenic agents. *J Clin Med*, 2020. **9**(3).
77. Huang, L., et al., KRAS mutation: from undruggable to druggable in cancer. *Signal Transduct Target Ther*, 2021. **6**(1): p. 386.
78. Bunn, P.A., Jr., et al., Small cell lung cancer: can recent advances in biology and molecular biology be translated into improved outcomes? *J Thorac Oncol*, 2016. **11**(4): p. 453–474.
79. Ma, L., et al., Liquid biopsy in cancer current: status, challenges and future prospects. *Signal Transduct Target Ther*, 2024. **9**(1): p. 336.
80. Li, C., et al., Circulating tumor DNA as liquid biopsy in lung cancer: biological characteristics and clinical integration. *Cancer Lett.*, 2023. **577**: p. 216365.
81. de Visser, K.E. and J.A. Joyce, The evolving tumor microenvironment: from cancer initiation to metastatic outgrowth. *Cancer Cell*, 2023. **41**(3): p. 374–403.
82. Jiang, X., et al., Role of the tumor microenvironment in PD-L1/PD-1-mediated tumor immune escape. *Mol Cancer*, 2019. **18**(1): p. 10.
83. Ratti, M., et al., MicroRNAs (miRNAs) and long non-coding RNAs (lncRNAs) as new tools for cancer therapy: first steps from bench to bedside. *Target Oncol*, 2020. **15**(3): p. 261–278.
84. Taherian, M., P. Bayati, and N. Mojtabavi, Stem cell-based therapy for fibrotic diseases: mechanisms and pathways. *Stem Cell Res Ther*, 2024. **15**(1): p. 170.
85. Sahu, M., et al., Chapter three—artificial intelligence and machine learning in precision medicine: a paradigm shift in big data analysis, in *Progress in molecular biology and translational science*, D.B. Teplow, Editor. 2022, Academic Press. P. 57–100.
86. Nikanjam, M., et al., Novel clinical trial designs emerging from the molecular reclassification of cancer. *CA Cancer J Clin*, 2025. **75**.
87. Lao, Y., et al., Immune checkpoint inhibitors in cancer therapy-how to overcome drug resistance? *Cancers (Basel)*, 2022. **14**(15).
88. Ai, L., A. Xu, and J. Xu, Roles of PD-1/PD-L1 pathway: signaling, cancer, and beyond. *Adv Exp Med Biol*, 2020. **1248**: p. 33–59.
89. Davies, M.J., PD-1/PD-L1 inhibitors for non-small cell lung cancer: incorporating care step pathways for effective side-effect management. *J Adv Pract Oncol*, 2019. **10**(Suppl 1): p. 21–35.
90. Dagar, G., et al., Harnessing the potential of CAR-T cell therapy: progress, challenges, and future directions in hematological and solid tumor treatments. *J Transl Med*, 2023. **21**(1): p. 449.
91. McGowan, E., et al., PD-1 disrupted CAR-T cells in the treatment of solid tumors: promises and challenges. *Biomed Pharmacother*, 2020. **121**: p. 109625.
92. Chen, L., et al., CAR-T cell therapy for lung cancer: potential and perspective. *Thorac Cancer*, 2022. **13**(7): p. 889–899.
93. Bremnes, R.M., et al., The role of tumor-infiltrating lymphocytes in development, progression, and prognosis of non-small cell lung cancer. *J Thorac Oncol*, 2016. **11**(6): p. 789–800.
94. Garg, P., et al., Next-generation immunotherapy: advancing clinical applications in cancer treatment. *J Clin Med*, 2024. **13**(21).
95. Vargas, A.J. and C.C. Harris, Biomarker development in the precision medicine era: lung cancer as a case study. *Nat Rev Cancer*, 2016. **16**(8): p. 525–537.
96. Ramos-Medina, R., et al., Technical challenges for CTC implementation in breast cancer. *Cancers (Basel)*, 2021. **13**(18).
97. Capuozzo, M., et al., Circulating tumor cells as predictive and prognostic biomarkers in solid tumors. *Cells*, 2023. **12**(22).
98. Li, W., et al., The signaling pathways of selected traditional Chinese medicine prescriptions and their metabolites in the treatment of diabetic cardiomyopathy: a review. *Front Pharmacol*, 2024. **15**: p. 1416403.
99. Bamodu, O.A., C.-C. Chung, and T.R. Pisanic, Harnessing liquid biopsies: exosomes and ctDNA as minimally invasive biomarkers for precision cancer medicine. *J Liq Biopsy*, 2023. **2**: p. 100126.
100. Raez, L.E., et al., Moving liquid biopsies to the front-line of lung cancer treatment decisions. *J Liq Biopsy*, 2023. **1**: p. 100006.
101. She, W., Y. Garitaonaindia, and Y. Lin, The latest advances in liquid biopsy for lung cancer-a narrative review. *Transl Lung Cancer Res*, 2024. **13**(11): p. 3241–3251.

102. Durham, A.L. and I.M. Adcock, The relationship between COPD and lung cancer. *Lung Cancer*, 2015. **90**(2): p. 121–127.
103. Katic, L., et al., The interplay between cardiovascular disease and lung cancer. *Cureus*, 2024. **16**(6): p. e62953.
104. Hardy, D., et al., Cardiac toxicity in association with chemotherapy and radiation therapy in a large cohort of older patients with non-small-cell lung cancer. *Ann Oncol*, 2010. **21**(9): p. 1825–1833.
105. Kolb, R., F.S. Sutterwala, and W. Zhang, Obesity and cancer: inflammation bridges the two. *Curr Opin Pharmacol*, 2016. **29**: p. 77–89.
106. Szablewski, L., Insulin resistance: the increased risk of cancers. *Curr Oncol*, 2024. **31**(2): p. 998–1027.
107. Islam, M.S., et al., The interplay of factors in metabolic syndrome: understanding its roots and complexity. *Mol Med*, 2024. **30**(1): p. 279.
108. Lahiri, A., et al., Lung cancer immunotherapy: progress, pitfalls, and promises. *Mol Cancer*, 2023. **22**(1): p. 40.
109. Avancini, A., et al., Physical activity and exercise in lung cancer care: will promises be fulfilled? *Oncologist*, 2020. **25**(3): p. e555–e569.
110. Hussen, B.M., et al., Revolutionizing medicine: recent developments and future prospects in stem-cell therapy. *Int J Surg*, 2024. **110**(12): p. 8002–8024.
111. Ikrama, M., et al., Pulmonary fibrosis: is stem cell therapy the way forward? J *Taibah Univ Med Sci*, 2024. **19**(1): p. 82–89.
112. Lai, S. and Z. Guo, Stem cell therapies for chronic obstructive pulmonary disease: mesenchymal stem cells as a promising treatment option. *Stem Cell Res Ther*, 2024. **15**(1): p. 312.
113. Joo, H., S. Min, and S.W. Cho, Advanced lung organoids for respiratory system and pulmonary disease modeling. *J Tissue Eng*, 2024. **15**: p. 20417314241232502.
114. Duan, W.W., et al., A TGF-β signaling-related lncRNA signature for prediction of glioma prognosis, immune microenvironment, and immunotherapy response. *CNS Neurosci Ther*, 2024. **30**(4): p. e14489.
115. Masucci, M., et al., Bridging the divide: a review on the implementation of personalized cancer medicine. *J Pers Med*, 2024. **14**(6).
116. Alowais, S.A., et al., Revolutionizing healthcare: the role of artificial intelligence in clinical practice. *BMC Med Educ*, 2023. **23**(1): p. 689.
117. Khalifa, M. and M. Albadawy, Artificial intelligence for clinical prediction: exploring key domains and essential functions. *Comput Methods Programs Biomed Update*, 2024. **5**: p. 100148.
118. Parihar, A., et al., 3D bioprinting for drug development and screening: recent trends towards personalized medicine. *Hybrid Adv*, 2024. **7**: p. 100320.
119. Toohey, K., et al., Exercise in cancer care for people with lung cancer: a narrative synthesis. *J Sci Med Sport*, 2025. **28**(1): p. 16–25.
120. Crawford, G.B., et al., Care of the adult cancer patient at the end of life: ESMO clinical practice guidelines☆. *ESMO Open*, 2021. **6**(4): p. 100225.
121. Feldstain, A., Psychosocial intervention in palliative care: what do psychologists need to know. *J Health Psychol*, 2024. **29**(7): p. 707–720.
122. Fan, S.-Y., et al., Psychosocial care provided by physicians and nurses in palliative care: a mixed methods study. *J Pain Symptom Manage*, 2017. **53**(2): p. 216–223.
123. Rome, R.B., et al., The role of palliative care at the end of life. *Ochsner J*, 2011. **11**(4): p. 348–352.

2 Epidemiology and Pathophysiology of Lung Cancer

Mohit Agarwal[1], *Anurag Mishra*[1*],
Sobhit Singh Rajput[2], *and Yogendra Singh*[3*]

[1] NIMS Institute of Pharmacy, NIMS University Rajasthan, Jaipur, Rajasthan, India

[2] MIPER, Mangalayatan University, Beswan, Aligarh, Uttar Pradesh, India

[3] Department of Pharmacology, Maharishi Arvind College of Pharmacy, Ambabari, Jaipur, Rajasthan, India

*Corresponding authors: Yogendra Singh, yogendra.singh119@gmail.com; Anurag Mishra, raag.mishra@gmail.com

2.1 INTRODUCTION

Worldwide medical experts have confirmed that lung cancer spread to epidemic levels because of cigarette-related injuries in the 1950s. Clinical evidence shows that lung cancer remains the leading cause of cancer-related mortality for both males and females worldwide (1). The worldwide statistics indicate that natural occurrences of lung cancer create 1.2 million new cases, followed by 1.1 million lethal cancer deaths each year (2).

In 2012 lung cancer led to the deaths of 160,340 Americans (3), with 234,000 new cases (4). Researchers have identified lung cancer death statistics at 1.8 million for 2020 as the primary cancer-related mortality worldwide according to the Global Cancer Statistics book (5). Mesothelioma results in just 3000 yearly deaths (6). The United States suffers the most deaths from cancer through lung cancer while breast together with prostate and colorectal cancers have proven less fatal. Two main types of lung cancer exist as small cell lung cancer (SCLC) as well as non–small cell lung cancer (NSCLC) based on their response to treatment and survival predictions (7, 8). The diagnosis of SCLC indicates an aggressive type of cancer which spreads through the entire body despite being called the "small-cell" lung cancer. The survival prognosis during untreated cases lasts between 2 and 4 months. The therapeutic care extends survival durations between 18 and 36 months.

The United States sees SCLC as one of the leading causes of new lung cancer cases with yearly occurrences between 15% and 20%. Cancer cells known as anaplastic cells exist as small fast-reproducing entities that characterize this condition. Surgery proves ineffective in treating this cancer since it spreads early, making it challenging to excise affected tissue. Chemotherapy administered through the bloodstream demonstrates a limited capacity to respond to the disease (9).

Standard treatment approaches for LS and ES illness include the etoposide combination therapy with cisplatin/carboplatin and optional radiation therapy (10). The recent medical advancements bringing disease-free intervals and survival benefits include preventative cranial irradiation as well as strengthened thoracic radiation therapy and its integration with systemic chemotherapy (11, 12).

The ACCP's current practice guidelines establish all functioning components of these measures (13). Professional performance of mediastinal evaluation allows doctors to offer surgery

DOI: 10.1201/9781003595830-2

recommendations to certain patients with stage I SCLC which raises their survival chances. The implementation rate of surgery in non–small cell lung cancer patient management stands at a higher level. Patients receive their highest likelihood of cure when treatment occurs throughout the early phases of disease because surgical intervention becomes available. NSCLC has three primary histopathologic forms containing squamous cell carcinoma and adenocarcinoma, which includes adenocarcinoma in situ along with bronchioloalveolar cell carcinoma (14).

2.2 EPIDEMIOLOGY OF LUNG CANCER

Statistics show that lung cancer occupies the leading position as the most occurring and fatal malignant tumor type in the world. Every year global disease reports place lung cancer as the top occurring illness. Exposure to tobacco smoke both indirectly and directly results in all forms of lung cancer diagnoses. Smoke consumption causes lung cancer risk to reach dangerous levels, which exceed nonsmoker numbers 15 to 50 times. Cigarette smokers who aim to minimize their lung cancer risk should stop smoking immediately. The risk rate decreases when smokers maintain abstinence for longer periods. Historical smokers represent 50% of people who develop lung cancer. The disease ranks as the sixth most deadly cancer form among nonsmokers even though it persists in 15–18% of people who never smoked.

The scanning cells of adenocarcinoma in situ contain specific mutations in tyrosine kinase receptors that affect EGFR and KRAS receptors located on the cell surface. Such somatic alterations allow the treatment of cancers with specific inhibitors through tyrosine kinase cellular pathways. Lung cancer–related mortality rates demonstrated positive progress during the 1990s after age-adjusted analysis but lung cancer deaths have increased throughout the entire US population since the 1950s because the majority of Americans survive past age 70 (15). Lung cancer statistics in the United States display two notable characteristics consisting of an adenocarcinoma prevalence increase and a female cancer death rate exceeding breast cancer mortality.

Through numerous case-control investigations which ran between the 1950s, British investigators generated the initial epidemiological basis linking cigarette smoking to lung cancer (16). Scientific results demonstrate that male lung cancer forms because of cigarette smoking activities (17). According to the 1964 Surgeon General's Report, scientists confirmed that lung cancer stems from tobacco smoke by merging accumulated data with established proper epidemiologic causality criteria which require consistency, strength, specificity, and temporal relationships and coherence of disease-to-variable associations. Humans can prevent lung cancer through avoiding exposure to cigarette smoke because it directly leads to this disease. Cigarette smoking drives 90% of male and 79% of female lung cancer cases, representing 80% of total US incidences (18). Genetic amplification approaches have delivered concrete evidence establishing the smoking-caused effects on p53 tumor suppressor gene loci. Studies reveal that cigarette smoke releases benzo[α]pyrene, which specifically damages three designated locations of the p53 tumor suppressor gene. Research confirms that aberrant loci exist within 60% of early-stage lung tumors diagnosed for the first time. Various environmental factors can strengthen the susceptibility towards developing lung cancer. Multiple innocent lung diseases as well as the factors of dietants and fat intake and exposure to radiation and haloethers and polycyclic aromatic hydrocarbons and nickel and genetic background and secondhand smoking contribute to lung cancer susceptibility (19).

The International Agency for Research on Cancer (IARC) issued recent estimations showing these worldwide cancer statistics based on global regions for the year 2022. A total of 20 million fresh cancer cases including NMSCs as well as 9.7 million cancer-related deaths occurred during 2022 (20). The diagnosis rate of cancer stands at 20% for all humans throughout their lifetimes. Cancer leads to the death of 9 males per 100 people while 12 females die from cancer annually. The medical community recorded lung cancer as the leading cancer diagnosis during

2022 worldwide. Among all new cancer diagnoses worldwide in 2022 the disease type affected 2.5 million patients which represented 8% of all cancer occurrences (12.4% of all cancers combined together). Breast cancer in female patients made up 11.6% of total cancers while colorectal cancer amounted to 9.6% and prostate cancer affected 7.3% of patients, whereas stomach cancer caused 4.9% of all diagnosed malignancies. Each year lung cancer stood as the main contributor to deaths because it resulted in the loss of 1.8 million individuals who comprised 18.7% of total fatalities. The prevalence ranking for female patients included colon cancer at 9.3% and liver cancer with 7.8% after breast cancer 6.9% but stomach cancer ranked second for male patients at 6.8%. Breast cancer led the statistics as the most prevalent cancer in women according to occurrence and death rates but among men lung cancer established itself as the primary cancer diagnosis. Global cancer incidence rates for NMSC and basal or squamous cell cancers spanned from 100 to about 500 per 100,000 population between locations across West Africa and Australia/New Zealand areas (507.9 per 100,00). South Central Asia had less than 100 cancer cases per 100,000 females but Australia/New Zealand reached above 400 (410.5 per 100,000) (20). The continuous rise of lung cancer cases in the United States since the 20th century appeared equally in both male and female populations. The number of smokers exploded among males 30 years ago before women began to adopt similar smoking habits 30 years afterward. Research data demonstrates that lung cancer occurrences have followed this pattern. The overall number of male lung cancer cases remains higher than female cases but these rates show a downward trend whereas female lung cancer counts keep increasing.

Ethnic groups show clear distinctions regarding disease occurrence. Black males currently have the highest mortality rate from lung cancer in America. The cancer risk between white and African American female populations remains equal while a mysterious element causes African American males to confront a significantly higher lung cancer threat compared to white men. Lung cancer poses fewer dangers to Latino people together with Native Americans and Asians/Pacific Islanders than it poses to white and African American populations (19). A person with either personal or family history of lung cancer shows higher chances of developing this disease. Multiple studies show that consuming a diet composed mainly of fruits and vegetables enables people to decrease their chances of getting lung cancer.

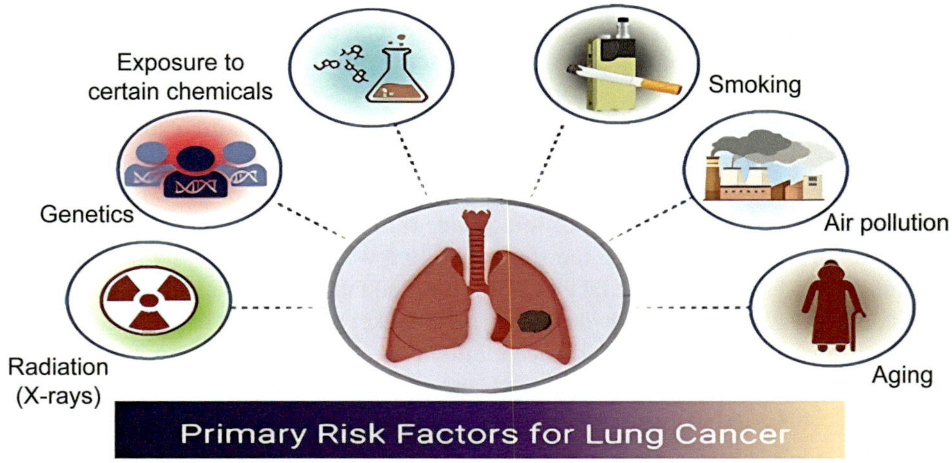

FIGURE 2.1 Factors associated with lung cancer

2.3 PATHOPHYSIOLOGY OF LUNG CANCER

The environmental factors cause a series of genetic changes which turn normal cells into lung cancer. Tumor growth with metastasis develops due to microenvironment changes in addition to mutations found in tumor oncogenes and tumor suppressors. A variety of intricate molecular pathways result in lung cancer development. At least one epigenetic change needs to occur during lung cancer development with DNA segment translocations among other point mutations, amplifications, and insertions and deletions being possible changes. The changes specifically work through both growth-promoting pathway activation combined with tumor suppressor pathway inactivation.

Genomic instability stands as a vital factor which initiates the development of malignancies. All body cells face impacts from this unpredictability. The disease has developed through multiple accumulated genetic anomalies. The DNA structure grows more susceptible to further mutations because it becomes weaker due to this chain of events. Mutations in proto-oncogenes along with tumor suppressor genes will produce cell cycle regulation abnormalities. All cancer formations initiate with cellular resistance to controlling signals through cell overproduction or neglect of cell growth control.

Squamous cell carcinoma and adenocarcinoma are the two most prevalent NSCLC forms; both begin as normal epithelium but progress to invasive cancer as a result of a cascade of mutations affecting tumor oncogenes and tumor suppressors. Atypical adenomatous alveolar hyperplasia is the initial histologically discernible alteration in adenocarcinoma. Subsequently, bronchoalveolar carcinoma and invasive adenocarcinoma are the stages that occur. The development of this advancement stems from various mutated genetic sequences which also modify *p53* and *p16* tumor suppressor genes and epidermal growth factor receptor (EGFR) and Kirsten rat sarcoma (K-RAS) viral oncogene homolog. Through membrane binding K-RAS genes create G proteins which establish links between both membrane-bound tyrosine kinase receptors and cytoplasmic signaling substances (21). An estimated 10–30% of lung adenocarcinomas have mutated K-RAS, which allows tumor cells to grow by constitutively activating guanosine triphosphatase. Women, nonsmokers, and patients with bronchoalveolar adenocarcinomas are more likely to have an EGFR mutation (22), which accounts for around 10–40% of adenocarcinomas with wild-type K-RAS. Amplification of tumor cells occurs when a mutated EGFR contacts ligands like EGF or transforming growth factor alpha and triggers signaling pathways that involve tyrosine kinases (23).

Three main tumor suppressor genes *p53*, *Rb*, and *p16* experience the majority of mutations which develop in lung cancer cells (24). Proto-oncogenes mutated in lung cancer patients belong to three primary groups: MYC, RAS, and HER families. Gene rearrangement of the ALK molecule occurs through a precise molecular mechanism (25).

Two critical tumor suppressor genes *p53* and *p16* play essential roles in developing adenocarcinoma. Through its transcription factor mechanism *p53* enables cells to stop dividing after DNA damage occurs and then leads them toward eventual cellular death. The ability of cancer cells to grow unrestricted emerges from *p53* mutations which block natural cell mortality mechanisms as well as stop cell proliferation (26). As a different tumor suppressor gene from *p53* the *p16* meditates activation of the retinoblastoma (Rb) gene to block cell cycle progression at the G1/S transition (27). The *p16* mutation affecting about 50% of lung adenocarcinomas finds loss of heterozygosity to be its most frequent variant that enables uncontrolled cell multiplication between tumor cells.

The development of invasive squamous cell carcinoma in the lung follows equivalent progression stages. The histologic pattern of squamous cell epithelium changes visually from normal epithelium to squamous hyperplasia during its first recognizable developmental stage. The cell development path leads toward invasive cancer or may result in dysplasia and carcinoma in situ and squamous metaplasia conditions (28). The developmental process enables mutations to occur in both oncogenes and tumor suppressors. The prevalence of *p53* mutations reaches 60–70% in patients diagnosed with squamous cell carcinoma similarly to what occurs in adenocarcinoma. The occurrence

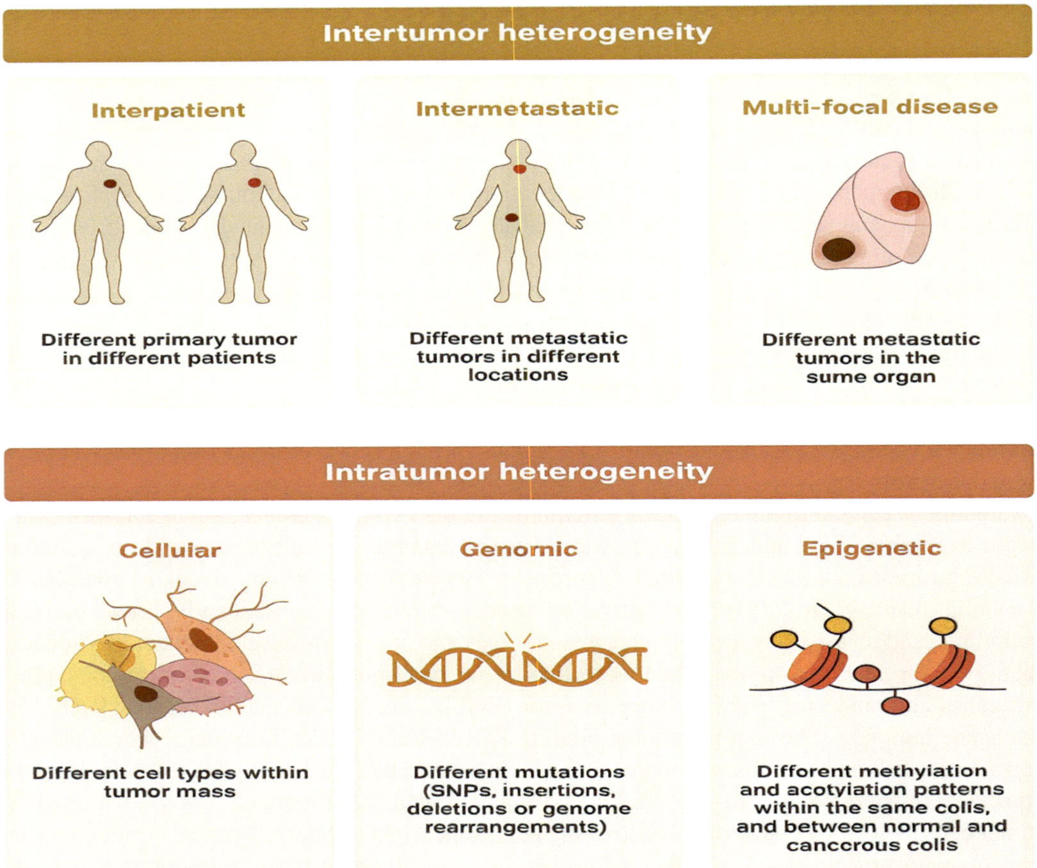

FIGURE 2.2 Types of tumor heterogeneity in lung cancer

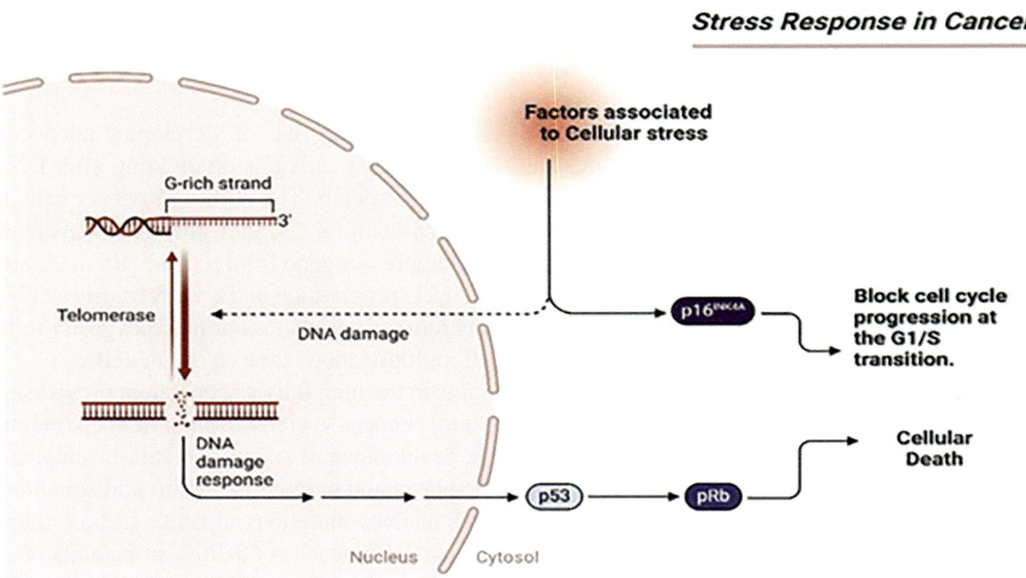

FIGURE 2.3 Tumor suppressor gene associated with epigenetic changes to lung cancer

of K-RAS mutations and EGFR mutations at less than 1% makes them exceptionally rare in squamous cell cancer cases (29, 30). The combination of p53 and Rb protein mutation occurs in approximately 90% of small cell lung carcinomas (31, 32). The Rb gene located downstream of the p16 pathway becomes mutated, which results in cancer cell proliferation because it influences cell cycle control (33). The Myc oncogene shows an increased version found in 30% of cases of small cell lung carcinoma (34).

Major signal transmitters that transmit outside information into cellular structures exist as cell receptors located on the external cell surfaces. NSCLC receives powerful positive growth input from the epidermal growth factor (EGF), which uses its receptor epidermal growth factor receptor (EGFR) for this function (35). Lung cancer development requires this vital stage, which leads to its dysfunction. This regulatory mechanism contrasts with aberrant angiogenesis because endothelial cells that undergo thousandfold cell divisions live longer durations while this process normally ends itself. VEGF stands as one of the strongest and dominant pro-angiogenic substances known to biology. Malignant tumors develop into tumors because the VEGF-mediated pathway stays perpetually activated while pro-angiogenic factors control over antiangiogenic factors prevails (36). A tumor needs its individual blood vessel network to expand beyond 3 mm in size. Tumor development requires the capability of vascular growth to supply blood vessels. The messenger RNA expression of VEGF remains constant in lung cancer cells. Among these two factors VEGFR shows increased expression levels in endothelial cells. This data points to the possibility of a paracrine action for VEGF in tumor cells. The production of VEGF by cancer cells leads to endothelial cell activation via their receptor facilities. Evidence demonstrates that tumor edges contain VEGF concentrations within 0.5 mm from the boundary (37). Physicians treating patients with this condition must be always alert for signs of lung cancer because hemoptysis frequently appears when blood vessel defects of tumors cause their fragility and organization issues and variable diameters. The failure of cells to perform proper apoptosis reactions to environmental triggers stands as a fundamental factor which helps lung cancer form. The cellular process of programmed suicide, which also goes as active death or physiological cell death, exists under multiple names to maintain body homeostasis.

The molecular model of lung cancer advancement demands multiple genetic abnormalities to become operational. The growing interest in molecular biology over the past few years has enabled scientists to observe cancer-causing genetic modifications with great precision. Carcinogenesis depends heavily on bidirectional *p53* gene mutations as one of its essential factors. Experimental evidence shows that neoplastic transformation becomes more likely when p53 protein failure happens simultaneously with mutations from *Rb*, *MYC*, *RAS*, and *EGFR* genes. The success of cancer treatment requires clinical mutation investigation of specific genes to provide precise diagnosis and enhanced treatment strategies. These genetic materials function as predictors at the molecular level. The clinical use of predictive tests which include all detection signals from this study remains uncommon. Parts of these genetic markers have entered treatment management algorithms while adding value to lung cancer detection. The process to receive EGFR TKI targeted treatment begins with establishing the EGFR gene mutation status of patients. ALK therapy depends on the detection of gene alterations in the ALK gene. Through their study the authors demonstrate optimism about molecular diagnostics testing advances which will lead to early non-invasive lung cancer detection to boost successful cancer interventions.

2.4 CONCLUSION

Lung cancer begins from tobacco exposure and environmental influences and stands among the leading preventable conditions leading to death. Cells perform multiplication and development together with metastasis through environmental stimuli which promote a succession of genomic alterations. Stage 4 lung cancer represents the most prevalent condition that patients exhibit at their initial treatment seeking stage. Clinical symptoms of patients depend on how far a tumor has spread and how large it is alongside where it has developed in the body structures. Tissue analysis of a tumor through pathological examination results in lung cancer diagnosis. A complete clinical staging of the disease incorporates X-rays alongside pathologic test results together with the medical records and physical examination outcomes from patients. Patients who receive a lung cancer diagnosis receive treatment through multiple different interventions. Radiotherapy serves as a backup treatment for surgery although surgery remains vital in local and regional disease cases. Chemotherapy represents the standard medical option for lung cancer patients when their tumor advances to different parts of the body. The diagnosis of lung cancer usually leads to unfavorable outcomes for patients. Research shows that lung cancer patients have a 16% chance of survival for five years after a diagnosis. The primary determinants of how long patients survive after receiving a lung cancer diagnosis are both their cancer type and the stage of pathologic progression.

REFERENCES

1. Siegel R, Naishadham D, Jemal A. Cancer statistics, 2013. *CA: A Cancer Journal for Clinicians.* 2013;63(1):11–30.
2. Ferlay J, Shin HR, Bray F, Forman D, Mathers C, Parkin DM. Estimates of worldwide burden of cancer in 2008: GLOBOCAN 2008. *International Journal of Cancer.* 2010;127(12):2893–917.
3. Ridge CA, McErlean AM, Ginsberg MS. Epidemiology of lung cancer. *Seminars in Interventional Radiology.* 2013;30(2):93–8.
4. Siegel RL, Miller KD, Jemal A. Cancer statistics, 2017. *CA: A Cancer Journal for Clinicians.* 2017;67(1):7–30.
5. Sung H, Ferlay J, Siegel RL, Laversanne M, Soerjomataram I, Jemal A, et al. Global cancer statistics 2020: GLOBOCAN estimates of incidence and mortality worldwide for 36 cancers in 185 countries. *CA: A Cancer Journal for Clinicians.* 2021;71(3):209–49.
6. Carbone M, Ly BH, Dodson RF, Pagano I, Morris PT, Dogan UA, et al. Malignant mesothelioma: facts, myths, and hypotheses. *Journal of Cellular Physiology.* 2012;227(1):44–58.
7. Lindeman NI, Cagle PT, Beasley MB, Chitale DA, Dacic S, Giaccone G, et al. Molecular testing guideline for selection of lung cancer patients for EGFR and ALK tyrosine kinase inhibitors: guideline from the college of American pathologists, international association for the study of lung cancer, and association for molecular pathology. *The Journal of Molecular Diagnostics: JMD.* 2013;15(4):415–53.
8. Lindeman NI, Cagle PT, Aisner DL, Arcila ME, Beasley MB, Bernicker EH, et al. Updated molecular testing guideline for the selection of lung cancer patients for treatment with targeted tyrosine kinase inhibitors: guideline from the college of American pathologists, the international association for the study of lung cancer, and the association for molecular pathology. *The Journal of Molecular Diagnostics: JMD.* 2018;20(2):129–59.
9. Murray N, Turrisi AT, 3rd. A review of first-line treatment for small-cell lung cancer. *Journal of Thoracic Oncology: Official Publication of the International Association for the Study of Lung Cancer.* 2006;1(3):270–8.
10. Okamoto H, Watanabe K, Kunikane H, Yokoyama A, Kudoh S, Asakawa T, et al. Randomised phase III trial of carboplatin plus etoposide vs split doses of cisplatin plus etoposide in elderly or poor-risk patients with extensive disease small-cell lung cancer: JCOG 9702. *British Journal of Cancer.* 2007;97(2):162–9.
11. Aupérin A, Arriagada R, Pignon JP, Le Péchoux C, Gregor A, Stephens RJ, et al. Prophylactic cranial irradiation for patients with small-cell lung cancer in complete remission: prophylactic cranial irradiation overview collaborative group. *The New England Journal of Medicine.* 1999;341(7):476–84.
12. Slotman B, Faivre-Finn C, Kramer G, Rankin E, Snee M, Hatton M, et al. Prophylactic cranial irradiation in extensive small-cell lung cancer. *The New England Journal of Medicine.* 2007;357(7):664–72.
13. Jett JR, Schild SE, Kesler KA, Kalemkerian GP. Treatment of small cell lung cancer: diagnosis and management of lung cancer, 3rd ed: American college of chest physicians evidence-based clinical practice guidelines. *Chest.* 2013;143(5 Suppl):e400S–19S.
14. Detterbeck FC, Boffa DJ, Tanoue LT. The new lung cancer staging system. *Chest.* 2009;136(1):260–71.
15. Jemal A, Murray T, Samuels A, Ghafoor A, Ward E, Thun MJ. Cancer statistics, 2003. *CA: A Cancer Journal for Clinicians.* 2003;53(1):5–26.
16. Doll R, Hill AB. Smoking and carcinoma of the lung; preliminary report. *British Medical Journal.* 1950;2(4682):739–48.
17. Doll R. Bronchial carcinoma: incidence and aetiology. *Medical World.* 1954;80(4):370–91.

18. Denissenko MF, Pao A, Tang M, Pfeifer GP. Preferential formation of benzo[a]pyrene adducts at lung cancer mutational hotspots in P53. *Science (New York, NY)*. 1996;274(5286):430–2.
19. Alberg AJ, Samet JM. Epidemiology of lung cancer*. *Chest*. 2003;123(1):21S–49S.
20. Bray F, Laversanne M, Sung H, Ferlay J, Siegel RL, Soerjomataram I, et al. Global cancer statistics 2022: GLOBOCAN estimates of incidence and mortality worldwide for 36 cancers in 185 countries. *CA: A Cancer Journal for Clinicians*. 2024;74(3):229–63.
21. Jänne PA, Engelman JA, Johnson BE. Epidermal growth factor receptor mutations in non-small-cell lung cancer: implications for treatment and tumor biology. *Journal of Clinical Oncology: Official Journal of the American Society of Clinical Oncology*. 2005;23(14):3227–34.
22. Quintanal-Villalonga Á, Chan JM, Yu HA, Pe'er D, Sawyers CL, Sen T, et al. Lineage plasticity in cancer: a shared pathway of therapeutic resistance. *Nature Reviews Clinical Oncology*. 2020;17(6):360–71.
23. Kim MP, Vaporciyan AA. Primary lung cancer. In: Yuh DD, Vricella LA, Yang SC, Doty JR, editors. *Johns Hopkins Textbook of Cardiothoracic Surgery*. New York, NY: McGraw-Hill Education; 2014.
24. Chen T, Ashwood LM, Kondrashova O, Strasser A, Kelly G, Sutherland KD. Breathing new insights into the role of mutant p53 in lung cancer. *Oncogene*. 2025;44(3):115–29.
25. Kontomanolis EN, Koutras A, Syllaios A, Schizas D, Mastoraki A, Garmpis N, et al. Role of oncogenes and tumor-suppressor genes in carcinogenesis: a review. *Anticancer Research*. 2020;40(11):6009–15.
26. Chen X, Zhang T, Su W, Dou Z, Zhao D, Jin X, et al. Mutant p53 in cancer: from molecular mechanism to therapeutic modulation. *Cell Death & Disease*. 2022;13(11):974.
27. Herbst RS, Heymach JV, Lippman SM. Lung cancer. *The New England Journal of Medicine*. 2008;359(13):1367–80.
28. Christopherson WM. Dysplasia, carcinoma in situ, and microinvasive carcinoma of the uterine cervix. *Human Pathology*. 1977;8(5):489–501.
29. Ladanyi M, Pao W. Lung adenocarcinoma: guiding EGFR-targeted therapy and beyond. *Modern Pathology*. 2008;21:S16–22.
30. Acker F, Stratmann J, Aspacher L, Nguyen NTT, Wagner S, Serve H, et al. KRAS Mutations in squamous cell carcinomas of the lung. *Frontiers in Oncology*. 2021;11:788084.
31. George J, Lim JS, Jang SJ, Cun Y, Ozretić L, Kong G, et al. Comprehensive genomic profiles of small cell lung cancer. *Nature*. 2015;524(7563):47–53.
32. Papavassiliou KA, Sofianidi AA, Gogou VA, Anagnostopoulos N, Papavassiliou AG. P53 and Rb aberrations in small cell lung cancer (SCLC): from molecular mechanisms to therapeutic modulation. *International Journal of Molecular Sciences*. 2024;25(5).
33. Agrelo R, Kishimoto H, Novatchkova M, Peraza V, Paolino M, Souabni A, et al. SATB1 collaborates with loss of p16 in cellular transformation. *Oncogene*. 2013;32(48):5492–500.
34. Cargill KR, Stewart CA, Park EM, Ramkumar K, Gay CM, Cardnell RJ, et al. Targeting MYC-enhanced glycolysis for the treatment of small cell lung cancer. *Cancer & Metabolism*. 2021;9(1):33.
35. Liu X, Wang P, Zhang C, Ma Z. Epidermal growth factor receptor (EGFR): a rising star in the era of precision medicine of lung cancer. *Oncotarget*. 2017;8(30):50209–20.
36. Vasudev NS, Reynolds AR. Anti-angiogenic therapy for cancer: current progress, unresolved questions and future directions. *Angiogenesis*. 2014;17(3):471–94.
37. Apte RS, Chen DS, Ferrara N. VEGF in signaling and disease: beyond discovery and development. *Cell*. 2019;176(6):1248–64.

3 Immunology of Lung Cancer

S. Afsar[a], Rahamat Unissa Syed[b], and Dra. Fulvia di Pillo[c]*

[a] Department of Virology, Sri Venkateswara University, Tirupathi, Andhra Pradesh 517502, India

[b] Department of Pharmaceutics, College of Pharmacy, University of Hail, Hail 81442, Saudi Arabia

[c] Centre for Medicinal Chemistry (CQMED) Av. André Tosello, 550—Térreo—Unicamp Campinas, SP, Brazil

*Corresponding author: Rahamat Unissa Syed, ru.syed@uoh.edu.sa.

3.1 FUNDAMENTALS OF IMMUNOLOGY AND LUNG CANCER

3.1.1 Introduction to Lung Cancer and Immunology

Lung cancer maintains the position as the primary cancer killer worldwide because patients usually receive a late-stage diagnosis coupled with insufficient treatment methods. The main epidemiological risk factor remains smoking because it leads to 85% of identified instances. Genetic predisposition along with occupational carcinogens (asbestos, radon, arsenic) and air pollution and chronic respiratory disorders cause substantial disease initiation along with smoking [1–4]. Lung cancer shows two major pathological categories: SCLC and NSCLC. SCLC represents a rapidly growing aggressive class with early metastasis behaviour and NSCLC contains three different distinctive groups: adenocarcinoma, squamous cell carcinoma, and large cell carcinoma. The immune system interacts with genetic and environmental factors to significantly affect the development of lung cancer pathophysiology [5]. Lung cancer cells use a variety of immune evasion strategies, such as upregulating immune checkpoints (PD-1/PD-L1, CTLA-4), secreting immunosuppressive cytokines, and recruiting regulatory immune cells like myeloid-derived suppressor cells (MDSCs) and tumour-associated macrophages (TAMs), which promote immune tolerance and tumour progression, even though immune surveillance mechanisms mediated by cytotoxic T cells, natural killer (NK) cells, and antigen-presenting cells can initially suppress tumour development [6–9]. The discovery of immune checkpoints enabled scientists to develop immune checkpoint inhibitors as new treatments for lung cancer where these medications restore the functionality of T-cells against tumour cells. Different response rates emerge because immune landscapes of tumours show heterogeneity as well as how well tumours present their antigens and what level of DNA damage exists (TMB) and require additional research into personalized immunotherapy methods. Maximizing requires a deeper understanding of lung cancer immunobiology because of its importance [10–13].

3.1.2 Tumour Immunology: The Role of the Immune System in Cancer Progression

Immune system regulation of defence between tumour immunity evasion and anti-tumour response determines cancer development. The tumour antigen recognition process is enabled by B cells and cytotoxic T lymphocytes, which establish targeted defence through adaptive immunity; however, innate immunity maintains its initial cellular defence through NK cells and dendritic cells and macrophages [14–16]. Tumour development remains under control by immune surveillance systems through cytotoxic activity and antigen presentation until cancer cells start using diverse evasion techniques for avoiding immune detection. Tumour cells use multiple defence

 DOI: 10.1201/9781003595830-3

mechanisms such as recruiting regulatory immune cells including TAMs and MDSCs and Tregs together with upregulating PD-L1 and CTLA-4 molecules and releasing TGF-β and IL-10 immunosuppressive cytokines [17]. These elements block the system's capability to initiate effective against tumours. Cancer metastasis gets accelerated through chronic inflammation, which along with angiogenesis and metabolic changes and immune suppression creates an enabling environment in the TME for tumour expansion. The revolutionary treatment of cancer happened due to immunological response revival through checkpoint inhibitors combined with CAR-T cell therapy and cancer vaccines but these treatments face barriers from tumour heterogeneity and immune resistance [18–20].

3.1.3 Lung Cancer and Immune Escape Mechanisms

Various escape mechanisms enable lung cancer to evade immune attacks, which develop an immunosuppressive operating environment that speeds up tumour development. LEVITATION of immunological checkpoints with emphasis on PD-1/PD-L1 and CTLA-4 pathways remains vital because it silences T-cell activation while leading to T-cell exhaustion that diminishes cytotoxic activity against tumour cells [21–24]. The depletion of effector T cells makes them perform worse at recognizing and fighting cancer cells so tumours have the chance to expand. The immune system uses T-cell proliferation suppression and regulatory T cell (Tregs) enhancement to create immunosuppressive chemicals including arginase-1, nitric oxide, and reactive oxygen species through myeloid-derived suppressor cells (MDSCs), which support immune evasion to weaken tumour immune defences [25]. Lung cancer cells develop because of tumour-associated macrophages (TAMs), which prefer M2 polarization to create immune suppression that stimulates angiogenesis and release cytokines like IL-10 and TGF-β to prevent cytotoxic immune responses. The relationship between cancer and neutrophils remains dual while TANs (tumour-associated neutrophils) in lung tumours tend to adopt pro-tumorigenic properties that accelerate metastasis and increase inflammatory responses while blocking effective anti-tumour immunity [26–29]. These immune escape pathways working together allow lung cancer cells to defend themselves from immune destruction thus both worsening the disease and making it tougher to treat. Lung cancer therapy has received substantial improvement because of immune checkpoint inhibitors targeting PD-1/PD-L1 and CTLA-4, which reactivate T-cell responses yet adaptive immune suppression combined with diverse TME heterogeneity remains a major challenge to therapy [30].

3.2 IMMUNE SIGNATURES AND BIOMARKERS IN LUNG CANCER

3.2.1 Inflammation and Immunomodulation in Lung Cancer

The creation of cancer and evasion of immune responses from chronic inflammation makes it one of the key contributors to both lung cancer initiation and its spreading mechanisms. Chronic inflammation produces DNA damage and genomic instability while also causing deregulated cell growth because of inflammatory factors, which include cigarette smoke together with environmental contaminants and infections [31–33]. The cancer propagation and survival of tumour cells and the development of new blood vessels alongside metastasis function through cytokines and chemokines including IL-6, TNF-α, IL-1β, and CXCL12, which simultaneously obstruct tumour immunity and support tumour cell development. These pro-inflammatory chemicals activate NF-κB signalling pathways, which are recognized as one of the main signalling pathways [34, 35]. Multiples genes involved in inflammation and metastasis and cell survival receive NF-κB activation, which establishes a positive feedback mechanism leading to tumour growth maintenance. The process of NF-κB activation induces immunological checkpoint molecules while developing (MDSCs), regulatory T cells (Tregs), and tumour-associated macrophages (TAMs) to sustain inhibition of tumour-targeted immune responses [36–38]. The combination of fastened tumour expansion and immunosuppressive conditions within tumour microenvironments leads to immune checkpoint inhibitor therapy resistance. The relationship between immunomodulation and inflammation comprehensively illustrates how difficult it is to treat lung cancer as well as the development of treatment resistance [39–41]. Active signalling pathways that promote tumours along with enhanced immune function emerge when medical experts investigate inflammatory mechanisms and related biological pathways, especially NF-κB. The newly developed anti-tumour treatment approaches focusing on inflammatory cytokines and chemokines with NF-κB pathway targets should enhance treatment results by reducing immune suppression and improving both anti-tumour immunity and current therapeutic resistance mechanisms [42].

3.2.2 Immunogenomics and Molecular Profiling

Knowledge of cancer immunology and treatment selection has advanced significantly because immunogenomics together with molecular profiling. The overall number of mutations within a tumour serves as an essential component that guides immune response. The production of mutant DNA-derived tumour-specific peptides referred to as neoantigens gets increasingly prevalent in cases with high TMB [43–45]. T-cells launch their immunological response when the immune system identifies these neoantigens as foreign substances. The immune evasion strategies operating within tumour microenvironments lead to reduced effectiveness of immune identification mechanism. The discovery of neoantigens alongside their validation plays a necessary role for determining tumour immunogenicity and immune checkpoint inhibitor susceptibility [46]. Single-cell sequencing approaches enable researchers to investigate distinct immune cells throughout the entire tumour microenvironment because they represent an innovative technique to elaborate immunological patterns present in cancers. The approach demonstrates how immune responses form and regulate inside tumours while showing the positions of immune cells and their relations with cancer cells [47–50]. The immune gene landscape of tumours becomes fully discernible when scientists unite TMB examination with neoantigen identification along with mapping individual immune cells throughout the tissue. The technique results in superior prediction capabilities for treatment outcomes. Together with genetic hereditary examinations immunological analytics enable healthcare providers to recognize which patients would best benefit from immunotherapy to develop unique therapeutic regimens [51]. The development of advanced cancer treatments requires a complete understanding of tumour genetic activity since it affects immuno-resistance mechanisms. Immunogenomics stands as the main foundation of precision oncology through which doctors establish specific treatments to target cancers while improving patient results [52–54].

3.2.3 Non-Invasive Immune Biomarkers for Early Detection

Modern diagnosis and treatment evaluation processes utilize non-invasive immunological biomarkers as they offer less violent yet more straightforward options than standard biopsy procedures. Liquid biopsy includes two important elements: circulating tumour cells (CTCs) and circulating tumour DNA (ctDNA) [55]. These elements make real-time genetic change detection possible for tumours and metastasis events. The analysis of ctDNA enables early detection of minimal residual illness while also helping identify tumour-specific mutations for disease diagnosis. CTCs provide valuable information about cancer development due to their source from major or metastatic tumours [56]. MiRNAs act as attractive biomarkers due to their ability to survive in body fluids and their capability of reflecting tumour-specific changes while existing as exosomal RNA molecules released from tumour cells. Immune system evasion occurs through miRNA regulation, which controls

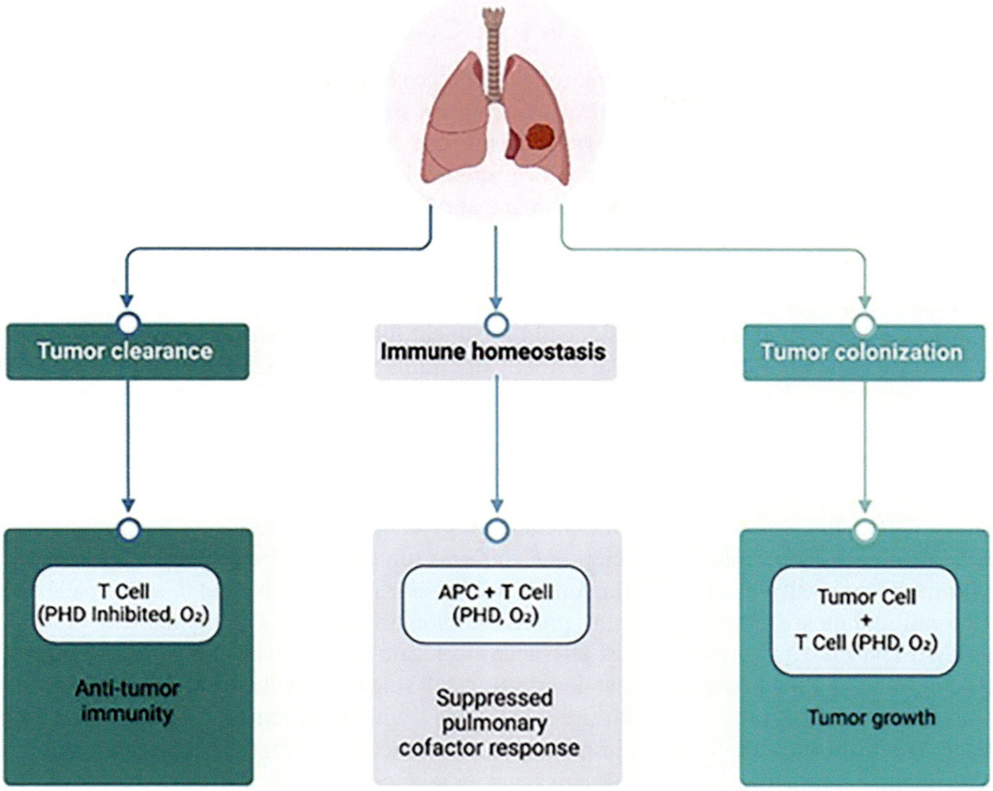

FIGURE 3.1 Lung cancer immune homeostasis through three distinct results, including tumour clearance and immune homeostasis, and tumour colonization

The procedure of tumour clearance depends on T cells with inhibited PHD activity to develop anti-tumour immunity. The immune balance operated by APCs together with T cells works to restrict pulmonary defence mechanisms. During tumour colonization, tumour cells interact with T cells while under the influence of PHD and oxygen to support the growth of tumours.

interactions between the tumour and immune system and the regulatory functions they perform [57–59]. Tumour antigens together with immune checkpoint molecules are immunogenic proteins found in circulation as biomarkers that help identify ideal immunotherapy approaches based on tumour immunological profiles. Liquid biopsy platforms using CTCs, ctDNA, exosomal miRNAs along with immunogenic proteins provide an all-inclusive method to monitor disease progression over time and detect early relapses and assess treatment performance including immunotherapy effectiveness [60]. Current technologies enable real-time tumour evolution and immune process evaluation through liquid biopsy; thus they help develop prediction models and support personal treatment plans. Fast therapy adjustments as well as the detection of treatment resistance become essential because of this invasive-free monitoring approach. Combined usage of non-invasive indicators creates opportunities to enhance patient results and precision oncology by providing better tumour detection methods and immunotherapy monitoring systems and personalized treatment approaches for cancer management [61–63].

3.3 IMMUNOTHERAPEUTIC STRATEGIES IN LUNG CANCER

3.3.1 Immune Checkpoint Inhibitors in Lung Cancer Therapy

The immune system function improved through immune checkpoint inhibitors (ICIs) produces better outcomes during lung cancer treatment. A treatment strategy must use blocking agents against the CTLA-4 and PD-1/PD-L1 pathways. Through this interaction between PD-1 expressed on T-cells and PD-L1 on tumour cells the immune response decreases, leading to tumour cells escaping immune system detection [64]. The activation and anti-tumour support of T-cells depend on inhibitors which target PD-1 (such as pembrolizumab and nivolumab) or PD-L1 (such as atezolizumab) via blocking their connection to reactivate T-cells. The use of ipilimumab and other CTLA-4 inhibitors helps T-cell activation by stopping the inhibitory signals of CTLA-4. Success with ICIs has not solved the persistent issue of resistance in treatment [65]. The resistance mechanisms include low TMB measures (tumour mutational burden) and immunological exclusion and the presence of immunosuppressive elements like regulatory T cells and myeloid-derived suppressor cells throughout the cancer microenvironment. The variables influence both acquired resistance through initial treatment success followed by relapse and primary resistance from complete treatment failure during the first course [66]. Researchers pursue combination therapy strategies to bypass these present barriers. The combination of ICIs with chemotherapy or targeted treatments or immunomodulators tries to enhance immune activation and improve tumour immunogenicity while countering the immunosuppressive elements in the tumour microenvironment. It is vital to select patients who would gain the most from ICIs by using predictive biomarkers [67]. The biomarkers TILs alongside TMB and PD-L1 expression assist physicians to locate patients who show potential treatment response while guiding better treatment decisions. Analysing exosomal microRNAs combined with immune cell screening provides healthcare professionals with targeted treatment strategies for each patient. The available data suggests immunotherapies function as effective treatments but patient results demand improved resistance systems together with combination strategies as well as predictive biomarkers and individual treatment protocols [68, 69].

3.3.2 Adoptive Cell Therapy and Personalized Immunotherapy

Personalised immunotherapy together with adoptive cell therapy (ACT) represents quickly evolving viable treatments which enhance tumour recognition by the immune system for lung cancer patients. Two primary cell treatment methods used in ACT consist of CAR-T (Chimeric Antigen Receptor T) and TCR-T (T-cell Receptor T) [70]. The CAR-T process consists of refined T-cells that receive specialized receptors that recognize particular tumour antigens thereby enabling them to strike cancer cells. The success of CAR-T in treating blood-based cancers is undermined by challenges within the tumour microenvironment and diverse tumour antigens which obstruct its application in solid tumours including lung cancer [71, 72]. The intracellular antigens unique to tumours in lung cancer can be targeted using TCR-T treatment to detect antigens presented by MHC molecules. Immuno-responsive cells named tumour-infiltrating lymphocytes (TILs) occur naturally within the human body. A new category of innovative medicines against cancer exists in the form of personalised vaccines. The vaccines specifically aimed at fighting cancer teach the immune system about tumour identifiers through patient-tailored designs based on individual tumour mutations [73, 74]. Targeted vaccinations made for neoantigens found exclusively in tumours generate immune responses capable of both tumour suppression and prevention of tumour recurrence. The therapeutic performance of lung cancer would significantly improve through the use of adoptive cell therapies with personalized immunotherapy approaches. Future oncology needs personalized solutions as an essential aspect because they develop long-term targeted treatments for patients with lung cancer by dealing with current therapeutic deficiencies like medication immuno-resistance and non-specific activity [75, 76].

3.3.3 The Role of Dendritic Cells and Cancer Vaccination

Dendritic cells function as the key component in cancer immunotherapy approaches since they serve as optimal initiators and controllers of immune response functions. The effective antigen-presenting abilities of dendritic cells are leveraged through immunotherapeutic treatments performed at the DC-based setting to activate immune responses against cancer cells [77]. Treatment success under these procedures requires scientists to extract DCs from patients to load them with tumour-linked antigens before returning them to patients. In vivo DC activation provides an additional strategy for improving T-cell antigen delivery by DCs, which enhances their anticancer immune response. The use of dendritic cell–based therapies for lung cancer treatment demonstrates good clinical potential because they enhance T-cell activity against tumours [78]. Researchers aim to develop peptide and mRNA vaccines as popular cancer vaccines that lead to effective immune responses against tumour-specific antigens. Tumour-associated anti-gens Short sequences are utilized in peptide vaccines to trigger T-cells for cancer cell recognition and elimination in antigen-expressing cancer cells. Immune response specificity increases when these vaccines target cancer antigens which are typically over-expressed in NSCLC and other lung cancers [79]. The implementation of mRNA vaccinations delivers tumour-targeted antigen carrier RNA molecules directly to cells of the patient. This is a newer method of vaccination. Once antigens are produced by the cells they start an immunological reaction. The mechanism enables fast adjustments because mRNA vaccines can easily be modified to detect specific antigens and tumour-related mutations. The second objective of neoantigen-based treatment involves focusing on antigens which develop solely from cancer cells' specific mutations [80]. The immune system detects and attacks neoantigens since they exclusively occur in tumours while being absent from regular body tissue. People who receive neoantigen-based vaccinations experience treatment that matches their individual needs because these vaccinations train their immune systems to identify tumour cells by detecting alterations to DNA. Clinical progress of lung cancer immunotherapy needs dendritic cell treatment integration with peptide and mRNA and neoantigen-based vaccines [81, 82]. Such treatment strategies work to enhance immune system efficiency in targeting cancer cells specifically while giving better therapeutic outcomes to patients through customized approaches. Different treatment approaches combined together provide hope for solving challenges with heterogeneity and immune cells, avoiding treatment for better lung cancer patient results [83].

TABLE 3.1

Exploration of Various Lung Cancer Immunology Elements Through Links Between Research Discoveries and Treatment Approaches and Forecasting Research Paths

Research Area	Key Findings	Mechanisms Involved	Therapeutic Strategies and Clinical Implications	Future Directions	References
Epidemiology and risk factors	Lung cancer is the leading cause of cancer death; smoking is the dominant risk factor, with contributions from environmental pollutants, genetic predisposition, and occupational exposures.	Tobacco carcinogens, asbestos, radon, genetic mutations	Early detection, public health interventions, risk reduction strategies	Improved screening and prevention programs tailored to high-risk populations	[84]

(Continued)

TABLE 3.1 (*Continued*)

Exploration of Various Lung Cancer Immunology Elements Through Links Between Research Discoveries and Treatment Approaches and Forecasting Research Paths

Research Area	Key Findings	Mechanisms Involved	Therapeutic Strategies and Clinical Implications	Future Directions	References
Histological classification	Lung cancer is mainly divided into NSCLC (adenocarcinoma, squamous cell carcinoma, large cell carcinoma) and SCLC (more aggressive with early metastasis).	Variations in tumour biology and genetic profiles	Histology-based treatment selection and targeted therapies	Refinement of molecular subtyping for personalized treatment	[85]
Immune evasion and TME	Tumour cells evade immune destruction via immune checkpoint upregulation and recruitment of immunosuppressive cells (MDSCs, TAMs, Tregs).	PD-1/PD-L1, CTLA-4 pathways; cytokine secretion; chronic inflammation	Use of immune checkpoint inhibitors (ICIs); combination therapies to reprogram the TME	Development of novel agents targeting multiple immune-suppressive pathways	[86, 87]
Chronic inflammation and immunomodulation	Persistent inflammatory stimuli (e.g., smoking, pollutants) promote tumorigenesis by causing DNA damage and creating an immunosuppressive environment.	NF-κB activation; cytokines (IL-6, TNF-α, IL-1β) release; chemokine-driven cell recruitment	Anti-inflammatory strategies; cytokine inhibitors; combinatorial immunotherapies	Targeting inflammatory pathways to enhance immunotherapy efficacy	[88, 89]
Immunogenomics and molecular profiling	High tumour mutational burden (TMB) leads to neoantigen formation, influencing immune recognition and treatment response.	Neoantigen presentation; single-cell sequencing; genomic profiling	Personalized immunotherapies based on biomarker-driven treatment selection	Integration of multi-omics data for precision immuno-oncology	[90]
Liquid biopsy and non-invasive biomarkers	Circulating tumour cells (CTCs), ctDNA, and exosomal miRNAs provide real-time insights into tumour dynamics and treatment response.	Genetic alterations, tumour-derived exosomes, immune profiling	Real-time monitoring, early detection of relapse, and therapy optimization	Development of robust liquid biopsy platforms for personalized management	[91]
Adoptive cell therapy and cancer vaccines	CAR-T, TCR-T, and TIL therapies, along with personalized vaccines, show promise in stimulating tumour-specific immune responses.	Genetic modification of T-cells; neoantigen-based vaccination; antigen presentation	Tailored cell therapies and vaccine platforms to overcome immune resistance	Advances in cell engineering and vaccine design for durable responses	[92]

TABLE 3.1 (*Continued*)
Exploration of Various Lung Cancer Immunology Elements Through Links Between Research Discoveries and Treatment Approaches and Forecasting Research Paths

Research Area	Key Findings	Mechanisms Involved	Therapeutic Strategies and Clinical Implications	Future Directions	References
Oncolytic viruses and nanotechnology	Oncolytic viruses selectively lyse tumour cells while stimulating systemic anti-tumour immunity; nanotechnology improves targeted drug delivery and immune modulation.	Viral-mediated tumour lysis; nanoparticle-based delivery systems; adjuvant effects	Combination of oncolytic virus therapy with ICIs; nanoparticle-enhanced immunotherapy	Integration of novel viral vectors and nanomaterials for precision delivery	[93]
Microbiome-immune interactions	The gut–lung axis and microbial dysbiosis influence lung cancer progression and response to immunotherapy.	Microbiota-induced modulation of immune checkpoints and cytokine networks	Use of probiotics, dietary interventions, and microbiome-targeted therapies	Personalized microbiome modulation to enhance immunotherapy outcomes	[94]
Resistance to immunotherapy and future prospects	Both primary and acquired resistance limit the efficacy of immunotherapies; combination strategies and biomarker-guided approaches are critical.	Adaptive immune suppression; TME heterogeneity; alternate checkpoint pathways	Combinatorial regimens integrating immunotherapy with chemotherapy, radiation, and targeted agents	Leveraging AI, next-gen immune therapeutics, and precision medicine for improved outcomes	[95, 96]

3.4 NOVEL AND EMERGING IMMUNOTHERAPIES

3.4.1 Oncolytic Viruses and Immune Stimulation in Lung Cancer

Oncolytic viruses represent a promising modern method of treating lung cancer because they attach to tumours for cell destruction while enhancing immune system response. The multiplication process of these viruses together with cell destruction creates tumour lysis through selective cancer cell targeting without harming regular tissue. After tumour-associated antigens are released through this process they activate immune system responses by drawing and activating T-cells and dendritic cells and natural killer (NK) cells [97, 98]. The immune response against tumours strengthens through oncolytic viral treatments that independently kill tumours as well as activate immune system defence mechanisms. Scientists have studied three different types of viral vectors for cancer therapy: vesicular stomatitis viruses, herpes simplex viruses, and adenoviruses. Scientists can modify the viruses with therapeutic genes that boost immune activation along with tumour control ability [99]. Research on oncolytic viruses in lung cancer treatment through clinical trials has identified encouraging effects that include partial tumour response with sustained therapeutic advantages. Despite progress there remain two significant challenges concerning tumour immunosuppression combined with developing efficient viral delivery methods into tumour tissues. The future of oncolytic virus therapy for lung cancer treatment will use combinations of immune checkpoint inhibitors together with chemotherapy along with targeted therapies to enhance their effectiveness [100]. Better patient selection criteria along with improved viral engineering will enhance the clinical effectiveness of this therapy. Continuing research will help develop viral strategies and identify therapeutic biomarkers to complete the therapeutic foundation of oncolytic viruses throughout future lung cancer clinical trial activities. The build-up of this approach will enable healthcare providers to produce customized treatment strategies that combine immune activators and tumour-destructing agents as part of sustained cancer protection approaches [101, 102].

3.4.2 Microbiome–Immune System Interactions in Lung Cancer

Medical experts now understand that treatment results along with lung cancer progression depend heavily on mutual relationships between the immune system and microbiome. The gut–lung axis operating as a two-way communication system between lungs and gut microbiota functions as the main mechanism through which immunological responses are affected [103, 104]. The development of lung cancer can be accelerated when dysbiosis activates immunological system modifications, leading to systemic inflammation processes. Studies suggest that the composition of gut microbiota determines immune system reactions to cancer because it controls T-cell functions and immuno-checkpoints. Several types of bacteria influence patient responsiveness to immune checkpoint inhibitor treatment therapies [4, 105]. Studies indicate that better therapeutic outcomes result from bacteria, which activate immune responses focused on tumour destruction while decreasing tumour-based immunosuppression in microenvironment. Unhealthy microbial balance in the body leads to decreased system immunity, which reduces treatment effectiveness of immunotherapy. Research indicates that lung cancer patients' microbiota plays a direct role in determining treatment effectiveness of immunotherapy [106]. Scientific studies analyse probiotics as potential elements for enhancing immune function together with restoring microbial equilibrium. Intake of probiotics improves immune function while reducing inflammation, which makes lung cancer treatments, particularly immunotherapies, more effective by supporting good gut bacterial composition. The management of gut–lung axis regulation might become possible through additional anti-tumour immune system enhancement. The research of lung cancer immune system–microbiota activity shows potential to create personalized therapeutic strategies [107]. Medical staff achieve optimum immune stimulation and enhance the effectiveness of immunotherapies along with better patient results through probiotic administration or dietary modifications that modify microbiome composition. The advancement of microbiome–immune system interaction research will probably produce secondary pharmaceutical options that surpass current lung cancer treatment strategies [108].

3.4.3 Nanotechnology and Immune Engineering in Lung Cancer

The treatment of lung cancer receives revolutionary transformations through nanotechnology, which enables better immune response enhancement as well as precise drug delivery and vaccination platform development. The application of nanoparticles in immune regulation provides the ability to either enhance or suppress immune responses in tumour microenvironments [109]. These carriers directly aim at cancer cells through immunomodulatory substances and boost immune activation while also decreasing systemic side effects such as cytokines and immune checkpoint inhibitors as well as vaccinations. This approach strengthens natural cancer defences and helps overcome immune evasion through direct modulation of tumour-associated immune cells like T cells, macrophages as well as dendritic cells [6, 110]. The development of immunotherapy for lung cancer increasingly depends on nanoparticles for their role in delivering targeted medications. The delivery of therapeutic materials or medications to cancer cells through nanoparticles becomes targeted and safe for healthy tissue because targeted ligands bind directly to the nanoparticles. Placing accurate dose amounts of drugs at the tumour site allows enhanced therapy outcomes together

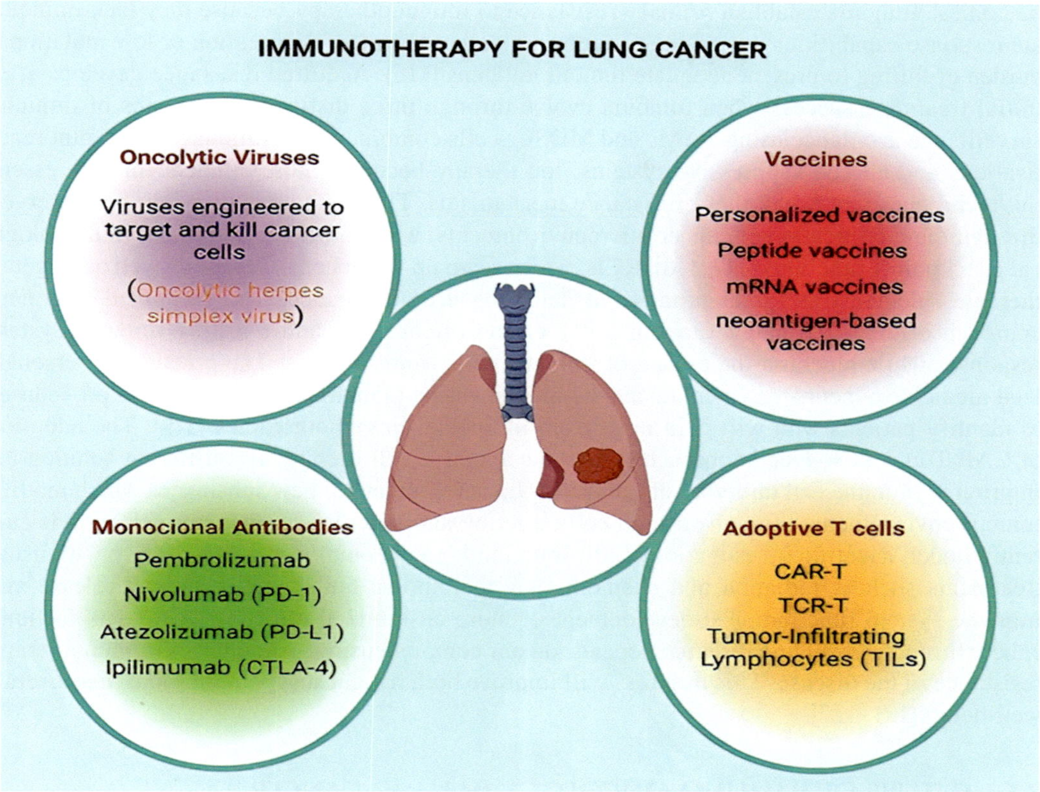

FIGURE 3.2 Multiple approaches for immunotherapeutic lung cancer treatment

Four vital therapeutic approaches, which include oncolytic viruses designed to target cancer cells as well as monoclonal antibodies that focus on PD-1, PD-L1, and CTLA-4 and different kinds of cancer vaccines alongside adoptive T-cell therapies such as CAR-T, TCR-T, and TILs. The central lung illustration demonstrates where the tumours would be situated.

with reduced systemic harm to patients [111]. The controlled mechanism which nanoparticles use for drug delivery results in extended therapeutic outcomes. Nanotechnology serves as a fundamental requirement for the development of lung cancer vaccination platforms made from biomaterials. These platforms contain nanoparticles and biocompatible materials that replicate a natural infection process to deliver tumour-associated antigens to the immune system for generating powerful enduring immune responses [112, 113]. Biomaterial-based vaccines activate T cells through dendritic cell stimulation while boosting antigen delivery thus leading to efficient presentation of tumour-specific immunity. Strong anti-tumour immune responses develop from combining vaccinations with adjuvants or immune-boosting chemicals that get encapsulated in nanoparticles. The effectiveness of cancer vaccinations increases through nanoparticle usage, which simultaneously serves as adjuvants to produce robust immunity. Nanotechnology achieves effective lung cancer treatment by developing vaccines along with target-driven medications and immune function modification systems to boost patient recovery success [114, 115].

3.5 CURRENT CHALLENGES AND FUTURE DIRECTIONS

3.5.1 Overcoming Resistance to Immunotherapy in Lung Cancer

The treatment of lung cancer faces significant obstacles from immunotherapy resistance because acquired along with primary resistance mechanisms reduce the effectiveness of treatment outcomes. Tumours establish primary resistance to immunotherapy because they lack immediate response capabilities due to either tumour microenvironment suppression or low mutational burden or failing to present adequate tumour antigens [116]. Acquired resistance develops after initial treatment success when tumours evolve through three distinct mechanisms of immune surveillance avoidance using TAMs and MDSCs cells combined with immune checkpoint reactivations and altered suppressive systems. Ion therapy becomes more efficient through essential techniques that break these resistance mechanisms. The immune system can be preserved through the modification of cancer microenvironments, while blocking additional immunological checkpoints functions [117, 118]. RCIs, which team up with chemotherapy as well as targeted therapies, demonstrate great promise in defeating drug resistance. The combined treatment approach offers three major functions: decreasing tumour size and boosting immune system responses while blocking the escape of immune cells from tumours. The practice of personalized medicine benefits treatment results by implementing biomarker-based selection procedures to identify patients who will gain most from individual immunotherapies [119]. The adoption of CAR-T and TCR-T cell approaches presents adoptive cell therapy as a different solution for improving immune cell targeting of tumours. The development of new substances which modify tumour environments through stromal cell targeting and vascular components selection is currently under scientific investigation [120]. Immunotherapy combines synergistically with usual treatments such as radiation and chemotherapy to improve both tumour antigen release and immune system function. The development of more customized effective treatments for lung cancer through advancing research depends on our comprehension and defeat of immunotherapy resistance in the disease. This progress will improve both patient survival rates and their overall well-being [121, 122].

3.6 FUTURE OF IMMUNO-ONCOLOGY IN LUNG CANCER

The advancement of therapeutic strategies with developing technologies drives immuno-oncology research of lung cancer towards new directions. Artificial intelligence (AI) uses major amounts of clinical, genomic, and imaging data to discover biomarkers and evaluate treatment responses for customized treatment planning in immunotherapy outcome prediction. AI algorithms using patient-specific prediction criteria enable physicians to pick suitable immunotherapy treatments that result in better clinical results [123]. Next-generation immune therapies have started a treatment revolution and simultaneously drive improvements in lung cancer therapy during this revolutionary period. The new generation of treatments implements innovative approaches such as bispecific antibodies and CAR-T cell therapies and oncolytic viruses as well as fresh immune checkpoint inhibitors that target Pacific-1/PD-L1 and CTLA-4 [124]. The modern approach of next-generation therapies modifies immune functions as well as tumour environment to reinforce immunological defences while helping immune system overcome resistance factors and boosting treatment outcomes. Flation may proceed with precision by analysing both tumour profiles and immune system characteristics and genetic variations in individual patients. Precision therapy approaches for immune therapy become achievable through genomic research because it provides information about tumour variety as well as immune resistance mechanisms [125]. Patient benefits come from using immunotherapy together with tumour-infiltrating lymphocytes and precision or chemotherapy treatments along with personalized cancer vaccines. Lung cancer patients will thrive in the upcoming years because of AI partnership with next-generation immune therapies and precision-medicine-developed treatments that provide individualized, efficient and extended outcomes. The potential of lung cancer immunotherapy cybersecurity will expand treatment success rates and offer better hope to patients through AI-based predictive analysis and development of novel treatment methods and precise medicine methodologies [126, 127].

3.7 CONCLUDING REMARKS AND PERSPECTIVES

Major advancements occurred in lung cancer immunology during recent years because immunotherapy treatments have brought new hope to patients. Exploitation of the immune system to fight cancer became possible due to experience gained from currently used PD-1/PD-L1 and CTLA-4 blocking treatments. The implementation of research-based clinical tools remains challenging even after reaching those significant achievements. The challenge of managing advanced lung cancer due to intricate factors involving immune system responses leads to ongoing hurdles which include heterogenous tumours and immune escape pathways together with resistance to immunotherapy treatments. New predictive biomarkers should be developed as they can select appropriate patients for immunotherapy and minimize drug-related side effects for non-responder patients. The development of combination treatments emerged from monotherapy shortcomings even though research on this field remains novel. Lung cancer immunology research aims to solve these barriers through modern technology implementation that provides superior prediction capabilities and individual patient treatment strategies. Tumour-specific antigen targeting together with enhanced treatment effectiveness are fundamental advantages that come from developing decades of next-generation immune solutions, which include personal cancer vaccines and CAR-T and TCR-T cellular therapies and bispecific antibodies. The clinical relevance of precision medicine increases when therapists apply individual patient-based tumour and immune profiles to develop prolonged effective treatment strategies. Advancements in lung cancer immunotherapy occur because researchers develop better understanding of tumour microenvironments combined with studies about microbial interactions and new immunomodulatory drug delivery systems. Clinical trials may become more adaptable through experimental tumour monitoring together with liquid biopsy testing that monitors resistance mechanisms during clinical trials. Patients worldwide will receive the most suitable and effective treatment for their lung cancer because development in immunology research has made medicines accessible to everyone.

REFERENCES

1. Florez, N., et al., Lung cancer in women: the past, present, and future. *Clin Lung Cancer*, 2024. **25**(1): p. 1–8.
2. Alberg, A.J., et al., Epidemiology of lung cancer: diagnosis and management of lung cancer, 3rd ed: American college of chest physicians evidence-based clinical practice guidelines. *Chest*, 2013. **143**(5 Suppl): p. e1S–e29S.
3. Thapa, R., et al., Epithelial-mesenchymal transition to mitigate age-related progression in lung cancer. *Ageing Res Rev*, 2024. **102**: p. 102576.
4. Thapa, R., et al., Influence of gut and lung dysbiosis on lung cancer progression and their modulation as promising therapeutic targets: a comprehensive review. *MedComm (2020)*, 2024. **5**(12): p. e70018.
5. Krpina, K., et al., Small cell lung carcinoma: current diagnosis, biomarkers, and treatment options with future perspectives. *Biomedicines*, 2023. **11**(7).
6. Zabeti Touchaei, A. and S. Vahidi, MicroRNAs as regulators of immune checkpoints in cancer immunotherapy: targeting PD-1/PD-L1 and CTLA-4 pathways. *Cancer Cell Int*, 2024. **24**(1): p. 102.
7. Zhang, H., et al., Regulatory mechanisms of immune checkpoints PD-L1 and CTLA-4 in cancer. *J Exp Clin Cancer Res*, 2021. **40**(1): p. 184.
8. Thapa, R., et al., New horizons in lung cancer management through ATR/CHK1 pathway modulation. *Future Med Chem*, 2023. **15**(19): p. 1807–1818.
9. Thapa, R., et al., A review of glycogen synthase kinase-3 (GSK3) inhibitors for cancers therapies. *Int J Biol Macromol*, 2023. **253**(Pt 7): p. 127375.
10. Sharma, P., et al., Immune checkpoint therapy—current perspectives and future directions. *Cell*, 2023. **186**(8): p. 1652–1669.
11. Barrueto, L., et al., Resistance to checkpoint inhibition in cancer immunotherapy. *Transl Oncol*, 2020. **13**(3): p. 100738.

12. Singh, S., et al., Emerging role of tumor suppressing microRNAs as therapeutics in managing non-small cell lung cancer. *Pathol Res Pract*, 2024. **256**: p. 155222.
13. Subramaniyan, V., et al., A review on epidermal growth factor receptor's role in breast and non-small cell lung cancer. *Chem Biol Interact*, 2022. **351**: p. 109735.
14. Deo, A.S., et al., Participation of T cells in generating immune protection against cancers. *Pathol Res Pract*, 2024. **262**: p. 155534.
15. Sharma, P., et al., Emerging trends in the novel drug delivery approaches for the treatment of lung cancer. *Chem Biol Interact*, 2019. **309**: p. 108720.
16. Shrivastava, G., et al., Targeting LIN28: a new hope in prostate cancer theranostics. *Future Oncol*, 2021. **17**(29): p. 3873–3880.
17. Kim, S.K. and S.W. Cho, The evasion mechanisms of cancer immunity and drug intervention in the tumor microenvironment. *Front Pharmacol*, 2022. **13**: p. 868695.
18. Ashique, S., et al., Multi drug resistance in colorectal cancer- approaches to overcome, advancements and future success. *Adv Cancer Biol Metastasis*, 2024. **10**: p. 100114.
19. Rohilla, S., et al., Recent developments and challenges in molecular-targeted therapy of non-small-cell lung cancer. *J Environ Pathol Toxicol Oncol*, 2023. **42**(1): p. 27–50.
20. Sadique Hussain, M., et al., Exosomal ncRNAs in liquid biopsies for lung cancer. *Clin Chim Acta*, 2025. **565**: p. 119983.
21. Dutta, S., et al., Targets of immune escape mechanisms in cancer: basis for development and evolution of cancer immune checkpoint inhibitors. *Biology (Basel)*, 2023. **12**(2).
22. Tian, Y., et al., Potential immune escape mechanisms underlying the distinct clinical outcome of immune checkpoint blockades in small cell lung cancer. *J Hematol Oncol*, 2019. **12**(1): p. 67.
23. Paudel, K.R., et al., Berberine-loaded liquid crystalline nanoparticles inhibit non-small cell lung cancer proliferation and migration in vitro. *Environ Sci Pollut Res Int*, 2022. **29**(31): p. 46830–46847.
24. Paudel, K.R., et al., Computational and biological approaches in repurposing ribavirin for lung cancer treatment: unveiling antitumorigenic strategies. *Life Sci*, 2024. **352**: p. 122859.
25. Labani-Motlagh, A., M. Ashja-Mahdavi, and A. Loskog, The tumor microenvironment: a milieu hindering and obstructing antitumor immune responses. *Front Immunol*, 2020. **11**: p. 940.
26. Zhang, W., et al., Macrophage polarization in the tumor microenvironment: emerging roles and therapeutic potentials. *Biomed Pharmacother*, 2024. **177**: p. 116930.
27. Zhou, Y., et al., The role of tumor-associated macrophages in lung cancer: from mechanism to small molecule therapy. *Biomed Pharmacother*, 2024. **170**: p. 116014.
28. Kokkinis, S., et al., Liposomal encapsulated curcumin attenuates lung cancer proliferation, migration, and induces apoptosis. *Heliyon*, 2024. **10**(19): p. e38409.
29. Lee, J.Y., et al., Molecular mechanisms underlying the regulation of tumour suppressor genes in lung cancer. *Biomed Pharmacother*, 2024. **173**: p. 116275.
30. Brahmer, J.R. and D.M. Pardoll, Immune checkpoint inhibitors: making immunotherapy a reality for the treatment of lung cancer. *Cancer Immunol Res*, 2013. **1**(2): p. 85–91.
31. Grivennikov, S.I., F.R. Greten, and M. Karin, Immunity, inflammation, and cancer. *Cell*, 2010. **140**(6): p. 883–899.
32. Hussain, M.S., et al., Groundbreaking mRNA lung cancer vaccine trials: a new dawn in cancer treatment. *Curr Cancer Drug Targets*, 2025. **25**.
33. Jha, S.K., et al., Cellular senescence in lung cancer: molecular mechanisms and therapeutic interventions. *Ageing Res Rev*, 2024. **97**: p. 102315.
34. Burkholder, B., et al., Tumor-induced perturbations of cytokines and immune cell networks. *Biochim Biophys Acta (BBA)—Rev Cancer*, 2014. **1845**(2): p. 182–201.
35. Chu, X., et al., Cancer stem cells: advances in knowledge and implications for cancer therapy. *Signal Transduct Target Ther*, 2024. **9**(1): p. 170.
36. Pavitra, E., et al., The role of NF-κB in breast cancer initiation, growth, metastasis, and resistance to chemotherapy. *Biomed Pharmacother*, 2023. **163**: p. 114822.
37. Hussain, M.S., et al., From carcinogenesis to therapeutic avenues: lncRNAs and mTOR crosstalk in lung cancer. *Pathol Res Pract*, 2024. **253**: p. 155015.
38. Hussain, M.S., et al., Unlocking the secrets: volatile organic compounds (VOCs) and their devastating effects on lung cancer. *Pathol Res Pract*, 2024. **255**: p. 155157.
39. Borgeaud, M., et al., Novel targets for immune-checkpoint inhibition in cancer. *Cancer Treat Rev*, 2023. **120**: p. 102614.
40. Hussain, M.S., et al., Long non-coding RNAs in lung cancer: unraveling the molecular modulators of MAPK signaling. *Pathol Res Pract*, 2023. **249**: p. 154738.

41. Hussain, M.S., et al., Unraveling NEAT1's complex role in lung cancer biology: a comprehensive review. *EXCLI J*, 2024. **23**: p. 34–52.
42. Cao, Y., et al., NF-κB signaling pathway in tumor microenvironment. *Front Immunol*, 2024. **15**: p. 1476030.
43. Wells, D.K., et al., Key parameters of tumor epitope immunogenicity revealed through a consortium approach improve neoantigen prediction. *Cell*, 2020. **183**(3): p. 818–834.e13.
44. Gupta, G., et al., A clinical update on metformin and lung cancer in diabetic patients. *Panminerva Med*, 2018. **60**(2): p. 70–75.
45. Hakami, M.A., et al., PVT1 lncRNA in lung cancer: a key player in tumorigenesis and therapeutic opportunities. *Pathol Res Pract*, 2024. **253**: p. 155019.
46. Xie, N., et al., Neoantigens: promising targets for cancer therapy. *Signal Transduct Target Ther*, 2023. **8**(1): p. 9.
47. Erfanian, N., et al., Immunotherapy of cancer in single-cell RNA sequencing era: a precision medicine perspective. *Biomed Pharmacother*, 2022. **146**: p. 112558.
48. Kuksin, M., et al., Applications of single-cell and bulk RNA sequencing in onco-immunology. *Eur J Cancer*, 2021. **149**: p. 193–210.
49. Dahiya, R., et al., Harnessing pyroptosis for lung cancer therapy: the impact of NLRP3 inflammasome activation. *Pathol Res Pract*, 2024. **260**: p. 155444.
50. Gupta, G., et al., Tumor suppressor role of miR-503. *Panminerva Med*, 2018. **60**(1): p. 17–24.
51. Anzar, I., et al., The interplay between neoantigens and immune cells in sarcomas treated with checkpoint inhibition. *Front Immunol*, 2023. **14**: p. 1226445.
52. Tsimberidou, A.M., et al., Review of precision cancer medicine: evolution of the treatment paradigm. *Cancer Treat Rev*, 2020. **86**: p. 102019.
53. Chan, Y., et al., Sugar-based nanoparticles for respiratory diseases: a new paradigm in the nanoworld. *Future Med Chem*, 2020. **12**(21): p. 1887–1890.
54. Chellappan, D.K., et al., Targeting the mitochondria in chronic respiratory diseases. *Mitochondrion*, 2022. **67**: p. 15–37.
55. Armakolas, A., M. Kotsari, and J. Koskinas, Liquid biopsies, novel approaches and future directions. *Cancers (Basel)*, 2023. **15**(5).
56. Zhang, L., et al., The interplay of circulating tumor DNA and chromatin modification, therapeutic resistance, and metastasis. *Mol Cancer*, 2019. **18**(1): p. 36.
57. Ortiz-Quintero, B., Extracellular microRNAs as intercellular mediators and noninvasive biomarkers of cancer. *Cancers (Basel)*, 2020. **12**(11).
58. Bhat, A.A., et al., MALAT1: a key regulator in lung cancer pathogenesis and therapeutic targeting. *Pathol Res Pract*, 2024. **253**: p. 154991.
59. Bhat, A.A., et al., Polysaccharide-based nanomedicines targeting lung cancer. *Pharmaceutics*, 2022. **14**(12).
60. Mazzeo, R., et al., Liquid biopsy in triple-negative breast cancer: unlocking the potential of precision oncology. *ESMO Open*, 2024. **9**(10): p. 103700.
61. Ma, L., et al., Liquid biopsy in cancer current: status, challenges and future prospects. *Signal Transduct Target Ther*, 2024. **9**(1): p. 336.
62. Ho, H.Y., et al., Liquid biopsy in the clinical management of cancers. *Int J Mol Sci*, 2024. **25**(16).
63. Alharbi, K.S., et al., PI3K/Akt/mTOR pathways inhibitors with potential prospects in non-small-cell lung cancer. *J Environ Pathol Toxicol Oncol*, 2022. **41**(4): p. 85–102.
64. Wojtukiewicz, M.Z., et al., Inhibitors of immune checkpoints-PD-1, PD-L1, CTLA-4-new opportunities for cancer patients and a new challenge for internists and general practitioners. *Cancer Metastasis Rev*, 2021. **40**(3): p. 949–982.
65. Shiravand, Y., et al., Immune checkpoint inhibitors in cancer therapy. *Curr Oncol*, 2022. **29**(5): p. 3044–3060.
66. Ma, L., et al., Targeting carnitine palmitoyl transferase 1A (CPT1A) induces ferroptosis and synergizes with immunotherapy in lung cancer. *Signal Transduct Target Ther*, 2024. **9**(1): p. 64.
67. Birnboim-Perach, R. and I. Benhar, Using combination therapy to overcome diverse challenges of immune checkpoint inhibitors treatment. *Int J Biol Sci*, 2024. **20**(10): p. 3911–3922.
68. Filipovic, A., G. Miller, and J. Bolen, Progress toward identifying exact proxies for predicting response to immunotherapies. *Front Cell Dev Biol*, 2020. **8**: p. 155.
69. Alotaibi, F., Exosomal microRNAs in cancer: potential biomarkers and immunotherapeutic targets for immune checkpoint molecules. *Front Genet*, 2023. **14**: p. 1052731.
70. Chang, Y., et al., Advancements in adoptive CAR immune cell immunotherapy synergistically combined with multimodal approaches for tumor treatment. *Bioact Mater*, 2024. **42**: p. 379–403.

71. Jogalekar, M.P., et al., CAR T-Cell-Based gene therapy for cancers: new perspectives, challenges, and clinical developments. *Front Immunol*, 2022. **13**: p. 925985.
72. Maalej, K.M., et al., CAR-cell therapy in the era of solid tumor treatment: current challenges and emerging therapeutic advances. *Mol Cancer*, 2023. **22**(1): p. 20.
73. Waldman, A.D., J.M. Fritz, and M.J. Lenardo, A guide to cancer immunotherapy: from T cell basic science to clinical practice. *Nat Rev Immunol*, 2020. **20**(11): p. 651–668.
74. Tsimberidou, A.M., et al., T-cell receptor-based therapy: an innovative therapeutic approach for solid tumors. *J Hematol Oncol*, 2021. **14**(1): p. 102.
75. Chakraborty, C., et al., The landscape of neoantigens and its clinical applications: from immunobiology to cancer vaccines. *Curr Res Biotechnol*, 2024. **7**: p. 100177.
76. Li, X., et al., Neoantigen cancer vaccines: a new star on the horizon. *Cancer Biol Med*, 2023. **21**(4): p. 274–311.
77. Hato, L., et al., Dendritic cells in cancer immunology and immunotherapy. *Cancers (Basel)*, 2024. **16**(5).
78. Najafi, S. and K. Mortezaee, Advances in dendritic cell vaccination therapy of cancer. *Biomed Pharmacother*, 2023. **164**: p. 114954.
79. Miao, L., Y. Zhang, and L. Huang, mRNA vaccine for cancer immunotherapy. *Mol Cancer*, 2021. **20**(1): p. 41.
80. Li, Y., et al., mRNA vaccine in cancer therapy: current advance and future outlook. *Clin Transl Med*, 2023. **13**(8): p. e1384.
81. Ruzzi, F., et al., Cancer vaccines: target antigens, vaccine platforms and preclinical models. *Mol Aspects Med*, 2025. **101**: p. 101324.
82. Zhang, Z., et al., Neoantigen: a new breakthrough in tumor immunotherapy. *Front Immunol*, 2021. **12**: p. 672356.
83. Garg, P., et al., Emerging therapeutic strategies to overcome drug resistance in cancer cells. *Cancers (Basel)*, 2024. **16**(13).
84. Dela Cruz, C.S., L.T. Tanoue, and R.A. Matthay, Lung cancer: epidemiology, etiology, and prevention. *Clin Chest Med*, 2011. **32**(4): p. 605–644.
85. Nicholson, A.G., et al., The 2021 WHO classification of lung tumors: impact of advances since 2015. *J Thorac Oncol*, 2022. **17**(3): p. 362–387.
86. Jiang, X., et al., Role of the tumor microenvironment in PD-L1/PD-1-mediated tumor immune escape. *Mol Cancer*, 2019. **18**(1): p. 10.
87. Li, Y.R., T. Halladay, and L. Yang, Immune evasion in cell-based immunotherapy: unraveling challenges and novel strategies. *J Biomed Sci*, 2024. **31**(1): p. 5.
88. Thompson, P.A., et al., Environmental immune disruptors, inflammation and cancer risk. *Carcinogenesis*, 2015. **36**(Suppl 1): p. S232–S253.
89. de Visser, K.E. and J.A. Joyce, The evolving tumor microenvironment: from cancer initiation to metastatic outgrowth. *Cancer Cell,* 2023. **41**(3): p. 374–403.
90. Wang, P., Y. Chen, and C. Wang, Beyond tumor mutation burden: tumor neoantigen burden as a biomarker for immunotherapy and other types of therapy. *Front Oncol*, 2021. **11**: p. 672677.
91. Wang, X., et al., Research progress of CTC, ctDNA, and Evs in cancer liquid biopsy. *Front Oncol*, 2024. **14**: p. 1303335.
92. Ott, P.A., et al., An update on adoptive T-cell therapy and neoantigen vaccines. *Am Soc Clin Oncol Educ Book*, 2019. **39**: p. e70–e78.
93. Zhang, Y., et al., Nanoengineering-armed oncolytic viruses drive antitumor response: progress and challenges. *MedComm (2020),* 2024. **5**(10): p. e755.
94. Lu, Y., et al., Gut microbiota influence immunotherapy responses: mechanisms and therapeutic strategies. *J Hematol Oncol*, 2022. **15**(1): p. 47.
95. Sharma, P., et al., Primary, adaptive, and acquired resistance to cancer immunotherapy. *Cell*, 2017. **168**(4): p. 707–723.
96. Tufail, M., et al., Hallmarks of cancer resistance. *iScience*, 2024. **27**(6): p. 109979.
97. Singh, P.K., et al., Oncolytic viruses & their specific targeting to tumour cells. *Indian J Med Res*, 2012. **136**(4): p. 571–584.
98. Kamta, J., et al., Advancing cancer therapy with present and emerging simer-oncology approaches. *Front Oncol*, 2017. **7**: p. 64.
99. Feola, S., et al., Oncolytic immunovirotherapy: a long history of crosstalk between viruses and immune system for cancer treatment. *Pharmacol Ther*, 2022. **236**: p. 108103.
100. Nia, G.E., et al., Advances in cell-based delivery of oncolytic viruses as therapy for lung cancer. *Mol Ther Oncol*, 2024. **32**(1): p. 200788.

101. Chen, Y., et al., Current status and research progress of oncolytic virus. *Pharm Sci Adv*, 2024. **2**: p. 100037.
102. Lin, D., Y. Shen, and T. Liang, Oncolytic virotherapy: basic principles, recent advances and future directions. *Signal Transduct Target Ther*, 2023. **8**(1): p. 156.
103. Zhao, Y., et al., Role of lung and gut microbiota on lung cancer pathogenesis. *J Cancer Res Clin Oncol*, 2021. **147**(8): p. 2177–2186.
104. Bou Zerdan, M., et al., The lung microbiota and lung cancer: a growing relationship. *Cancers (Basel)*, 2022. **14**(19).
105. Liu, X., et al., The role of gut microbiota in lung cancer: from carcinogenesis to immunotherapy. *Front Oncol*, 2021. **11**: p. 720842.
106. Wu, M., et al., The role of gut microbiota in tumor immunotherapy. *J Immunol Res*, 2021. **2021**: p. 5061570.
107. Jiang, S., et al., An emerging strategy: probiotics enhance the effectiveness of tumor immunotherapy via mediating the gut microbiome. *Gut Microbes*, 2024. **16**(1): p. 2341717.
108. Ma, T., et al., Targeting gut microbiota and metabolism as the major probiotic mechanism—an evidence-based review. *Trends Food Sci Technol*, 2023. **138**: p. 178–198.
109. Karahmet Sher, E., et al., Nanotechnology in medicine revolutionizing drug delivery for cancer and viral infection treatments. *Int J Pharm*, 2024. **660**: p. 124345.
110. Kong, B., et al., mRNA: A promising platform for cancer immunotherapy. *Adv Drug Deliv Rev*, 2023. **199**: p. 114993.
111. Wang, J., et al., Application of nanoparticles in the treatment of lung cancer with emphasis on receptors. *Front Pharmacol*, 2021. **12**: p. 781425.
112. Hsu, C.-Y., et al., An overview of nanoparticles in drug delivery: properties and applications. *S Afr J Chem Eng*, 2023. **46**: p. 233–270.
113. Ashrafizadeh, M., et al. (Nano)platforms in breast cancer therapy: drug/gene delivery, advanced nanocarriers and immunotherapy. *Med Res Rev*, 2023. **43**(6): p. 2115–2176.
114. Sheykhhasan, M., et al., Revolutionizing cancer treatment: the power of dendritic cell-based vaccines in immunotherapy. *Biomed Pharmacother*, 2025. **184**: p. 117858.
115. Calmeiro, J., et al., Biomaterial-based platforms for in situ dendritic cell programming and their use in antitumor immunotherapy. *J Immunother Cancer*, 2019. **7**(1): p. 238.
116. Boyero, L., et al., Primary and acquired resistance to immunotherapy in lung cancer: unveiling the mechanisms underlying of immune checkpoint blockade therapy. *Cancers (Basel)*, 2020. **12**(12).
117. Hou, A., et al., Targeting myeloid-derived suppressor cell, a promising strategy to overcome resistance to immune checkpoint inhibitors. *Front Immunol*, 2020. **11**: p. 783.
118. Cui, J.-W., et al., Tumor immunotherapy resistance: revealing the mechanism of PD-1/PD-L1-mediated tumor immune escape. *Biomed Pharmacother*, 2024. **171**: p. 116203.
119. Bayat Mokhtari, R., et al., Combination therapy in combating cancer. *Oncotarget*, 2017. **8**(23): p. 38022–38043.
120. Liu, Y., et al., TCR-T immunotherapy: the challenges and solutions. *Front Oncol*, 2021. **11**: p. 794183.
121. Jagodinsky, J.C., P.M. Harari, and Z.S. Morris, The promise of combining radiation therapy with immunotherapy. *Int J Radiat Oncol Biol Phys*, 2020. **108**(1): p. 6–16.
122. Garg, P., et al., Next-generation immunotherapy: advancing clinical applications in cancer treatment. *J Clin Med*, 2024. **13**(21).
123. Zhang, H., et al., The application of artificial intelligence in lung cancer: a narrative review. *Transl Cancer Res*, 2021. **10**(5): p. 2478–2487.
124. Al-Haideri, M., et al., CAR-T cell combination therapy: the next revolution in cancer treatment. *Cancer Cell Int*, 2022. **22**(1): p. 365.
125. Ghemrawi, R., et al., Revolutionizing cancer treatment: recent advances in immunotherapy. *Biomedicines*, 2024. **12**(9).
126. Dailah, H.G., et al., Potential role of immunotherapy and targeted therapy in the treatment of cancer: a contemporary nursing practice. *Heliyon*, 2024. **10**(2): p. e24559.
127. Oli, A.N., et al., Tumour immunotherapy and applications of immunological products: a review of literature. *J Immunol Res*, 2024. **2024**: p. 8481761.

4 Nanotechnology in the Diagnosis of Lung Cancer

Bhawana Sharma[1], *Piush Sharma*[2], *Bhushan Rane*[3], *Anurag Mishra*[4], *Sobhit Singh Rajput*[5], *and Yogendra Singh*[6*]

[1] Department of Pharmaceutics, Maharishi Arvind College of Pharmacy, Jaipur, Rajasthan, India

[2] Department of Pharmacognosy, Maharishi Arvind College of Pharmacy, Jaipur, Rajasthan, India

[3] Department of Pharmaceutics, Shri D Vispute College of Pharmacy and Research Centre, Panvel, Navi Mumbai, Maharashtra 410206, India

[4] NIMS Institute of Pharmacy, NIMS University Rajasthan, Jaipur, India

[5] MIPER, Mangalayatan University, Beswan, Aligarh, Uttar Pradesh, India

[6] Department of Pharmacology, Maharishi Arvind College of Pharmacy, Ambabari, Jaipur, Rajasthan, India

*Corresponding author: Yogendra Singh, yogendra.singh119@gmail.com.

4.1 INTRODUCTION

The most current statistics show that lung cancer accounts for the majority of cancer-related fatalities worldwide. The second most prevalent cancer-related death in 2022 was colorectal cancer, while lung cancer was 2.5 million times more common, killing almost 1.8 million individuals. Based on projections from the American Cancer Society (ACS), lung cancer was expected to account for 127,070 fatalities in 2023, making it the top cause of cancer-related mortality overall. A total of 120,790 women and 117,550 men will be affected by the 238,340 new cases of lung cancer. Lung cancer affects 1.16% of men and 1.17% of women; the lifetime incidence rates for these two genders are 6.2% and 5.8%, respectively. Since these probabilities are based on the total lung cancer prevalence in the population, individuals with a history of smoking are at a higher risk (1).

Despite the terrible impact this disease has on people's lives, nobody talks about how preventable the majority of lung cancer cases are. Among all causes of lung cancer, smoking is the leading culprit, accounting for 85% of cases according to data from the WHO. They not only are one of the leading causes of cancer-related deaths overall, but also account for a disproportionately high number of deaths among nonsmokers (2). Tobacco smoke, welding fumes, asbestos-containing materials, automobile combustion, air pollution from both indoors and outdoors, and other contaminants are common causes. Lung cancer is typically fatal since it is often discovered at a late stage, when treatment has already failed. Signs of lung cancer include being extremely lethargic, not wanting to eat, having trouble breathing, experiencing chest discomfort, and having a chronic cough. Lung cancer occurs when abnormal cells proliferate uncontrollably in lung tissue, a process accelerated by exposure to carcinogens such as smoke from cigarettes, airborne contaminants, or certain chemicals (3).

The tremendous strides in molecular cancer research have not yet translated into clinically applicable translational improvements, indicating that our knowledge of the illness is outpacing our capacity to treat it. This is due to the fact that lung cancers are often detected after they are already

 DOI: 10.1201/9781003595830-4

advanced, and it is challenging to provide therapy regimens for late-stage tumors at high-enough concentrations to be effective without harming healthy tissue in the process. Imagine a perfect therapeutic scenario where a curative medicine is administered to tumor cells in their early phases of malignant transformation in a very selective manner.

The concept of the "magic bullet" paradigm initially emerged in the work of Paul Ehrlich (4), an immunologist. The advancement of nanotechnology has opened the door for this concept to be considered seriously. More and more people are getting CT scans, and early identification of lung cancer has become increasingly common, all because people are paying more attention to their health these days (5, 6). A number of navigational techniques are available for use with peripheral lung lesions. The list goes on and on, and it includes things like virtual bronchoscopic navigation, electromagnetic navigation bronchoscopy, advanced bronchoscopic trans parenchymal nodule access, Lungpro, and X-ray line monitoring for lung biopsy (7).

The recent years have seen a proliferation of tools for the detection of lung cancer. Among these methods are imaging examinations such as CT scans, bone scans, PET scans, and MRI, as well as cytology of pleural fluid and sputum, AFB, EUS, and EBUS (8–10). Cancer biomarkers, including specific proteins, nucleic acids, cells, and aromatic chemical substances, can be detected in various bodily fluids and tissues, including urine, blood, and sputum, exhaled breath, and tissues. Because gaseous molecules adsorb to solid surfaces, these tests aren't very sensitive, specific, complicated, or useful (11, 12).

While undergoing imaging, patients run the risk of incurring substantial radiation exposure and incur a hefty financial burden (13). A lack of specificity in the photographic developer or an inability of the beam to penetrate the tissue effectively can also lead to false-negative findings (14).

In actuality, X-ray chest imaging provides a simple way to identify lung cancer without revealing any interior structures. The decreased radiation exposure and improved safety of low dose computed tomography (LDCT) make it an attractive option for screening for lung cancer (15, 16).

There is a significant sensitivity bottleneck with these detection methods: sputum cytology (66%), pleural fluid cytology (70%), PET scan (88%), CT scan (55%), and bone scan (77%) (17). Developing sensitive, practical, non-invasive, and cost-effective detection tools for lung cancer is of the utmost importance. Fiber optic bronchoscopy has a low positive rate since it can't reach lesions in the periphery of the lung. The capacity of nanotechnology to identify cancer cells in their early stages has led many to anticipate that it will play a significant role in cancer diagnosis. The study and application of synthetic materials, devices, and systems is known as nanotechnology. However, the spectrum of size of between 100 and 1 nm in at least one dimension is not consistently used in scientific publications and is not particularly relevant for solving unfulfilled medical needs.

These materials are ideally suited for usage in biological and medical settings due to their intrinsic physicochemical properties, which are a result of their size and composition. One topic of nanomedicine that has garnered a lot of interest, particularly in the past two decades, is nanoparticle (NP) research. NPs may totally revolutionize the way cancer and other illnesses are detected and treated. The term "nanoparticle" describes the vast variety of nano-vector formations in terms of size and shape (Figure 4.1).

NPs can adapt their surface chemistry and chemical makeup to various environmental and internal influences. All of these factors contribute to their popularity. Utilizing their many characteristics, NPs may be able to traverse the chemical and biological obstacles encountered by the human body. Improved localization and efficacy of diagnostic and therapeutic procedures, with less invasiveness and more biocompatibility, can improve patients' quality of life.

However, the full translational potential of NPs has not been realized yet, even if preclinical studies are intriguing. Problems with repeatability, issues with producing on a big scale, and the potential for miscalculation, neglect, or undervaluation of toxicological and safety concerns are the primary reasons for this. Elongated and bio-persistent NPs used in preclinical studies to carry medications to the lungs have been shown to induce fibrosis, highlighting the importance of conducting long-term safety assessments of nanomaterials (18) and an asbestos-like scar ring of the lung (19).

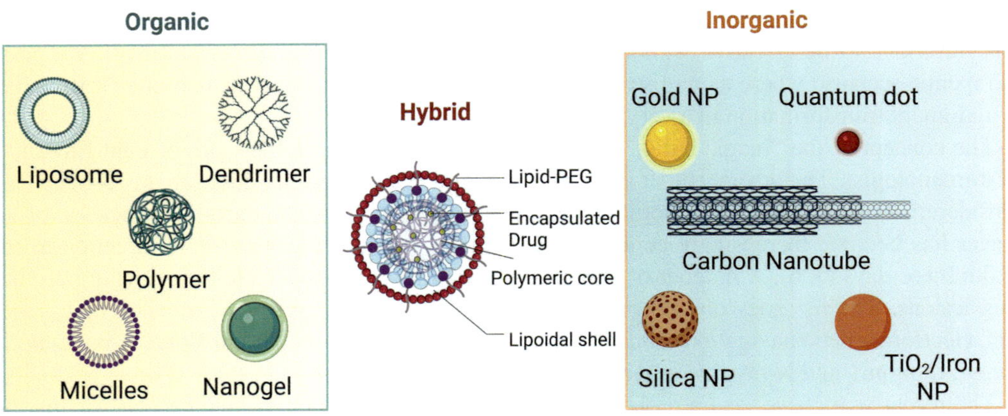

FIGURE 4.1 Nanoparticles utilized for cancer detection

The composition of nanoparticles may be described in three ways: organic (including polymer compounds, lipids, dendrimers, and proteins), inorganic (which incorporates metallic substances, rare earthy substances, and carbonaceous), and hybrid.

Currently, certain NPs are proving to be useful in cancer diagnoses, allowing doctors to scan the exact areas of the body where cancer is detected. Extensive use of nanomaterials in cancer detection and treatment has been facilitated by their remarkable flexibility, scalability, and controllability of thermal, magnetic, and optical characteristics (20). Surface modification can be used to create functional NPs that stabilize drugs, as well as those that target specificities, respond to stimuli, increase detection sensitivity, and improve biocompatibility (21).

Additionally, it has been shown that a novel concept may distinguish between healthy volunteers and individuals with non–small cell lung cancer. To do this, we compare the two sets of blood proteins using the biomolecular corona that surrounds the NPs. The creation of new, very efficient, and specialized diagnostic processes has been spurred by the integration of nanotechnology with diagnostic goals. It is believed that these methods will lead to more effective early identification, better tracking of development patterns, and more precise medicines for the follow-up treatment of lung cancer (22).

4.2 NANOTECHNOLOGY IN LUNG CANCER DIAGNOSIS

Getting a correct diagnosis is critical since it influences the treatment strategy. Locating the tumor precisely and quantifying its spread are prerequisites for optimizing the efficacy of some treatment techniques, such as radiation. Cancer diagnostics primarily consists of two subfields: imaging of tumor bodies and identification of chemicals linked with tumors. Nanoscale material science provides an opportunity to improve upon established diagnostic methods in each of these fields. The unique properties of materials at the nanoscale are responsible for these enhancements. Improved signal strength and image quality, together with reduced medicine toxicity, are advantages to the patient due to the more focused administration (23). Also, cancer markers like certain metabolites or CT scans can be detected early using nano-based sensors due to their increased sensitivity and specificity. Because of this, cancer can be detected earlier, which leads to better treatment choices.

4.2.1 CT Scan Imaging Agents

Contemporary CT scans rely on iodine molecules as their imaging agents; however, the kidneys rapidly excrete these molecules from the circulation due to their diminutive size. Iodine encapsulated in a NP is one way to circumvent this problem; doing so increases the agent's circulation length while decreasing the dosage. Metal NPs, especially those of tantalum, gold, and bismuth, are common in contemporary CT scan contrast agents, the reason being their surface chemistry, high atomic number, and malleability, making them ideal for attaching biomolecules such as aptamers, antibodies, or peptides, which help in tumor targeting (24). Gold NPs are the most appealing option because of their exceptional biocompatibility, distinctive optical characteristics, and ease of altering their surface chemical study and physicochemical properties. Body imaging of cancers like lung cancer has shown their promise as better contrast agents.

4.2.2 MRI Imaging Agents

Lung cancer diagnosis by MRI is uncommon due to the organ's complex electromagnetic area, weak proton magnitude, and encroachment from respiration and the cardiac rhythm. Nevertheless, chemistry and technology have addressed some of these challenges, and recent research by Koyama et al. (25) demonstrated that MRI can identify malignant lung nodules with comparable accuracy to multi-detector CT scans.

Researchers have looked into NPs formulations of gadolinium to see whether they may increase contrast, decrease dose, and reduce toxicity while also changing gadolinium's biodistribution and pharmacokinetics in a way that is beneficial. NPs of semiconductor polymer, nanodiamond conjugates, nanoporous silica, and ultra-small, rigid plat forms fabricated from a polysiloxane matrix ringed by gadolinium chelates grafted to the matrix are all components of these formulations (26). When contrasted with a commercially available gadolinium agent, these platforms showed greater signal intensity at lower doses and excellent colocalization with the tumor as shown by MRI.

4.2.3 PET Isotopes

Although PET imaging has progressed, the diagnostic value of common PET tracers such 2-[18F] fluoro-2-deoxy-D-glucose is restricted for some types of lung cancer. Researchers are investigating the use of nanotechnology to study longer-lived PET isotopes with the goal of improving the accumulation of PET tracers inside malignancies. Petersen et al. (27) found that clinically significant intratumoral accumulation occurred in vivo when 64Cu and 177Lu were encased in PEGylated liposomes, which modified radionuclide pharmacokinetics and biodistribution.

Using bovine serum albumin as a scaffold, a peptide targeting luteinizing-hormone-releasing hormone was synthesized into ultrasmall 64Cu nanoclusters not long ago. In a mouse orthotopic A549 lung cancer model, nanoclusters that were engineered to target peptides exhibited a tumor absorption rate four times higher than nontargeted nanoclusters. In vivo comparisons between PET and near-infrared (NIR) fluorescence imaging showed that PET might better see tumors, suggesting that nanoscale structures could be useful in PET imaging (28).

4.2.4 Quantum Dots

As NIR fluorophores, semiconductor nanocrystals called quantum dots (QDs) may have their emission wavelength and color tuned by controlling their size. In order to target the $\alpha v \beta 3$ integrin receptor, Tang and colleagues made silver sulfide QDs that were ornamented with cyclic arginine-glycine-aspartic acid-(D)-phenylalanine-lysine peptides.

The QDs exhibited a significant bias toward tumor uptake as opposed to uptake into other in vivo organs, such as the liver or spleen (29). A similar strategy was employed to target mucin 1, an overexpressed protein seen in several malignancies (including lung adenocarcinoma), using zinc doped cadmium:tellurium QDs that were functionalized with aptamers. Whether in a lab or in living organisms, these dots were able to locate and capture images of A549 cells (30).

4.2.5 Use of Polystyrene Nanobeads

Another approach focused on the use of polystyrene nanobeads linked with a synthetic "self" peptide or human CD47. The results showed that tumor NIR fluorescence intensity was higher in mice given human lung xenografts, as tumor uptake was enhanced while splenic clearance and macrophage uptake were reduced relative to nontargeted nanobeads. More and more, doctors are turning to multimodal imaging techniques to glean more data than a single method might provide. Examples of cancers where PET-CT combination has proven to be more helpful than PET alone in terms of correct staging include NSCLC. As a CT and MRI contrast agent, nanoscale holmium-based NPs that are PEGylated have demonstrated encouraging in vivo characteristics (31).

Sun et al. (32) created gold NPs by conjugating glycol chitosan polymers; these NPs were further functionalized with a peptide that serves as a substrate for matrix metalloproteinase second enzyme. Furthermore, a black hole quencher and an ANIR fluorescent dye (Cy5.5) were attached to the peptides. The peptide cleaves in MMP-positive tumors, resulting in selective fluorescence. This enables the use of NIR and CT imaging modalities for the detection of malignancies.

A trimodal nano formulation was functionalized with the radioisotope fluorine-18 (18F-CLIO). This formulation consisted of an SPIO core, a cross-linked aminated dextran coating that had been tagged with an NIR fluorochrome. Using MRI and PET/CT to identify 18F-CLIO, the scientists reasoned that NIR fluorochrome may be employed for NIR imaging detection as well. The PET/CT imaging of living organisms showed a signal-to-noise ratio of over 200, was well tolerated, and was biocompatible. It could detect 18F-CLIO at a level 200 times lower than MRI (33).

4.2.6 Use of Nanomaterials

Nanoscale biosensors have the potential to improve upon existing diagnostic tools by making cancer biomarker detection more precise. In this case, the use of nanomaterials has improved molecular selectivity and reduced the detection limit to the level of individual molecules. For example, by taking use of carbon nanotubes' special characteristics, Heller et al. were able to detect single molecules of ROS like hydrogen peroxide and to measure the amounts of alkylating chemotherapeutic medicines in real time in living human cells (34).

4.2.7 Use of Gold Nanoparticles

Research on the potential sensory and imaging uses of gold NPs has been conducted. Coupled with oligonucleotide probes that match the mutations in EGFR exons 19 and 21, gold NPs were used. In order to build Chemi resistive sensors, a cross-reactive array was made by combining 5-nm gold NPs with distinct functionalizations. This array was subsequently subjected to breath samples from both healthy and lung cancer patients in order to identify vapor organic chemical indicators that had previously been identified. The array was a powerful diagnostic tool, able to differentiate between lung cancer and healthy breath samples with a sensitivity of less than 1 ppb, a high degree of repeatability, and a low limit of detection (35).

4.2.8 Use of Nanopore Sensors

Discoveries of molecular markers linked to cancer and other disorders have been made possible through the utilization of nanotechnology sensors. Distributed among these sensors are nanopores with the conformational ability to capture particular molecules. In order to better understand and quantify the molecule electrically, this change in conductance is necessary. Wang et al. used this method to detect and quantify over 100 microRNAs in the blood, particularly miR-155, a microRNA that has been found to be abnormal in cancer of the lungs. By utilizing the nanopore sensor and quantitative reverse transcription PCR, the researchers discovered that miR-155 levels were noticeably higher in lung cancer patients in comparison to healthy controls. There was decreased intra-assay variability and improved accuracy with nanopore sensors (36).

Researchers have looked into the use of QD-based microarrays to identify circulating microRNAs in blood as a potential biomarker or signature for resectable NSCLC patients (37).

To add to that, QDs have found utility in imaging and detecting vectors. QDs of cadmium and zinc measuring 3.3 nm were attached to HER2 antibodies by a PEGylation process. Instead of using conventional monoclonal antibodies coupled with a fluorophore, this method significantly increased HER2 detection and fluorescence in A549 and H596 lung cancer cells, which only weakly express the protein.

4.2.9 Microfluidic Devices

In order to create bifunctional nanocomposites with magnetic and fluorescent properties, cadmium:tellurium QDs were attached to SPIONs covered with silica. The nanocomposites' magnetic properties allowed them to enrich lung adenocarcinoma SPCA-1 cells from suspension, which was a process before their discovery. The presence of fluorescence, a marker of CEA binding, made them evident as well. It is possible that our technology discovers cancer cells in pleural effusions from patients with lung cancer (38).

To build on this study, 27 patients with NSCLC had their CTCs tested for EGFR mutations after they were drawn from their blood. Eleven of the 19 samples with EGFR mutations contained the replacement mutation T790M, which is associated with resistance to tyrosine kinase inhibitors; the total number of EGFR mutations was 19. This information is vital for making decisions about diagnosis, prognosis, and therapy, and it is usually obtained by invasive biopsies.

A microfluidic-based volumetric bar chart chip was constructed utilizing an ELISA substrate made of nanoporous glass. It allows for the rapid and precise measurement of the three blood tumor biomarkers in lung cancer patients: carcinoma of squamous cells antigen, cytokeratin 19 fragment, and CEA. These research are only a small sample of the numerous ways nanotechnology is changing the face of respiratory cancer diagnostics and, by implication, human disease in general (39).

4.3 THERANOSTIC STRATEGIES IN LUNG CANCER

The adaptability of nanotechnology has made it feasible to develop single agents with both diagnostic and therapeutic capabilities, greatly improving their efficacy. Nanotechnology has expanded the scope of classical theranostics to encompass the co-delivery of diagnostic and pharmacological platforms, allowing for contemporaneous imaging prior, during, and after therapy. This idea has been around for a while, but it is much more feasible now.

The fact that both parties agree that the medications can't work until they build up a significant amount at the site of the disease makes this task easier (40). More and more promising preclinical research are popping up, even if translational development remains in its infancy.

Hou et al. were able to create NPs by mixing modified DOX cis-aconitic anhydride with D-α-tocopherol polyethylene glycol 1000 succinate, a vitamin E derivative that has been authorized by the FDA. The hydrophilic shell is provided by the vitamin E derivative, and the DOX-centered NPs are manufactured by self-assembly in water using the amphiphilic structure. To further facilitate near-infrared fluorescence imaging and photodynamic therapy by producing ROS in response to photoexcitation, the NPs were encapsulated with the FDA-approved photosensitizing drug Chlorin e6 (Ce6). A pH-sensitive amide linker in this design releases DOX and eliminates Ce6's self-quenching limits when exposed to low pH levels, as is the case in tumors or lysosomes (41).

The theranostic potential of up-conversion NPs (UCNPs) was the subject of another study. The formulation of Ce6-CPT-UCNPs was achieved by loading mPEG-COOH, Ce6, and ROS-cleavable thioketal conjugated camptothecin onto UCNPs. A common kind of UCNP is a nanocrystal doped with lanthanides. Fluorescence resonance energy transfer allows UCNPs to convert NIR light at

980 nm into light between 645 and 675 nm, which may then activate Ce6. Besides serving as the foundation for photodynamic treatment, ROS generated by active Ce6 have the ability to release camptothecin by cleaving the thioketal linker. The imaging-friendly fluorescence that Ce6 produces falls within the range of 650 to 750 nm is an added bonus. The H460 orthotopic mouse model was used to characterize Ce6-CPT-UCNPs before in vivo studies. Fluorescence imaging revealed that Ce6-CPT-UCNPs accumulated in tumors, liver, and lungs, in contrast to the little accumulation of free Ce6. Despite the presence of Ce6-CPT UCNPs in the tumor for a minimum of 24 hours, no free Ce6 was detected throughout this period. Every other therapy failed to reduce tumor growth as much as the PBS/laser group. The development of tumors was entirely stopped by laser-irradiated Ce6-CPT-UCNPs (42).

4.4 THE ROLE OF NANOPARTICLES IN CANCER DIAGNOSIS/PROGRESSION

These alterations at the cellular or molecular level are frequently present throughout cancer's genesis and progression and can serve as unique diagnostic markers. Diagnostic methods in niduses that involve imaging evaluation and biomarker identification to find particular targets are common.

4.4.1 In Vivo Examination

For a visual representation of the disease, EUS, AFB, PET, MRI, and CT are the primary in vivo techniques. Despite MRI's ultrahigh resolution, allowing one to observe the tumor's anatomical structure and boundary range, the lengthier imaging duration and absence of molecular imaging data limited its application. However, PET imaging has limitations such as low spatial resolution and the potential for false positives due to inadequate specificity, despite its high sensitivity and ability to detect systemic lesions. While low dosage CT scans can reveal the location, size, and form of lymph node cancer cells, these images are severely distorted due to things like tattoos and internal organ movements, which limits the scans' sensitivity and specificity.

4.4.2 Single-Modal Imageological Examination: MRI

Using large-scale, high-resolution imaging, MRI enables the observation of tumors, anatomical features, and subtleties. Because of issues with mobility, different susceptibility gradients, and low proton density, direct lung detection via MRI is yet in the future (43). Optimized proton MRI sequences derived from UTE can be used in conjunction with UTE-MRI to provide images of lung tissue (44). Furthermore, whereas healthy tissues may excrete gadolinium, malignant ones may accumulate it preferentially (45). Accurate and thorough detection information may be obtained by using two MRI contrast agents concurrently (46).

Due to the strong magnetic coupling between the paramagnetic T1 and T2 contrast media, the relaxing effect of the former may be compromised, leading to an unintended reduction in signal strength and quenching of the MR signal (47). Consequently, a smart MRI contrast agent was developed, consisting of a silica shell containing the anticancer drug DOX and water-soluble Mn-porphyrin (T1 contrast medium) and a core of Fe_3O_4 NPs (T2 contrast medium). Once cancer cells absorbed the contrast media, the acidic tumor microenvironment stimulated porphyrin release, reviving the quenched signal caused by Fe_3O_4 and Mn-porphyrin used together (48).

4.4.3 Imaging Technique for a Single-Modal Image: PET

PET is a type of diagnostic imaging technique that can scan through an endless amount of tissue, is very sensitive, has high temporal resolution, and quantitatively analyzes pictures in a systematic way (49). The physicochemical characteristics of NPs utilized in PET imaging are significantly affected by the poor in vivo stability of radiometal-chelator complexes. ^{64}Cu nanoclusters were produced using a simple one-pot chemical reduction method without chelators to enhance stability and accumulation. The PET lung cancer detection framework utilized bovine serum albumin (BSA) (50). Despite extensive investigation and some success, future research on lung diagnostics will still center on the limitations of Cu-based radionuclide, such as its instability and higher accumulation in the liver.

4.4.4 Multi-Modal Imageological Examination

Imaging using more than one contrast medium can be more helpful in cancer detection than imaging with only one. This is because multi-modal imaging uses a variety of contrast agents. When many contrast agents are given at once, the imaging signal could be severely degraded. In order to create USRPs-Cy5.5, nebulized gadolinium NPs were covalently conjugated with cyanine 5.5. This opens the door to the non-invasive use of UTE-MRI and fluorescence tomography for the diagnosis of lung cancer (51). A combination of an electrophilic substitution reaction and a chelation reaction allows melanin NPs photoacoustic imaging (PAI) to transport 124I (a PET contrast agent) and Mn^{2+} (an MRI contrast agent) simultaneously. Particularly well suited for tri-mode imaging, which considerably enhances the precision of early-stage lung cancer diagnosis, their exceptional qualities make them ideal for this procedure.

4.4.5 In Vitro Detection

Biomarkers that have demonstrated utility in cancer analysis and diagnosis encompass proteins such as CEA, CYFRA21-1, NSE, DKK1, nucleic acids including DNA and RNA, tumor-associated macrophages, and volatile organic compounds such as aldehydes (52). The substantial inter-tumor heterogeneity is ascribed to poor specificity, although biomarker identification allows non-invasive repeated sampling (53). The use of NPs has the potential to significantly improve specificity and sensitivity (Table 4.1).

TABLE 4.1
Nanoparticles for the Detection of Lung Cancer Biomarkers

Nanoparticles	Cancer Biomarkers	Limit of Detection	Methods
Quantum dots	CYFRA21-1	0.3 pg/mL	Electro chemiluminescent immunoassay
	HER2	—	Flow cytometry, Western blot, confocal icroscopy, ELISA,
Fe0 nanoparticles in a liposome	TKTL1, TTF1	—	BIOCHIP TIRF and TLN
	ctDNA	0.1 pg/mL	Assistive plasma mass spectrometry
Nanocomposites of Fe3O4, Au, and Ag	Adenosine	—	Optimized Raman scattering on surfaces
Gold nanoparticles	EGFR, CK, Nap	—	Fluorescence and optimized Raman scattering on surfaces
	CEA, CYFRA21–1, NSE, Dkk1	<1000 pg/mL	NP-and microarray-based co-detection approach
	MiR-205	—	Targeted surface plasmon resonance chronoamperometric method
	MUC1	8 cells/mL	
	Aldehydes	10 ppb	Optimized Raman scattering on surfaces
Fluorescent nanoparticles	EGFR	—	Flow cytometry, Western blot, confocal microscopy
Carbon nanomaterials	CEA/EGFR	14 cells/mL	A method of research entitled differential pulse voltammetry

4.4.6 Gold Nanoparticles

Gold NPs are optimal for identifying cancer biomarkers owing to their superior biocompatibility, controlled particle size and volume, and surface plasmon resonance impact. Ultra-sensitive cytosensing can dramatically improve the specific detection signal of lung cancer. A control method was initiated by adding an enzyme to commence the detection.

We utilized a peptide linker that was designed utilizing a cathepsin B enzyme cleavage site to bind the gold NPs to the dual-functional Raman active luciferin. When combined with these antibodies—cytokeratin-19, napkin-A, and epidermal growth factor receptor—and cathepsin B enzyme, the nano probes have the potential to multi-target lung cancer and perform enzyme-driven fluorescence imaging in cells (54). Enhanced sensitivity for the detection of circulating ctDNA was achieved by combining two NPs. The fragile ctDNA may be separated and enriched with the help of the amorphous Fe0 nanomaterials' magnetic characteristics and high monodispersity, and the strong amplification detecting signal is supplied by the Au NPs. One possible use of Fe-Au NPs is the detection of gene mutations at concentrations as low as 0.1 pg/mL; this might pave the way for an initial diagnosis of KRAS mutations. The possibility of test variance due to traditional PCR amplification was eradicated by this approach (51).

Immobilizing gold NPs onto microarrays and then immersing them in a solution of $HauCl_4$ and H_2O_2 can enhance the detection signal. Four different types of markers—CEA, CYFRA21–1, NSE, and DKK1—can be detected simultaneously using microarray technology according to this approach (52). Gold NPs have the ability to self-assemble into gold super particles, which can improve the Raman signal. NPs have a very high sensitivity for non-invasive lung cancer diagnostics in vitro; however, their potential toxicity should still be taken into account (55).

4.4.7 Quantum Dots

For several reasons, including their size-dependent electrochemical properties, extended fluorescence lifespan, excellent photostability, and the flexibility to tune ECL parameters, QDs find widespread use in the production of electrochemical luminescence (ECL) sensors (56). The detection limit of 0.3 pg/mL for QD-based methods was much lower than the 0.65 pg/mL limit of other detection approaches. A sandwich structure was established between the magnetic beads and the QDs when the suspension interacted with the target proteins through particular antibody-antigen interactions.

In principle, modifying the QDs' surface-bound lung cancer-specific antibody might allow for target identification. Some problems with the QDs' surface-bound benign lung-cancer-specific antibody and the bigger complexes formed between them and the antibody led to lower response for intrinsic tumor biomarkers identification, which were problems in real-world applications. Molecular weight reduction via single domain antibodies (sdAbs) is a boon in situations when conventional antibodies are not feasible (57). Although surface modification can be used to boost biocompatibility and minimize biological toxicity to cells, QD security and targeting specificity remain key difficulties.

4.4.8 Carbon Nanomaterials

Electrochemical biosensors made of carbon nanostructures have a lot of promise because of their great electrical and thermal conductivity, huge surface area, high mechanical strength, and exceptional electron transport rate. A two-step electro-polymerization procedure may be used to deposit graphene oxide and ordered mesoporous carbon onto nano-carrier surfaces, increasing the electrochemical effective surface area and electrical conductivity (58).

One possible mechanism by which electrochemical process cytosensors based on 3D carbon nanospheres produced by microwave-hydrothermal methods can detect CEA is the self-assembly of gold NPs on their surfaces. As a result of enhanced electron transport brought about by the combination of gold NPs with monodisperse colloidal carbon nanospheres, the detection sensitivity was enhanced and the limit of detection was reduced to 14 cells/mL.

The purpose of this research was to develop biosensors that could more accurately detect the lung cancer biomarker toluene in a patient's exhaled breath by utilizing single-wall carbon nanotubes coated with platinum-group transition metals. Due to the near proximity of the d and p orbitals of the metal and carbon atoms in the toluene benzene ring, the successful identification (59).

4.4.9 Other Nanomaterials

In addition, several other nanomaterials have been studied and utilized as biomarkers for lung cancer. The lipid bilayer, similar to the cell membrane, consists of phospholipids and cholesterol (60). Liposomes outperform other synthetic materials in terms of biosafety and biocompatibility. Liposomes, which contain the transcription factors thyroid transcription factor 1 (TTF1) and transketolase 1 (TKTL1), can measure the RNA concentration in the blood of people with lung cancer by forming larger nanoscale complexes out of extracellular vesicles that are electrostatically attracted to them.

4.5 CONCLUSIONS

Much effort has been devoted to describing the challenges of using nanomaterials for lung cancer detection. Although many NPs have demonstrated efficacy in detecting lung cancer and made significant progress in enhancing sensitivity, decreasing detection time, increasing detection limit, and streamlining detection procedures, there are still obstacles to overcome in areas like specificity and biosafety, detection effectiveness, detection expenditure, and patients' tolerance. Nanomedicines can only be as effective as the knowledge we now have on tumor heterogeneity, the tumor microenvironment, nano-bio interactions, host-tumor immunological interplay, tumor transport to tumors, and the enhanced permeability and retention effect.

One potential way nanomaterials might help overcome these obstacles is by binding to biomarkers of lung cancer in vitro or by enhancing the absorption of diagnostic reagents by lung cancer cells rather than normal cells. You have complete command over their dimensions, form, and physicochemical properties, which encompass thermal, magnetic, and optical features, as well as those of surfaces that may be altered. The goal is to create more accurate preclinical models by following appropriate manufacturing practices while producing NPs on a big scale and subsequently characterizing them thoroughly. Metabolomic findings and the microbiome's function may provide key clues for nanomedicine research and development.

Nanotechnology, according to most people, might revolutionize the medical field and the cancer treatment industry in particular. It is well known that nanomedicine has challenges in both diagnostic and therapeutic domains. Array, chip, and nanotechnology integration offers hope for a future that is less expensive and more efficient in detecting lung cancer.

Increasing the efficacy of detection without adding further damage requires more effort into creating non-invasive detection methods. To overcome these current roadblocks in early lung cancer detection, we need to bring together clinical diagnostics, biology, material science, and nanotechnology.

REFERENCES

1. Thun MJ, Carter BD, Feskanich D, Freedman ND, Prentice R, Lopez AD, et al. 50-year trends in smoking-related mortality in the United States. *The New England Journal of Medicine*. 2013;368(4):351–64.
2. Siegel RL, Miller KD, Wagle NS, Jemal A. Cancer statistics, 2023. *CA: A Cancer Journal for Clinicians*. 2023;73(1):17–48.

3. Pandi A, Mamo G, Getachew D, Kitila F, Kalappan V, Dhiravidamani S. A brief review on lung cancer. 2016.
4. Strebhardt K, Ullrich A. Paul Ehrlich's magic bullet concept: 100 years of progress. *Nature Reviews Cancer*. 2008;8(6):473–80.
5. Cui EN, Yu T, Shang SJ, Wang XY, Jin YL, Dong Y, et al. Radiomics model for distinguishing tuberculosis and lung cancer on computed tomography scans. *World Journal of Clinical Cases*. 2020;8(21):5203–12.
6. Jaine R, Kvizhinadze G, Nair N, Blakely T. Cost-effectiveness of a low-dose computed tomography screening programme for lung cancer in New Zealand. *Lung Cancer (Amsterdam, Netherlands)*. 2020;144:99–106.
7. Cicenia J, Avasarala SK, Gildea TR. Navigational bronchoscopy: a guide through history, current use, and developing technology. *Journal of Thoracic Disease*. 2020;12(6):3263–71.
8. Gould MK, Donington J, Lynch WR, Mazzone PJ, Midthun DE, Naidich DP, et al. Evaluation of individuals with pulmonary nodules: when is it lung cancer? Diagnosis and management of lung cancer, 3rd ed: American college of chest physicians evidence-based clinical practice guidelines. *Chest*. 2013;143(5 Suppl):e93S–120S.
9. Lu H, Zhang H, Wei Y, Chen H. Ambient mass spectrometry for the molecular diagnosis of lung cancer. *The Analyst*. 2020;145(2):313–20.
10. Khanmohammadi A, Aghaie A, Vahedi E, Qazvini A, Ghanei M, Afkhami A, et al. Electrochemical biosensors for the detection of lung cancer biomarkers: a review. *Talanta*. 2020;206:120251.
11. Capuano R, Catini A, Paolesse R, Di Natale C. Sensors for lung cancer diagnosis. *Journal of Clinical Medicine*. 2019;8(2):235.
12. Nakhleh MK, Amal H, Jeries R, Broza YY, Aboud M, Gharra A, et al. Diagnosis and classification of 17 diseases from 1404 subjects via pattern analysis of exhaled molecules. *ACS Nano*. 2017;11(1):112–25.
13. Gao M, Su H, Lin G, Li S, Yu X, Qin A, et al. Targeted imaging of EGFR overexpressed cancer cells by brightly fluorescent nanoparticles conjugated with cetuximab. *Nanoscale*. 2016;8(32):15027–32.
14. Lee LJ, Yang Z, Rahman M, Ma J, Kwak KJ, McElroy J, et al. Extracellular mRNA detected by tethered lipoplex nanoparticle biochip for lung adenocarcinoma detection. *American Journal of Respiratory and Critical Care Medicine*. 2016;193(12):1431–3.
15. Tartarone A, Lerose R, Aieta M. Focus on lung cancer screening. *Journal of Thoracic Disease*. 2020;12(7):3815–20.
16. Chen KN. The diagnosis and treatment of lung cancer presented as ground-glass nodule. *General Thoracic and Cardiovascular Surgery*. 2020;68(7):697–702.
17. Mottaghitalab F, Farokhi M, Fatahi Y, Atyabi F, Dinarvand R. New insights into designing hybrid nanoparticles for lung cancer: diagnosis and treatment. *Journal of Controlled Release: Official Journal of the Controlled Release Society*. 2019;295:250–67.
18. Wang H, Zheng L, Peng C, Shen M, Shi X, Zhang G. Folic acid-modified dendrimer-entrapped gold nanoparticles as nanoprobes for targeted CT imaging of human lung adencarcinoma. *Biomaterials*. 2013;34(2):470–80.
19. Poland CA, Duffin R, Kinloch I, Maynard A, Wallace WA, Seaton A, et al. Carbon nanotubes introduced into the abdominal cavity of mice show asbestos-like pathogenicity in a pilot study. *Nature Nanotechnology*. 2008;3(7):423–8.
20. Yang C, Huang W, Gao Y, Liu Z, An N, Mu W, et al. Phototherapy ablation of rabbit orthotopic tumors by non-stoichiometric BiPO4-x nanoparticles. *Chemical Engineering Journal*. 2020;386:123961.
21. Mir TA, Yoon JH, Gurudatt NG, Won MS, Shim YB. Ultrasensitive cytosensing based on an aptamer modified nanobiosensor with a bioconjugate: detection of human non-small-cell lung cancer cells. *Biosensors & Bioelectronics*. 2015;74:594–600.
22. Card JW, Zeldin DC, Bonner JC, Nestmann ER. Pulmonary applications and toxicity of engineered nanoparticles. *American Journal of Physiology Lung Cellular and Molecular Physiology*. 2008;295(3):L400–11.
23. Toy R, Bauer L, Hoimes C, Ghaghada KB, Karathanasis E. Targeted nanotechnology for cancer imaging. *Advanced Drug Delivery Reviews*. 2014;76:79–97.
24. Shilo M, Reuveni T, Motiei M, Popovtzer R. Nanoparticles as computed tomography contrast agents: current status and future perspectives. *Nanomedicine (London, England)*. 2012;7(2):257–69.
25. Koyama H, Ohno Y, Kono A, Takenaka D, Maniwa Y, Nishimura Y, et al. Quantitative and qualitative assessment of non-contrast-enhanced pulmonary MR imaging for management of pulmonary nodules in 161 subjects. *European Radiology*. 2008;18(10):2120–31.

26. Hashim Z, Green M, Chung PH, Suhling K, Protti A, Phinikaridou A, et al. Gd-containing conjugated polymer nanoparticles: bimodal nanoparticles for fluorescence and MRI imaging. *Nanoscale*. 2014;6(14):8376–86.
27. Petersen AL, Henriksen JR, Binderup T, Elema DR, Rasmussen PH, Hag AM, et al. In vivo evaluation of PEGylated ^{64}Cu-liposomes with theranostic and radiotherapeutic potential using micro PET/CT. *European Journal of Nuclear Medicine and Molecular Imaging*. 2016;43(5):941–52.
28. Gao F, Cai P, Yang W, Xue J, Gao L, Liu R, et al. Ultrasmall [64Cu]Cu nanoclusters for targeting orthotopic lung tumors using accurate positron emission tomography imaging. *ACS Nano*. 2015;9(5):4976–86.
29. Tang R, Xue J, Xu B, Shen D, Sudlow GP, Achilefu S. Tunable ultrasmall visible-to-extended near-infrared emitting silver sulfide quantum dots for integrin-targeted cancer imaging. *ACS Nano*. 2015;9(1):220–30.
30. Zhang C, Ji X, Zhang Y, Zhou G, Ke X, Wang H, et al. One-pot synthesized aptamer-functionalized CdTe:Zn2+ quantum dots for tumor-targeted fluorescence imaging in vitro and in vivo. *Analytical Chemistry*. 2013;85(12):5843–9.
31. Ni D, Zhang J, Bu W, Zhang C, Yao Z, Xing H, et al. PEGylated NaHoF4 nanoparticles as contrast agents for both X-ray computed tomography and ultra-high field magnetic resonance imaging. *Biomaterials*. 2016;76:218–25.
32. Sun I-C, Eun D-K, Koo H, Ko C-Y, Kim H-S, Yi DK, et al. Tumor-targeting gold particles for dual computed tomography/optical cancer imaging. *Angewandte Chemie International Edition*. 2011;50(40):9348–51.
33. Devaraj NK, Keliher EJ, Thurber GM, Nahrendorf M, Weissleder R. 18F labeled nanoparticles for in vivo PET-CT imaging. *Bioconjugate Chemistry*. 2009;20(2):397–401.
34. Heller DA, Jin H, Martinez BM, Patel D, Miller BM, Yeung TK, et al. Multimodal optical sensing and analyte specificity using single-walled carbon nanotubes. *Nature Nanotechnology*. 2009;4(2):114–20.
35. Peng G, Tisch U, Adams O, Hakim M, Shehada N, Broza YY, et al. Diagnosing lung cancer in exhaled breath using gold nanoparticles. *Nature Nanotechnology*. 2009;4(10):669–73.
36. Wang H, Zheng L, Peng C, Guo R, Shen M, Shi X, et al. Computed tomography imaging of cancer cells using acetylated dendrimer-entrapped gold nanoparticles. *Biomaterials*. 2011;32(11):2979–88.
37. Fan L, Qi H, Teng J, Su B, Chen H, Wang C, et al. Identification of serum miRNAs by nano-quantum dots microarray as diagnostic biomarkers for early detection of non-small cell lung cancer. *Tumour Biology: The Journal of the International Society for Oncodevelopmental Biology and Medicine*. 2016;37(6):7777–84.
38. Nagrath S, Sequist LV, Maheswaran S, Bell DW, Irimia D, Ulkus L, et al. Isolation of rare circulating tumour cells in cancer patients by microchip technology. *Nature*. 2007;450(7173):1235–9.
39. Li Y, Xuan J, Song Y, Qi W, He B, Wang P, et al. Nanoporous glass integrated in volumetric bar-chart chip for point-of-care diagnostics of non-small cell lung cancer. *ACS Nano*. 2016;10(1):1640–7.
40. Xie J, Lee S, Chen X. Nanoparticle-based theranostic agents. *Adv Drug Deliv Rev*. 2010;62(11): 1064–79.
41. Hou W, Zhao X, Qian X, Pan F, Zhang C, Yang Y, et al. pH-Sensitive self-assembling nanoparticles for tumor near-infrared fluorescence imaging and chemo-photodynamic combination therapy. *Nanoscale*. 2016;8(1):104–16.
42. Yue C, Zhang C, Alfranca G, Yang Y, Jiang X, Yang Y, et al. Near-infrared light triggered ROS-activated theranostic platform based on Ce6-CPT-UCNPs for simultaneous fluorescence imaging and chemo-photodynamic combined therapy. *Theranostics*. 2016;6(4):456–69.
43. Sim AJ, Kaza E, Singer L, Rosenberg SA. A review of the role of MRI in diagnosis and treatment of early stage lung cancer. *Clinical and Translational Radiation Oncology*. 2020;24:16–22.
44. Wild JM, Marshall H, Bock M, Schad LR, Jakob PM, Puderbach M, et al. MRI of the lung (1/3): methods. *Insights into Imaging*. 2012;3(4):345–53.
45. Bianchi A, Dufort S, Lux F, Fortin PY, Tassali N, Tillement O, et al. Targeting and in vivo imaging of non-small-cell lung cancer using nebulized multimodal contrast agents. *Proceedings of the National Academy of Sciences of the United States of America*. 2014;111(25):9247–52.
46. Guldris N, Argibay B, Kolen'ko YV, Carbó-Argibay E, Sobrino T, Campos F, et al. Influence of the separation procedure on the properties of magnetic nanoparticles: gaining in vitro stability and T1–T2 magnetic resonance imaging performance. *Journal of Colloid and Interface Science*. 2016;472:229–36.
47. Yoo D, Lee J-H, Shin T-H, Cheon J. Theranostic magnetic nanoparticles. *Accounts of Chemical Research*. 2011;44(10):863–74.

48. Huang X, Yuan Y, Ruan W, Liu L, Liu M, Chen S, et al. pH-responsive theranostic nanocomposites as synergistically enhancing positive and negative magnetic resonance imaging contrast agents. *Journal of Nanobiotechnology*. 2018;16(1):30.
49. Teramoto A, Yamada A, Tsukamoto T, Imaizumi K, Toyama H, Saito K, et al. Decision support system for lung cancer using PET/CT and microscopic images. *Advances in Experimental Medicine and Biology*. 2020;1213:73–94.
50. Kim H-Y, Li R, Ng TSC, Courties G, Rodell CB, Prytyskach M, et al. Quantitative imaging of tumor-associated macrophages and their response to therapy using 64Cu-labeled Macrin. *ACS Nano*. 2018;12(12):12015–29.
51. Fu L, Wang R, Yin L, Shang X, Zhang R, Zhang P. CYFRA21–1 tests in the diagnosis of non-small cell lung cancer: a meta-analysis. *The International Journal of Biological Markers*. 2019;34(3):251–61.
52. Hu P, Zhang S, Wu T, Ni D, Fan W, Zhu Y, et al. Fe-Au nanoparticle-coupling for ultrasensitive detections of circulating tumor DNA. *Advanced Materials (Deerfield Beach, Fla)*. 2018;30(31):e1801690.
53. Maharjan N, Thapa N, Tu J. Blood-based biomarkers for early diagnosis of lung cancer: a review article. *JNMA; Journal of the Nepal Medical Association*. 2020;58(227):519–24.
54. Saranya G, Joseph MM, Karunakaran V, Nair JB, Saritha VN, Veena VS, et al. Enzyme-driven switchable fluorescence-SERS diagnostic nanococktail for the multiplex detection of lung cancer biomarkers. *ACS Applied Materials & Interfaces*. 2018;10(45):38807–18.
55. Qiao X, Su B, Liu C, Song Q, Luo D, Mo G, et al. Selective surface enhanced raman scattering for quantitative detection of lung cancer biomarkers in superparticle@MOF structure. *Advanced Materials*. 2018;30(5):1702275.
56. Ren D, Wang B, Hu C, You Z. Quantum dot probes for cellular analysis. *Analytical Methods*. 2017;9(18):2621–32.
57. Singh RD, Shandilya R, Bhargava A, Kumar R, Tiwari R, Chaudhury K, et al. Quantum dot based nano-biosensors for detection of circulating cell free miRNAs in Lung carcinogenesis: from biology to clinical translation. *Frontiers in Genetics*. 2018;9:616.
58. Shoja Y, Kermanpur A, Karimzadeh F. Diagnosis of EGFR exon21 L858R point mutation as lung cancer biomarker by electrochemical DNA biosensor based on reduced graphene oxide/functionalized ordered mesoporous carbon/Ni-oxytetracycline metallopolymer nanoparticles modified pencil graphite electrode. *Biosensors & Bioelectronics*. 2018;113:108–15.
59. Aasi A, Aghaei SM, Panchapakesan B. A density functional theory study on the interaction of toluene with transition metal decorated carbon nanotubes: a promising platform for early detection of lung cancer from human breath. *Nanotechnology*. 2020;31(41):415707.
60. Matczuk M, Ruzik L, Aleksenko SS, Keppler BK, Jarosz M, Timerbaev AR. Analytical methodology for studying cellular uptake, processing and localization of gold nanoparticles. *Analytica Chimica Acta*. 2019;1052:1–9.

5 Metal and Metal Oxide Nanoparticles for the Management of Lung Cancer Diseases

Vaishali Wadhwa[1], *Shivam Pandey*[1*], *and Nishesh Sharma*[2]
[1] Department of Chemistry, School of Applied and Life Sciences, Uttaranchal University, Dehradun 248007, India
[2] Department of Biotechnology, School of Applied and Life Sciences, Uttaranchal University, Dehradun 248007, India
*Corresponding author: Dr. Shivam Pandey, pandeyshivam547@gmail.com.

5.1 INTRODUCTION

Lung disorders, a major and expanding public health issue, impact millions of people globally and provide difficulties for healthcare systems. Globally, lung diseases, which include a variety of ailments like asthma, lung cancer, and chronic obstructive pulmonary disease (COPD), present serious obstacles to public health. These conditions not only undermine respiratory function but also lead to debilitating symptoms that can drastically reduce quality of life. Environmental factors including pollution and occupational dangers frequently make these diseases worse, which raises the morbidity and mortality rates in the afflicted populations. Many diseases with distinct pathophysiological mechanisms are included in the classification of lung diseases. Asthma and COPD are two well-known examples of obstructive lung illnesses, which restrict airflow because of inflammation and constricted airways. In particular, asthma is not a single illness but rather a collection of distinct disease entities called asthma endotypes, each of which reflects a distinct underlying cause of respiratory failure (Custovic et al. 2023).

Restrictive lung illnesses, such as pulmonary fibrosis, cause decreased lung capacity and reduced gas exchange. Furthermore, malignancies, particularly lung cancer, represent considerable health hazards, with symptoms including chronic cough and shortness of breath typically serving as important diagnostic markers. Predominantly, smoking remains a critical risk factor, significantly contributing to the development of COPD and lung cancer. Economic assessments indicate that smoking-related conditions, particularly lung cancer, incur substantial healthcare costs, further straining public health resources (Sutrisna et al. 2009). For instance, while smoking accounts for a large percentage of lung cancer fatalities, other factors play a substantial role as well. In a study addressing lung cancer mortality, it was noted that the probabilistic estimates of cases attributable to each risk factor could exceed the observed total, highlighting uncertainties in exposure and dose-response relationships (Ha-Duong et al. 2004). Similarly, behavioural factors such as diet and inactivity have been linked to lung health, emphasizing the multifaceted nature of risk assessment (Lazarus et al. 2022).

One of the main environmental risk factors associated with respiratory diseases is exposure to particulate matter, especially particulate matter (PM), that is PM 2.5. Research has demonstrated

DOI: 10.1201/9781003595830-5

that PM 2.5 induces oxidative stress, which is a precursor to various respiratory and cardiovascular diseases. Additionally, lifestyle choices such as smoking, poor dietary habits, and physical inactivity play critical roles in lung health. According to recent studies, modifiable risk factors, including health behaviours and social circumstances, account for up to 60% of disease risk (Yamashita & Millar 2017).

5.2 CLASSIFICATION OF LUNG DISEASES

The classification of lung diseases primarily falls into two categories: obstructive and restrictive diseases, which exhibit distinct physiological characteristics. One more prominent category of lung diseases is lung cancer termed under neoplastic lung diseases. Figure 5.1 shows different classification of lung diseases.

5.2.1 Obstructive Lung Disease

One major type of respiratory illness is obstructive lung disease, which is often described by the blockages in airway. The smaller bronchi and larger bronchioles narrow (obstruct) in these obstructive illnesses as a result of excessive smooth muscle contraction. Inflammatory and readily collapsible airways along with airflow obstruction that impede airflow often lead to symptoms like wheezing and breathlessness. Chronic obstructive pulmonary disease, cystic fibrosis, bronchitis, and asthma are examples of types of obstructive lung diseases. Here's a detailed explanation about same.

5.2.1.1 Asthma

Asthma is a chronic lung disease that mostly affects the peripheral airways, obstructing airflow in a variable and reversible manner (Hamid et al. 1997). This commonly occurs when allergens, pollen, dust, or other particles are breathed into the lungs and the bronchioles constrict and generate mucus, which in turn limits the supply of oxygen to the alveoli. Symptoms include occurrences of wheezing, coughing, chest tightness, and breathlessness. Asthma poses a significant global health and economic burden and its prevalence is rising in many regions of the world that have embraced elements of the western way of life (Holgate et al. 2015). Asthma in adults and children has historically been categorized according to the degree of disease control achieved through a step-by-step management process or the intensity of symptoms. These symptoms are then used to decide the need for controller medication treatment.

5.2.1.2 Bronchitis

Bronchitis is a respiratory disease usually defined by inflammation of the airways that carry air into and out of the lungs. It is further classified into two categories: acute and chronic bronchitis. Acute bronchitis is a brief inflammation of the bronchial tubes that is typically caused by viral infections such as common cold or flu. It usually occurs a few days to a few weeks following an upper respiratory illness (Kinkade & Long 2016). Its typical symptoms include a persistent cough that may generate mucus, chest pain, wheezing, dyspnoea, exhaustion and a low-grade fever. Chronic bronchitis is a long-term respiratory condition that involves inflammation of the bronchial tubes causing increased mucus secretion and a cough lasting for months. It is generally caused by long-term exposure to harmful substances such as dust, chemicals, cigarette smoke, and air pollution (Fletcher & Tinker 1961).

5.2.1.3 Chronic Obstructive Pulmonary Disease

The progressive lung condition known as COPD leads to airway blockage and breathing difficulties (Hussain et al. 2021). Extended exposure to irritants such as dust, gaseous pollutants, and so forth,

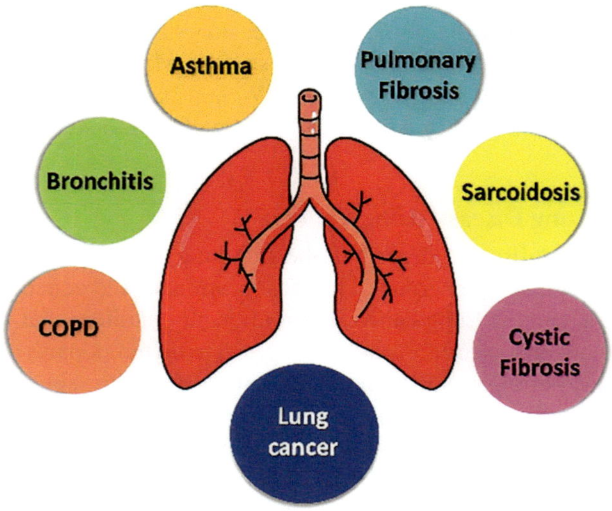

FIGURE 5.1 Lung-related diseases

along with cigarette smoking and air pollution are the primary causes of COPD. Patients with any form of COPD display a wide range of symptoms with breathlessness or breath shortness during daily activities being the significant indication that breaks down over time. The lung's capacity to interact with gases is diminished in such conditions, which impede airflow. This lowers blood oxygen levels (hypoxemia) and eventually leads to respiratory failure (Joshi 2018). Common symptoms of any type of COPD include fatigue, wheezing, tightness in the chest, and a persistent cough (with or without mucus); these symptoms are occasionally misdiagnosed as age related. It's possible for some people to show no symptoms at all until their condition gets worse. Despite advancements in therapeutic options to help control symptoms and prevent complications, COPD is incurable (Shukla et al. 2020).

5.2.2 Restrictive Lung Diseases

Restrictive lung diseases limit lung expansion, which results in a reduction in lung volume and compromised gas exchange. These disorders mainly impair lung compliance or the lung's capacity to expand and stretch, which makes it challenging for people to take deep breaths. Restrictive lung illnesses can be brought on by extra-parenchymal conditions (extrinsic reasons) or by inflammation, toxins, and mechanisms that are still unknown (intrinsic conditions), which cause the distal lung parenchyma to be destroyed (Martinez-Pitre et al. 2023).

5.2.2.1 Pulmonary Fibrosis

Pulmonary fibrosis is a condition that seeds lung scarring, wherein fibrotic tissue prevents oxygen from entering the bloodstream from the lungs' small air sacs. Because of stiff scar tissue and low oxygen levels especially when walking and exercising, people with pulmonary fibrosis have dyspnoea. Such lung disorders are a subset of the interstitial lung diseases (ILD), a broader category of diseases that encompasses all conditions characterized by inflammation and/or lung scarring (Noble et al. 2012). As a result of scarring, the lungs are less able to expand normally, which lowers oxygen exchange and causes chronic breathlessness, dry cough, tiredness, and in more severe cases respiratory failure. Although there is no cure, antifibrotic drugs are used to delay the disease's progression. One of the most widespread kind of pulmonary fibrosis is idiopathic pulmonary fibrosis (IPF). Millions of people worldwide suffer with IPF and its frequency rises sharply with age. Weight loss, fever, and arthralgias are common signs of IPF that should lead to a secondary cause of pulmonary fibrosis study (Martinez et al. 2017).

5.2.2.2 Neoplastic Lung Disease (Lung Cancer)

The term "neoplastic lung disease" describes aberrant lung growths which can be either benign (non-cancerous) or malignant (cancerous). The most prominent and concerning form of neoplastic lung disease is lung cancer. Lung cancer is a malignant condition that occurs when aberrant cells in the lungs expand uncontrollably and may spread to different parts of the body. Lung cancer is the primary cause of mortality in the United States for both men and women. More than a million people worldwide die from lung cancer every year. It is often associated with smoking and environmental exposures and is one of the leading causes of cancer deaths worldwide (Witschi 2001).

Lung cancer is clearly a burdensome and widespread disease with a major public health impact. There are two main types of lung cancer: small cell lung cancer (SCLC), which tends to be more aggressive and spreads more rapidly than the other major type, non–small cell lung cancer (NSCLC), which accounts for about 85% of cases (Cersosimo 2002). The type of the tumour and extent of metastases may also influence signs and symptoms (Collins et al. 2007).

Contingent on the phase and kind of lung cancer, medication options include immunotherapy, radiation therapy, chemotherapy, surgery, and targeted therapy. While prevention measures like quitting smoking and limiting exposure to dangerous chemicals might reduce the risk of getting the disease, early detection through screening, particularly in high-risk patients, greatly increases survival rates (Alberg et al. 2013).

5.3 MANAGEMENT OF LUNG DISEASES

All the categorized lung diseases can be caused by numerous elements like ecological, genetic, infectious, and lifestyle-related elements. Contact with air pollution, workplace risks such as dust and chemicals, and allergens like pollen and mould also contribute to all these respiratory problems. Among the major causes, smoking and tobacco use greatly raises the risk of lung diseases. The majority of patients with obstructive lung diseases arrive with coughing and dyspnoea, occasionally with wheezing or the production of sputum. The degree of airflow restriction and the underlying condition determines the physical symptoms in patients with obstructive lung diseases (Ryu & Scanlon 2001).

Therapeutics for respiratory diseases are to control symptoms, enhance lung function, and treat the underlying causes of illnesses such as pulmonary fibrosis, pneumonia, asthma, and COPD. Pharmacological therapies include corticosteroids to lower inflammation in diseases like asthma and COPD and bronchodilators (like beta-agonists and anticholinergics) to ease airway muscles and improve airflow. Pneumonia, influenza, and other respiratory diseases are treated with antibiotics and antiviral drugs. Combination inhalers and other inhaled drugs offer localized treatment with less systemic side effects (Wu et al. 2024).

Long-term oxygen therapy is used to treat hypoxemia in chronic illnesses and pulmonary therapy include exercises to improve lung function and quality of life. Advanced treatments that target certain disease mechanisms include biologics (omalizumab for severe asthma) and antifibrotic medications (pirfenidone for pulmonary fibrosis). According to current guidelines, the medication schedule should be tailored to the patient's symptoms and risk of exacerbations, which is determined by looking at prior instances of severe or frequent exacerbations (Seshadri & Ramamurthi 2018).

5.4 NANO-APPROACHES IN THERAPEUTICS FOR LUNG DISEASE MANAGEMENT

A new period of precise and accurate detection and management of pulmonary inflammatory diseases has been made possible by nanotechnology in medicine, that is nanomedicine. To overcome the drawbacks of existing medical therapies and get past biological barriers, nanoparticles are created and used in drug delivery devices (Figure 5.2). Compared to the conventional method of administering therapeutic agents, the use of nanotechnology in pulmonary inflammatory disorders offers several advantages (Doroudian et al. 2023).

Since lungs have a huge absorptive surface area, a decent blood supply and a relatively thin absorptive mucosal barrier, they are an appealing target for drug administration. This has led to a recent surge in interest in pulmonary delivery of nanoparticle-based therapeutics (Shah et al. 2019). Magnetic nanoparticles have also emerged as a novel approach for the targeted treatment of lung cancer. By delivering medications to the tumour site specifically, these nanoparticles can reduce harm to healthy tissues and increase the effectiveness of treatment (Mangal et al. 2017).

Metal and metal oxide nanoparticles also provide a number of benefits in comparison to conventional cancer treatments. To improve their selectivity and effectiveness, they can be functionalized with other substances. Additionally, they can serve as multipurpose agents that combine medicinal and diagnostic properties. Metal and metal oxide nanoparticles have shown promising results in preclinical studies for lung cancer management. They can be used as drug delivery agents, radiosensitizers, and photothermal agents, enhancing the efficacy and specificity of cancer therapy (Taghavizadeh et al. 2023). Numerous nanoparticles (NPs) are utilized in treatment of lung related diseases which can be seen in Table 5.1.

5.4.1 Metal Nanoparticles

Drug delivery agents include a variety of organic (solid lipid, liposomes, and polymers) and inorganic (quantum dots with various mineral compositions, silica, gold, and silver) NPs. Recent

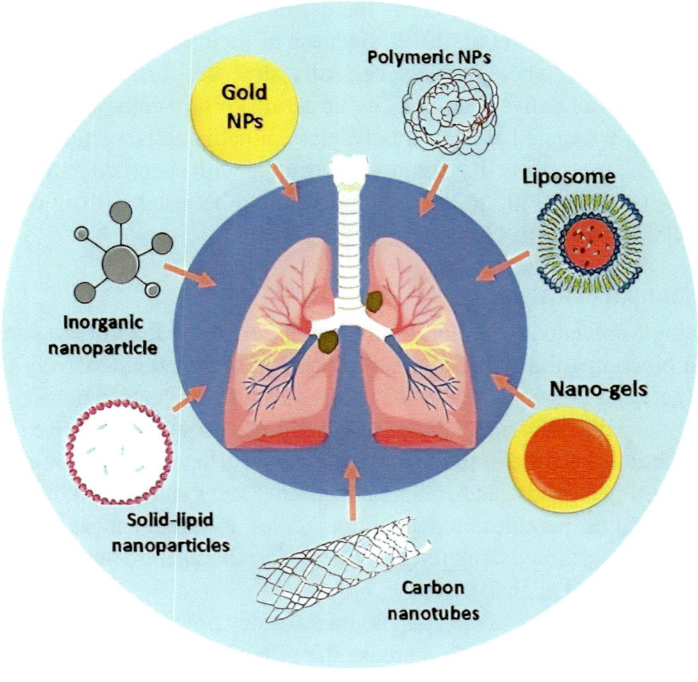

FIGURE 5.2 Use of nanoparticles in lung disease management

TABLE 5.1
Nanoparticles and Their Applications in the Treatment of Different Lung Diseases

Nanoparticles	Application in Treatment	Reference
Polymeric	COPD	Varshosaz et al. 2013
ZnO and Ag	Tuberculosis	Ellis et al. 2018
Liposome	COPD	Moro et al. 2019
Nano-SBS	Asthma	Guo et al. 2013
Polymeric/Silica	Lung Cancer	Bregoli et al. 2016
Selenium	Tuberculosis	Pi et al. 2020
Gold	Lung Cancer	Liu et al. 2019
Polymeric	Cystic fibrosis	Konstan et al., 2004

TABLE 5.2
Metal Nanoparticles and Their Applications in the Management of Diseases

Metal	Applications	References
Gold (Au)	Photodynamic therapy, Cellular imaging	Raji et al. 2020
Iron (Fe)	Nanoezymes, Targeted drug delivery	Kuzmov & Minko 2015
Silver (Ag)	Chemotherapeutic medicines	Leo et al. 2013
Zinc (Zn)	Biosensing, Biological imaging	Singh et al. 2020
Silica (Si)	Chemotherapy, Gene therapy	Garcia-Fernandez et al. 2021

advancements in nanotechnology, particularly the use of metal NPs, present promising avenues for overcoming challenges (Wadhwa et al. 2019). Because of their use in cancer treatment and microbial resistance management, where they successfully slow the formation of resistant infections, these NPs can improve drug delivery systems by targeting certain cells while reducing toxicity. By enabling tailored therapy targeted at aberrant alveolar epithelial cells, metal NPs present a possible alternative to traditional treatments that frequently struggle with specificity and side effects and off-target consequences (Dudhat et al. 2020). Common metal NPs used in drug delivery include gold (Au), silver (Ag) NPs and so forth (Table 5.2).

5.4.1.1 Gold Nanoparticles

Because of their high-precision drug and biomolecule delivery capabilities, simplicity of surface modification, and biocompatibility, gold NPs are being researched extensively (Shah et al. 2019). Gold is a highly inert and non-toxic metal that possesses higher electron density. Gold NPs can be used to treat respiratory diseases such lung cancer, asthma, chronic obstructive pulmonary disease, and pulmonary infections because of their small size, high surface-to-volume ratio, and adaptability to medications, targeting ligands, and imaging medium. They can be modified with various particles, including antibodies, peptides, and small molecules, to target cancer cells specifically. They facilitate controlled and sustained drug release, improving therapeutic outcomes while minimizing systemic side effects (Bansal et al. 2020).

For pulmonary drug delivery, Au NPs can be made as dry powders or aerosols, offering targeted therapy with less systemic exposure. To increase the effectiveness of radiation, gold NPs can also be employed in chemotherapy, and in radiotherapy as radiosensitizers. They have the ability to scatter and absorb X-rays, which increases the radiation dose that cancer cells receive. When these Au

NPs come into contact with visible light, they can react by producing heat. As a result, they are also employed as photothermal and labelling agents, and by generating enough heat, they destroy nearby tumour cells (Kumari et al. 2023).

5.4.1.2 Silver Nanoparticles

Studies describing Ag NPs as drug delivery agents suggest that Ag^+ is the only stable nanoscale entity that achieves effective anticancer and immunomodulatory expression (Malik & Mukherjee 2018). The strong antibacterial, anti-inflammatory, and antioxidant qualities of silver NPs have drawn a lot of interest in the treatment of lung conditions. They offer intriguing options for treating a range of respiratory disorders, such as infections, inflammation, and even lung cancer, because of their substantial surface area, distinct physicochemical characteristics, and broad-spectrum antibacterial action. In fact, this contributes to Ag NPs' increased anticancer potential. Through mitochondrial damage, Ag NPs cause apoptosis, or programmed cell death, which has cytotoxic effects on lung cancer cells.

Ag NPs have been effective in treating pathogen-induced respiratory infections by generation of reactive oxygen species (ROS), which damage the outer membrane of bacteria or disrupt microbial DNA. For a longer-lasting therapeutic effect, silver NPs can act as carriers for the targeted delivery of medications such chemotherapeutic medicines, antibiotics, and anti-inflammatory therapies straight to the lungs. Some well-known uses in drug delivery include the transfer of anti-TB medications to infected lung tissues and the encapsulation of corticosteroids for the treatment of asthma and COPD.

5.4.2 Metal Oxide Nanoparticles

Because of their special qualities, engineered metal oxide NPs (MONPs) are one of the most widely used produced nanomaterials. MONPs have become attractive options for the diagnosis and therapy of lung disorders because of their special physicochemical characteristics, such as their large surface area, adjustable reactivity, and capacity to generate ROS (Chavali & Nikolova 2019). The physicochemical characteristics of NPs, such as their surface area, diffusibility, and electrophilic nature, regulate the quantity of ROS generated (Raghunath & Perumal 2017). MONPs often have electron-donating surfaces in addition to producing ROS through the metal ions they release from their surfaces. The ROS generated following exposure to oxygen and/or UV light cause the cell membrane to rupture, permeability to decrease, proteins and DNA to be damaged, and enzymes to become damaged (Sawai et al. 1996). The most widely used MONPs in the detection and management of lung disorders are iron oxide (Fe_3O_4), zinc oxide (ZnO), cerium oxide (CeO_2), titanium dioxide (TiO_2), and so forth.

5.4.2.1 Iron Oxide

Recent developments in nanotechnology, especially iron oxide NPs, have made it possible for new approaches to accurately identify and diagnose inflammatory lung diseases (Ansari et al. 2024). The term "nanozyme" characterizes the distinctive characteristics of iron oxide nanoparticles, which include their inherent enzyme-like activities. These nanozymes have proven to be effective in lowering reactive oxygen species, which are known to be important in the ethology of a number of lung conditions, including lung cancer, asthma, and chronic obstructive pulmonary disease (Gao et al. 2017). Furthermore, iron oxide NPs have been used widely in targeted drug delivery and imaging in treatment of lung cancer (Ahmad et al. 2023).

5.4.2.2 Zinc Oxide

The versatility of ZnO NPs is attributed to their ability to be synthesized through various methods, including physical, chemical, and green synthesis techniques (Kandari & Pandey 2024). Nano particles have demonstrated encouraging outcomes in a number of fields related to lung diseases, such as drug delivery, biosensing, and biological imaging. ZnO NPs have been shown to produce reactive oxygen species through their pH-dependent dissolution into Zn^{2+} ions, which can cause cytotoxicity in cancer cells. As a result, they may be used as a treatment for lung cancer (Singh et al. 2020). ZnO NPs' easy synthesis from inexpensive precursors, biocompatibility, and efficient cellular internalization through endocytosis have made them a viable candidate for targeted drug delivery.

5.4.2.3 Cerium Oxide

Cerium oxide NPs are an effective anticancer agent because of their cytoprotectant properties, which cause the cancer cells to produce ROS (Perez et al. 2008). The anti-invasive characteristic of cerium oxide NPs regulates antioxidant enzymes, which in turn controls the amount of ROS and offers normal cells radioprotection. Therefore, by shielding healthy cells from radiation damage, this could be implemented as a possible radiation sensitizer for cancer treatment (Alili et al. 2011). They can scavenge ROS and lower oxidative stress in conditions like pulmonary fibrosis and asthma because of their regenerative antioxidant qualities. These nanomaterials can reduce lung tissue damage and inflammation (Wason et al. 2013).

5.5 MECHANISM OF ACTION IN TARGETING LUNG TISSUES AND REDUCING INFLAMMATION

Metal and metal oxide NPs target lung tissues and reduce inflammation through intricate and varied mechanisms of action. These nanomaterials interact with cellular components in the lungs by utilizing their distinct physicochemical characteristics, which has a major impact on inflammatory pathways (Horvat et al. 2024). In contrast to their massive equivalents, NPs are more effective at inhibiting inflammation boosters such as cytokines and enzymes that aid inflammation due to their high surface area to volume ratio. Damage to cells and tissues brought on by an imbalance in the inflammatory process's regulatory signals is what causes inflammation.

Metal NPs can be administered via the mouth, nostrils, or skin. They may readily penetrate most biological barriers, such as mucosal linings, and even reach the sense organs due to their small size (Agarwal et al. 2019). Metal NPs enter the bloodstreams immediately upon entering the body. The course of action can be seen in Figure 5.3. Au NPs are recognized to act as possible anti-inflammatory drugs by reducing ROS generation in phagocytes in a dose-dependent manner. Additionally, metal NPs such as silver NPs have been shown to possess inherent antimicrobial properties, which are critical in combating respiratory infections and minimizing biofilm formation on lung surfaces (Ahamed et al. 2008).

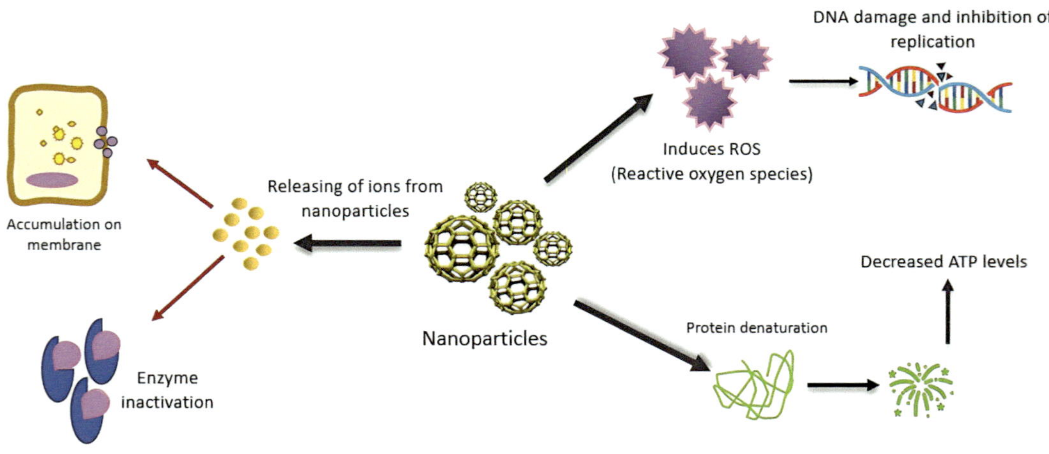

FIGURE 5.3 Mechanism of action of nanoparticles

ZnO NPs have several means of reducing inflammation. They are known to reduce the formation of thymic stromal lymphopoietin (TSLP). In reaction to pathogenic microorganisms, internal or external damage, trapped foreign particles, and pre-existing inflammatory cytokines, epithelial cells release TSLP (Brenner & Krakauer 2003). Similar to ZnO NPs, Ag NPs act as an anti-inflammatory agent. Ag NPs in lung tissues prevent the overproduction of mucus glycoproteins (mucins), which impairs pulmonary function by obstructing airways and causing chronic inflammation (Jang et al. 2012).

According to studies, there is a concentration increase in NPs in the plasma that is dependent on the dosage over a 24-hour period. Within 72 hours of injection, the liver, lung, and kidney receive the majority of the NPs. Smaller NPs are removed through the urine itself, whereas the biliary and faecal pathways remove the bigger ones (Baek et al. 2012). On a broader scale, the innovative designs of biocompatible nanomaterials enhance delivery systems to affected areas, improving therapeutic effectiveness while reducing systemic adverse effects. Thus, the careful engineering of these NPs is promising for future nanomedical applications in lung disease management (Damasco et al. 2020).

5.6 CONCLUSION

The exploration of metal and metal oxide NPs presents promising avenues for managing lung diseases, highlighting their distinctive characteristics and properties possessed for therapeutic approach. These nanomaterials can enhance drug delivery systems, allowing for targeted therapy that minimizes systemic exposure while maximizing local impact in the lungs. For instance, the unique properties of transition metal oxide NPs can be fine-tuned to improve their interaction with biological tissues, potentially reducing cytotoxic effects and enhancing therapeutic efficacy. Consequently, studying the properties of these NPs is essential not only for developing effective treatments but also for mitigating potential adverse effects. Through meticulous characterization and understanding of their behaviours, researchers can harness their potential for innovative therapeutic strategies in lung disease management. However, potential challenges remain, such as the complex interactions of NPs with biological systems, including immunogenic responses and toxicological concerns. As researchers navigate these obstacles, interdisciplinary collaboration and rigorous testing methodologies will be essential to fully harness the benefits of NPs, paving the way for innovative therapies in managing chronic respiratory conditions and other lung diseases.

REFERENCES

Agarwal, H., Nakara, A., & Shanmugam, V. K. (2019). Anti-inflammatory mechanism of various metal and metal oxide nanoparticles synthesized using plant extracts: A review. *Biomedicine & Pharmacotherapy*, *109*, 2561–2572.

Ahamed, M., Karns, M., Goodson, M., Rowe, J., Hussain, S. M., Schlager, J. J., & Hong, Y. (2008). DNA damage response to different surface chemistry of silver nanoparticles in mammalian cells. *Toxicology and Applied Pharmacology*, *233*(3), 404–410.

Ahmad, W., Joshi, H. C., Pandey, S., Kumar, V., & Verma, M. (2023). An overview of green methods for Fe2O3 nanoparticle synthesis and their applications. *International Nano Letters*, *13*(2), 117–130.

Alberg, A. J., Brock, M. V., Ford, J. G., Samet, J. M., & Spivack, S. D. (2013). Epidemiology of lung cancer: Diagnosis and management of lung cancer: American college of chest physicians evidence-based clinical practice guidelines. *Chest*, *143*(5), e1S–e29S.

Alili, L., Sack, M., Karakoti, A. S., Teuber, S., Puschmann, K., Hirst, S. M., . . . & Brenneisen, P. (2011). Combined cytotoxic and anti-invasive properties of redox-active nanoparticles in tumor–stroma interactions. *Biomaterials*, *32*(11), 2918–2929.

Ansari, S. R., Mahajan, J., & Teleki, A. (2024). Iron oxide nanoparticles for treatment and diagnosis of chronic inflammatory diseases: A systematic review. *Wiley Interdisciplinary Reviews: Nanomedicine and Nanobiotechnology*, *16*(3), e1963.

Baek, M., Chung, H. E., Yu, J., Lee, J. A., Kim, T. H., Oh, J. M., . . . & Choi, S. J. (2012). Pharmacokinetics, tissue distribution, and excretion of zinc oxide nanoparticles. *International Journal of Nanomedicine*, 3081–3097.

Bansal, S. A., Kumar, V., Karimi, J., Singh, A. P., & Kumar, S. (2020). Role of gold nanoparticles in advanced biomedical applications. *Nanoscale Advances*, *2*(9), 3764–3787.

Bregoli, L., Movia, D., Gavigan-Imedio, J. D., Lysaght, J., Reynolds, J., & Prina-Mello, A. (2016). Nanomedicine applied to translational oncology: A future perspective on cancer treatment. *Nanomedicine: Nanotechnology, Biology and Medicine*, *12*(1), 81–103.

Brenner, P. S., & Krakauer, T. (2003). Regulation of inflammation: A review of recent advances in anti-inflammatory strategies. *Current Medicinal Chemistry-Anti-Inflammatory & Anti-Allergy Agents*, *2*(3), 274–283.

Cersosimo, R. J. (2002). Lung cancer: A review. *American Journal of Health-System Pharmacy*, *59*(7), 611–642.

Chavali, M. S., & Nikolova, M. P. (2019). Metal oxide nanoparticles and their applications in nanotechnology. *SN Applied Sciences*, *1*(6), 607.

Collins, L. G., Haines, C., Perkel, R., & Enck, R. E. (2007). Lung cancer: Diagnosis and management. *American Family Physician*, *75*(1), 56–63.

Custovic, A., de Moira, A. P., Murray, C. S., & Simpson, A. (2023). Environmental influences on childhood asthma: Allergens. *Pediatric Allergy and Immunology*, *34*(2), e13915.

Damasco, J. A., Ravi, S., Perez, J. D., Hagaman, D. E., & Melancon, M. P. (2020). Understanding nanoparticle toxicity to direct a safe-by-design approach in cancer nanomedicine. *Nanomaterials*, *10*(11), 2186.

Doroudian, M., Armstrong, M. E., & Donnelly, S. C. (2023). Nano-based therapies for acute and chronic lung diseases. In *Biotechnology Applied to Inflammatory Diseases: Cellular Mechanisms and Nanomedicine* (pp. 271–286). Singapore: Springer Nature Singapore.

Dudhat, K. R., & Patel, H. V. (2020). Novel nanoparticulate systems for idiopathic pulmonary fibrosis: a review. *Asian Journal of Pharmaceutical and Clinical Research*, 3–11.

Ellis, T., Chiappi, M., Garcia-Trenco, A., Al-Ejji, M., Sarkar, S., Georgiou, T. K., . . . & Ryan, M. P. (2018). Multimetallic microparticles increase the potency of rifampicin against intracellular Mycobacterium tuberculosis. *ACS Nano*, *12*(6), 5228–5240.

Fletcher, C. M., & Tinker, C. M. (1961). Chronic bronchitis. *British Medical Journal*, *1*(5238), 1491.

Gao, Z., Ye, H., Tang, D., Tao, J., Habibi, S., Minerick, A., . . . & Xia, X. (2017). Platinum-decorated gold nanoparticles with dual functionalities for ultrasensitive colorimetric in vitro diagnostics. *Nano Letters*, *17*(9), 5572–5579.

Garcia-Fernandez, A., Sancenon, F., & Martinez-Manez, R. (2021). Mesoporous silica nanoparticles for pulmonary drug delivery. *Advanced Drug Delivery Reviews*, *177*, 113953.

Guo, D., Xie, G., & Luo, J. (2013). Mechanical properties of nanoparticles: Basics and applications. *Journal of Physics D: Applied Physics*, *47*(1), 013001.

Ha-Duong, M., Casman, E., & Morgan, G. M. (2004). A strategy for bounding attributable risk: A lung cancer example. *Risk Analysis*, *24*(5), 1093–1095.

Hamid, Q., Song, Y., Kotsimbos, T. C., Minshall, E., Bai, T. R., Hegele, R. G., & Hogg, J. C. (1997). Inflammation of small airways in asthma. *Journal of Allergy and Clinical Immunology*, *100*(1), 44–51.

Holgate, S. T., Wenzel, S., Postma, D. S., Weiss, S. T., Renz, H., & Sly, P. D. (2015). Asthma. *Nature Reviews: Disease Primers*, *1*(1), 15025.

Horvat, N. K., Chocarro, S., Marques, O., Bauer, T. A., Qiu, R., Diaz-Jimenez, A., . . . & Muckenthaler, M. U. (2024). Superparamagnetic iron oxide nanoparticles reprogram the tumor microenvironment and reduce lung cancer regrowth after crizotinib treatment. *ACS Nano*, *18*(17), 11025–11041.

Hussain, M. S., Sharma, P., Dhanjal, D. S., Khurana, N., Vyas, M., Sharma, N., . . . & Sohal, S. S. (2021). Nanotechnology based advanced therapeutic strategies for targeting interleukins in chronic respiratory diseases. *Chemico-Biological Interactions*, *348*, 109637.

Kandari, D., & Pandey, S. (2024). Comparative study of photo catalytic activity of Tinospora Cordifolia derived nanoparticles for degradation of indigo carmine. *Desalination and Water Treatment*, *320*, 100870.

Jang, S., Park, J. W., Cha, H. R., Jung, S. Y., Lee, J. E., Jung, S. S., . . . & Park, H. S. (2012). Silver nanoparticles modify VEGF signaling pathway and mucus hypersecretion in allergic airway inflammation. *International Journal of Nanomedicine*, 1329–1343.

Joshi, N. (2018). Nanotechnology enabled inhalation of bio-therapeutics for pulmonary diseases: Design considerations and challenges. *Current Pathobiology Reports*, *6*, 225–231.

Kinkade, S., & Long, N. A. (2016). Acute bronchitis. *American Family Physician*, *94*(7), 560–565.

Konstan, M. W., Davis, P. B., Wagener, J. S., Hilliard, K. A., Stern, R. C., Milgram, L. J., . . . & Gedeon, C. R. (2004). Compacted DNA nanoparticles administered to the nasal mucosa of cystic fibrosis subjects are safe and demonstrate partial to complete cystic fibrosis transmembrane regulator reconstitution. *Human Gene Therapy*, *15*(12), 1255–1269.

Kumari, V., Vishwas, S., Kumar, R., Kakoty, V., Khursheed, R., Babu, M. R., . . . & Singh, S. K. (2023). An overview of biomedical applications for gold nanoparticles against lung cancer. *Journal of Drug Delivery Science and Technology*, 104729.

Kuzmov, A., & Minko, T. (2015). Nanotechnology approaches for inhalation treatment of lung diseases. *Journal of Controlled Release*, *219*, 500–518.

Lazarus, J. V., Ortiz, A., Tyrovolas, S., Fernández, E., Guy, D., White, T. M., . . . & Soriano, J. B. (2022). A GBD 2019 study of health and sustainable development goal gains and forecasts to 2030 in Spain. *Scientific Reports*, *12*(1), 21154.

Leo, B. F., Chen, S., Kyo, Y., Herpoldt, K. L., Terrill, N. J., Dunlop, I. E., . . . & Ryan, M. P. (2013). The stability of silver nanoparticles in a model of pulmonary surfactant. *Environmental Science & Technology*, *47*(19), 11232–11240.

Liu, B., Cao, W., Qiao, G., Yao, S., Pan, S., Wang, L., . . . & Cui, D. (2019). Effects of gold nanoprism-assisted human PD-L1 siRNA on both gene down-regulation and photothermal therapy on lung cancer. *Acta Biomaterialia*, *99*, 307–319.

Malik, P., & Mukherjee, T. K. (2018). Recent advances in gold and silver nanoparticle-based therapies for lung and breast cancers. *International Journal of Pharmaceutics*, *553*(1–2), 483–509.

Mangal, S., Gao, W., Li, T., & Zhou, Q. T. (2017). Pulmonary delivery of nanoparticle chemotherapy for the treatment of lung cancers: challenges and opportunities. *Acta Pharmacologica Sinica*, *38*(6), 782–797.

Martinez, F. J., Collard, H. R., Pardo, A., Raghu, G., Richeldi, L., Selman, M., . . . & Wells, A. U. (2017). Idiopathic pulmonary fibrosis. *Nature Reviews: Disease Primers*, *3*(1), 1–19.

Martinez-Pitre, P. J., Sabbula, B. R., & Cascella, M. (2023). Restrictive lung disease. In *StatPearls [Internet]*. StatPearls Publishing.

Moro, M., Di Paolo, D., Milione, M., Centonze, G., Bornaghi, V., Borzi, C., . . . & Ponzoni, M. (2019). Coated cationic lipid-nanoparticles entrapping miR-660 inhibit tumor growth in patient-derived xenografts lung cancer models. *Journal of Controlled Release*, *308*, 44–56.

Noble, P. W., Barkauskas, C. E., & Jiang, D. (2012). Pulmonary fibrosis: patterns and perpetrators. *The Journal of Clinical Investigation*, *122*(8), 2756–2762.

Perez, J. M., Asati, A., Nath, S., & Kaittanis, C. (2008). Synthesis of biocompatible dextran-coated nanoceria with pH-dependent antioxidant properties. *Small (Weinheim an der Bergstrasse, Germany)*, *4*(5), 552–556.

Pi, J., Shen, L., Yang, E., Shen, H., Huang, D., Wang, R., . . . & Cai, J. (2020). Macrophage-targeted isoniazid-selenium nanoparticles promote antimicrobial immunity and synergize bactericidal destruction of tuberculosis bacilli. *Angewandte Chemie*, *132*(8), 3252–3260.

Raghunath, A., & Perumal, E. (2017). Metal oxide nanoparticles as antimicrobial agents: A promise for the future. *International Journal of Antimicrobial Agents*, *49*(2), 137–152.

Raji, V., Pal, K., Zaheer, T., Kalarikkal, N., Sabu, T., de Souza, F. G., & Asiya, S. I. (2020). Gold nanoparticles against respiratory diseases: Oncogenic and viral pathogens review. *Therapeutic Delivery*, *11*(8), 521.

Ryu, J. H., & Scanlon, P. D. (2001, November). Obstructive lung diseases: COPD, asthma, and many imitators. In *Mayo Clinic Proceedings* (Vol. 76, No. 11, pp. 1144–1153). Elsevier.

Sawai, J., Kawada, E., Kanou, F., Igarashi, H., Hashimto, A., Kokugan, T., & Shimizu, M. (1996). Detection of active oxygen generated from ceramic powders having antibacterial activity. *Journal of Chemical Engineering of Japan*, *29*(4), 627–633.

Seshadri, D. R., & Ramamurthi, A. (2018). Nanotherapeutics to modulate the compromised micro-environment for lung cancers and chronic obstructive pulmonary disease. *Frontiers in Pharmacology*, *9*, 340406.

Shah, M. R., Imran, M., & Ullah, S. (2019). *Nanocarriers for Cancer Diagnosis and Targeted Chemotherapy*. Elsevier.

Shukla, S. D., Vanka, K. S., Chavelier, A., Shastri, M. D., Tambuwala, M. M., Bakshi, H. A., . . . & O'Toole, R. F. (2020). Chronic respiratory diseases: An introduction and need for novel drug delivery approaches. In *Targeting Chronic Inflammatory Lung Diseases Using Advanced Drug Delivery Systems* (pp. 1–31). Elsevier.

Singh, T. A., Das, J., & Sil, P. C. (2020). Zinc oxide nanoparticles: A comprehensive review on its synthesis, anticancer and drug delivery applications as well as health risks. *Advances in Colloid and Interface Science*, *286*, 102317.

Sutrisna, B., Surtidewi, L., Jusuf, A., Hudoyo, A., Kusmana, D., Setianto, B., & Purwaningsih, E. (2009). Estimating the annual cost of smoking-related diseases in Indonesia. *Media Medika Indonesiana*, *43*(5), 247–253.

Taghavizadeh Yazdi, M. E., Qayoomian, M., Beigoli, S., & Boskabady, M. H. (2023). Recent advances in nanoparticle applications in respiratory disorders: A review. *Frontiers in Pharmacology*, *14*, 1059343.

Varshosaz, J., Ghaffari, S., Mirshojaei, S., Jafarian, A., Atyabi, F., Kobarfard, F., & Azarmi, S. (2013). Biodistribution of amikacin solid lipid nanoparticles after pulmonary delivery. *BioMed Research International, 2013*.

Wadhwa, R., Aggarwal, T., Thapliyal, N., Chellappan, D. K., Gupta, G., Gulati, M., . . . & Maurya, P. K. (2019). Nanoparticle-based drug delivery for chronic obstructive pulmonary disorder and asthma: Progress and challenges. *Nanotechnology in Modern Animal Biotechnology*, 59–73.

Wason, M. S., Colon, J., Das, S., Seal, S., Turkson, J., Zhao, J., & Baker, C. H. (2013). Sensitization of pancreatic cancer cells to radiation by cerium oxide nanoparticle-induced ROS production. *Nanomedicine: Nanotechnology, Biology and Medicine*, *9*(4), 558–569.

Witschi, H. (2001). A short history of lung cancer. *Toxicological Sciences*, *64*(1), 4–6.

Wu, P., Jiang, Y. Q., Si, F. L., Wang, H. Y., Song, X. B., Sheng, C. F., . . . & Zhang, J. (2024). Pharmaceutical treatment status of patients with COPD in the community based on medical Internet of Things: A real-world study. *NPJ Primary Care Respiratory Medicine*, *34*(1), 10.

Yamashita, T., & Millar, R. (2017). Disease prevalence and health determinants in Nevada. In Dmitri N. Shalin (ed.), *The Social Health of Nevada: Leading Indicators and Quality of Life in the Silver State* (pp. 1–28). UNLV, Center for Democratic Culture Publications.

6 Nanomicelles for the Management of Lung Cancer

Aditi Singhal[1*], *Verma Mansi*[1*], *and Silpi Chanda*[2]

[1] Meerut Institute of Engineering & Technology, Department of Pharmaceutical Technology, Meerut

[2] Amity Institute of Pharmacy, Department of Pharmacognosy, Amity University Lucknow campus, Uttar Pradesh, Sector 125, Noida 201313, India

*Corresponding author:

6.1 INTRODUCTION

Lung cancer has the highest fatality rate among all cancers and is the second most common malignant tumor worldwide [1]. It is classified into two main types: small cell lung cancer (SCLC) and non–small cell lung cancer (NSCLC), with NSCLC making up about 85% of cases globally. The five-year survival rate for lung cancer patients varies between 4% and 17%, depending on the stage and geographic location. Unfortunately, most lung cancers are diagnosed at an advanced stage, making surgery no longer a viable option due to tumor invasion or distant metastases [2–4]. Chemotherapy remains a key treatment for lung cancer, but its lack of selectivity leads to severe side effects. For example, while doxorubicin (DOX) is one of the most powerful anticancer drugs, its clinical use is significantly limited due to its toxic effects on the heart and bone marrow [5].

Nanoparticles (NPs) have emerged as promising drug delivery systems, attracting significant attention in cancer treatment. Among them, polymer micelles—self-assembling nanoparticles with a core-shell structure—have gained popularity. Studies have shown that polymer micelles can efficiently encapsulate hydrophobic drugs, improving their water solubility [6, 7]. They also offer excellent biocompatibility, extend drug circulation time in the bloodstream, accumulate in tumor tissues through the enhanced permeability and retention (EPR) effect, and can be modified to actively target tumor cells. Additionally, they provide an environmentally responsive delivery system, making them a promising tool for more effective lung cancer treatment [8, 9].

The EPR effect is a key mechanism that explains how nanoparticles accumulate in tumors. Unlike normal blood vessels, tumor vessels have larger gaps between their endothelial cells and poor lymphatic drainage, allowing nanoparticles to enter and stay within the tumor [10]. Polyamines such as spermidine, spermine, and putrescine are naturally occurring small molecules found in mammalian cells. They play a vital role in cell growth and differentiation. Research has shown that many cancer cells have a highly active polyamine transport system (PTS), which absorbs polyamines to support rapid tumor growth [11, 12]. Because of this, PTS is considered a promising target for cancer therapy. Studies have demonstrated that when chemotherapeutic drugs are linked to polyamines, their ability to enter cancer cells increases significantly, leading to enhanced drug effectiveness [13, 15]. This makes polyamines a strong candidate for developing targeted drug delivery systems. However, despite advancements in nano-targeted drug delivery—including active targeting systems and environmentally sensitive targeting systems—current targeting efficiency remains low. Only about 0.7% of an administered drug dose actually reaches solid tumors. One major challenge is that NPs

DOI: 10.1201/9781003595830-6

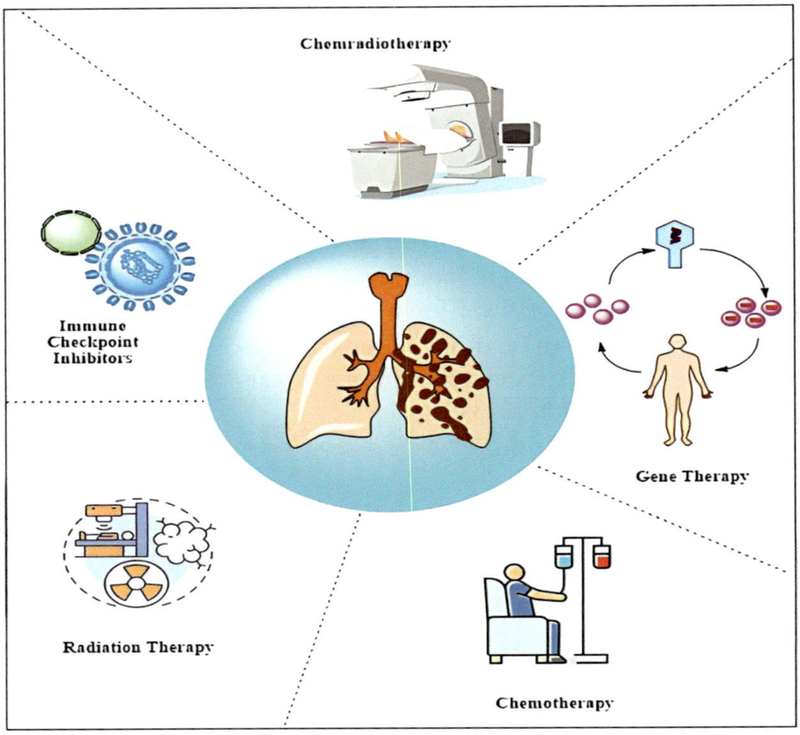

FIGURE 6.1 Nanomicelles for the management of lung cancer

are rapidly cleared by the reticuloendothelial system (RES), which removes foreign particles from the bloodstream before they can reach tumor cells. While active targeting systems are designed to direct drugs to tumors, they still struggle to avoid detection and elimination by the immune system. To improve drug delivery, the first priority should be preventing rapid clearance by the RES so that nanoparticles can accumulate near tumor tissues. By ensuring that nano-targeted drug delivery systems reach the tumor site effectively, they will have a better chance of delivering therapeutic drugs directly to cancer cells [16, 17].

6.2 NANOPARTICLES IN LUNG CANCER

6.2.1 Magnetic NPs

Over the past two decades, magnetic nanoparticles (MNPs) have been extensively studied for their potential in diagnosing and treating various diseases. Several MNP formulations—such as LumirenVR, GastromarkVR, Feridex I.V.VR, and EndoremVR—have received clinical approval for medical use [18, 19]. Their unique properties, including a high surface-to-volume ratio and the ability to be controlled by external magnetic fields, make them valuable tools in biomedical research. Recent studies suggest that MNPs could enable more precise and earlier detection of lung cancer, improving diagnostic accuracy [20]. One of the key advantages of MNPs is their ability to localize drug delivery directly to tumors while minimizing harm to healthy tissues. However, their application is somewhat limited due to potential toxicity concerns, which are closely tied to their biocompatibility. Several factors influence MNP biocompatibility, including size, shape, concentration, surface properties, solubility, biodegradability, and pharmacokinetics [21].

For MNPs to remain in circulation long enough to be effective, they must be larger than 10 nm to avoid rapid kidney filtration, yet smaller than 200 nm to prevent clearance by the liver and spleen [22–24]. Additionally, neutral surface charge helps extend their circulation time by reducing opsonization, making them less likely to be absorbed by the immune system. However, MNPs tend to clump together, so researchers are actively developing surface coatings to improve their stability and ensure uniform dispersion in biological fluids. Common biocompatible coatings include proteins, polymers, liposomes, and inorganic materials. Without proper biocompatibility, MNPs can have harmful effects on cells, disrupting metabolism and triggering toxicity [25]. Studies have linked MNP exposure to oxidative stress and the production of reactive oxygen species, which can damage cells. Additionally, when iron oxide NPs break down, they release excess free iron, leading to another form of cellular toxicity. Understanding and addressing these challenges is crucial for the safe and effective use of MNPs in lung cancer detection and treatment [26].

The human body tightly regulates iron levels, so an excess of iron particles can be harmful. Overloading the system with too many MNPs—especially in high or repeated doses—can disrupt the function of delicate organs like the heart, liver, and pancreatic beta cells [27, 28]. To prevent exceeding the body's regulatory capacity, MNP dosages must be carefully controlled. Research has linked excess MNP exposure to various toxic effects, including cell cycle disruption, cytoskeletal damage, mitochondrial dysfunction, and reduced cell viability [29–34]. Although this study primarily focuses on MNPs for lung cancer treatment, these nanoparticles also play a significant role in diagnostics. When coated with carboxymethyl dextran and oleic acid, superparamagnetic iron oxide nanoparticles (SPIONs) can be linked to mouse anti-CD44v6 monoclonal antibodies, which are specific markers for metastatic cancer [35–39]. These SPIONs can then be used as T2 contrast agents in MRI imaging, enhancing the detection of lung cancer metastases. Additionally, fluorescent particles combined with magnetic particles have been explored for imaging applications. For instance, galactose-displaying core-shell nanospheres have shown strong targeting ability toward $\alpha v\beta 6$-positive H2009 cells, a peptide linked to lung adenocarcinoma [40].

Another promising imaging technique involves nebulized MNPs, which could improve lung imaging precision. Beyond diagnostics, MNPs also show potential in thermal therapies. There are two key heat-based treatment strategies [41]:

- **Hyperthermia:** A controlled temperature increase between 40°C and 45°C, which can trigger cancer cell apoptosis [42].
- **Thermoablation:** A higher temperature increase that causes instant cancer cell death [43].

Studies have explored magnetically induced hyperthermia for various cancers, including glioblastoma, oral cancer, osteosarcoma, pancreatic cancer, breast cancer, melanoma, and prostate cancer. For lung cancer, iron oxide MNPs have been integrated into magnetic nanocomposite microparticles using spray drying techniques. When exposed to an alternating magnetic field, these particles generate heat remotely, which can help stimulate drug movement, activate treatments, and induce hyperthermia as a therapeutic strategy. MNPs can also serve as carriers for chemotherapy drugs. For example, hydroxyapatite, a highly biocompatible compound, has a natural affinity for certain chemotherapy medications. By combining iron and platinum with hydroxyapatite, researchers have developed a dual-action chemo-hyperthermia agent, which may become a future lung cancer treatment [44–50].

Further studies have shown that polyacrylic acid–coated MNP clusters can effectively slow tumor growth and sensitize lung cancer cells (NCI-H460) to radiation therapy when heated in an alternating magnetic field. Additionally, exposure to a pulsed electromagnetic field has been found to trigger lung cancer cell death [51–53]. Another promising hyperthermia-based treatment involves ferucarbotran, an MRI contrast agent that generates heat in an alternating magnetic field. Researchers have also evaluated doxorubicin-loaded porous MNPs with a polyethylene glycol (PEG) coating for their ability to combine hyperthermia with chemotherapy. Studies on human lung adenocarcinoma (A495) cells have shown encouraging results, highlighting the potential of MNP-based drug delivery and thermal therapy in lung cancer treatment [54, 55].

6.2.2 Polymer NPs

The use of polymeric NPs in cancer therapy has been widely studied due to their unique physical and chemical properties [56]. These nanoparticles are particularly attractive for delivering anticancer drugs because they allow precise control over particle size, surface charge, and drug encapsulation. They can also be designed in various functional forms, such as capsules, dendrimers, colloids, or micelles. Both natural and synthetic polymers are used in NP development [57]. Natural polymers, including polypeptides, albumin, gelatin, and chitosan, are favored for their biocompatibility and biodegradability. Synthetic polymers, such as PEG, polylactic acid (PLA), polylactic-co-glycolic acid (PLGA), and polycaprolactone (PCL), also play a crucial role in nanoparticle-based drug delivery [58]. One major advantage of polymeric NPs is their versatile administration methods, as they can be delivered intravenously, orally, nasally, or through topical absorption [59]. They help address key challenges in cancer drug delivery, particularly by reducing side effects and enhancing drug effectiveness [60].

Polymeric NPs are often created through self-assembly of block copolymers, which form nanoparticles due to differences in hydrophobicity. This allows for the high-concentration encapsulation of hydrophobic drugs, improving chemotherapy outcomes by extending drug circulation time and ensuring targeted delivery [61]. Abraxane®, an FDA-approved albumin-based nanocarrier loaded with paclitaxel, has been successfully used for treating NSCLC and metastatic breast cancer [62].

PEG-modified nanoparticles combined with taxanes have demonstrated improved effectiveness in chemoradiation for NSCLC, both in lab studies and in A549 lung cancer cell lines [63]. Hu et al. (2017) explored how circadian rhythms influence tumor growth, showing that paclitaxel-loaded PCL nanoparticles were more effective when combined with chronomodulated therapy (treatment timed with the body's natural rhythms) [64]. Wang et al. recently demonstrated how mesenchymal stem cells can act as carriers for drug-loaded NPs, enhancing drug delivery directly to cancer cells, which led to suppressed tumor growth in vivo [65]. Gold nanorods coated with a polymer shell offer a promising photothermal therapy approach. When activated by a near-infrared laser, they generate heat, which increases doxorubicin concentration at the tumor site while controlling dosage, duration, and drug distribution. These advancements highlight how polymeric nanoparticles improve cancer treatment by increasing drug effectiveness, reducing toxicity, and enhancing targeted delivery, making them a key focus in the fight against lung cancer [66].

6.2.3 Liposomes and Solid Lipid NPs

Liposomes are tiny, spherical vesicles made of phospholipid bilayers, which makes them effective carriers for chemotherapy drugs [67]. These vesicles can encapsulate hydrophilic (water-soluble) drugs in their core and hydrophobic (fat-soluble) drugs within their bilayer membrane, allowing for versatile drug delivery. One challenge with conventional liposomes is their tendency to fuse together, making them less stable in the body and more susceptible to being broken down by the RES [68, 69]. To overcome this, researchers have modified their surface by coating them with inert hydrophilic polymers like PEG. This PEGylation helps stabilize liposomes, prevents them from being recognized by the immune system, and extends their circulation time in the bloodstream. A prime example of this approach is PEGylated liposomal doxorubicin, which has shown promising results in multiple Phase III trials for ovarian cancer. Beyond liposomes, solid lipid nanoparticles

(SLNPs) have emerged as another promising drug delivery system. These submicron colloidal carriers offer advantages such as the following [70]:

- The ability to transport both water-soluble and fat-soluble drugs
- Greater stability in the bloodstream
- Longer drug circulation time
- Potential for using biocompatible ingredients

Liposomal formulations have gained interest in lung cancer treatment because they improve drug pharmacokinetics while reducing side effects. In locally advanced NSCLC, platinum-based chemotherapy is the standard treatment, but 20–40% of patients experience severe nephrotoxicity (kidney damage) from cisplatin [71]. To reduce these side effects, scientists developed Lipoplatin, a liposomal version of cisplatin [72]. Clinical trials have shown that Lipoplatin significantly reduces toxic side effects such as kidney damage, nerve toxicity, and hearing loss while maintaining effectiveness. In a phase III trial comparing Lipoplatin + paclitaxel versus traditional cisplatin + paclitaxel in NSCLC patients, the Lipoplatin group had a significantly higher response rate [73, 74].

6.2.4 Hybrid Nanoparticles and Subcellular Targeting

The new approaches are also being explored for controlled drug delivery to tumors. One such development is lipid–polymer hybrid nanoparticles, which combine the stability of polymers with the biocompatibility of lipids. For example, lipid–chitosan hybrid nanoparticles loaded with cisplatin have shown promising results in cancer therapy [75, 76].

Subcellular drug targeting is another emerging field. Researchers have designed a lipid-based platinum complex that self-assembles into nanoparticles and releases cisplatin based on pH levels in the tumor environment. This enhances the anticancer effect while reducing toxicity [77].

Taxanes (such as paclitaxel) are another common class of chemotherapy drugs, but their poor solubility makes them difficult to administer. Traditionally, solubilizing agents like Cremophor EL and Polysorbate 80 have been used to dissolve paclitaxel, but these cause severe side effects, including nerve damage, immune reactions, and bone marrow suppression. To reduce these risks, researchers developed liposomal paclitaxel, which has shown higher tolerated doses and improved effectiveness compared to traditional formulations. In a phase I clinical trial, liposomal paclitaxel was tested in NSCLC patients with malignant pleural effusions, showing positive safety and toxicity outcomes [78].

Ongoing clinical trials are now evaluating liposomal paclitaxel as a first-line treatment for advanced NSCLC, often in combination with cisplatin and other drugs. Beyond liposomes, researchers are also working on micelle-based drug formulations to enhance chemotherapy effectiveness while reducing side effects [79]. Drugs like doxorubicin and SN-38 have been successfully encapsulated in micelles to improve their delivery and minimize toxicity. For advanced NSCLC, current studies are comparing the effectiveness of traditional paclitaxel injections (with Cremophor EL) versus paclitaxel micelles combined with cisplatin, showing encouraging results for safer and more efficient treatment options [80]. Liposomes, solid lipid nanoparticles, and hybrid nanosystems are revolutionizing cancer treatment by making chemotherapy drugs more effective, longer lasting, and less toxic. These advancements, particularly in lung cancer therapy, hold great promise for improving patient outcomes [81–85].

6.2.5 Metal NPs

Metallic nanoparticles (MeNPs) are highly versatile materials with a wide range of applications in medicine, including thermal ablation, radiation therapy enhancement [86] and diagnostic imaging [87, 88]. Their effectiveness comes from their strong electromagnetic properties, unique optical characteristics, ease of synthesis, and ability to be customized with surface modifications [89, 90]. For enhancing drug delivery and gene therapy, one major advantage of MeNPs is that their surface can be coated with biocompatible polymers like PEG. This coating helps with the following [91, 92]:

- Prolong drug circulation time, ensuring a longer therapeutic effect
- Improve targeted drug delivery
- Enhance gene therapy applications, such as gene silencing

MeNPs contribute to cancer treatment in two key ways [93–95]:

1. **Intrinsic anti-tumor effects**—Some MeNPs naturally exhibit antioxidant properties, which can slow tumor growth.
2. **Extrinsic effects**—Their physicochemical properties allow them to be used in various cancer treatments, including drug delivery, imaging, and targeted therapy.

Several metallic nanoparticles have been explored for their anticancer potential, including the following [96–99]:

- **Iron nanoparticles**—Used for targeted drug delivery and magnetic hyperthermia
- **Titanium dioxide nanoparticles**—Known for their photocatalytic properties in cancer treatment
- **Zinc oxide nanoparticles**—Have anticancer and antibacterial effects
- **Cerium nanoparticles**—Studied for their antioxidant properties
- **Silver nanoparticles**—Effective in destroying cancer cells due to their antimicrobial and cytotoxic properties
- **Gold nanoparticles**—The most widely studied due to their unique optical, electrical, and surface properties

Gold nanoparticles (Au NPs) stand out in cancer research because of their biocompatibility, easy surface modification, and tunable size and shape (ranging from 1 to 150 nm). Their distinct optical properties make them ideal for the following:

- **Drug delivery**—Au NPs can be coated with chemotherapy drugs to enhance their effectiveness.
- **Photothermal therapy**—This uses near-infrared light to heat and destroy tumor cells.
- **Imaging and diagnostics**—AuNPs improve MRI and CT scan visibility, helping doctors detect and monitor cancer.

Because of their flexibility, safety, and effectiveness, Au NPs have become a leading choice in cancer nanomedicine [100–105].

6.3 VIRUS NPS

Viruses, including bacteriophages, plant viruses, and mammalian viruses, have gained attention for their ability to penetrate cells efficiently and deliver therapeutic agents inside their protein shells. These viral-based nanoparticles (VNPs) have been explored for cancer treatment, particularly in targeted drug delivery [106–108].

There are two main types of virus-based nanoparticles:

- **Viral Nanoparticles**—naturally occurring, containing a viral genome
- **Virus-Like Particles (VLPs)**—genome-free, making them non-infectious and potentially safer for medical applications

Virus-based drug delivery for lung cancer has been widely used to deliver platinum-based chemotherapy, a common treatment for lung cancer. Another key chemotherapy drug, DOX, used for small cell lung cancer, has been successfully paired with virus-based nanoparticles like the following [109–115]:

- Potato virus X (PVX)
- Hepatitis B virus-like particles (HB VLPs)
- Cucumber mosaic virus (FA-CMV-Dox)

A promising new delivery system involves physalis mottle virus (PhMV). Scientists have loaded its empty core with a prodrug form of DOX-EMCH and coated the surface with PEG to enhance biocompatibility. This formulation showed greater stability and effectiveness in a mouse breast cancer model, opening the door for further research [116].

6.3.1 Virus-Based Nanoparticles in Cancer Diagnosis

These viruses are not just useful for drug delivery—they can also help detect cancer cells. The T4 bacteriophage, tagged with fluorescent dyes, has been used to detect A549 lung cancer cells. The Cowpea mosaic virus, when conjugated with a FA-PEG molecule, can specifically target tumor cells that overexpress folate receptors, found in cancers of the kidneys, ovaries, and lungs. These virus-based approaches could enhance precision in both cancer diagnosis and treatment, reducing side effects and improving patient outcomes [117–119].

6.4 QUANTUM DOTS

Quantum dots (QDs) are tiny nanocrystals (3–30 nm) made from semiconductor materials. They typically have a core-shell or alloy structure and are composed of elements from the II–VI or III–V groups of the periodic table [120–122]. Some common examples include the following:

- **Group II–VI QDs**: Cadmium telluride (CdTe), cadmium selenide (CdSe), zinc selenide (ZnSe), and zinc sulfide (ZnS)
- **Group III–V QDs**: Gallium arsenide (GaAs), gallium nitride (GaN), indium arsenide (InAs), and indium phosphide (InP) [123, 124]

QDs have shown potential in drug delivery and imaging, but their solubility and bioavailability can be problematic. To address this challenge, polymer coatings are often applied to improve their solubility and absorption in biological systems. Despite their promising applications, the biggest challenge in using QDs for clinical purposes is their potential toxicity to cells. Research is ongoing to develop safer, biocompatible alternatives that minimize the risk of cytotoxic effects [125]. QDs have the potential to revolutionize cancer imaging and treatment, but their toxicity remains a major concern. This is because they contain toxic metal atoms, which can enhance the colloidal effect and generate photon-induced free radicals, leading to cell damage [126].

- **Applications in cancer research:** Despite these challenges, QDs are valuable in genetic profiling, cellular imaging, and cancer diagnosis. Their unique optical properties and ability to be functionalized with proteins make them ideal for multimodal theranostics—technologies that combine therapy and diagnosis [127].
- **Recent studies and innovations—doxorubicin-conjugated QDs**: A study by Ranjbar-Navazi et al. explored the use of indium phosphide/zinc sulfide (InP/ZnS) QDs, conjugated with doxorubicin, folic acid (FA), and D-glucosamine (GA) in OVCAR-3 ovarian cancer and A549 lung cancer cells. Results showed that FA and GA improved QD absorption by cancer cells, making chemotherapy drugs more effective. QDs also acted as a

chemosensitizer, reducing cell toxicity while enhancing drug delivery. Multispectral analysis detected a narrow fluorescence emission at 570 nm when stimulated at 400 nm, highlighting their potential for real-time cancer imaging and treatment.

- **pH-responsive drug delivery with ZnO QDs:** A study by Cai et al. introduced a zinc oxide (ZnO) QD–based drug delivery system that responds to pH changes for controlled drug release inside cancer cells. DOX was effectively loaded onto PEG-functionalized ZnO QDs using metal-drug interactions and covalent bonds. This technology enables targeted drug release within tumor environments, improving treatment precision while minimizing damage to healthy cells [128].

These advancements suggest that QD-based drug delivery and imaging systems hold great promise for early cancer detection and targeted therapy, though further research is needed to address toxicity concerns. Researchers have developed hyaluronic acid (HA)-conjugated zinc oxide (ZnO) QDs designed to selectively target CD44, a glycoprotein overexpressed in many cancer cells. Once absorbed by cancer cells, these pH-sensitive QDs dissolve in the acidic environment of the endosome/lysosome, releasing zinc ions (Zn^{2+}) and the chemotherapy drug DOX. This controlled drug release allows for a synergistic anti-tumor effect, combining the cytotoxic properties of both Zn^{2+} and DOX, leading to enhanced cancer cell destruction [129]. Erlotinib hydrochloride is a first-generation tyrosine kinase inhibitor used to treat NSCLC. Its active metabolite, desmethyl erlotinib (OSI-420), has similar anticancer properties. A study by Kulkarni et al. investigated the impact of OSI-420 conjugated with QDs on A549 lung cancer cells. The results showed that QD-conjugated OSI-420 was significantly more effective than the standalone drug, demonstrating higher intracellular drug delivery and cytotoxicity across multiple cancer cell lines. These studies suggest that QD-based drug delivery can improve targeting, enhance drug effectiveness, and reduce side effects. By leveraging QD bioconjugation, researchers are developing more efficient alternatives to traditional cancer treatments, paving the way for next-generation nanomedicine [130].

6.5 CLINICAL STATUS OF NANOPARTICLES IN LUNG CANCER

For patients with advanced lung cancer, the standard treatment typically involves a combination of chemotherapy and radiation, often alongside immunotherapy or personalized medicine. However, drug resistance and the complexities of combining different treatments remain significant challenges. To address these issues, numerous clinical trials are now exploring the potential of NP-based therapies for lung cancer treatment. Nab-Paclitaxel has demonstrated effectiveness as first-line chemotherapy for NSCLC in some early studies [131–133]. However, its efficacy in later-stage disease or in combination with other treatments is still under investigation (NCT00729612, NCT00553462). BIND-014, a docetaxel NP injectable, is being tested as a second-line treatment for metastatic lung cancer, following promising results in other cancer types [134, 135]. However, no published clinical data are currently available (NCT01792479, NCT02283320). CRLX101, an NP-based drug containing camptothecin linked to a cyclodextrin-based polymer, has shown promising early results (NCT01380769). This therapy aims to enhance tumor exposure to camptothecin while reducing side effects. As research progresses, NP-based drug delivery has the potential to improve treatment efficacy, minimize side effects, and overcome drug resistance in lung cancer. Ongoing clinical trials will determine how these innovative therapies can be integrated into standard cancer care [136].

6.6 LIPID-BASED NANOCARRIERS

The most promising drug delivery systems (DDSs) today use nanoparticles (NPs) made from natural polymers such as proteins, phospholipids, and polysaccharides. These natural NPs are considered superior to synthetic polymers due to their following properties:

- Higher drug-loading capacity
- Better biocompatibility
- Reduced immune system detection (lower RES opsonization)
- Safer degradation, producing fewer toxic byproducts [137–139]

Since they are easier to produce and safer for human use, natural polymer-based NPs present a promising alternative for drug delivery technologies. One major challenge in chemotherapy is delivering hydrophobic drugs, which do not dissolve easily in water. Lipid-based nanocarriers help solve this problem because they mimic cell membranes, making them easier to absorb and reduce side effects compared to other drug carriers [140–143]

A key advantage of lipid-based nanocarriers is their dual functionality:

- Hydrophilic (water-loving) targeting agents (e.g., antibodies, polysaccharides) attach to the outer surface.
- Hydrophobic (water-repelling) chemotherapy drugs are encapsulated inside.

The targeting molecules attached to these carriers play a crucial role in directing anticancer drugs specifically to tumor cells, improving treatment precision. Lipid-based nanocarriers can be delivered in multiple ways, including the following:

- Oral (most common)—Non-invasive, cost effective, and lower risk of side effects [144]
- Parenteral (injections)
- Ocular (eye drops)
- Intranasal (nose sprays, inhalers)
- Dermal/transdermal (patches, creams)

Although intranasal drug delivery is considered a promising method for lung cancer treatment, its effectiveness has not yet been fully proven in clinical studies. The shape and composition of lipid-based NPs impact how effectively they carry and release drugs. By fine-tuning formulation techniques, researchers aim to enhance drug delivery efficiency and improve treatment success.

6.7 DIAGNOSTIC AGENTS USED IN LUNG CANCER

Early detection of cancer significantly increases survival rates, making it a critical area of scientific research. Detecting tumors in their early stages can greatly improve the chances of complete recovery, which has led to a growing demand for advanced diagnostic techniques that can identify lung nodules sooner. Studies have shown that early detection is linked to better survival rates, emphasizing the need for accurate staging of lung cancer to ensure the most effective treatment plans.

Current lung cancer diagnostic methods:

Several traditional methods are used to diagnose and stage lung cancer, including the following:

- Sputum cytology
- Computed tomography (CT scans)
- Bronchial biopsy
- Polymerase chain reaction (PCR) testing
- Fluorescence bronchoscopy
- Chest radiography (CXR)

However, many of these techniques have limitations, particularly when it comes to early-stage tumor detection. A key concern is the risk of "missing lung cancer" in imaging tests, which can allow the disease to progress undetected to an advanced stage [152].

TABLE 6.1
The most widely utilized lipid-based nanocarriers as drug delivery systems in the treatment of lung cancer

Lipid-based nanocarriers	Pros	Cons	Drug: Target cancer Phase: Status (Source: FDA)
Liposomes	Enhanced therapeutic effectiveness against infections; decreased toxicity; enhanced pharmacokinetics and pharmacodynamics; and enhanced drug-target selectivity [145, 146]	Low stability, limited shelf life, and high production costs [145]	SCLC; phase 2; Hiring irinotecan hydrochloride liposome injection (LY01610)
Lipid micelles	Prolongs the duration of a drug's blood circulation; dissolves hydrophobic medications; has distinct diameters and particle sizes [147]	Theirdissociation may result from a change in equilibrium caused by dilution toward the unimer state. [147]	Genexol (Pacitaxel) Drug: Genexol-PM®, a paclitaxel-loaded polymeric micelle; NSCLC; phase 2; Finalized
Solid lipid nanoparticles (SLNPs)	Protects the medication against severe environmental conditions and has great biodegradability and biocompatibility [148, 149]	Drug evacuation as a result of storage conditions, crystalline structure, and poor drug loading efficiency. [148]	Not available
Lipid nanoparticle (LNP)	Simple synthesis with the use of microfluidic methods; enhanced drug loading that is based on charge, like nucleic acid	There aren't many instances of hepatotoxicity or other adverse consequences from using polyethylene glycol [150, 151]	Not accessible

The potential of nanotechnology in cancer imaging:

Advancements in nanotechnology are transforming cancer detection, staging, and treatment. NPs have unique physical and chemical properties that enable the creation of highly sensitive imaging tools with the following:

- Stronger signal intensity
- Better contrast and magnification
- Improved dispersion for clearer imaging [153, 154]

Because nanoparticles can specifically target tumors, they offer a promising solution to overcome the limitations of traditional diagnostic methods.

6.8 NANOMEDICINE

Nanomedicines are a game changer in cancer treatment. Nanomedicine has revolutionized cancer therapy by enabling the precise delivery of both chemotherapeutic agents and diagnostic molecules. Nanocarriers, which include both metallic and non-metallic nanoparticles, play a crucial role in enhancing imaging and treatment effectiveness. Among these, noble metal nanoparticles (such as gold and silver NPs) have gained attention due to their exceptional thermal, electrical, and optical properties. Their biocompatibility, stability, and low toxicity make them an ideal starting point for developing next-generation drug delivery systems [155–157].

6.8.1 Polymeric Nanocarriers

The biocompatibility, biodegradability, and non-toxic nature of polymeric nanocarriers, along with their diverse physicochemical properties, make them highly valuable for both cancer diagnosis and treatment.

6.8.2 Biologically Derived Nanocarriers

These nanocarriers are increasingly used in medical applications due to their natural compatibility with the body, ability to be easily modified, and flexibility to be coated with different active drugs for targeted therapy.

6.9 VASCULAR ENDOTHELIAL–DERIVED GROWTH FACTOR RECEPTORS

Lung cancer cells and endothelial cells tend to overproduce vascular endothelial growth factor receptors (VEGFRs) due to genetic mutations and low oxygen levels in tissues (hypoxia). This leads to an increase in vascular endothelial growth factor (VEGF), regulated by hypoxia-inducible factor-1α (HIF-1α) and matrix metalloproteinase (MMP) [158]. VEGF plays a key role in tumor growth and spread by activating VEGFR signaling pathways, which increase blood vessel permeability, promote new blood vessel formation (angiogenesis), and attract endothelial cells. Additionally, VEGF contributes to lung cancer progression and drug resistance by triggering the production of survivin, a protein that helps cancer cells evade death [159, 160].

The VEGF family includes PGF, VEGF-A, VEGF-B, VEGF-C, VEGF-D, and VEGF-E, while its receptors include VEGFR-1 (Flt-1), VEGFR-2 (Flk-1/KDR), VEGFR-3 (Flt-4), neuropilin-1 (NRP-1), and neuropilin-2 (NRP-2), with VEGFR-2 being the most important subtype. VEGFRs consist of a tyrosine kinase-containing cytoplasmic domain, a transmembrane domain, and an extracellular ligand-binding domain. According to Frezzetti et al., NRPs have a short intracellular domain lacking catalytic function, an extracellular domain, and a single transmembrane domain [161]. Because VEGFRs are highly expressed in both tumor cells and tumor-associated blood vessels, they are considered ideal targets for dual tumor and vascular therapy [162]. Specifically, VEGFRs are prime targets for precision drug delivery [163].

Nanoparticle-based targeting of VEGFR in lung cancer:

Jin et al. (2018) designed a new type of liposomal nanoparticle with active targeting properties for treating NSCLC. These nanoparticles carried parthenolide and ginsenoside compound K (CK) and were coated with tumor-homing peptide tLYP-1 (CGNKRTR) within a hydrophilic PEG shell. The nanoparticles measured 188 nm in size, with encapsulation efficiencies of 83.4% for parthenolide and 70.7% for CK, and drug-loading efficiencies of 14.8% and 2.9%, respectively. The PEG shell helped increase solubility, improve stability, and extend circulation time. More importantly, tLYP-1 enabled the nanoparticles to specifically target lung cancer cells by binding to the overexpressed NRP-1 receptor, allowing them to penetrate tumor tissues more effectively. Once inside the cells, the nanoparticles triggered mitochondrial swelling, increased ROS, and induced apoptosis, ultimately enhancing the anti-tumor effect while minimizing side effects. This suggests that combining low-toxic natural compounds could further refine this innovative cancer treatment strategy [163].

Similarly, Liu et al. developed a nanostructured lipid carrier for targeted delivery of docetaxel (DTX) [163]. They analyzed its biological distribution, tumor cell uptake, and overall therapeutic efficiency. By binding to VEGFR-2 (Flk-1 receptor) on both tumor and endothelial cells, the nanoparticles demonstrated strong anti-tumor activity. The study found that the PEG coating prevented unwanted interactions with serum proteins, improving stability and selectivity. This approach enhanced tumor inhibition while reducing adverse effects, though further research is needed to fully understand its impact on metastasis. Since tumor blood vessel formation (neovascularization) not only nourishes the tumor but also facilitates its spread, developing VEGFR-targeting nanoparticles represents a promising precision medicine approach. Future research will likely introduce new NP-based therapies into clinical treatment, providing lung cancer patients with personalized options that are more effective and have fewer side effects.

6.10 αVβ3 INTEGRIN

Integrins are transmembrane glycoproteins composed of 18 α and 8 β subunits, forming heterodimers that help regulate cell adhesion, migration, differentiation, proliferation, and survival [164]. Among the 24 known subtypes, 11 specifically bind to the Arg-Gly-Asp (RGD) sequence, a common feature in extracellular matrix proteins such as collagen, fibronectin, fibrinogen, laminin, and osteopontin. When integrins bind to these proteins, they undergo conformational changes that affect how cells interact with their surroundings. The αvβ3 integrin subtype is particularly overexpressed in lung cancer and endothelial cells, playing a key role in tumor growth, blood vessel formation (angiogenesis), and metastasis. Since RGD peptides can selectively bind to αvβ3 integrins, they have been widely explored for targeted drug delivery. Researchers have modified RGD peptides to enhance their stability, improve their biological properties, and increase their ability to attach to lung cancer cells [165].

6.11 RGD-MODIFIED NANOPARTICLES FOR LUNG CANCER TREATMENT

6.11.1 RGD-Modified PLGA-Chitosan Nanoparticles for Paclitaxel Delivery

Babu et al. developed an RGD-modified polylactic acid-glycolic acid (PLGA)-chitosan nanoparticle system (CSNP-RGD) for delivering paclitaxel (PTX) to NSCLC cells. These nanoparticles were primarily PLGA based and coated with positively charged chitosan, which improved stability, adhesion, and controlled drug release. The linear RGD peptide (GRGDSP) was added to enhance targeting of αvβ3 and α5β1 integrins. The nanoparticles had an average size of 217 nm, a drug loading rate of 6.5%, and a PTX entrapment efficiency of 93.7%.

Key findings [166]:

- The nanoparticles showed no toxicity when drug free and maintained good physical and chemical stability.
- For NSCLC cell lines (A549 and H1299) overexpressing αvβ3 integrin, the RGD-modified nanoparticles exhibited high selectivity.
- Once inside the cells, they induced apoptosis and blocked tumor cell division at the G2/M phase, while having minimal impact on healthy bronchial epithelial cells.
- In vivo, the nanoparticles reduced side effects and significantly inhibited tumor growth, making them a promising targeted drug delivery approach.

6.11.2 RGD-Modified Lipid Polymer Nanoparticles for Combined Chemotherapy

Wang et al. explored a nanoparticle system (RGD-ss-PTX-CDDP LPNs) that combined PTX and cisplatin (CDDP) for synergistic lung cancer treatment [167]. These nanoparticles were composed of soybean lecithin-coated PLGA loaded with PTX and CDDP, connected by PEG disulfide bonds. The high glutathione (GSH) levels in tumor cells triggered intracellular drug release, enhancing targeted chemotherapy. The nanoparticles had an average size of 191.3 nm, a zeta potential of –37.2 mV, and stable drug loading rates (PTX: 5.4%, CDDP: 12.3%).

Key Findings:

- The nanoparticles were stable in culture medium for five days *and in* phosphate-buffered saline (PBS) for 30 days.
- Minimal accumulation in healthy tissues *and* no significant toxicity at 100 μg/mL.
- RGD-modification improved tumor selectivity, allowing for controlled drug release *and* enhanced anti-tumor activity.
- This method could be a potential redox-sensitive, RGD-modified nanoparticle system *for* combined lung cancer treatment.

6.11.3 RGD-Linked Polymer Nanoparticles for Doxorubicin Delivery

Zou et al. (2018) developed cRGD-PS-DOX, a cyclic RGD peptide-disulfide cross-linked polymer, *for* targeted DOX delivery in NSCLC [168]. The system used PEG-P (TMC-DTC) as a ligand carrier, with DOX loaded at a rate of 15.2% and an average nanoparticle size of 96 nm. Drug release studies showed low DOX release (<15%) at normal conditions (37°C, pH 7.4, low GSH levels) but higher release in tumor environments with increased GSH concentrations.

Key Findings:

- The nanoparticles demonstrated strong stability in circulation and efficient drug transport into cancer cells.
- RGD modification improved the selective uptake of DOX into $\alpha v\beta 3$-overexpressing lung cancer cells (A549), while minimizing effects on non-cancerous cells.
- Flow cytometry analysis confirmed that cRGD-PS-DOX had higher selectivity for NSCLC compared to other formulations.
- Pharmacokinetics studies showed that the nanoparticles accumulated more in tumors than in normal tissues, enhancing treatment efficiency.

6.12 CHALLENGES AND FUTURE DIRECTIONS

Despite the promising results, there are still limitations to RGD-based nanoparticle therapies, including the following [168]:

- Low specificity and affinity of some RGD peptides
- Variability in drug release rates and nanoparticle stability
- Potential off-target effects in normal cells

To overcome these challenges, future research should focus on enhancing RGD peptide selectivity to improve binding to lung cancer cells while avoiding normal tissues. Combining RGD-modified nanoparticles with additional targeting ligands increases precision and efficacy and explores new biodegradable and biocompatible nanocarrier materials to reduce toxicity and improve drug delivery efficiency.

RGD-modified nanoparticles show great potential for targeted lung cancer therapy by selectively delivering chemotherapy drugs directly to tumor cells, reducing systemic toxicity, and enhancing treatment effectiveness. With further refinements in design and specificity, these nanoparticles could become a key component of precision medicine for lung cancer treatment [168].

6.13 EPIDERMAL GROWTH FACTOR RECEPTOR

The epidermal growth factor receptor (EGFR) belongs to the class I transmembrane receptor tyrosine kinase superfamily, which includes EGFR/ERBB1/HER1, ERBB2/HER2, ERBB3/HER3, and

ERBB4/HER4. EGFR plays a crucial role in cell growth, proliferation, and differentiation and consists of three main regions: an extracellular ligand-binding domain, a transmembrane region, and an intracellular kinase domain. It is present on the surface of various cell types, including epithelial cells, fibroblasts, glial cells, and keratinocytes and is often overexpressed in multiple cancers. While SCLC does not exhibit EGFR overexpression, it is found in over 60% of NSCLC [169].

EGFR can bind to seven different ligands: betacellulin (BTC), amphiregulin (AREG), epiregulin (EREG), transforming growth factor-alpha (TGF-α), heparin-binding EGF-like growth factor (HB-EGF), epidermal growth factor (EGF), and epigen (EPGN). Among these, EGF, TGF-α, HB-EGF, and BTC exhibit high affinity, while AREG, EREG, and EPGN bind with lower affinity. According to Singh et al., these ligands activate EGFR signaling pathways, which play key roles in tumor progression, including invasion, angiogenesis, metastasis, and proliferation [170].

EGFR tyrosine kinase inhibitors (EGFR-TKIs), such as erlotinib, gefitinib, and afatinib, are commonly used to target these pathways [169]. Another promising approach involves anti-EGFR aptamers (APT), which selectively target EGFR [171]. Kim and Huang explored a peptide-based therapy designed to disrupt intracellular EGFR protein interactions, delivering these peptides via a targeted lipid nanoparticle system to human cancer cells [172].

To enhance EGFR-targeted lung cancer therapy, Zhang et al. developed a redox- and pH-responsive nanotheranostic agent (ECMI) for both treatment and molecular imaging. ECMI consists of mesoporous silica nanoparticles loaded with ZnO QDs and indocyanine green, coated with chitosan modified with a cross-linked disulfide ligand for improved targeting [173]. The nanoparticles measured approximately 220 nm in diameter, with a polydispersity index (PDI) of 0.41, a zeta potential of −18.5 mV, a drug loading capacity of 15 μg/mg, and an ER conjugation efficiency of 35 μg/mg. Both in vitro and in vivo studies confirmed that ECMI effectively accumulated in EGFR-mutant lung cancer cells (such as PC-9 and H1975) while maintaining high biocompatibility and biodegradability. The fluorescence intensity in PC-9 cells was significantly higher with ECMI compared to non-targeted controls, demonstrating enhanced targeting efficiency.

Another approach by Zhang et al. involved the development of PEG-PLGA-PLL-based nanoparticles (EGF-PEAL NPs) modified with EGF ligands. These nanoparticles were loaded with either DOX or Bcl-2 siRNA to enhance tumor cell apoptosis. The nanoparticles measured around 204 nm in size, with zeta potentials of 3.5 mV and 2.1 mV for DOX- and Bcl-2-loaded versions, respectively [174]. These particles exhibited slow and sustained drug release, increased nanoparticle uptake in lung cancer cells, and enhanced drug distribution in tumor tissues while minimizing accumulation in healthy organs such as the liver, spleen, and kidneys. Annexin V/PI staining suggested that the combination of DOX-EGF-PEAL NPs and Bcl-2-EGF-PEAL NPs could have the most substantial therapeutic impact.

Lv et al. further explored EGFR-targeted strategies by developing nanoparticles modified with an anti-EGFR aptamer, polyamine dendrimer as the core, and loaded with erlotinib and survivin-shRNA. These nanoparticles exhibited an average size of 383 nm, a zeta potential of 12.1 mV, an erlotinib loading rate of 10.4%, and an encapsulation efficiency of 18.7%. The results showed that these anti-EGFR APT-modified nanoparticles could effectively bind to EGFR-overexpressing NSCLC cells, enhance tumor inhibition, and induce apoptosis with minimal side effects. Furthermore, when combined with chloroquine (CQ), they improved tumor microcirculation, enhanced gene transfection efficiency, and increased drug delivery effectiveness, making them particularly promising for treating EGFR-resistant NSCLC. With ongoing advancements in tumor biology and targeted drug delivery, EGFR remains a crucial target in lung cancer therapy. However, challenges such as specificity, resistance, and effective nanoparticle design still need to be addressed for further clinical success [171].

6.14 σ RECEPTOR

Lung cancer is categorized into two types, σ1 and σ2, with the σ receptor being a key target for treatment. Although these receptors have similar functions, they are not genetically related. Recent studies indicate that the σ1 receptor is a single-channel transmembrane protein, while the σ2 receptor, also known as TMEM97, is a four-channel transmembrane protein. While a small portion of the σ receptor is found on the cell membrane, the majority is expressed within the intracellular membrane system [175].

Interestingly, the σ receptor—especially the σ2 receptor—is rapidly overexpressed in actively proliferating normal cells and various cancer cells, including malignant melanoma, gliomas, breast cancer, prostate cancer, SCLC, and NSCLC. A remarkable finding is that σ ligands selectively target and kill cancer cells [176]. Σ2 ligands contribute to cancer cell death through autophagy, interference with the cell cycle, and apoptosis, whereas σ1 ligands can reduce tumor invasiveness [177].

Σ2 ligands can be classified into three main types: amine derivatives with rigid conformations, indole analogs related to Siramesine, and amine derivatives with flexible conformations. On the other hand, σ1 ligands are categorized into the Gilligan and Glennon/Ablordeppey models [178]. One of the most commonly used σ ligands is anisamide (AA), a small benzamide derivative.

Yang et al. developed lipid-calcium-phosphate (LCP) NPs targeting NSCLC using AA as the ligand. These nanoparticles were designed to simultaneously deliver three different siRNAs (targeting HDM2, c-myc, and VEGF in a 1:1:1 weight ratio). The core of the NPs consisted of calcium phosphate (CaP)-siRNA stabilized with dioleoylphosphatidic acid (DOPA) and coated with DOTAP. The final structure was modified with DSPE-PEG2000-AA to enhance targeting efficiency [179]. These nanoparticles had a size of approximately 38.6 nm and a zeta potential of 29.1 mV.

The results showed that AA-LCP NPs had strong affinity for nucleic acids, were highly biodegradable and biocompatible, and degraded quickly in acidic environments. More importantly, they efficiently delivered siRNA to NSCLC cells overexpressing the σ receptor while minimizing uptake by the RES. The addition of AA significantly enhanced siRNA delivery—up to nine times more effective—leading to suppressed expression of HDM2, c-myc, and VEGF, reduced angiogenesis, inhibited tumor growth, and induced cancer cell death. This pooled siRNA approach within targeted LCP nanoparticles presents a promising strategy for NSCLC treatment by simultaneously blocking multiple targets and limiting tumor progression.

In another study, Xiong et al. developed a cationic liposome-based nanocomposite using AA as a ligand to deliver metformin and cisplatin (CDDP) to NSCLC cells [180]. The nanoparticle core consisted of an electrostatic complex of the cationic polymer metformin (Polymet) and anionic polyglutamic acid, coated with polyglycolic cationic LP and linked to AA via DSPE-PEG.

These nanoparticles had a zeta potential of 49 mV and a size of 145 nm. The results demonstrated that these nanoparticles were highly stable, biodegradable, and biocompatible. Unlike free CDDP, which is often hindered by PGA and faces reduced therapeutic efficacy, these NPs significantly lowered RES uptake and enhanced tumor targeting. The combination of AA with CDDP and Polymet resulted in nanoparticles that were just as effective—or even more effective—than free CDDP while dramatically reducing systemic toxicity and drug resistance. Additionally, the σ2 receptor, which is overexpressed in NSCLC, has been targeted using the ligand haloperidol (Hal). Varshosaz et al. developed haloperidol-targeted bovine serum albumin nanoparticles for pulmonary delivery of DOX. These nanoparticles exhibited an entrapment efficiency of 89%, a zeta potential of –25.4 mV, a size of 218 nm, and a drug release rate of 56% within two hours [181].

Despite the promising potential of σ receptor ligands in cancer therapy, research in this area is still in its early stages, with AA being the most widely studied ligand so far. Further exploration of these receptors and their ligands could lead to more effective and targeted lung cancer treatments [181]

6.15 FOLATE RECEPTOR

FA, also known as vitamin B_9, is a water-soluble vitamin essential for the body's use of amino acids and carbohydrates in protein synthesis, nucleic acid production, and cell division. In the body, FA functions as tetrahydrofolate acid, working alongside vitamin B_{12} to aid in protein metabolism, red blood cell formation, and the synthesis of purine and pyrimidine nucleotides. Because of its low molecular weight, FA is easily taken up by tumor cells, making it a promising tool for targeted therapy. Its benefits include rapid blood clearance, strong tissue penetration, minimal immune response, and fast targeting capabilities [182, 183].

FA is widely used in clinical settings and is particularly valuable as a tumor-targeting agent. The folate receptor is a 38 kDa glycoprotein that binds with high affinity to FA, FA-linked drugs, or FA-modified nanoparticles. By regulating the endocytosis of FR, FA is transported into cells. FR is one of the most studied tumor markers due to its selective overexpression in certain cancers while being nearly absent in normal tissues. This has driven significant interest in FRs for targeted drug delivery, cancer treatment, and immunotherapy for diseases like rheumatoid arthritis. There are three known FR isomers: FRα, FRβ, and FRγ [184].

FRα is highly expressed in epithelial cancers such as lung adenocarcinoma, colorectal cancer, and ovarian cancer. FRβ is predominantly found on tumor cells originating from non-epithelial tissues, such as sarcomas and acute myeloid leukemia. FRγ, which is not found in normal blood, has potential as a serum biomarker for lymphoma and a target for FR-mediated leukemia treatments [185].

The presence and activity of FRs on the surface of tumor cells—including ovarian, colorectal, breast, lung, and renal carcinomas—are significantly higher than in normal cells. FR is anchored to the cell membrane via phosphatidylinositol, and it can be detached using specific phospholipase enzymes. This understanding has laid the foundation for research into FA-based tumor-targeting strategies. Since FR expression is particularly high in many epithelial tumors, including lung, colon, stomach, and gynecological cancers, targeting FR can improve drug specificity, reducing side effects while increasing drug potency. However, current methods for evaluating FR-targeted drug delivery mainly rely on in vitro tests measuring cytotoxicity and cellular uptake. More animal studies are needed to assess the therapeutic potential in humans.

New advances suggest that FA can also enhance the effectiveness of photodynamic therapy (PDT), a minimally invasive approach for treating lung cancer and metastatic lymph nodes. Kato et al. [186] developed folate receptor 1 (FOLR1)-targeting nanoparticles (NPs) using FA and porphysomes. These unique nanoparticles, made from natural lipids and chlorophyll, can be used for various bio-photon applications. They act like tiny, multicolored water balloons that deliver drug-loaded NPs directly to tumors. Lovell et al. [187] highlighted that porphysome-based NPs exhibit strong photothermal and photoacoustic properties, a broad absorption spectrum, and structure-dependent fluorescence quenching.

Dendrimer (Den) nanoparticle technology targeting FRα has also been explored for lung cancer treatment. Amreddy et al. developed Den-PEI-cisplatin (CDDP)-siRNA FA NPs, which are well-dispersed, spherical particles smaller than 10 nm in size, with a zeta potential of +17.2 mV and a cisplatin encapsulation efficiency of 40.52 ± 4.18% [188]. Similarly, Muralidharan et al. [189] demonstrated the effectiveness of FRα-targeted DOTAP cholesterol lipid nanoparticles loaded with HuR siRNA (HuR-FNP) in treating lung cancer. These nanoparticles measured 303.3 nm in diameter and exhibited significantly higher uptake in lung cancer cells (H1299) compared to normal lung fibroblasts (CCD 16), indicating a receptor dose-dependent effect. The zeta potential of HuR-FNP was 4.3 mV, and about 39% of the siRNA was released within the first hour in an acidic buffer [189].

Another promising approach was explored by He et al. [190], who developed FA-modified amphiphilic PEG-PLGA copolymer nanoparticles loaded with cisplatin (CDDP) and PTX. These co-FA-NPs (CDDP: PTX in a 2:1 ratio) had a zeta potential of 21.50 ± 0.88 mV, a polydispersity index (PDI) of 0.133 ± 0.008, and a particle size of 171.36 ± 8.67 nm. Such targeted nanoparticles hold promise for delivering both paclitaxel and cisplatin in the treatment of NSCLC. Overall, FA-targeted drug delivery represents a promising strategy for treating various cancers, particularly epithelial tumors, by enhancing drug efficacy while reducing systemic toxicity. However, further research is needed to validate its effectiveness in clinical settings [190].

6.16 TRANSFERRIN RECEPTOR

Human transferrin (TF) is an iron-binding protein with a molecular weight of 79.57 kDa. It plays a crucial role in transporting iron to cells by binding to the transferrin receptor (TFR). This process is essential for red blood cell production, as TF carries iron from both dietary sources and red blood cell breakdown to the bone marrow in the form of the TFR-Fe3+ complex. Additionally, TF helps deliver iron to cells that express TFR on their surface. Because TF is biodegradable, non-toxic, and non-immunogenic, it has potential as a targeted drug delivery tool through TFR-mediated pathways [191].

While all normal nucleated cells express low levels of TFR, its expression significantly increases in rapidly proliferating cells, particularly in tumor cells. High TFR levels are often found in cancers such as liver cancer, chronic lymphocytic leukemia, and lung cancer [192–195]. This phenomenon, known as "iron addiction," occurs because iron is essential for cell growth and proliferation [192]. Due to the overexpression of TFR in many tumor cells, researchers have explored TF/TFR-mediated mechanisms to deliver therapeutic drugs directly to cancerous tissues [196, 197].

According to Zhu et al., measuring TFR expression in lung cancer tissues could help assess how responsive tumors are to gambogic acid treatment in clinical settings. In small cell lung cancer (SCLC), multidrug resistance is a major challenge [198]. However, research by Sadava et al. suggests that combining TFR with artemisinin may help overcome treatment resistance. Their findings indicate that after pretreatment, low concentrations of artemisinin can effectively kill SCLC cells [199]. Additionally, Upadhyay et al. found that transferrin-modified nanoparticles loaded with thymoquinone (a natural compound) can enhance apoptosis and cell death in highly aggressive NSCLC cells. Unlike conventional chemotherapy combinations, this approach effectively activates two different miRNA pathways, limiting cancer cell migration without significant toxicity. In summary, the transferrin receptor plays a crucial role in the altered iron metabolism of lung cancer cells. While the exact mechanisms are not yet fully understood, targeting TFR offers a promising strategy to disrupt cancer cells' iron intake. By interfering with their iron metabolism, we can slow their growth and potentially open new doors for lung cancer treatment. As research advances, TFR-based therapies could become a powerful tool in combating cancer [200].

6.17 CD44

CD44 is a membrane-bound glycoprotein involved in the progression of malignant tumors. It has been identified as a key marker for tumor-initiating cells (TICs) and cancer stem cells (CSCs) [201–203]. According to the CSC theory, tumors contain subpopulations of cells that behave like stem or progenitor cells. These cells are highly resistant to conventional chemotherapy and can drive tumor growth and differentiation through various mechanisms [204].

Research by Matheolabakis et al. has shown that CD44 is overexpressed in several solid cancers, including lung, breast, and pancreatic cancers [205]. On the cell surface, CD44 interacts with components of the extracellular matrix, such as hyaluronic acid (HA), to promote cancer cell invasion and metastasis. This interaction can lead to changes in the cytoskeleton of cancer cells, making them more aggressive [206]. Additionally, CD44 plays a role in multiple signaling pathways that enhance tumor progression. It acts as a co-receptor for epidermal growth factor receptor (EGFR) and other ErbB family receptor tyrosine kinases, indirectly stimulating cell proliferation [207]. It also helps tumor cells survive by activating anti-apoptotic pathways, such as Bcl-2 and Bcl-xl transcription, and the PI3K/AKT signaling cascade [207, 208].

In small cell lung cancer (SCLC), CD44-mediated signaling can increase the expression of urokinase plasminogen activator (uPA)/uPA receptor (uPAR) and the multidrug resistance gene (MDR1). This enhances the invasiveness of lung cancer cells and contributes to their resistance to multiple drugs [209]. According to Jeannot et al., the size of HA nanoparticles significantly influences their cellular uptake and biodistribution [210]. Smaller nanoparticles show promising targeting effects in tumors that overexpress CD44, making them ideal candidates for drug delivery. One commonly used biodegradable polymer for such drug delivery systems is poly-ε-caprolactone (PCL). Capsaicin (CAP), a compound known for its anticancer properties, has been widely studied for its effects on lung cancer cells [211, 212]. Parashar et al. demonstrated that HA-PCL-CAP nanoparticles enhance CAP's cytotoxic, anti-proliferative, and apoptotic effects in A549 lung cancer cells, showing potential for treating NSCLC [213]. Moving forward, cancer therapies should focus on developing targeted drug delivery strategies based on CD44 expression in CSCs and TICs [214]. As research progresses, understanding the molecular roles of CD44 isoforms, its transcriptional regulation, and alternative splicing will be crucial for improving CD44-targeted treatments [215, 216].

6.18 NANOCANCER TARGETING STRATEGY

For a successful nanoparticle-based drug delivery system, it's essential to understand how nanomaterials interact with the body. Before these nanoparticles (NPs) reach a tumor, they must go through three key steps. First, they travel through the bloodstream, where they are often detected and cleared by macrophages in the RES, mainly in the liver and spleen [217]. Next, the NPs must penetrate the tumor and deliver the drug. Finally, they accumulate at the tumor site, where they exert their therapeutic effects. However, before reaching cancer cells, NPs are more likely to interact with immune cells like macrophages and neutrophils, leading to their clearance from the body via the urinary system [218]. While it's important to prevent excessive accumulation in the RES to prolong circulation, longer circulation times also increase the likelihood of immune interactions. One way to enhance NP delivery to tumors is by modifying their surface, which can improve their absorption and distribution in tumor tissues. A common strategy is to coat nanoparticles with PEG, which improves the solubility of hydrophobic drugs, extends their circulation time, and enhances their tumor-targeting ability through the EPR effect [219]. There are two main types of PEG modification: first-generation PEGylation (random coating) and second-generation PEGylation (targeted coating). PEGylation allows both passive and active targeting of drugs, helping to develop nanoparticles with improved pharmacokinetics (PK). These PEG-coated nanoparticles are often called "stealth" NPs because they are better at evading detection and clearance by the immune system compared to unmodified nanoparticles [220]. For lung cancer treatment, nanoparticle-based drug delivery offers clear advantages, but if anticancer drugs are released uncontrollably, they can harm healthy lung tissue. Targeted drug delivery systems ensure that the drug is released only where it is needed, minimizing toxicity to normal cells. Additionally, these advanced drug formulations help prolong drug retention and improve effectiveness. As a result, innovative small-molecule drug delivery systems are becoming essential for the effective treatment of lung cancer.

6.19 STRATEGIES ARE GENERALLY CLASSIFIED INTO THREE CATEGORIES

(6.19.1) receptor-based targeting
(6.19.2) passive targeting, and
(6.19.3) stimuli responsive system-based targeting

6.19.1 Receptor-Based Targeting

By modifying the surface of drug-carrying nanoparticles, receptor-mediated endocytosis ensures that medication is delivered directly to tumor cells. This approach improves targeting efficiency using molecules like folic acid, transferrin, albumin, hyaluronic acid, RGD (arginine-glycine-aspartic acid), and Herceptin. Researchers have developed receptor-based targeting strategies that enhance drug delivery by increasing tumor penetration, cellular uptake, and internalization.

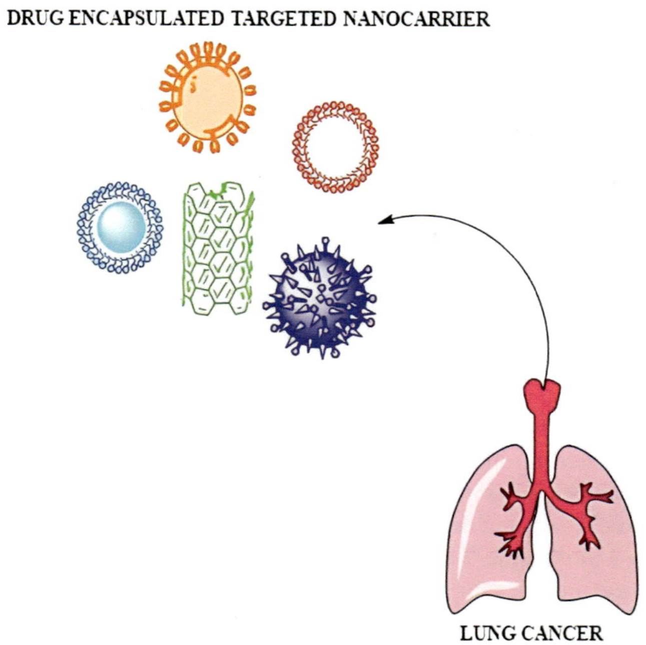

FIGURE 6.2 Nanocancer targeting strategy

6.19.1.1 EGFR-Targeted Drug Delivery

Receptor-based drug targeting can be achieved using Herceptin, which recognizes and binds to HER2 receptors that are overexpressed in breast cancer cells. To improve drug effectiveness, researchers have modified nanocarriers containing PTX and DTX with Herceptin. These modifications help prevent premature drug loss before the medication reaches the tumor. Herceptin-modified nanocarriers are created through an adsorption process at room temperature with gentle shaking overnight, resulting in enhanced cytotoxic effects, better drug release, and improved absorption by cancer cells [221].

6.19.1.2 Targeting Transferrin and Folic Acid Receptors

Apo-transferrin has been integrated into paclitaxel nanocarriers to enable targeted drug delivery. Another method involves modifying the paclitaxel nanocarrier's surface with transferrin, which has been shown to significantly improve tumor suppression. Additionally, the tumor-targeting peptide XQ1 (HAIYPRHGGGF) enhances the anticancer effect of camptothecin-loaded nanocarriers by binding to transferrin receptors, demonstrating effectiveness against HeLa and A549 cell lines [221].

6.19.1.3 Targeting Vascular Endothelial Growth Factor Receptors

Studies using immunohistochemistry analysis have shown that oral administration of bovine lactoferrin reduces VEGF overexpression in lung cancer cells. Since VEGF plays a key role in tumor growth, invasion, and metastasis by promoting blood vessel formation, BLF's ability to inhibit angiogenesis presents a promising strategy for treating cancer [222].

6.19.1.4 Luteinizing Hormone-Releasing Receptor Targeting

Tartula et al. developed a lung-targeted drug delivery system using nanostructured lipid carriers (NLCs) modified with luteinizing hormone-releasing receptors (LHRH). Their findings revealed that inhalable LHRH-modified paclitaxel nanoparticles were preferentially delivered to tumor sites and induced apoptosis at a rate 16 times higher than the standard drug's, demonstrating their potential for more effective lung cancer treatment [223].

6.19.1.5 CD44 Receptor

Lapatinib-loaded nanocarriers coated with hyaluronic acid are designed to actively target the CD44 receptor, which is highly expressed in cancer cells. Hyaluronic acid naturally binds to CD44, a cell surface receptor that is upregulated in certain cancers, including breast and stomach cancer. By incorporating hyaluronic acid into paclitaxel nanocarriers, researchers have developed a way to selectively target tumors. This adhesion molecule interacts with receptors involved in hyaluronic acid-mediated motility, ensuring precise drug delivery to cancerous tissues [224, 225].

Targeting integrin adhesion molecule αvβ3:

Nanoparticles engineered with RGD peptides (arginine-glycine-aspartic acid) can bind to αvβ3, an integrin protein commonly found on tumor cells. Once attached, these RGD-modified nanocrystals (NCs) are internalized through receptor-mediated endocytosis. A specific example is RGD-modified paclitaxel nanocapsules (RGD-PTX NC), which can be synthesized via a Schiff base reaction to enhance αvβ3-targeted drug delivery.

Targeting sigma-1 and sigma-2 receptors:

Sigma receptors are involved in various cellular processes, including ion channel regulation and tumor proliferation. Ligands such as [18F] FTC-146, [18F] fluspidine, and donepezil target the sigma-1 receptor, while [11C]RHM-1, [18F]ISO-1, siramesine, WC-26, DTG (1,3-ditolyl guanidine), and SV119.112 target sigma-2. Drugs that interact with these receptors have shown anticancer properties by influencing ion channels and suppressing tumor growth. This receptor-based targeting approach makes sigma receptor modulators a promising option for treating lung cancer [226, 227].

Targeting ROS1 Proto-oncogene receptors:

Crizotinib was the first drug proven effective against advanced NSCLC in patients with ROS1 gene fusions. Other drugs that inhibit ROS1 (c-ros oncogene 1) include Entrectinib, Lorlatinib, DS-6051b, and Ceritinib (LDK378). These targeted therapies block ROS1-driven cancer cell growth, offering an effective treatment strategy for ROS1-positive lung cancer patients [228–232].

6.19.1.6 Other Endocytosis

Drugs can be delivered directly into cancer cells through specific endocytosis pathways, such as caveolar or clathrin-mediated endocytosis. To enhance drug targeting, paclitaxel nanocarriers are modified by combining triphenyl phosphonium cation with albumin-absorbed paclitaxel. This modification improves cellular uptake and increases drug delivery efficiency, ensuring the medication reaches its intended target more effectively.

6.19.2 Passive Targeting

Passive targeting techniques take advantage of the EPR effect, allowing drugs to accumulate more effectively in tumors compared to normal tissues. This method improves drug infiltration in tumor-associated blood vessels by 20–30%, increasing the chances of the drug reaching cancer cells [233, 234]. Tumors typically grow in harsh conditions like low oxygen (hypoxia) or inflammation, which can block both newly formed and existing blood vessels. However, this environment also allows nanosystems to penetrate more effectively, improving drug retention in tumor tissues. The downside is that tiny drug molecules may be quickly cleared through lymphatic drainage, reducing their effectiveness. Controlling this drainage is essential to prolong drug retention. Key factors that enhance EPR-based targeting include charge, particle size, and surface chemistry. Even with these improvements, standard drugs often struggle to consistently reach every cancer cell. To overcome this challenge, researchers are developing advanced nanocarriers designed to better interact with tumor cells, ensuring more efficient drug delivery. These next-generation drug delivery systems help improve the drug's pharmacokinetics, extend retention time, and reduce side effects by enhancing tumor selectivity [235].

6.19.3 Stimuli-Responsive System-Based Targeting

Stimuli-responsive drug delivery systems provide a precise and efficient way to target cancer cells. Tumor cells have a unique environment that differs from normal cells, with variations in temperature, pH levels, oxygen availability (hypoxia), and glutathione concentration.

To take advantage of these differences, researchers have developed nanocarrier-based drug delivery systems that react to specific tumor conditions. For example, certain polymers can either be introduced from the surrounding environment or be designed to release drugs when triggered by specific stimuli within the tumor microenvironment. This approach ensures that the drug is delivered exactly where it is needed, improving treatment effectiveness while minimizing damage to healthy cells.

6.19.3.1 pH-Responsive Systems

Researchers have explored drug release strategies that specifically target cancer cells by taking advantage of their acidic environment. Tumor microenvironments typically have a lower pH (4.5–6.5), as do lysosomes (5.5–6.8), making them different from normal tissues. To improve drug delivery, Liang et al. developed a method using a pH-responsive peptide combined with mannitol in a spray-dried powder inhalation system. This approach ensures that the drug remains stable in lung fluids while effectively reaching tumor sites with lower pH levels, enhancing nucleic acid delivery and treatment effectiveness. In cancer treatment, controlling drug release based on temperature changes has proven to be highly effective, as tumors often exhibit higher temperatures than normal tissues. Thermoresponsive polymers can be used to modify the surface of nanocrystals to take advantage of this feature. Wu et al. utilized thermoresponsive PNIPAM [poly(N-isopropylacrylamide)] brushes grafted onto cellulose nanocrystals to encapsulate the chemotherapy drug 5-fluorouracil. Their study showed that when the temperature increased from 25°C to 37–40°C, PNIPAM significantly boosted drug release. This approach can also be integrated with stimulus-guided methods, where drug release is triggered by external signals for more precise treatment [236, 237].

6.19.3.2 Enzyme-Responsive Systems

Certain enzymes present at tumor sites can be used to trigger the controlled release of drugs from enzyme-responsive systems, ensuring targeted delivery. Tagami et al. developed a hybrid liposomal system made of phospholipids and poloxamer 188, loaded with doxorubicin, which responds to the PLA2 (phospholipase A2) enzyme found in cancer cells. Their study showed that the presence of PLA2 increased drug release by eight times. For localized cancer treatment, enzyme-responsive nanocarriers can encapsulate drug nanocrystals and deliver them through the pulmonary route [238].

6.19.3.3 Ligand-Attached Stimuli-Responsive Systems

Combining receptor-based targeting with stimuli-responsive drug delivery can improve drug absorption inside lung cancer cells. Tseng et al. created aerosolized gelatin nanoparticles targeting the EGFR receptor and delivered them directly into the lungs of mice. This method significantly increased drug accumulation within lung cancer cells. Similarly, a dextran nanogel coated with a pulmonary surfactant and decorated with folate was used to enhance siRNA stability and prevent its premature extracellular release. This approach led to higher intracellular siRNA concentrations, improving the effectiveness of gene silencing therapies for lung cancer [239].

6.19.3.4 Redox-Responsive Systems

These systems show great potential for cancer therapy, offering a targeted approach to delivering anticancer drugs directly to tumor cells. These systems take advantage of the elevated intracellular

ROS levels found in tumors, particularly in cancer cells, to enhance drug delivery. Chen and colleagues developed a liposome-based nanosystem designed for the co-delivery of paclitaxel and anti-survival siRNA. This system was composed of cholesterol, soybean phosphatidylcholine (PC), and a redox-sensitive cationic oligopeptide lipid with a proton sponge effect. The liposomes effectively released both the drug and siRNA into the cytosol via endo-lysosomal leakage, leading to increased apoptosis and reduced survival protein expression in cancer cells.

6.19.3.5 Magnetic Field Responsive Drug Delivery

Magnetic fields can also be used as an external trigger to direct therapeutic agents toward lung cancer cells, either inside the cytosol or in extracellular tumor regions. Park et al. developed a magnetite nanocrystal cluster coated with PEG-polyethylenimine-DOPA (dopamine). This system successfully facilitated the magnetically induced intracellular transport of siRNA, enhancing drug delivery and therapeutic effectiveness in cancer cells [240, 241].

6.19.3.6 Multi-Stimuli-Responsive Systems

Several advanced delivery techniques have been developed to transport biomolecules to cancer cells through the lungs. Qu et al. designed a dual-responsive system that reacts to both redox conditions and pH changes, ensuring precise drug release within the tumor environment for enhanced anticancer effects. This method relies on a pH-sensitive component and a disulfide linker that responds to glutathione (GSH) levels [242–244].

6.20 FORMULATION DRAWBACKS

Before reaching clinical trials, it is crucial to develop a stable and highly reproducible formulation to address the complexities of malignant tumors [245]. The foundation of new anticancer drug delivery strategies lies in designing multifunctional NPs with targeted delivery and imaging-enhancing properties. However, incorporating specific targeting molecules into nanocarriers increases production complexity, raises costs, and leads to intricate interactions in the body, complicating regulatory approval. Targeted therapies are promising and often seen as the ideal solution to the challenges of traditional chemotherapy. However, their development encompassing synthesis, purification, and selecting the most effective ligand receptor pair can be difficult and risky sometimes yielding impractical products. Even minor structural modifications can alter binding properties, potentially leading to changes in shape, steric hindrance, and reduced effectiveness. Additionally, the presence of multiple active compounds in nanomedicines further complicates their pharmacological profile [246–248]. To overcome these challenges, novel bioconjugation techniques are being explored. One approach, click chemistry, simplifies nanoparticle design, making drug synthesis more straightforward and cost effective while enabling the rapid creation of large libraries of new compounds. This strategy has been gaining renewed interest [249, 250]. Moreover, ensuring a consistent and scalable manufacturing process is essential for clinical approval. Unfortunately, nanoparticles used in preclinical research are often produced in small batches, making it difficult to scale up for clinical studies. Beyond pharmacological and toxicological concerns, the complexity of characterizing nanomedicines necessitates clear regulatory guidelines and a streamlined approval process. These challenges remain significant barriers to the advancement of nanomedicines for therapeutic applications [251].

6.21 NEW APPROACHES FOR THE USE OF NANOMEDICINE IN THE TREATMENT OF RESISTANT LUNG CANCER

Innovative receptor-targeting techniques that leverage the unique characteristics of the tumor microenvironment may help overcome challenges associated with nanoparticle-based drug delivery while also reducing chemotherapy resistance. Traditional methods rely on environmental signals to direct nanoparticles, whereas some new approaches focus on controlling their movement externally. For instance, researchers have demonstrated that applying a magnetic field gradient directly over a tumor can guide iron-based nanoparticles carrying anticancer drugs to the targeted site. Alexiou et al. used commercially available ferrofluids as intravenous drug carriers, showing that an external magnetic field applied after injection successfully directed the nanoparticles to the tumor, reducing systemic toxicity. Additionally, magnetic nanoparticles can deliver multiple anticancer agents, including radionuclides, genes, and cancer-specific antibodies [252]. Another promising approach is PDT, which has gained traction in the last decade for cancer treatment. Since most photosensitizers are hydrophobic and difficult to deliver efficiently, researchers have developed nanocarriers to transport them. In vitro studies on lung cancer cell lines have shown promising results, demonstrating that photosensitizer-loaded nanocarriers can induce mitochondrial dysfunction by targeting specific cellular structures. A study by Shi et al. used a liposomal nanomedicine (L@BP) loaded with PTX and a mitochondria-anchored photosensitizer (Cy-Br), showing effective drug accumulation at the tumor site and improved therapeutic outcomes in PTX-resistant lung cancer cells. This approach could pave the way for treating multidrug-resistant cancers. Another cutting-edge strategy involves using nanoparticles to deliver small interfering RNAs (siRNAs) to silence drug resistance-related genes [253–257]. By delivering multiple siRNAs, several resistance-associated genes can be targeted simultaneously. However, siRNAs struggle to cross cell membranes due to their physicochemical properties, limiting their cellular uptake. Therefore, new RNA interference (RNAi) technologies and advanced carriers are needed [258]. Genome editing techniques could also provide a foundation for more effective solutions. A promising approach to improving drug delivery systems for cancer treatment involves integrating biocompatible materials. For example, Reshma et al. proposed using a biopolymer derived from tamarind seeds (PST) to create PTX-loaded nanoparticles via epichlorohydrin crosslinking. Their study indicated that PST-PTX nanoparticles could down-regulate multidrug resistance proteins like P-gp and BCRP in resistant cancer cells, highlighting their potential as MDR inhibitors [259].

6.22 FUTURE PERSPECTIVES

Researchers are actively exploring a wide range of applications for novel drug delivery systems, which have shown promise in improving treatment precision, effectiveness, and safety. These advanced technologies not only enhance early detection of metastases but also address key limitations of traditional treatments, such as poor targeting, low therapeutic efficiency, unwanted side effects, and drug resistance. By using nanoparticles linked with peptides, PEGylation, or receptor ligands, medications can be delivered specifically to targeted tissues, achieving optimal drug concentrations with lower doses and reduced toxicity. Studies have shown that embedding chemotherapy drugs in nanocarriers significantly lowers cytotoxicity compared to administering free drug solutions.

Nano-based drug delivery systems are considered a potential breakthrough in lung cancer treatment. However, several challenges must be addressed before they gain formal recognition. Poor drug release from nanocarriers can contribute to drug resistance, but strategies such as multifunctional nanocarriers for precise tumor targeting, endosomal disruption for immediate drug release, and combination therapies using multiple drugs can help overcome this issue. Additionally, as particles shrink from micro to nano size, their surface area and chemical reactivity increase, making their behavior harder to predict. This heightened reactivity can lead to the production of ROS, causing oxidative stress, DNA damage, and inflammation, ultimately contributing to cytotoxicity. Despite the promise of nanoparticles, the FDA has approved only a limited number of liposome- and nanoparticle-based formulations. One major drawback is the instability of liposomes and lipid-based nanoparticles, which tend to aggregate and degrade over time, reducing their effectiveness. Since most lung cancers are diagnosed at an advanced stage, an ideal delivery system should effectively target both local and distant metastatic tumors, reaching central and peripheral lung tissues as well as the bloodstream. More research is needed to develop controlled-release nanocarriers that respond to specific biological stimuli.

A growing area of focus is cancer theranostics—an approach that combines diagnostics with targeted therapy. In the future, multifunctional nanocarriers carrying both imaging agents and chemotherapy drugs could play a critical role in cancer management. Future research should aim to develop nanoparticles with greater stability, better dispersion, and improved tissue penetration. Adjusting physical properties such as particle size and porosity, along with chemical modifications like amphiphilic coatings, targeted ligands, and stable polymers, could help prevent agglomeration and enhance drug delivery. Different types of nanocarriers—including liposomes, dendrimers, and micelles—can be used to encapsulate drugs, enhancing both local and systemic distribution. However, a major challenge in nanomedicine is the lack of standardized protocols for manufacturing, testing, and regulatory approval. Nanocarrier-based therapies must meet the same stringent requirements as newly developed drugs, including reproducible manufacturing, comprehensive characterization, rigorous pharmacological testing, and proof of safety and efficacy through preclinical and clinical trials. Unlike conventional drugs that typically contain a single active ingredient, nanoparticles often carry multiple active compounds, making their pharmacokinetics and safety evaluation more complex. This complexity calls for updated regulatory guidelines to assess bioequivalence, pharmacokinetics, and safety. Regulatory agencies need to establish comprehensive testing protocols and streamlined approval pathways to facilitate the development of nanomedicines. Moving forward, continued research in NDDSs is expected to transform drug delivery, replacing traditional dosing methods with more efficient and patient-friendly treatment strategies.

6.23 CONCLUSION

Lung cancer remains one of the deadliest cancers, making early detection and advanced treatment approaches essential for improving patient outcomes. Theranostic agents, which combine therapy and imaging, use nanotechnology-based delivery systems to target cancer cells more precisely. By modifying their surface properties, nanostructures can enhance the solubility and stability of chemotherapy drugs, addressing common challenges in traditional treatments. Nanotechnology continues to revolutionize lung cancer treatment, with researchers constantly exploring new applications. Various nanocarriers have been developed to improve targeting—either passively or actively—while reducing harm to healthy cells. Targeting overexpressed cancer markers significantly enhances treatment effectiveness, and the large surface area of nanoparticles allows for a higher therapeutic payload. Several types of nanocarriers are being explored for theranostic applications, including liposomes, SLNPs, micelles, dendrimers, MNPs, carbon nanotubes, QDs, hydrogels, nanoemulsions, and polymeric NPs. The quality of these nanocarriers plays a crucial role in ensuring effective drug delivery to tumors by improving drug penetration and retention in cancerous tissues. Beyond drug delivery, theranostic agents also aid in early cancer detection and monitoring treatment response with high precision. These advancements, along with the development of more effective drug delivery technologies, offer hope to lung cancer patients, improving both treatment success rates and overall quality of life.

REFERENCES

1. Bade, B.C.; Dela Cruz, C.S. Lung Cancer 2020: Epidemiology, Etiology, and Prevention. *Clin. Chest Med.* **2020**, 41, 1–24.
2. Ramalingam, S.S.; Owonikoko, T.K.; Khuri, F.R. Lung Cancer: New Biological Insights and Recent Therapeutic Advances. *CA Cancer J. Clin.* **2011**, 61, 91–112.
3. Rodriguez-Canales, J.; Parra-Cuentas, E.; Wistuba, I.I. Diagnosis and Molecular Classification of Lung Cancer. *Cancer Treat. Res.* **2016**, 170, 25–46.
4. Nasim, F.; Sabath, B.F.; Eapen, G.A. Lung Cancer. *Med. Clin. N. Am.* **2019**, 103, 463–473.

5. Jyoti, K.; Pandey, R.S.; Kush, P.; Kaushik, D.; Jain, U.K.; Madan, J. Inhalable Bioresponsive Chitosan Microspheres of Doxorubicin and Soluble Curcumin Augmented Drug Delivery in Lung Cancer Cells. *Int. J. Biol. Macromol.* **2017**, 98, 50–58.
6. Smejkalova, D.; Nesporova, K.; Hermannova, M.; Huerta-Angeles, G.; Cozikova, D.; Vistejnova, L.; Safrankova, B.; Novotny, J.; Kucerik, J.; Velebny, V. Paclitaxel Isomerisation in Polymeric Micelles Based on Hydrophobized Hyaluronic Acid. *Int. J. Pharm.* **2014**, 466, 147–155.
7. Zhao, D.; Wu, J.; Li, C.; Zhang, H.; Li, Z.; Luan, Y. Precise Ratiometric Loading of PTX and DOX Based on Redox-Sensitive Mixed Micelles for Cancer Therapy. *Colloids Surf. B Biointerfaces* **2017**, 155, 51–60.
8. Diao, Y.Y.; Li, H.Y.; Fu, Y.H.; Han, M.; Hu, Y.L.; Jiang, H.L.; Tsutsumi, Y.; Wei, Q.C.; Chen, D.W.; Gao, J.Q. Doxorubicin-Loaded PEG-PCL Copolymer Micelles Enhance Cytotoxicity and Intracellular Accumulation of Doxorubicin in Adriamycin-Resistant Tumor Cells. *Int. J. Nanomed.* **2011**, 6, 1955–1962.
9. Zhu, Y.; Zhang, J.; Meng, F.; Deng, C.; Cheng, R.; Feijen, J.; Zhong, Z. cRGD-Functionalized Reduction-Sensitive Shell-Sheddable Biodegradable Micelles Mediate Enhanced Doxorubicin Delivery to Human Glioma Xenografts in Vivo. *J. Control. Release* **2016**, 233, 29–38.
10. Pegg, A.E.; Casero, R.A., Jr. Current Status of the Polyamine Research Field. *Methods Mol Biol.* **2011**, 720, 3–35.
11. Abdulhussein, A.A.; Wallace, H.M. Polyamines and Membrane Transporters. *Amino Acids* **2014**, 46, 655–660.
12. Kruczynski, A.; Vandenberghe, I.; Pillon, A.; Pesnel, S.; Goetsch, L.; Barret, J.M.; Guminski, Y.; Le Pape, A.; Imbert, T.; Bailly, C.; et al. Preclinical Activity of F14512, Designed to Target Tumors Expressing an Active Polyamine Transport System. *Investig. New Drugs* **2011**, 29, 9–21.
13. Alliot, J.; Theodorou, I.; Duconge, F.; Gravel, E.; Doris, E. Polyamine Transport System-Targeted Nanometric Micelles Assembled from Epipodophyllotoxin-Amphiphiles. *Chem. Commun. Camb.* **2019**, 55, 14968–14971.
14. Kruczynski, A.; Pillon, A.; Creancier, L.; Vandenberghe, I.; Gomes, B.; Brel, V.; Fournier, E.; Annereau, J.P.; Currie, E.; Guminski, Y.; et al. F14512, a Polyamine-Vectorized Anti-Cancer Drug, Currently in Clinical Trials Exhibits a Marked Preclinical Anti-Leukemic Activity. *Leukemia* **2013**, 27, 2139–2148.
15. Li, J.; Mao, J.; Tang, J.; Li, G.; Fang, F.; Tang, Y.; Ding, J. Surface Spermidine Functionalized PEGylated Poly(Lactide-Co-Glycolide) Nanoparticles for Tumor-Targeted Drug Delivery. *RSC Adv.* **2017**, 7, 22954–22963.
16. Tavares, A.J.; Poon, W.; Zhang, Y.N.; Dai, Q.; Besla, R.; Ding, D.; Ouyang, B.; Li, A.; Chen, J.; Zheng, G.; et al. Effect of Removing Kupffer Cells on Nanoparticle Tumor Delivery. *Proc. Natl. Acad. Sci. USA* **2017**, 114, E10871–E10880.
17. Zhang, Y.N.; Poon, W.; Tavares, A.J.; McGilvray, I.D.; Chan, W.C.W. Nanoparticle-Liver Interactions: Cellular Uptake and Hepatobiliary Elimination. *J. Control. Release* **2016**, 240, 332–348.
18. Avval, Z.M.; Malekpour, L.; Raeisi, F.; Babapoor, A.; Mousavi, S.M.; Hashemi, S.A.; et al. Introduction of Magnetic and Supermagnetic Nanoparticles in New Approach of Targeting Drug Delivery and Cancer Therapy Application. *Drug Metab. Rev.* **2020**, 52, 157–184.
19. Wu, K.; Su, D.; Liu, J.; Saha, R.; Wang, J.P. Magnetic Nanoparticles in Nanomedicine: A Review of Recent Advances. *Nanotechnol.* **2019**, 30, 502003.
20. Woodman, C.; Vundu, G.; George, A.; Wilson, C.M. Applications and Strategies in Nanodiagnosis and Nanotherapy in Lung Cancer. *Semin. Cancer Biol.* **2021**, 69, 349–364.
21. Saadat, M.; Manshadi, M.K.; Mohammadi, M.; Zare, M.J.; Zarei, M.; Kamali, R.; et al. Magnetic Particle Targeting for Diagnosis and Therapy of Lung Cancers. *J. Control Release* **2020**, 328, 776–791.
22. Singh, N.; Jenkins, G.J.; Asadi, R.; Doak, S.H. Potential Toxicity of Superparamagnetic Iron Oxide Nanoparticles (SPION). *Nano Rev.* **2010**, 1, 5358.
23. Ankamwar, B.; Lai, T.C.; Huang, J.H.; Liu, R.S.; Hsiao, M.; Chen, C.H.; et al. Biocompatibility of Fe3O4 Nanoparticles Evaluated Byin Vitrocytotoxicity Assays Using Normal, Glia and Breast Cancer Cells. *Nanotechnol.* **2010**, 21, 075102.
24. Kandasamy, G.; Maity, D. Recent Advances in Superparamagnetic Iron Oxide Nanoparticles (SPIONs) for *In Vitro* and *In Vivo* Cancer Nanotheranostics. *Int. J. Pharm.* **2015**, 496, 191–218.
25. Aggarwal, P.; Hall, J.B.; McLeland, C.B.; Dobrovolskaia, M.A.; McNeil, S.E. Nanoparticle Interaction with Plasma Proteins as It Relates to Particle Biodistribution, Biocompatibility and Therapeutic Efficacy. *Adv. Drug Deliv. Rev.* **2009**, 61, 428–437.
26. Lu, A.H.; Salabas, E.L.; Schüth, F. Magnetic Nanoparticles: Synthesis, Protection, Functionalization, and Application. *Angew. Chem. Int. Ed. Engl.* **2007**, 46, 1222–1244.

27. Li, X.; Wei, J.; Aifantis, K.E.; Fan, Y.; Feng, Q.; Cui, F.Z.; et al. Current Investigations into Magnetic Nanoparticles for Biomedical Applications. *J. Biomed. Mater. Res. Part A* **2016**, 104, 1285–1296.
28. Oxide Nanoparticles by Biocompatible Polymers for Tissue Imaging and Targeting. *Biotechnol. Adv.* **2013**, 31, 1224–1236.
29. Malvindi, M.A.; De Matteis, V.; Galeone, A.; Brunetti, V.; Anyfantis, G.C.; Athanassiou, A.; et al. Toxicity Assessment of Silica Coated Iron Oxide Nanoparticles and Biocompatibility Improvement by Surface Engineering. *PLOS One* **2014**, 9, e85835.
30. Dwivedi, S.; Siddiqui, M.A.; Farshori, N.N.; Ahamed, M.; Musarrat, J.; Al Khedhairy, A.A. Synthesis, Characterization and Toxicological Evaluation of Iron Oxide Nanoparticles in Human Lung Alveolar Epithelial Cells. *Colloids Surf. B Biointerfaces* **2014**, 122, 209–215.
31. Soenen, S.J.H.; De Cuyper, M. Assessing Cytotoxicity of (Iron Oxide Based) Nanoparticles: An Overview of Different Methods Exemplified with Cationic Magnetoliposomes. *Contrast Media Mol. Imaging* **2009**, 4, 207–219.
32. Kunzmann, A.; Andersson, B.; Vogt, C.; Feliu, N.; Ye, F.; Gabrielsson, S.; et al. Efficient Internalization of Silica-Coated Iron Oxide Nanoparticles of Different Sizes by Primary Human Macrophages and Dendritic Cells. *Toxicol. Appl. Pharmacol.* **2011**, 253, 81–93.
33. Eaton, J.W.; Qian, M. Molecular Bases of Cellular Iron Toxicity. *Free Radic. Biol. Med.* **2002**, 32, 833–840.
34. Wu, J.; Sun, J. Investigation on Mechanism of Growth Arrest Induced by Iron Oxide Nanoparticles in PC12 Cells. *J. Nanosci. Nanotechnol.* **2011**, 11, 11079–11083.
35. Wu, W.; Chen, B.; Cheng, J.; Wang, J.; Xu, W.; Liu, L.; et al. Biocompatibility of Fe3O4/DNR Magnetic Nanoparticles in the Treatment of Hematologic Malignancies. *Int. J. Nanomed.* **2010**, 5, 1079–1084.
36. Zhu, M.T.; Wang, Y.; Feng, W.Y.; Wang, B.; Wang, M.; Ouyang, H.; et al. Oxidative Stress and Apoptosis Induced by Iron Oxide Nanoparticles in Cultured Human Umbilical Endothelial Cells. *J. Nanosci. Nanotechnol.* **2010**, 10, 8584–8590.
37. Wang, D.; Wang, L.H.; Zhao, Y.; Lu, Y.P.; Zhu, L. Hypoxia Regulates the Ferrous Iron Uptake and Reactive Oxygen Species Level Via Divalent Metal Transporter 1 (DMT1) Exon1B by Hypoxia Inducible Factor-1. *IUBMB Life* **2010**, 62, 629–636.
38. Wan, X.; Song, Y.; Song, N.; Li, J.; Yang, L.; Li, Y.; et al. The Preliminary Study of Immune Superparamagnetic Iron Oxide Nanoparticles for the Detection of Lung Cancer in Magnetic Resonance Imaging. *Carbohydr. Res.* **2016**, 419, 33–40.
39. Pfaff, A.; Schallon, A.; Ruhland, T.M.; Majewski, A.P.; Schmalz, H.; Freitag, R.; et al. Magnetic and Fluorescent Glycopolymer Hybrid Nanoparticles for Intranuclear Optical Imaging. *Biomacromolecules* **2011**, 12, 3805–3811.
40. Huang, G.; Zhang, C.; Li, S., Khemtong, C.; Yang, S.G.; Tian, R.; et al. A Novel Strategy for Surface Modification of Superparamagnetic Iron Oxide Nanoparticles for Lung Cancer Imaging. *J. Mater. Chem.* **2009**, 19, 6367–6372.
41. Nishimoto, K.; Mimura, A.; Aoki, M.; Banura, N.; Murase, K. Application of Magnetic Particle Imaging to Pulmonary Imaging Using Nebulized Magnetic Nanoparticles. *Open J. Med. Imag.* **2015**, 5, 49.
42. Chatterjee, D.K.; Diagaradjane, P.; Krishnan, S. Nanoparticle Mediated Hyperthermia in Cancer Therapy. *Ther. Deliv.* **2011**, 2, 1001–1014.
43. Gupta, R.; Sharma, D. Evolution of Magnetic Hyperthermia for Glioblastoma Multiforme Therapy. *ACS Chem. Neurosci.* **2019**, 10, 1157–1172.
44. Legge, C.J.; Colley, H.E.; Lawson, M.A.; Rawlings, A.E. Targeted Magnetic Nanoparticle Hyperthermia for the Treatment of Oral Cancer. *J. Oral Pathol. Med.* **2019**, 48, 803–809.
45. Liang, B.; Zuo, D.; Yu, K.; Cai, X.; Qiao, B.; Deng, R.; et al. Multifunctional Bone Cement for Synergistic Magnetic Hyperthermia Ablation and Chemotherapy of Osteosarcoma. *Mater. Sci. Eng. C Mater. Biol. Appl.* **2020**, 108, 110460.
46. Brero, F.; Albino, M.; Antoccia, A.; Arosio, P.; Avolio, M.; Berardinelli, F.; et al. Hadron Therapy, Magnetic Nanoparticles and Hyperthermia: A Promising Combined Tool for Pancreatic Cancer Treatment. *Nanomaterials (Basel)* **2020**, 10, 1919.
47. Cędrowska, E.; Pruszyński, M.; Gawęda, W.; Żuk, M.; Krysiński, P.; Bruchertseifer, F.; et al. Trastuzumab Conjugated Superparamagnetic Iron Oxide Nanoparticles Labeled with 225Ac as a Perspective Tool for Combined α-Radioimmunotherapy and Magnetic Hyperthermia of HER2-Positive Breast Cancer. *Molecules* **2020**, 25, 1025.
48. Duval, K.E.; Vernice, N.A.; Wagner, R.J.; Fiering, S.N.; Petryk, J.D.; Lowry, G.J.; et al. Immunogenetic Effects of Low Dose (CEM43 30) Magnetic Nanoparticle Hyperthermia and Radiation in Melanoma Cells. *Int. J. Hyperthermia* **2019**, 36, 37–46.

49. Attaluri, A.; Kandala, S.K.; Wabler, M.; Zhou, H.; Cornejo, C.; Armour, M.; et al. Magnetic Nanoparticle Hyperthermia Enhances Radiation Therapy: A Study in Mouse Models of Human Prostate Cancer. *Int. J. Hyperthermia* **2015**, 31, 359–374.
50. Stocke, N.A.; Meenach, S.A.; Arnold, S.M.; Mansour, H.M.; Hilt, J.Z. Formulation and Characterization of Inhalable Magnetic Nanocomposite Microparticles (MnMs) for Targeted Pulmonary Delivery Via Spray Drying. *Int. J. Pharm.* **2015**, 479, 320–328.
51. Tseng, C.L.; Chang, K.C.; Yeh, M.C.; Yang, K.C.; Tang, T.P.; Lin, F.H. Development of a Dual-Functional PtFeHAP Magnetic Nanoparticles Application for Chemo-Hyperthermia Treatment of Cancer. *Ceram. Int.* **2014**, 40, 511727.
52. Ma, J.; Zhang, Z.; Zhang, Z.; Huang, J.; Qin, Y.; Li, X.; et al. Magnetic Nanoparticle Clusters Radiosensitise Human Nasopharyngeal and Lung Cancer Cells After Alternating Magnetic Field Treatment. *Int. J. Hyperthermia* **2015**, 31, 800–812.
53. Baskar, G.; Ravi, M.; Panda, J.J.; Khatri, A.; Dev, B.; Santosham, R.; et al. Efficacy of Dipeptide-Coated Magnetic Nanoparticles in Lung Cancer Models Under Pulsed Electromagnetic Field. *Cancer Invest.* **2017**, 35, 431–442.
54. Araya, T.; Kasahara, K.; Nishikawa, S.; Kimura, H.; Sone, T.; Nagae, H.; et al. Antitumor Effects of Inductive Hyperthermia Using Magnetic Ferucarbotran Nanoparticles on Human Lung Cancer Xenografts in Nude Mice. *Onco. Targets Ther.* **2013**, 6, 237–242.
55. Dabbagh, A.; Hedayatnasab, Z.; Karimian, H.; Sarraf, M.; Yeong, C.H.; Hosseini, H.R.; et al. Polyethylene Glycol-coated Porous Magnetic Nanoparticles for Targeted Delivery of Chemotherapeutics Under Magnetic Hyperthermia Condition. *Int. J. Hyperthermia* **2019**, 36, 104–114.
56. Hoffman, A. The Origins and Evolution of "Controlled" Drug Delivery Systems. *J Control. Release* **2008**, 132, 153–163.
57. Kong, W.; Sung, D.; Shim, Y.H.; Bae, K.H.; Dubois, P.; Park, T.G.; et al. Efficient Intracellular siRNA Delivery Strategy through Rapid and Simple Two Steps Mixing Involving Noncovalent Post-PEGylation. *J. Control. Release* **2009**, 138, 141–147.
58. Chan, J.M.; Valencia, P.M.; Zhang, L.; Langer, R.; Farokhzad, O.C. Polymeric Nanoparticles for Drug Delivery. *Methods Mol. Biol.* **2010**, 624, 163–175.
59. Liu, J.; Liu, J.; Chu, L.; Wang, Y.; Duan, Y.; Feng, L.; et al. Novel Peptide-Dendrimer Conjugates as Drug Carriers for Targeting Nonsmall Cell Lung Cancer. *Int. J. Nanomedicine* **2010**, 6, 59–69.
60. Singh, S.; Nalwa, H.S. Nanotechnology and Health Safety Toxicity and Risk Assessment of Nanostructured Material Son Human Health. *J. Nanosci. Nanotechnol.* **2007**, 7, 3048–3070.
61. Allouche, J. Synthesis of Organic and Bioorganic Nanoparticles: An Overview of the Preparation Methos. In: *Nanomaterials: A Danger or a Promise*. **2013**. P. 27–74.
62. Ma, P.; Mumper, R.J. Paclitaxel Nano-Delivery Systems: A Comprehensive Review. *J. Nanomed. Nanotechnol.* **2013**, 4, 100164.
63. Jung, J.; Park, S.J.; Chung, H.K.; Kang, H.W.; Lee, S.W.; Seo, M.H.; et al. Polymeric Nanoparticles Containing Taxanes Enhance Chemoradiotherapeutic Efficacy in Non-Small Cell Lung Cancer. *Int. J. Radiat. Oncol. Biol. Phys.* **2012**, 84, e77–e83.
64. Hu, J.; Fu, S.; Peng, Q.; Han, Y.W.; Xie, J.; Zan, N.; et al. Paclitaxel-Loaded Polymeric Nanoparticles Combined with Chronomodulated Chemotherapy on Lung Cancer: *In Vitro* and *In Vivo* Evaluation. *Int. J. Pharm.* **2017**, 516, 313–322.
65. Wang, X.; Chen, H.; Zeng, X.; Guo, W.; Jin, Y.; Wang, S.; et al. Efficient Lung Cancer-Targeted Drug Delivery Via a Nanoparticle/MSC System. *Acta Pharm. Sin. B* **2019**, 9, 167–176.
66. Zhang, Z.; Wang, J.; Nie, X.; Wen, T.; Ji, Y.; Wu, X.; et al. Near Infrared Laser-Induced Targeted Cancer Therapy Using Thermoresponsive Polymer Encapsulated Gold Nanorods. *J. Am. Chem. Soc.* **2014**, 136, 7317–7326.
67. Abu Lila, A.S.; Ishida, T. Liposomal Delivery Systems: Design Optimization and Current Applications. *Biol. Pharm. Bul. Biol. Pharm. Bull.* **2017**, 40, 1–10.
68. Haluska, C.K.; Riske, K.A.; Marchi-Artzner, V.; Lehn, J.M.; Lipowsky, R.; Dimova, R. Time Scales of Membrane Fusion Revealed by Direct Imaging of Vesicle Fusion with High Temporal Resolution. *Proc. Natl. Acad. Sci.* **2006**, 103, 15841–15846.
69. Torchilin, V.P. Recent Advances with Liposomes as Pharmaceutical Carriers. *Nat. Rev. Drug Discov.* **2005**, 4, 145–160.
70. Wagner, U.; Marth, C.; Largillier, R.; Kaern, J.; Brown, C.; Heywood, M.; et al. Final Overall Survival Results of Phase III GCIG CALYPSO Trial of Pegylated Liposomal Doxorubicin and Carboplatin vs Paclitaxel and Carboplatin in Platinum-Sensitive Ovarian Cancer Patients. *Br. J. Cancer* **2012**, 107, 588–591.

71. Paliwal, R.; Paliwal, S.R.; Kenwat, R.; Kurmi, B.D.; Sahu, M.K. Solid Lipid Nanoparticles: A Review on Recent Perspectives and Patents. *Expert Opin. Ther. Pat.* **2020**, 30, 179–194.
72. Aupérin, A.; Péchoux, C.L.; Rolland, E.; Curran, W.J.; Furuse, K.; Fournel, P.; et al. Meta-Analysis of Concomitant Versus Sequential Radio Chemotherapy in Locally Advanced Non-Small-Cell Lung Cancer. *J. Clin. Oncol.* **2010**, 28, 2181–2190.
73. Hamroun, A.; Lenain, R.; Bigna, J.J.; Speyer, E.; Bui, L.; Chamley, P.; et al. Prevention of Cisplatin-Induced Acute Kidney Injury: A Systematic Review and Meta-Analysis. *Drugs* **2019**, 79, 1567–1582.
74. Boulikas, T. Clinical Overview on Lipoplatin™: A Successful Liposomal Formulation of Cisplatin. *Expert Opin. Investig. Drugs* **2009**, 18, 1197–1218.
75. Giuberti, C.D.; Reis, E.C.; Rocha, T.G.; Leite, E.A.; Lacerda, R.G.; Ramaldes, G.A.; et al. Study of the Pilot Production Process of Long-Circulating and pH-Sensitive Liposomes Containing Cisplatin. *J. Liposome Res.* **2010**, 21, 60–69.
76. Stathopoulos, G.P.; Antoniou, D.; Dimitroulis, J.; Stathopoulos, J.; Marosis, K.; Michalopoulou, P. Comparison of Liposomal Cisplatin Versus Cisplatin in Non-Squamous Cell Non-Small-Cell Lung Cancer. *Cancer Chemother. Pharmacol.* **2011**, 68, 945–950.
77. Khan, M.M.; Madni, A.; Torchilin, V.; Filipczak, N.; Pan, J.; Tahir, N.; et al. Lipid-Chitosan Hybrid Nanoparticles for Controlled Delivery of Cisplatin. *Drug Deliv.* **2019**, 26, 765–772.
78. Paraskar, A.S.; Soni, S.; Chin, K.T.; Chaudhuri, P.; Muto, K.W.; Berkowitz, J.; et al. Harnessing Structure-Activity Relationship to Engineer a Cisplatin Nanoparticle for Enhanced Antitumor Efficacy. *Proc. Natl. Acad. Sci.* **2010**, 107, 12435–12440.
79. Khurana, R.K.; Mahajan, M.; Teenu, Kapoor, S.; Jain, S.; Singh, B. The Sojourn from Parenteral to Oral Taxanes Using Nanocarrier Systems: A Patent Review. *Recent Pat. Drug Deliv. Formul.* **2016**, 10, 44–58.
80. Koudelka, Š.; Turánek, J. Liposomal Paclitaxel Formulations. *J. Control. Release* **2012**, 163, 322–334.
81. Wang, X.; Zhou, J.; Wang, Y.; Zhu, Z.; Lu, Y.; Wei, Y.; et al. A Phase I Clinical and Pharmacokinetic Study of Paclitaxel Liposome Infused in Non-Small Cell Lung Cancer Patients with Malignant Pleural Effusions. *Eur. J. Cancer* **2010**, 46, 1474–1480.
82. Paclitaxel Liposome for Squamous Non-Small Cell Lung Cancer Study (LIPUSU) Full Text View.
83. Matsumura, Y.; Hamaguchi, T.; Ura, T.; Muro, K.; Yamada, Y.; Shimada, Y.; et al. Phase I Clinical Trial and Pharmacokinetic Evaluation of NK911, a Micelle-Encapsulated Doxorubicin. *Br. J. Cancer* **2004**, 91, 1775–1781.
84. Hamaguchi, T.; Doi, T.; Eguchi-Nakajima, T.; Kato, K.; Yamada, Y.; Shimada, Y.; et al. Phase I Study of NK012, a Novel SN-38-Incorporating Micellar Nanoparticle, in Adult Patients with Solid Tumors. *Clin. Cancer Res.* **2010**, 16, 5058–5066.
85. Paclitaxel Micelles for Injection/Paclitaxel Injection in Combination with Cisplatin for First Line Therapy of Advanced NSCLC.
86. Yavuz, M.S.; Cheng, Y.; Chen, J.; Cobley, C.M.; Zhang, Q.; Rycenga, M.; et al. Gold Nanocages Covered by Smart Polymers for Controlled Release with Near-Infrared Light. *Nat. Mater.* **2009**, 8, 935–939.
87. Hainfeld, J.F.; Dilmanian, F.A.; Slatkin, D.N.; Smilowitz, H.M. Radiotherapy Enhancement with Gold Nanoparticles. *J. Pharm. Pharmacol.* **2008**, 60, 977–985.
88. Selvan, S.T.; Tan, T.T.; Yi, D.K.; Jana, N.R. Functional and Multifunctional Nanoparticles for Bioimaging and Biosensing. *Langmuir* **2010**, 26, 11631–11641.
89. Xie, J.; Lee, S.; Chen, X. Nanoparticle-Based Theranostic Agents. *Adv. Drug Deliv. Rev.* **2010**, 62, 1064–1079.
90. Sperling, R.A.; Parak, W. Surface Modification, Functionalization and Bioconjugation of Colloidal Inorganic Nanoparticles. *Philos. Trans. R Soc. Lond. A Math. Phys. Eng. Sci.* **2010**, 368, 1333–1383.
91. Sau, T.K.; Rogach, A.L.; Jäckel, F.; Klar, T.A.; Feldmann, J. Properties and Applications of Colloidal Nonspherical Noble Metal Nanoparticles. *Adv. Mater.* **2010**, 22, 1805–1825.
92. Sperling, R.A.; Gil, P.R.; Zhang, F.; Zanella, M.; Parak, W.J. Biological Applications of Gold Nanoparticles. *Chem. Soc. Rev.* **2008**, 37, 1896–1908.
93. Porcel, E.; Liehn, S.; Remita, H.; Usami, N.; Kobayashi, K.; Furusawa, Y.; et al. Platinum Nanoparticles: A Promising Material for Future Cancer Therapy? *Nanotechnol.* **2010**, 21, 85103.
94. Danhier, F.; Feron, O.; Préat, V. To Exploit the Tumor Microenvironment: Passive and Active Tumor Targeting of Nanocarriers for Anti-Cancer Drug Delivery. *J. Control. Release* **2010**, 148, 135–146.
95. Silva, A.C.; Oliveira, T.R.; Mamani, J.B.; Malheiros, S.M.; Malavolta, L.; Pavon, L.F.; et al. Application of Hyperthermia Induced by Superparamagnetic Iron Oxide Nanoparticles in Glioma Treatment. *Int. J. Nanomed.* **2011**, 6, 591–603.

96. Maier-Hauff, K.; Ulrich, F.; Nestler, D.; Niehoff, H.; Wust, P.; Thiesen, B.; et al. Efficacy and Safety of Intratumoral Thermotherapy Using Magnetic Iron-Oxide Nanoparticles Combined with External Beam Radiotherapy on Patients with Recurrent Glioblastoma Multiforme. *J. Neurooncol.* **2011**, 103, 317–324.
97. Hou, Z.; Zhang, Y.; Deng, K.; Chen, Y.; Li, X.; Deng, X.; et al. UV-Emitting Upconversionbased TiO2 Photosensitizing Nanoplatform: Near-Infrared Light Mediated *In Vivo* Photodynamic Therapy Via Mitochondria Involved Apoptosis Pathway. *ACS Nano* **2015**, 9, 2584–2599.
98. Baskar, G.; Chandhuru, J.; Fahad, K.S.; Praveen, A.S.; Chamundeeswari, M.; Muthukumar, T. Anticancer Activity of Fungal l-Asparaginase Conjugated with Zinc Oxide Nanoparticles. *J. Mater. Sci. Mater. Med.* **2015**, 26, 5380.
99. Ali, D.; Alarifi, S.; Alkahtani, S.; AlKahtane, A.A.; Almalik, A. Cerium Oxide Nanoparticles Induce Oxidative Stress and Genotoxicity in Human Skin Melanoma Cells. *Cell Biochem. Biophys.* **2015**, 71, 1643–1651.
100. Zhang, X.F.; Liu, Z.G.; Shen, W.; Gurunathan, S. Silver Nanoparticles: Synthesis, Characterization, Properties, Applications, and Therapeutic Approaches. *Int. J. Mol. Sci.* **2016**, 17, 1534.
101. Haume, K.; Rosa, S.; Grellet, S.; Śmiałek, M.A.; Butterworth, K.T.; Solov'yov, A.V.; et al. Gold Nanoparticles for Cancer Radiotherapy: A Review. *Cancer Nano* **2016**, 7, 8.
102. Lopez-Campos, F.; Candini, D.; Carrasco, E.; Francés, M.A. Nanoparticles Applied to Cancer Immunoregulation. *Rep. Pract. Oncol. Radiother.* **2019**, 24, 47–55.
103. Zhao, N.; Pan, Y.; Cheng, Z.; Liu, H. Gold Nanoparticles for Cancer Theranostics a Brief Update. *J. Innov. Opt. Health Sci.* **2016**, 9, 1630004.
104. Zhao, P.; Li, N.; Astruc, D. State of the Art in Gold Nanoparticle Synthesis. *Coord. Chem. Rev.* **2013**, 257, 638–665.
105. Singh, P.; Pandit, S.; Mokkapati, V.R.; Garg, A.; Ravikumar, V.; Mijakovic, I. Gold Nanoparticles in Diagnostics and Therapeutics for Human Cancer. *Int. J. Mol. Sci.* **2018**, 19, 1979.
106. Guthi, J.S.; Yang, S.G.; Huang, G.; Li, S.; Khemtong, C.; Kessinger, C.W.; et al. MRI-Visible Micellar Nanomedicine for Targeted Drug Delivery to Lung Cancer Cells. *Mol. Pharm.* **2010**, 7, 32–40.
107. Wang, C.; Beiss, V.; Steinmetz, N.F. Cowpea Mosaic Virus Nanoparticles and Empty Virus-Like Particles Show Distinct but Overlapping Immuno Stimulatory Properties. *J. Virol.* **2019**, 93, e00129.19.
108. Johnstone, T.C.; Suntharalingam, K.; Lippard, S.J. The Next Generation of Platinum Drugs: Targeted Pt(II) Agents, Nanoparticle Delivery, and Pt(IV) Prodrugs. *Chem. Rev.* **2016**, 116, 3436–3486.
109. Chung, Y.H.; Cai, H.; Steinmetz, N.F. Viral Nanoparticles for Drug Delivery, Imaging, Immunotherapy, and Theranostic Applications. *Adv. Drug Deliv. Rev.* **2020**, 156, 214–235.
110. Ren, Y.; Wong, S.M.; Lim, L.Y. Folic Acid-Conjugated Protein Cages of a Plant Virus: A Novel Delivery Platform for Doxorubicin. *Bioconjug. Chem.* **2007**, 18, 836–843.
111. Zeng, Q.; Wen, H.; Wen, Q.; Chen, X.; Wang, Y.; Xuan, W.; et al. Cucumber Mosaic Virus as Drug Delivery Vehicle for Doxorubicin. *Biomaterials* **2013**, 34, 4632–4642.
112. Aljabali, A.A.; Shukla, S.; Lomonossoff, G.P.; Steinmetz, N.F.; Evans, D.J. CPMV-DOX Delivers. *Mol. Pharm.* **2013**, 10, 3–10.
113. Cao, J.; Guenther, R.H.; Sit, T.L.; Opperman, C.H.; Lommel, S.A.; Willoughby, J.A. Loading and Release Mechanism of Red Clover Necrotic Mosaic Virus Derived Plant Viral Nanoparticles for Drug Delivery of Doxorubicin. *Small* **2014**, 10, 5126–5136.
114. Le, D.H.; Lee, K.L.; Shukla, S.; Commandeur, U.; Steinmetz, N.F. Potato Virus X, a Filamentous Plant Viral Nanoparticle for Doxorubicin Delivery in Cancer Therapy. *Nanoscale* **2017**, 9, 2348–2357.
115. Shan, W.; Zhang, D.; Wu, Y.; Lv, X.; Hu, B.; Zhou, X.; et al. Modularized Peptides Modified HBc Virus-Like Particles for Encapsulation and Tumor-Targeted Delivery of Doxorubicin. *Nanomedicine* **2018**, 14, 725–734.
116. Hu, H.; Steinmetz, N.F. Doxorubicin-Loaded Physalis Mottle Virus Particles as a pH-Responsive Prodrug for Cancer Therapy. *Biotechnol. J.* **2020**, 12, e2000077.
117. Robertson, K.L.; Soto, C.M.; Archer, M.J.; Odoemene, O.; Liu, J.L. Engineered T4 Viral Nanoparticles for Cellular Imaging and Flow Cytometry. *Bioconjug. Chem.* **2011**, 22, 595–604.
118. Destito, G.; Yeh, R.; Rae, C.S.; Finn, M.G.; Manchester, M. Folic Acid-Mediated Targeting of Cowpea Mosaic Virus Particles to Tumor Cells. *Chem. Biol.* **2007**, 14, 1152–1162.
119. Beatty, P.H.; Lewis, J.D. Cowpea Mosaic Virus Nanoparticles for Cancer Imaging and Therapy. *Adv. Drug Deliv. Rev.* **2019**, 145, 130–144.
120. Lewinski, N.; Colvin, V.; Drezek, R. Cytotoxicity of Nanoparticles. *Small* **2008**, 4, 26–49.
121. de Mello Donegá, C. Synthesis and Properties of Colloidal Heteronanocrystals. *Chem. Soc. Rev.* **2011**, 40, 1512–1546.

122. Ghasemi, Y.; Peymani, P.; Afifi, S. Quantum Dot: Magic Nanoparticle for Imaging, Detection and Targeting. *Acta Biomed.* **2009**, 80, 156–165.
123. Jin, S.; Hu, Y.X.; Gu, Z.J.; Liu, L.; Wu, H.C. Application of Quantum Dots in Biological Imaging. *J. Nanomater.* **2011**, 2011, 834139.
124. Hardman, R. A Toxicologic Review of Quantum Dots: Toxicity Depends on Physicochemical and Environmental Factors. *Environ. Health Perspect.* **2006**, 114, 165–172.
125. Smith, A.M.; Nie, S. Next-Generation Quantum Dots. *Nat. Biotechnol.* **2009**, 27, 732–733.
126. Singh, R.D.; Shandilya, R.; Bhargava, A.; Kumar, R.; Tiwari, R.; Chaudhury, K.; et al. Quantum Dot Based Nano-Biosensors for Detection of Circulating Cell Free miRNAs in Lung Carcinogenesis: From Biology to Clinical Translation. *Front. Genet.* **2018**, 9, 616.
127. Mashinchian, O.; Johari-Ahar, M.; Ghaemi, B.; Rashidi, M.; Barar, J.; Omidi, Y. Impacts of Quantum Dots in Molecular Detection and Bioimaging of Cancer. *Bioimpacts* **2014**, 4, 149–166.
128. Ranjbar-Navazi, Z.; Eskandani, M.; Johari-Ahar, M.; Nemati, A.; Akbari, H.; Davaran, S.; et al. Doxorubicin Conjugated D-Glucosamine-and Folate-bi-Functionalised InP/ZnS Quantum Dots for Cancer Cells Imaging and Therapy. *J. Drug Target.* **2018**, 26, 267–277.
129. Cai, X.; Luo, Y.; Zhang, W.; Du, D.; Lin, Y. pH-Sensitive ZnO Quantum Dots-Doxorubicin Nanoparticles for Lung Cancer Targeted Drug Delivery. *ACS Appl. Mater. Interfaces* **2016**, 8, 22442–22450.
130. Kulkarni, N.S.; Parvathaneni, V.; Shukla, S.K.; Barasa, L.; Perron, J.C.; Yoganathan, S.; et al. Tyrosine Kinase Inhibitor Conjugated Quantum Dots for Non-Small Cell Lung Cancer (NSCLC) Treatment. *Eur. J. Pharm. Sci.* **2019**, 133, 145–159.
131. Vynnychenko, I.; Okamoto, I.; et al. Weekly Nab-Paclitaxel in Combination with Carboplatin Versus Solvent-Based Paclitaxel Plus Carboplatin as First-Line Therapy in Patients with Advanced Non-Small Cell Lung Cancer: Final Results of a Phase III Trial. *J. Clin. Oncol.* **2012**, 30, 2055–2062.
132. Yang, J.J.; Huang, C.; Chen, G.Y.; Song, Y.; Cheng, Y.; Yan, H.H.; et al. A Randomized Phase II Clinical Trial of Nabpaclitaxel and Carboplatin Compared with Gemcitabine and Carboplatin as First-line Therapy in Locally Advanced or Metastatic Squamous Cell Carcinoma of Lung. *BMC Cancer* **2014**, 14, 684.
133. Langer, C.J.; Hirsh, V.; Ko, A.; Renschler, M.F.; Socinski, M.A. Weekly Nab-Paclitaxel in Combination with Carboplatin as First-line Therapy in Patients with Advanced Nonsmall Cell Lung Cancer: Analysis of Safety and Efficacy in Patients with Renal Impairment. *Clin. Lung Cancer* **2015**, 16, 112–120.
134. Autio, K.A.; Dreicer, R.; Anderson, J.; Garcia, J.A.; Alva, A.; Hart, L.L.; et al. Safety and Efficacy of BIND-014, a Docetaxel Nanoparticle Targeting Prostate-Specific Membrane Antigen for Patients With Metastatic Castration-Resistant Prostate Cancer: A Phase 2 Clinical Trial. *JAMA Oncol.* **2018**, 4, 1344–1351.
135. Von Hoff, D.D.; Mita, M.M.; Ramanathan, R.K.; Weiss, G.J.; Mita, A.C.; LoRusso, P.M.; et al. Phase I Study of PSMA Targeted Docetaxel-Containing Nanoparticle BIND-014 in Patients with Advanced Solid Tumors. *Clin. Cancer Res.* **2016**, 22, 3157–3163.
136. Chen, Y.F.; Wang, Y.H.; Lei, C.S.; Changou, C.A.; Davis, M.E.; Yen, Y. Host Immune Response to Anti-Cancer Camptothecin Conjugated Cyclodextrin-Based Polymers. *J. Biomed. Sci.* **2019**, 26, 85.
137. Abarca-Cabrera, L.; Fraga-García, P.; Berensmeier, S. Bio-Nano Interactions: Binding Proteins, Polysaccharides, Lipids and Nucleic Acids Onto Magnetic Nanoparticles. *Biomater. Res.* **2021**, 25(1), 12.
138. Duan, Y.; Dhar, A.; Patel, C.; Khimani, M.; Neogi, S.; Sharma, P.; et al. A Brief Review on Solid Lipid Nanoparticles: Part and Parcel of Contemporary Drug Delivery Systems. *RSC Adv.* **2020**, 10(45), 26777–26791.
139. Reddy, M.S.B.; Ponnamma, D.; Choudhary, R.; Sadasivuni, K.K. A comparative review of natural and synthetic biopolymer composite scaffolds. *Polymers* **2021**, 13(7), 1105.
140. Din, F.U.; Aman, W.; Ullah, I.; Qureshi, O.S.; Mustapha, O.; Shafique, S.; et al. Effective Use of Nanocarriers as Drug Delivery Systems for the Treatment of Selected Tumors. *Int. J. Nanomedicine* **2017**, 12, 7291–7309.
141. Alavi, M.; Karimi, N.; Safaei, M. Application of Various Types of Liposomes in Drug Delivery Systems. *Adv. Pharma. Bull.* **2017**, 7(1), 3–9.
142. Attama, A.A.; Momoh, M.A.; Builders, P.F. Lipid Nanoparticulate Drug Delivery Systems: A Revolution in Dosage form Design and Development. *Recent Adv. Novel Drug Carrier Syst.* **2012**, 5, 107–140.35.
143. Bulbake, U.; Doppalapudi, S.; Kommineni, N.; Khan, W. Liposomal Formulations in Clinical Use: An Updated Review. *Pharmaceutics* **2017**, 9(2), 12.
144. Shrestha, H.; Bala, R.; Arora, S. Lipid-Based Drug Delivery Systems. *J. Pharma.* **2014**, 2014, 801820.
145. Yadav, D.; Sandeep, K.; Pandey, D.; Dutta, R.K. Liposomes for Drug Delivery. *J. Biotechnol. Biomater.* **2017**, 7(4), 276.

146. Rommasi, F.; Esfandiari, N. Liposomal Nanomedicine: Applications for Drug Delivery in Cancer Therapy. *Nanoscale Res. Lett.* **2021**, 16(1), 95.
147. Wakaskar, R.R. General Overview of Lipid-Polymer Hybrid Nanoparticles, Dendrimers, Micelles, Liposomes, Spongosomes and Cubosomes. *J. Drug Target* **2018**, 26(4), 311–318.
148. Ghasemiyeh, P.; Mohammadi-Samani, S. Solid Lipid Nanoparticles and Nanostructured Lipid Carriers as Novel Drug Delivery Systems: Applications, Advantages and Disadvantages. *Res. Pharm. Sci.* **2018**, 13(4), 288–303.
149. Yoon, G.; Park, J.W.; Yoon, I.-S. Solid Lipid Nanoparticles (SLNs) and Nanostructured Lipid Carriers (NLCs): Recent Advances in Drug Delivery. *J. Pharma. Investig.* **2013**, 43(5), 353–362.
150. Shiraishi, K.; Yokoyama, M. Toxicity and Immunogenicity Concerns Related to PEGylated-Micelle Carrier Systems: A Review. *Sci. Technol. Adv. Mater.* **2019**, 20(1), 324–336.
151. Chen, B.-M.; Cheng, T.-L.; Roffler, S.R. Polyethylene Glycol Immunogenicity: Theoretical, Clinical, and Practical Aspects of Anti-Polyethylene Glycol Antibodies. *ACS Nano* **2021**, 15(9), 14022–14048.
152. Shah, P.K.; Austin, J.H.; White, C.S.; Patel, P.; Haramati, L.B.; Pearson, G.D.; et al. Missed Non-Small Cell Lung Cancer: Radiographic Findings of Potentially Resectable Lesions Evident Only in Retrospect. *Radiology* **2003**, 226, 235–241.
153. Weissleder, R.; Pittet, M.J. Imaging in the Era of Molecular Oncology. *Nature* **2008**, 452, 580–589.
154. Chapman, S.; Dobrovolskaia, M.; Farahani, K.; Goodwin, A.; Joshi, A.; Lee, H.; et al. Nanoparticles for Cancer Imaging: The Good, the Bad, and the Promise. *Nano Today* **2013**, 8, 454–460.
155. Safdar, M.H.; Hussain, Z.; Abourehab, M.A.; Hasan, H.; Afzal, S.; Thu, H.E. New Developments and Clinical Transition of Hyaluronic Acid-Based Nanotherapeutics for Treatment Of Cancer: Reversing Multidrug Resistance, Tumour-Specific Targetability and Improved Anticancer Efficacy. *Artif. Cells Nanomed. Biotechnol.* **2018**, 46, 1967–1980.
156. Arshad, R.; Kiani, M.H.; Rahdar, A.; Sargazi, S.; Barani, M.; Shojaei, S.; et al. Nano-Based Theranostic Platforms for Breast Cancer: A Review of Latest Advancements. *Bioengineering* **2022**, 9, 320.
157. Ali, I.; Rahis-Uddin; Salim, K.; Rather, M.A.; Wani, W.A.; Haque, A. Advances in Nano Drugs for Cancer Chemotherapy. *Curr. Cancer Drug Targets* **2011**, 11, 135–146.
158. Prabhu, R.H.; Patravale, V.B.; Joshi, M.D. Polymeric Nanoparticles for Targeted Treatment in Oncology: Current Insights. *Int. J. Nanomedicine* **2015**, 10, 1001–1018.
159. Tran, J.; Master, Z.; Yu, J.L.; Rak, J.; Dumont, D.J.; Kerbel, R.S. A Role for Survivin in Chemoresistance of Endothelial Cells Mediated by VEGF. *Proc. Natl. Acad. Sci. U S A.* **2002**, 99(7), 4349–4354.
160. Herbst, R.S.; Onn, A.; Sandler, A. Angiogenesis and Lung Cancer: Prognostic and Therapeutic Implications. *J. Clin. Oncol.* **2005**, 23(14), 3243–3256.
161. Frezzetti, D.; Gallo, M.; Maiello, M.R.; D'Alessio, A.; Esposito, C.; Chicchinelli, N., et al. VEGF as a Potential Target in Lung Cancer. *Expert Opin. Ther. Targets* **2017**, 21(10), 959–966.
162. Liu, D.; Liu, F.; Liu, Z.; Wang, L.; Zhang, N. Tumor Specific Delivery and Therapy by Double-Targeted Nanostructured Lipid Carriers with Anti-VEGFR-2 Antibody. *Mol. Pharm.* **2011**, 8(6), 2291–2301.
163. Jin, X.; Zhou, J.; Zhang, Z.; Lv, H. The Combined Administration of Parthenolide and Ginsenoside CK in Long Circulation Liposomes with Targeted tLyp-1 Ligand Induce Mitochondria-Mediated Lung Cancer Apoptosis. *Artif. Cell Nanomed. Biotechnol.* **2018**, 46(Suppl. 3), S931–S942.
164. Aksorn, N.; Chanvorachote, P. Integrin as a Molecular Target for Anticancer Approaches in Lung Cancer. Anticancer Res. **2019**, 39(2), 541–548.
165. Marelli, U.K.; Rechenmacher, F.; Sobahi, T.R.; Mas-Moruno, C.; Kessler, H. Tumor Targeting via Integrin Ligands. *Front. Oncol.* **2013**, 3, 222.
166. Babu, A.; Amreddy, N.; Muralidharan, R.; Pathuri, G.; Gali, H.; Chen, A.; et al. Chemodrug Delivery Using Integrin-Targeted PLGA-Chitosan Nanoparticle for Lung Cancer Therapy. *Sci. Rep.* **2017**, 7(1), 14674.
167. Wang, G.; Wang, Z.; Li, C.; Duan, G.; Wang, K.; Li, Q.; et al. RGD Peptide-Modified, Paclitaxel Prodrug-Based, Dual-Drugs Loaded, and Redox-Sensitive Lipid-Polymer Nanoparticles for the Enhanced Lung Cancer Therapy. *Biomed. Pharmacother.* **2018**, 106, 275–284.
168. Zou, Y.; Wei, J.; Xia, Y.; Meng, F.; Yuan, J.; Zhong, Z. Targeted Chemotherapy for Subcutaneous and Orthotopic Non-small Cell Lung Tumors with Cyclic RGD-Functionalized and Disulfide-Crosslinked Polymersomal Doxorubicin. *Signal. Transduct Target. Ther.* **2018**, 3, 32.
169. Zhang, Z.; Stiegler, A.L.; Boggon, T.J.; Kobayashi, S.; Halmos, B. EGFR-mutated Lung Cancer: A Paradigm of Molecular Oncology. *Oncotarget* **2010**, 1(7), 497–514.
170. Singh, B.; Carpenter, G.; Coffey, R.J. EGF Receptor Ligands: Recent Advances. *F1000Res.* **2016**, 5, 2270.
171. Lv, T.; Li, Z.; Xu, L.; Zhang, Y.; Chen, H.; Gao, Y. Chloroquine in Combination with Aptamer-Modified Nanocomplexes for Tumor Vessel Normalization and Efficient erlotinib/Survivin shRNA Co-delivery to

Overcome Drug Resistance in EGFR-Mutated Non-Small Cell Lung Cancer. *Acta Biomater.* **2018**, 76, 257–274.
172. Kim, S.K.; Huang, L. Nanoparticle Delivery of a Peptide Targeting EGFR Signaling. *J. Control. Release* **2012**, 157(2), 279–286.
173. Zhang, Y.; Zhang, L.; Lin, X.; Ke, L.; Li, B.; Xu, L.; et al. Dual-Responsive Nanosystem for Precise Molecular Subtyping and Resistant Reversal of EGFR Targeted Therapy. *Chem. Eng. J.* **2019**, 372, 483–495.
174. Zhang, X.; Wang, Q.; Qin, L.; Fu, H.; Fang, Y.; Han, B., et al. EGF-Modified mPEG-PLGA-PLL Nanoparticle for Delivering Doxorubicin Combined with Bcl-2 siRNA as a Potential Treatment Strategy for Lung Cancer. *Drug Deliv.* **2016**, 23(8), 2936–2945.
175. Schmidt, H.R.; Kruse, A.C. The Molecular Function of σ Receptors: Past, Present, and Future. *Trends Pharmacol. Sci.* **2019**, 40(9), 636–654.
176. van Waarde, A.; Rybczynska, A.A.; Ramakrishnan, N.K.; Ishiwata, K.; Elsinga, P.H.; Dierckx, R.A. Potential Applications for Sigma Receptor Ligands in Cancer Diagnosis and Therapy. *Biochim. Biophys. Acta* **2015**, 1848(10 Pt B), 2703–2714.
177. Ramzy, L.; Nasr, M.; Metwally, A.A.; Awad, G.A.S. Cancer Nanotheranostics: A Review of the Role of Conjugated Ligands for Overexpressed Receptors. *Eur. J. Pharm. Sci.* **2017**, 104, 273–292.
178. Georgiadis, M.O.; Karoutzou, O.; Foscolos, A.S.; Papanastasiou, I. Sigma Receptor (σR) Ligands with Antiproliferative and Anticancer Activity. *Molecules* **2017**, 22(9).
179. Yang, Y.; Hu, Y.; Wang, Y.; Li, J.; Liu, F.; Huang, L. Nanoparticle Delivery of Pooled siRNA for Effective Treatment of Non-Small Cell Lung Cancer. *Mol. Pharm.* **2012**, 9(8), 2280–2289.
180. Xiong, Y.; Zhao, Y.; Miao, L.; Lin, C.M.; Huang, L. Co-Delivery of Polymeric Metformin and Cisplatin by Self-Assembled Core-Membrane Nanoparticles to Treat Non-Small Cell Lung Cancer. *J. Control. Release* **2016**, 244(Pt A), 63–73.
181. Varshosaz, J.; Hassanzadeh, F.; Mardani, A.; Rostami, M. Feasibility of Haloperidol-Anchored Albumin Nanoparticles Loaded with Doxorubicin as Dry Powder Inhaler for Pulmonary Delivery. *Pharm. Dev. Technol.* **2015**, 20(2), 183–196.
182. Prabhu, R.H.; Patravale, V.B.; Joshi, M.D. Polymeric Nanoparticles for Targeted Treatment in Oncology: Current Insights. *Int. J. Nanomedicine* **2015**, 10, 1001–1018.
183. Sharma, P.; Mehta, M.; Dhanjal, D.S., Kaur, S.; Gupta, G.; Singh, H.; et al. Emerging Trends in the Novel Drug Delivery Approaches for the Treatment of Lung Cancer. *Chem. Biol. Interact* **2019**, 309, 108720.
184. Bazak, R.; Houri, M.; El Achy, S.; Kamel, S.; Refaat, T. Cancer Active Targeting by Nanoparticles: A Comprehensive Review of Literature. *J. Cancer Res. Clin. Oncol.* **2015**, 141(5), 769–784.
185. Jin, X.; Zhou, J.; Zhang, Z.; Lv, H. The Combined Administration of Parthenolide and Ginsenoside CK in Long Circulation Liposomes with Targeted tLyp-1 Ligand Induce Mitochondria-Mediated Lung Cancer Apoptosis. *Artif. Cell Nanomed. Biotechnol.* **2018**, 46(Suppl. 3), S931–S942.
186. Kato, T.; Jin, C.S.; Ujiie, H.; Lee, D.; Fujino, K.; Wada, H.; et al. Nanoparticle Targeted Folate Receptor 1-Enhanced Photodynamic Therapy for Lung Cancer. *Lung Cancer* **2017**, 113, 59–68.
187. Lovell, J.F., Jin, C.S., Huynh, E., Jin, H., Kim, C., Rubinstein, J.L., et al. Porphysome Nanovesicles Generated by Porphyrin Bilayers for Use as Multimodal Biophotonic Contrast Agents. *Nat. Mater.* **2011**, 10(4), 324–332.
188. Amreddy, N.; Babu, A.; Panneerselvam, J.; Srivastava, A.; Muralidharan, R.; Chen, A.; et al. Chemo-Biologic Combinatorial Drug Delivery Using Folate Receptor-Targeted Dendrimer Nanoparticles for Lung Cancer Treatment. *Nanomedicine* **2018**, 14(2), 373–384.
189. Muralidharan, R.; Babu, A.; Amreddy, N.; Basalingappa, K.; Mehta, M.; Chen, A.; et al. Folate Receptor-Targeted Nanoparticle Delivery of HuR-RNAi Suppresses Lung Cancer Cell Proliferation and Migration. *J. Nanobiotechnology* **2016**, 14(1), 47.
190. He, Z., Shi, Z., Sun, W.,Ma, J., Xia, J., Zhang, X., et al. Hemocompatibility of Folic-Acid-Conjugated Amphiphilic PEG-PLGA Copolymer Nanoparticles for Co-Delivery of Cisplatin and Paclitaxel: Treatment Effects for Non-Small-cell Lung Cancer. *Tumour Biol.* **2016**, 37(6), 7809–7821. https://doi.org/10.1007/s13277-015-4634-1
191. Vaidya, B.; Vyas, S.P. Transferrin Coupled Vesicular Systemfor Intracellular Drug Delivery for the Treatment of Cancer: Development and Characterization. *J. Drug Target.* **2012**, 20(4), 372–380.
192. Feelders, R.A., Kuiper-Kramer, E.P.; van Eijk, H.G. Structure, Function and Clinical Significance of Transferrin Receptors. *Clin. Chem. Lab. Med.* **1999**, 37(1), 1–10.
193. Daniels, T.R., Delgado, T., Rodriguez, J.A., Helguera, G.; Penichet, M.L. The Transferrin Receptor Part I: Biology and Targeting with Cytotoxic Antibodies for the Treatment of Cancer. *Clin. Immunol.* **2006**, 121(2), 144–158.

194. Cheng, Z.; Liu, Y.F.; Song, Y.N.; Dai, L.L.; Kang, Y.; Xia, J.; et al. Regulating Action of Iron Regulatory Protein-2 in Iron Metabolism of Lung Cancer. *Zhonghua Yi Xue Za Zhi* **2011**, 91(28), 1992–1995.
195. Dev, S.; Babitt, J.L. Overview of Iron Metabolism in Health and Disease. *Hemodial Int.* **2017**, 21(Suppl 1), S6–S20.
196. Manz, D.H.; Blanchette, N.L.; Paul, B.T.; Torti, F.M.; Torti, S.V. Iron and Cancer: Recent Insights. *Ann. N. Y Acad. Sci.* **2016**, 1368(1), 149–161.
197. Tros de Ilarduya, C.; Düzgüneş, N. Delivery of Therapeutic Nucleic Acids via Transferrin and Transferrin Receptors: Lipoplexes and Other Carriers. *Expert Opin. Drug Deliv.* **2013**, 10(11), 1583–1591.
198. Zhu, X.; Zhang, H.; Lin, Y.; Chen, P.; Min, J.; Wang, Z., et al. Mechanisms of Gambogic Acid-Induced Apoptosis in Non-Small Cell Lung Cancer Cells in Relation to Transferrin Receptors. *J. Chemother.* **2009**, 21(6), 666–672.
199. Sadava, D.; Phillips, T.; Lin, C.; Kane, S.E. Transferrin Overcomes Drug Resistance to Artemisinin in Human Small-Cell Lung Carcinoma Cells. *Cancer Lett.* **2002**, 179(2), 151–156.
200. Upadhyay, P.; Sarker, S.; Ghosh, A.; Gupta, P.; Das, S.; Ahir, M.; et al. Transferrin-Decorated Thymoquinone-Loaded PEG-PLGA Nanoparticles Exhibit Anticarcinogenic Effect in Non-Small Cell Lung Carcinoma via the Modulation of miR-34a and miR-16. *Biomater. Sci.* **2019**, 7(10), 4325–4344.
201. Fasano, M.; Sabatini, M.T.; Wieczorek, R.; Sidhu, G.; Goswami, S.; Jagirdar, J. CD44 and Its V6 Spliced Variant in Lung Tumors: A Role in Histogenesis? *Cancer* **1997**, 80(1), 34–41.
202. Du, L.; Wang, H.; He, L.; Zhang, J.; Ni, B.; Wang, X.; et al. CD44 Is of Functional Importance for Colorectal Cancer Stem Cells. *Clin. Cancer Res.* **2008**, 14(21), 6751–6760.
203. Ghosh, S.C.; Neslihan Alpay, S.; Klostergaard, J. CD44: A Validated Target for Improved Delivery of Cancer Therapeutics. *Expert Opin. Ther. Targets* **2012**, 16(7), 635–650.
204. Leung, E.L.; Fiscus, R.R.; Tung, J.W.; Tin, V.P.; Cheng, L.C.; Sihoe, A.D.; et al. Non-Small Cell Lung Cancer Cells Expressing CD44 Are Enriched for Stem Cell-like Properties. *PLOS One* **2010**, 5(11), e14062.
205. Mattheolabakis, G.; Milane, L.; Singh, A.; Amiji, M.M. Hyaluronic Acid Targeting of CD44 for Cancer Therapy: From Receptor Biology to Nanomedicine. *J. Drug Target.* **2015**, 23(7–8), 605–618.
206. Marhaba, R.; Klingbeil, P.; Nuebel, T.; Nazarenko, I.; Buechler, M.W.; Zoeller, M. CD44 and EpCAM: Cancer-Initiating Cell Markers. *Curr. Mol. Med.* **2008**, 8(8), 784–804.
207. Bourguignon, L.Y.; Gilad, E.; Peyrollier, K. Heregulin-Mediated ErbB2-ERK Signaling Activates Hyaluronan Synthases Leading to CD44-Dependent Ovarian Tumor Cell Growth and Migration. *J. Biol. Chem.* **2007**, 282(27), 19426–19441.
208. Madjd, Z.; Mehrjerdi, A.Z.; Sharifi, A.M.; Molanaei, S.; Shahzadi, S.Z.; Asadi-Lari, M. CD44+ Cancer Cells Express Higher Levels of the Antiapoptotic Protein Bcl-2 in Breast Tumours. *Cancer Immun.* **2009**, 9, 4.
209. Gutova, M.; Najbauer, J.; Gevorgyan, A.; Metz, M.Z.; Weng, Y.; Shih, C.C., et al. Identification of uPAR-Positive Chemoresistant Cells in Small Cell Lung Cancer. *PLOS One* **2007**, 2(2), e243.
210. Jeannot, V.; Mazzaferro, S.; Lavaud, J.; Vanwonterghem, L.; Henry, M.; Arboléas, M.; et al. Targeting CD44 Receptor-Positive Lung Tumors Using Polysaccharide-Based Nanocarriers: Influence of Nanoparticle Size and Administration Route. *Nanomedicine* **2016**, 12(4), 921–932.
211. Amreddy, N.; Babu, A.; Muralidharan, R.; Munshi, A.; Ramesh, R. Polymeric Nanoparticle-Mediated Gene Delivery for Lung Cancer Treatment. *Top. Curr. Chem. (Cham)* **2017**, 375(2), 35.
212. Galano, A.; Martínez, A. Capsaicin, a Tasty Free Radical Scavenger: Mechanism of Action and Kinetics. *J. Phys. Chem. B* **2012**, 116(3), 1200–1208.
213. Parashar, P.; Tripathi, C.B.; Arya, M.; Kanoujia, J.; Singh, M.; Yadav, A.; et al. A Facile Approach for Fabricating CD44-Targeted Delivery of Hyaluronic Acid-Functionalized PCL Nanoparticles in Urethane-Induced Lung Cancer: Bcl-2, MMP-9, Caspase-9, and BAX as Potential Markers. *Drug Deliv. Transl Res.* **2019**, 9(1), 37–52.
214. Ghosh, S.C.; Neslihan Alpay, S.; Klostergaard, J. CD44: A Validated Target for Improved Delivery of Cancer Therapeutics. *Expert Opin. Ther. Targets* **2012**, 16(7), 635–650.
215. Orian-Rousseau, V.; Ponta, H. Perspectives of CD44 Targeting Therapies. *Arch. Toxicol.* **2015**, 89(1), 3–14.
216. Yan, Y.; Zuo, X.; Wei, D. Concise Review: Emerging Role of CD44 in Cancer Stem Cells: A Promising Biomarker and Therapeutic Target. *Stem Cell Transl. Med.* **2015**, 4(9), 1033–1043.
217. Zhang, M.; Gao, S.; Yang, D.; Fang, Y.; Lin, X.; Jin, X.; Liu, Y.; Liu, X.; Su, K.; Shi, K. Influencing Factors and Strategies of Enhancing Nanoparticles into Tumors *In Vivo*. *Acta Pharm. Sin. B* **2021**, 11, 2265–2285.
218. Kim, S.W.; Khang, D. Multiple Cues on the Physiochemical, Mesenchymal, and Intracellular Trafficking Interactions with Nanocarriers to Maximize Tumor Target Efficiency. *Int. J. Nanomed.* **2015**, 10, 3989.

219. Shukla, R.; Handa, M.; Lokesh, S.B.; Ruwali, M.; Kohli, K.; Kesharwani, P. Conclusion and Future Prospective of Polymeric Nanoparticles for Cancer Therapy. *Polymeric Nanoparticles as a Promising Tool for Anti-cancer Therapeutics* **2019**, 389–408.
220. Choi, J.S.; Park, J.S. Surface Modification of Docetaxel Nanocrystals with HER2 Antibody to Enhance Cell Growth Inhibition in Breast Cancer Cells. *Colloids Surf. B* **2017**, 159, 139–150.
221. Zhan, H.; Jagtiani, T.; Liang, J.F. A New Targeted Delivery Approach by Functionalizing Drug Nanocrystals Through Polydopamine Coating. *Eur. J. Pharm. Biopharm.* **2017**, 114, 221–229.
222. Tung, Y.T.; Chen, H.L.; Yen, C.C.; Lee, P.Y.; Tsai, H.C.; Lin, M.F.; Chen, C.M. Bovine Lactoferrin Inhibits Lung Cancer Growth Through Suppression of Both Inflammation and Expression of Vascular Endothelial Growth Factor. *J. Dairy Sci.* **2013**, 96, 2095–2106.
223. Taratula, O.; Kuzmov, A.; Shah, M.; Garbuzenko, O.B.; Minko, T. Nanostructured Lipid Carriers as Multifunctional Nanomedicine Platform for Pulmonary Co-Delivery of Anticancer Drugs and siRNA. *J. Control. Release* **2013**, 171, 349–357.
224. Agrawal, S.; Dwivedi, M.; Ahmad, H.; Chadchan, S.B.; Arya, A.; Sikandar, R.; et al. CD44 Targeting Hyaluronic Acid Coated Lapatinib Nanocrystals Foster the Efficacy Against Triple-Negative Breast Cancer. *Nanomed.: Nanotechnol. Biol. Med.* **2018**, 14, 327–337.
225. Sohn, J.S.; Yoon, D.S.; Sohn, J.Y.; Park, J.S.; Choi, J.S. Development and Evaluation of Targeting Ligands Surface Modified Paclitaxel Nanocrystals. *Mater. Sci. Eng.* **2017**, 72, 228–237.
226. Huang, Z.G.; Lv, F.M.; Wang, J.; Cao, S.J.; Liu, Z.P.; Liu, Y.; Lu, W.Y. RGD-Modified PEGylated Paclitaxel Nanocrystals with Enhanced Stability and Tumor-Targeting Capability. *Int. J. Pharm.* **2019**, 556, 217–225.
227. Georgiadis, M.O.; Karoutzou, O.; Foscolos, A.S.; Papanastasiou, I. Sigma Receptor (σR) Ligands with Antiproliferative and Anticancer Activity. *Molecules* **2017**, 22, 1408.
228. van Waarde, A.; Rybczynska, A.A.; Ramakrishnan, N.K.; Ishiwata, K.; Elsinga, P.H.; Dierckx, R.A. Potential Applications for Sigma Receptor Ligands in Cancer Diagnosis and Therapy. *Biochim. Biophys. Acta Biophys. Incl. Photosynth.* **2015**, 1848, 2703–2714.
229. Aydar, E.; Palmer, C.P.; Djamgoz, M.B. Sigma Receptors and Cancer: Possible Involvement of Ion Channels. *Cancer Res.* **2004**, 64, 5029–5035.
230. Sehgal, K.; Patell, R.; Rangachari, D.; Costa, D.B. Targeting ROS1 Rearrangements in Non-Small Cell Lung Cancer with Crizotinib and Other Kinase Inhibitors. *Transl. Cancer Res.* **2018**, 7, S779.
231. Mehta, P.; Kadam, S.; Pawar, A.; Bothiraja, C. Dendrimers for Pulmonary Delivery: Current Perspectives and Future Challenges. *New J. Chem.* **2019**, 43, 8396–8409.
232. Akamine, T.; Toyokawa, G.; Tagawa, T.; Yamazaki, K.; Seto, T.; Takeo, S.; et al. Lorlatinib for the Treatment of Patients with Non-Small Cell Lung Cancer. *Drugs Today (Barc)* **2019**, 55, 107–116.
233. Katayama, R.; Gong, B.O.; Togashi, N.; Miyamoto, M.; Kiga, M.; Iwasaki, S.; et al. The New-Generation Selective ROS1/NTRK Inhibitor DS-6051b Overcomes Crizotinib Resistant ROS1-G2032R Mutation in Preclinical Models. *Nat. Commun.* **2019**, 10, 3604.
234. Facchinetti, F.; Friboulet, L. Profile of Entrectinib and Its Potential in the Treatment of ROS1-Positive NSCLC: Evidence to Date. *Lung Cancer Targets Ther.* **2019**, 10, 87.
235. Albanese, A.; Tang, P.S.; Chan, W.C. The Effect of Nanoparticle Size, Shape, and Surface Chemistry on Biological Systems. *Annu. Rev. Biomed. Eng.* **2012**, 14, 1–16.
236. Attia, M.F.; Anton, N.; Wallyn, J.; Omran, Z.; Vandamme, T.F. An Overview of Active and Passive Targeting Strategies to Improve the Nanocarriers Efficiency to Tumour Sites. *J. Pharm. Pharmacol.* **2019**, 71, 1185–1198.
237. Huynh, N.T.; Roger, E.; Lautram, N.; Benoit, J.P.; Passirani, C. The Rise and Rise of Stealth Nanocarriers for Cancer Therapy: Passive Versus Active Targeting. *Nanomedicine* **2010**, 5, 1415–1433.
238. Lombardo, D.; Kiselev, M.A.; Caccamo, M.T. Smart Nanoparticles for Drug Delivery Application: Development of Versatile Nanocarrier Platforms in Biotechnology and Nanomedicine. *J. Nanomater.* **2019**, 2019, 1–26.
239. Lu, X.Y.; Wu, D.C.; Li, Z.J.; Cheng, G.Q. Polymer Nanoparticles. In *Progress in Molecular Biology and Translational Science*; Villaverde, A., Ed.; Academic Press, **2011**; pp 299–323.
240. Casalini, T.; Rossi, F.; Castrovinci, A.; Perale, G. A Perspective on Polylactic Acid-Based Polymers Use for Nanoparticles Synthesis and Applications. *Front. Bioeng. Biotechnol.* **2019**, 7, 259.
241. Lu, Y.; Low, P.S. Folate-Mediated Delivery of Macromolecular Anticancer Therapeutic Agents. *Adv. Drug Delivery Rev.* **2002**, 54, 675–693.
242. Marasini, N.; Haque, S.; Kaminskas, L.M. Polymer-Drug Conjugates as Inhalable Drug Delivery Systems: A Review. *Curr. Opin. Colloid Interface Sci.* **2017**, 31, 18–29.

243. Harush-Frenkel, O.; Debotton, N.; Benita, S.; Altschuler, Y. Targeting of Nanoparticles to the Clathrin-Mediated Endocytic Pathway. *Biochem. Biophys. Res. Commun.* **2007**, 353, 26–32.
244. Vasir, J.K.; Labhasetwar, V. Quantification of the Force of Nanoparticle-Cell Membrane Interactions and Its Influence on Intracellular Trafficking of Nanoparticles. *Biomaterials* **2008**, 29, 4244–4252.
245. Anselmo, A.C.; Mitragotri, S. Nanoparticles in the Clinic. *Bioeng. Transl. Med.* **2016**, 1, 10–29.
246. Kolb, H.C.; Finn, M.G.; Sharpless, K.B. Click Chemistry: Diverse Chemical Function from a Few Good Reactions. *Angew. Chem. Int. Ed.* **2001**, 40, 2004–2021.
247. Lallana, E.; Sousa-Herves, A.; Fernandez-Trillo, P.; Riguera, R.; Fernandez-Megia, E. Click Chemistry for Drug Delivery Nanosystems. *Pharm. Res.* **2011**, 29, 1–34.
248. Shoba, G.; Joy, D.; Joseph, T.; Majeed, M.; Rajendran, R.; Srinivas, P.S. Influence of Piperine on the Pharmacokinetics of Curcumin in Animals and Human Volunteers. *Planta Med.* **1998**, 64, 353–356.
249. Cheng, Z.; Al Zaki, A.; Hui, J.Z.; Muzykantov, V.R.; Tsourkas, A. Multifunctional Nanoparticles: Cost Versus Benefit of Adding Targeting and Imaging Capabilities. *Science* **2012**, 338, 903–910. [CrossRef]
250. Fleck, L.M. The Costs of Caring: Who Pays? Who Profits? Who Panders? *Hast. Cent. Rep.* **2006**, 36, 13–17.
251. Desai, N. Challenges in Development of Nanoparticle-Based Therapeutics. *AAPS J.* **2012**, 14, 282–295.
252. Alexiou, C.; Schmid, R.J.; Jurgons, R.; Kremer, M.; Wanner, G.; Bergemann, C.; et al. Targeting Cancer Cells: Magnetic Nanoparticles as Drug Carriers. *Eur. Biophys. J.* **2006**, 35, 446–450.
253. Chatterjee, D.K.; Fong, L.S.; Zhang, Y. Nanoparticles in Photodynamic Therapy: An Emerging Paradigm. *Adv. Drug Deliv. Rev.* **2008**, 60, 1627–1637.
254. Luo, M.; Li, H.; Han, D.; Yang, K.; Kang, L. Inhibition of Autophagy Enhances Apoptosis Induced by Ce6-Photodynamic Therapy in Human Colon Cancer Cells. *Photodiagn. Photodyn. Ther.* **2021**, 36, 102605.
255. Deng, X.; Song, Q.; Zhang, Y.; Liu, W.; Hu, H.; Zhang, Y. Tumour Microenvironment-Responsive Nanoplatform Based on Biodegradable Liposome-Coated Hollow MnO2 for Synergistically Enhanced Chemotherapy and Photodynamic Therapy. *J. Drug Target.* **2021**, 14, 1–23.
256. Crous, A.; Abrahamse, H. Aluminium (III) Phthalocyanine Chloride Tetrasulphonate Is an Effective Photosensitizer for the Eradication of Lung Cancer Stem Cells. *R. Soc. Open Sci.* **2021**, 8, 210148.
257. Yu, T.-T.; Sang, X.-Y.; Han, N.; Peng, X.-C.; Li, Q.-R.; Xu, X.; et al. Macrophages Mediated Delivery of Chlorin e6 and Treatment of Lung Cancer by Photodynamic Reprogramming. *Int. Immunopharmacol.* **2021**, 100, 108164.
258. Shi, C.; Huang, H.; Zhou, X.; Zhang, Z.; Ma, H.; Yao, Q.; et al. Reversing Multidrug Resistance by Inducing Mitochondrial Dysfunction for Enhanced Chemo-Photodynamic Therapy in Tumor. *ACS Appl. Mater. Interfaces* **2021**, 13, 45259–45268.
259. Reshma, P.; Unnikrishnan, B.; Preethi, G.; Syama, H.; Archana, M.; Remya, K.; et al. Overcoming Drug-Resistance in Lung Cancer Cells by Paclitaxel Loaded Galactoxyloglucan Nanoparticles. *Int. J. Biol. Macromol.* **2019**, 136, 266–274.

7 Carbon Nanotubes for the Management of Lung Cancer

Dr. Sarath Chandran C.[1], Rahul. K.[1], Abhijith. K. Shaji[1], Krishnameera Sajayan[2], Gopika Nath C.[1], Hiba K.[1], Zakir Hussain K P[3], and Dr. Rajesh Sreedharan Nair[4]*

[1] College of Pharmaceutical Sciences, Government Medical College, Kannur, India

[2] Department of Medical Devices, NIPER, Kolkata, India

[3] Department of Pharmacy Practice, JSS College of Pharmacy Ooty, Tamil Nadu, India

[4] School of Pharmacy, University of Nottingham, Malaysia Jalan Broga, 43500 Semenyih Selangor, Malaysia

*Corresponding author: Dr. Sarath Chandran C., scschenoy@gmail.com.

7.1 INTRODUCTION

Carbon nanotubes (CNTs) are a type of nanomaterial consisting of carbon atoms arranged in a two-dimensional hexagonal lattice structure. They are curved in a single direction and assembled to form hollow cylindrical structures (1). CNTs are carbon allotropes in the dimensional range of fullerenes (0-dimensional) and graphene (2-dimensional). They belong to the fullerene family, which also includes buckyballs. Unlike spherical buckyballs, CNTs are cylindrical. Some CNTs have open ends, whereas others feature closed ends with complete fullerene caps. The quasi-1-dimensional nanotube form of carbon was first reported by Iijima et al. (1991), who discovered multiwalled carbon nanotubes (MWCNTs) in carbon soot generated using the arc discharge method (2). Although CNTs were produced and observed under various conditions before 1991, the report by Iijima et al. was considered particularly significant for bringing CNTs to the attention of the broader scientific community (1). In 1993, Ichihashi reported the synthesis of single-walled carbon nanotubes (SWCNTs) with a diameter of 1 nm. Radushkevich and Lukyanovich are recognized for discovering that carbon filaments can be hollow and possess nanoscale diameters (3). In 1979, John Abrahamson described CNTs as carbon fibers formed on a carbon anode during arc discharge and outlined the properties of these fibers. Considering their mechanical and electrochemical properties, CNTs were initially used as additives in various structural materials, including electronics, optics, plastics, and other nanotechnology applications (4). Since the early 21st century, CNTs have shown exceptional potential as drug delivery systems capable of directly penetrating cells while maintaining drug integrity and preventing metabolism during transport in the body (5).

CNTs are primarily categorized as single-walled or multiwalled, with other less common variants including fullerites, torus, and nanoknots (6). SWCNTs are formed by rolling a single graphene sheet into a seamless cylindrical structure with a diameter of approximately 1 nm (5). They comprise two distinct regions, the sidewall and the end cap, each possessing unique physical and chemical properties. Owing to their distinctive electrical properties, SWCNTs demonstrate greater efficacy than MWCNTs (7). MWCNTs consist of multiple concentric SWCNTs with several layers of graphite rolled into a tubular shape with varying diameters. The interlayer spacing in the MWCNTs is approximately 3.3 Å, which is comparable to the distance between the graphene layers

DOI: 10.1201/9781003595830-7

in graphite. A notable variant, double-walled carbon nanotubes (DWCNTs), shares morphology and properties with SWCNTs, rendering them particularly significant.

The primary applications of CNTs in medicine include drug delivery, biomolecule and gene delivery to cells or organs, tissue regeneration, biosensor diagnostics, and analysis. Functionalized CNTs facilitate safe and efficient drug delivery by traversing cytoplasmic and nuclear membranes without toxicity, rendering them more efficacious than conventional drug formulations. CNTs present potential solutions for addressing the resistance of infectious agents to standard drugs and vaccine inefficacies. Cationic CNT plasmid DNA complexes enhance gene therapy compared with naked DNA. Furthermore, functionalized SWCNTs and MWCNTs are effective delivery systems for treating neurodegenerative diseases and brain tumors. CNTs represent an emerging field in biosensing nanotechnology, with significant potential for therapeutic monitoring and diagnostics. CNTs also demonstrate promise in targeted cancer therapy, where their unique properties enable their selective accumulation in tumor tissues and the controlled release of therapeutic agents. Moreover, the high surface-area-to-volume ratio of CNTs facilitates the loading of multiple drug molecules, thereby enhancing the efficacy of combination therapies. The versatility of CNTs extends their potential applications in theranostics by combining diagnostic imaging capabilities with therapeutic delivery for personalized medical approaches (5).

7.2 METHODS OF PREPARATION OF CARBON NANOTUBES

Several methods have been adopted to develop CNTs, including chemical vapor deposition, arc discharge, laser ablation, and thermal synthesis (Figure 7.1).

Among these, chemical vapor deposition is one of the most widely used and economical methods. This process involves the thermal dehydrogenation of gaseous carbon sources, such as methane, ethane, ethylene, acetylene, and xylene, in the presence of a transition-metal catalyst (8).

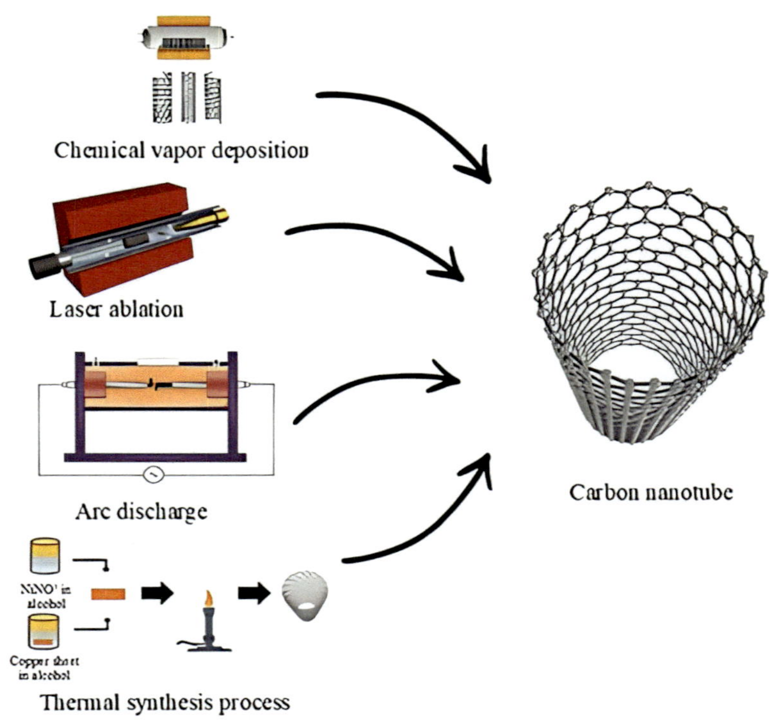

FIGURE 7.1 Methods of preparation of carbon nanotubes

The growth of CNTs depends on the volume and concentration of the gas-phase intermediates used, temperature and pressure conditions provided, and reaction duration (2). The arc discharge method is another commonly employed method that converts carbon atoms into plasma at high temperatures (approximately 3000°C). An electric arc was generated between the two closely positioned graphite rods in an inert atmosphere under a pressure of 30–700 mbar. CNTs can be synthesized through two methods: evaporation of graphite alone or graphite evaporation combined with a metal catalyst. Flexible CNTs with minimal defects can be produced using this method. Gasification using oxygen or carbon dioxide is often employed to purify CNTs. Plasma generated from argon, helium, nitrogen, carbon, cobalt, and nickel is commonly used in synthesis (8). Similar to arc discharge, the laser ablation method operates at high temperatures in the presence of background gas (7). A continuous laser beam is employed to vaporize the graphite target to form CNT. Gasification was applied to eliminate carbonaceous impurities. This method is known to produce high-purity SWCNTs with uniform diameters (5–20 nm) and lengths ranging from 10 to 100 μm (2). In contrast to arc discharge and laser ablation, the thermal synthesis method operates at lower temperatures not exceeding 1200°C. This technique utilizes iron, nickel, and cobalt as catalysts with molybdenum and ruthenium promoters to enhance the CNT yield. In addition to these methods, flame synthesis, high-pressure carbon monoxide, plasma-enhanced chemical vapor deposition, ball milling, and annealing have been used to develop CNTs (7). However, chemical vapor deposition remains the most widely preferred method for biomedical applications, including drug delivery, owing to its adaptability, scalability, and cost efficiency.

7.3 CARBON NANOTUBES IN LUNG CANCER THERAPY

Lung cancer therapy has traditionally relied on chemotherapy, radiation, and surgical interventions, each of which is associated with significant limitations, such as non-specific targeting, systemic toxicity, and drug resistance. The introduction of CNTs into biomedical applications has presented unprecedented opportunities for addressing diseases such as cancer, including lung cancer, which remains a leading cause of cancer-related mortality worldwide. By utilizing the structural and functional attributes of CNTs, researchers have endeavored to address these challenges by developing innovative therapeutic platforms.

7.4 DRUG DELIVERY SYSTEMS IN LUNG CANCER TREATMENT

The success of CNTs in lung cancer therapy is because of their remarkable capability to function as targeted drug delivery systems. Owing to their unique structural and physicochemical properties, CNTs have emerged as effective drug delivery systems for precisely delivering chemotherapeutic agents (9–14). Their large surface area enables significantly high drug loading, either by non-covalent interactions, such as π–π stacking and hydrophobic forces, or by encapsulation within their hollow cores, protecting therapeutic agents from premature degradation. Selective delivery to tumor cells with minimal off-target effects is facilitated by the ability to functionalize their surfaces with targeting ligands such as peptides, antibodies, or small molecules that bind cancer-specific signals. Additionally, modification of CNTs with pH-sensitive or stimuli-responsive coatings can release medication in an acidic tumor microenvironment or respond to specific stimuli such as heat, light, or enzymatic activity (9–14). This controlled release ensures sustained therapeutic effects and reduces systemic toxicity. Through the enhanced permeability and retention (EPR) effect, CNTs preferentially accumulate in tumor tissues and demonstrate exceptional cellular uptake efficiency. They can penetrate the cell membranes via direct needle-like insertions or receptor-mediated endocytosis. In addition to drug delivery, CNTs can be coupled with imaging agents for real-time diagnostics and monitoring, or utilized with photodynamic and photothermal therapies to enhance their therapeutic potential (9–14). Despite challenges such as potential toxicity and biodegradability concerns, ongoing advancements in functionalization and coating strategies are addressing these limitations and positioning CNTs as transformative tools for cancer treatment (9–14).

Studies have expanded the scope of CNT research by exploring the functionalization of MWCNTs with diverse bioactive molecules. These molecules include chemotherapeutic drugs, surfactants, diagnostic agents, antibodies, and targeting ligands designed to engage overexpressed receptors on cancer cells. These conjugation strategies promise to improve the selectivity and effectiveness of lung cancer therapies and enable multimodal diagnostic and therapeutic applications, underscoring the versatility of CNT-based oncology platforms (15–20). Sheikhpour et al. critically reviewed applications of CNTs in the diagnosis and treatment of lung cancer, emphasizing that their side effects manifested during therapy, such as inflammation, fibrosis, and carcinogenesis, can be partially mitigated by functionalization of CNTs to appropriate dimensions, including increased length, width, and curvature (14). Recent advancements have validated the efficacy of CNT as an effective drug delivery system for lung cancer treatment. Based on these findings, researchers have explored novel functionalization strategies to enhance the biocompatibility and targeting capabilities of CNTs. These efforts have led to the development of "smart" CNT-based delivery systems that can respond to specific stimuli within the tumor microenvironment, such as pH changes or enzymatic activity. Additionally, integrating CNTs with other nanomaterials, such as liposomes or polymeric nanoparticles, has shown promise for creating hybrid systems that combine the strengths of multiple nanocarriers to improve therapeutic outcomes in lung cancer treatment. For instance, the conjugation of MWCNTs with transferrin has been investigated and has demonstrated improved targeting of docetaxel to adenocarcinoma human alveolar basal epithelial A549 lung cancer cells by engaging transferrin receptors overexpressed on the cancer cells (21). This CNT formulation exhibited superior therapeutic efficiency during *in vitro* investigation, with increased cytotoxicity and enhanced safety profiles *in vivo* compared to commercially available docetaxel formulations (Docetaxel™). Furthermore, combining multiple strategies has shown additional promise, as evidenced by Li et al., who developed a pH-responsive SWCNT system conjugated with arginyl glycyl aspartic acid (RGD) peptide (22), which selectively binds to surface receptors on A549 cells, facilitating the transport of docetaxel-loaded SWCNTs. This system facilitated a higher release of docetaxel in the acidic tumor microenvironment, simulating *in vivo* conditions. Animal models have revealed that this approach enhances therapeutic outcomes and minimizes systemic toxicity, highlighting the potential of pH-responsive CNT platforms for lung cancer treatment.

The persistent challenges posed by multidrug resistance (MDR) and incomplete eradication of deeply embedded lung tumors remain significant obstacles in lung cancer therapy, particularly for small-cell lung cancer (SCLC). Park et al. demonstrated that CNT-conjugated doxorubicin nanodrugs enhance intracellular drug retention and selectively accumulate within MDR tumor cells, such as the H69AR cell line. By circumventing efflux pump mechanisms, these nano drugs prolonged intracellular drug presence by 24 hours, disrupted mitochondrial function, inhibited ATP production, and induced significant cytotoxicity. The apoptosis rates were four times higher for CNT-conjugated doxorubicin than for free drugs. *In vivo* studies have revealed a marked morphological collapse of tumor tissue, effectively eliminating MDR tumors and demonstrating the efficacy of CNT-based nanodrugs (23). The enhanced intracellular retention and selective accumulation of CNT-conjugated doxorubicin in MDR tumor cells suggest a promising approach for improving treatment efficacy. The observed disruption of mitochondrial function and increased apoptosis rates indicate that the CNT formulation may offer a more potent therapeutic option for resistant lung cancer cells. These findings highlight the potential of CNT-based drug delivery systems to overcome multidrug resistance in small-cell lung cancer. Several studies have explored the impact of CNT-based nanodrugs on the tumor microenvironment, revealing significant alterations in extracellular matrix composition and reduced angiogenesis. These changes contribute to improved drug penetration and distribution within the tumor tissue, enhancing overall therapeutic efficacy. Researchers observed a notable decrease in the expression of efflux pumps responsible for drug resistance, suggesting that CNT-conjugated drugs may help circumvent the traditional mechanisms of MDR in lung cancer cells.

CNT conjugation is a promising avenue for lung cancer therapy, as demonstrated by Kamazani et al. Their investigation of the conjugation of bromocriptine to MWCNTs revealed a significant induction of apoptosis in lung cancer cells at lower doses compared to the drug alone. The study demonstrated that a nanofluid containing CNT-conjugated bromocriptine exhibited significantly higher antitumor activity in the human lung cancer cell lines A549 and QU-DB than in the normal control cell line MRC5. Their findings demonstrated that treatment with CNT-conjugated bromocriptine nanofluid significantly increased cell death in both A549 and QU-DB lung cancer cell lines within 48 hours. Total cell death reached 90.67% in A549 cells (with 89.85% attributed to apoptosis) and 69% in QU-DB cells (63.7% apoptosis), compared to only 14.5% total cell death (11.77% apoptosis) observed in the normal MRC5 cell line ($p < 0.0001$). The observed differential cytotoxicity suggests that CNT-based formulations may offer selective targeting of cancer cells while minimizing damage to healthy tissues. This approach could improve the treatment efficacy and reduce the side effects of lung cancer therapy. These results highlight the potential of CNT-based formulations to enhance therapeutic outcomes by achieving greater cancer cell lethality at a reduced dosage (10).

Functionalization of polyethylene glycol (PEG) with CNTs has been investigated to enhance drug delivery in lung cancer. PEGylation of CNTs can improve their biocompatibility and circulation time in the bloodstream, potentially increasing their drug delivery efficacy in lung cancer cells. This modification may also address challenges associated with CNT-based drug delivery systems, such as their aggregation and rapid clearance from the body. PEGylated CNTs conjugated with ABT737 facilitated mitochondrial apoptosis and reactive oxygen species (ROS) generation in non–small cell lung cancer (NSCLC) cells. These conjugates primarily enter cells via clathrin-mediated endocytosis, leading to early endosome formation in A549 lung cancer cell lines. Upon release, they disrupt the mitochondrial membrane potential and trigger Bcl-2-mediated apoptosis. Compared to unconjugated ABT737, these conjugates elicited significantly higher levels of mitochondria-mediated cell death, demonstrating the enhanced efficacy of PEGylated CNT-based drug delivery systems (24). Studies have reported that the combination of PEGylated CNTs with targeted ligands or specific antibodies can further enhance the selective delivery of therapeutic agents to lung cancer cells, potentially improving treatment outcomes while minimizing off-target effects (24–32).

Various studies indicated that cisplatin-loaded MWCNTs may be a promising approach for overcoming drug resistance in NSCLC treatment (33–38). The utilization of advanced characterization techniques provides valuable insights into the structural and physicochemical properties of nanocarriers, which can inform future optimization. Comprehensive *in vitro* evaluations across multiple cell lines and assays offer a robust foundation for understanding how cisplatin-loaded MWCNTs exert their anticancer effects and potentially reverse drug resistance. Qi et al. synthesized cisplatin-loaded MWCNTs, characterized by advanced X-ray diffraction and scanning electron microscopy (SEM). The anticancer efficacy of cisplatin-loaded MWCNTs was evaluated using cisplatin-sensitive (A549) and cisplatin-resistant (A549/DDP) NSCLC cell lines through *in vitro* assays, including the MTT assay for cell viability, transwell assays for invasion, and Western blotting for apoptosis and epithelial-mesenchymal transition (EMT)-related markers. These nanotubes exhibited enhanced intracellular accumulation of cisplatin, resulting in increased apoptosis of cisplatin-resistant A549/DDP cells. Western blot analyses revealed upregulation of pro-apoptotic proteins (e.g., Bax, Bim, Bid, Caspase-3, and Caspase-9) and suppression of EMT markers such as vimentin and N-cadherin. *In vivo* studies have confirmed the improved therapeutic efficacy and tolerability of cisplatin-loaded MWCNTs, with reduced tumor burden and minimal systemic toxicity. The observed upregulation of pro-apoptotic proteins and suppression of EMT markers suggest that CNTs enhance drug delivery and modulate cellular signaling pathways to overcome cisplatin resistance. These findings indicate a potential dual mechanism of action wherein nanotubes function as drug carriers and therapeutic agents (33). Investigating the molecular interactions between CNTs and cellular components could provide valuable insights for the development of more effective targeted therapies for cisplatin-resistant cancers (39–44).

Biopolymers have played a pivotal role in the pharmaceutical field in recent years, extending their applications to various areas such as targeted drug delivery, solubility enhancement, controlled release, and numerous other applications. The incorporation of biopolymers, such as chitosan, has significantly improved the solubility and functionalization of CNTs (11, 22, 45). Chitosan-modified SWCNTs loaded with etoposide demonstrated selective targeting of A549 cells through epidermal growth factor receptor (EGFR)-dependent and EGFR-independent pathways. This system exhibited pH-sensitive drug release and enhanced cytotoxicity against cancer cells while minimizing off-target effects. Chitosan modification also facilitates higher drug-loading capacity and efficient cellular uptake, further optimizing therapeutic outcomes (11, 45). The versatility of biopolymers in drug delivery systems extends beyond chitosan and other natural polymers, such as alginate, gelatin, and cellulose derivatives, and shows promise for improving CNT-based formulations. These biopolymers can be tailored to achieve specific release profiles, enhance biocompatibility, and target particular cellular receptors, offering a wide range of possibilities for personalized medicine. Furthermore, the combination of biopolymers with CNTs opens new avenues for theranostic applications, where imaging capabilities can be integrated with therapeutic functions to enable the real-time monitoring of drug delivery and treatment efficacy (33, 34, 46–48).

In a recent study, naringenin-functionalized MWCNTs were developed as targeted drug delivery systems for lung cancer treatment. Naringenin is a flavonoid found in citrus fruits with antioxidant, anti-inflammatory, and anticancer properties. Characterization using transmission electron microscopy (TEM) confirmed successful functionalization, with an increased average diameter of 9.6 ± 1.4 nm for the composite. Fourier transform infrared spectroscopy (FTIR) and X-ray photoelectron spectroscopy confirmed the presence of non-covalent interactions, including hydrogen bonding and π–π stacking, between naringenin and MWCNTs. The system exhibited pH-sensitive drug release, with 83.5% release at pH 7.4 and 60.8% at pH 5.5 after 9 hours ($p < 0.05$), suggesting its suitability for the tumor microenvironment. Cytotoxicity studies revealed that functionalized nanotubes (CC_{50}:82.6 ± 7.3 μg/mL) exhibited enhanced selectivity toward A549 lung cancer cells compared to free naringenin (CC_{50}:129.3 ± 6.0 μg/mL) while showing reduced toxicity to non-malignant cells (hFB) at concentrations below 125 μg/mL. This approach demonstrates the significant potential of functionalized MWCNTs for improving therapeutic outcomes in lung cancer with reduced off-target effects, highlighting their capacity for pH-sensitive release, improved anticancer activity, and reduced systemic toxicity. Such innovations provide a robust foundation for advancing nanomedicine in oncology, particularly for enhancing the therapeutic index of existing anticancer agents (45).

The transformative potential of CNTs as effective drug delivery systems for lung cancer treatment has been investigated using various drug molecules and polymers. By addressing key challenges such as multidrug resistance, low drug solubility, and unintended effects, the ability to modify CNTs with targeting ligands, anticancer drugs, and diagnostic molecules has facilitated the development of highly targeted, efficient, and adaptable approaches for combating both NSCLC and SCLC. By incorporating pH-sensitive and receptor-targeting methodologies, CNTs have demonstrated significant potential for advancing lung cancer therapy, providing a robust foundation for future clinical applications.

7.5 PHOTOTHERMAL AND PHOTODYNAMIC THERAPY IN LUNG CANCER

Building on advancements in utilizing CNTs for drug delivery, researchers are increasingly recognizing their potential in phototherapy applications, particularly in photothermal therapy (PTT) and photodynamic therapy (PDT) (49–51). These therapeutic modalities exploit the distinctive optical properties of CNTs to generate localized heat or ROS upon light exposure. The applications of CNTs in multifaceted cancer treatments emphasize their capacity to precisely target cancer cells while minimizing damage to the surrounding healthy tissues (49, 52–54). Although significant progress has been made in the treatment of various cancers, including triple-negative breast cancer, colorectal cancer (50, 55, 56), and melanoma, research on CNT-based phototherapy for lung cancer is still in its nascent stages. Nevertheless, preliminary studies have shown promising results (50, 55–59).

Zhang et al. investigated the potential of multifunctional Fe_3O_4-carbon quantum dot (CQD)-coated SWCNTs for combined PTT, PDT, and chemotherapy under 808 nm laser irradiation. Utilizing a PEG-modified platform, they synthesized SWCNTs-PEG-Fe_3O_4-CQDs of Doxorubicin conjugated with an sgc8c aptamer for targeted therapy. The results demonstrated a photothermal conversion efficiency of 38.9%, accompanied by significant ROS generation, as evidenced by the degradation of methylene blue dye. The nanocomposites exhibited rapid pH-sensitive doxorubicin release upon laser irradiation, achieving targeted cancer cell ablation. This approach has demonstrated potential for the treatment of lung and other cancers that require precise drug delivery (60). Similarly, Arya et al. examined a novel strategy combining SWCNTs, graphene oxide, and paclitaxel for lung cancer therapy. Their study demonstrated enhanced ROS-mediated cytotoxicity in the NSCLC cell lines A549 and NCI-H460, with significantly increased cell death observed in combination treatments compared to paclitaxel alone. The synergistic effect was ROS dependent, as confirmed by its abrogation with *N*-acetylcysteine. CNTs enhance mitogen-associated protein kinase activation, highlighting their potential as co-therapeutic agents with Paclitaxel (15). The PTT action of CNTs on tumor metastasis was investigated by Wang et al. through the development of manganese oxide-coated MWCNTs for dual-modality mapping and PTT of metastatic lymph nodes in lung cancer. Using A549 cell lines and nude mouse models, they demonstrated low toxicity, efficient photothermal conversion, and significant tumor ablation without damaging the adjacent tissues. *In vivo* studies have elucidated their capability for lymphatic tracing and metastatic tumor eradication, indicating their potential for clinical application in lung cancer staging and therapy (61).

Gene therapy using CNTs has also been used in lung cancer therapy. Recently, Zhao et al. established a new foundation for future research. Zhao et al. combined PTT with gene therapy using peptide lipids, sucrose laurate-coated SWCNTs, and MWCNTs. These bifunctional drug delivery systems achieved photothermal conversion efficiencies of 59.3% (SWCNTs) and 57.8% (MWCNTs), which were approximately 15% higher than those previously reported. Irradiation at 1.0 W/cm² for 10 minutes resulted in the optimal thermal doses for cancer ablation. Additionally, CNTs delivered anti-luciferase siRNA to A549 cells, achieving 75–80% luciferase silencing. This study highlights the potential of CNT-based platforms for systemic delivery, synergistic PTT, and gene therapy (62). Another study by Zhang et al. provided an innovative approach for combining PTT and chemotherapy. The nanoconjugates achieved a photothermal conversion efficiency of 45.3%. These nanoconjugates enabled near-infrared (NIR)-triggered doxorubicin release in an acidic TME. *In vivo* studies using A549 cells demonstrated 100% tumor elimination without recurrence at a laser power density of 2 W/cm², illustrating the efficacy of integrating PTT and chemotherapy for a comprehensive cancer treatment (63).

Emerging evidence underscores the transformative potential of CNTs in PTT and PDT for lung cancer, despite limited research compared with other cancers. CNT-based platforms offer innovative multimodal therapies that integrate photothermal and photodynamic effects with conventional chemotherapy. Their capacity for selective cancer cell ablation, minimal collateral damage, and enhanced therapeutic efficacy positions CNTs as promising candidates for future clinical applications in lung cancer treatment. Further investigation is necessary to optimize CNT-based PTT and PDT for lung cancer, including determining the optimal dosages, elucidating long-term effects, and improving targeting mechanisms. Clinical trials are essential to validate the safety and efficacy of these nanoconjugates in humans.

7.6 IMAGING AND THERANOSTICS APPLICATIONS IN LUNG CANCER

CNTs exhibit considerable potential across various imaging modalities, including photoacoustic, fluorescence imaging, and Raman imaging. These approaches utilize the optical properties of CNTs to enhance the resolution and specificity of cancer imaging, providing a robust platform for detecting and managing diseases, such as lung cancer. In response to the increasing interest in utilizing CNTs for theranostic applications, Minati et al. demonstrated the potential of doxorubicin-loaded CNT/gold hybrids in A549 lung cancer cells. Doxorubicin was adsorbed onto oxidized CNTs via π–π stacking interactions with a high loading efficiency. These CNT–gold hybrids exhibited broad absorption bands in the red and NIR regions, which significantly enhanced their cytotoxicity in A549 cells. TEM confirmed the presence of gold nanostructures around the CNTs, and laser scanning confocal microscopy (LSCM) demonstrated efficient doxorubicin delivery and imaging capabilities (64, 65). Datir et al. developed hyaluronic acid-functionalized MWCNTs for dual imaging and targeted therapy. Owing to hyaluronic acid-mediated endocytosis, doxorubicin-loaded hyaluronic acid-MWCNTs exhibited 3.2-fold higher cytotoxicity than free doxorubicin in A549 cells. The NIR fluorescent dye Alexa-Fluor-647 (AF-647) and radiotracer Technetium-99m (99mTc) were used for optical and scintigraphy imaging, respectively. Organ distribution studies in an Ehlrich ascites tumor mouse model revealed that tumor-specific localization was significantly higher for 99mTc-MWCNT-hyaluronic acid-doxorubicin compared to free doxorubicin. Notably, the platform demonstrated enhanced tumor inhibition with no detectable cardiotoxicity, hepatotoxicity, or nephrotoxicity (17, 64).

Similarly, in another study, Das et al. engineered folic acid-conjugated, methotrexate-loaded MWCNTs decorated with fluorochromes (Alexa Fluor 488/647) and radionuclides (Technetium-99m, 99mTc), enabling multimodal imaging and targeted drug delivery. Optical fluorescence allowed real-time cellular tracking, whereas radiolabeling with 99mTc facilitated whole-body biodistribution studies via scintigraphy. Selective internalization into folate receptor-positive A549 and MCF-7 cells was achieved through folic acid-mediated endocytosis, demonstrating tumor-specific accumulation 19-fold and 8.6-fold higher than that of free methotrexate and non-targeted MWCNTs, respectively. Scintigraphic imaging of Ehrlich ascites tumor (EAT)-xenografted mice confirmed enhanced tumor accumulation, allowing real-time monitoring of drug localization and therapeutic response. Pharmacokinetic studies have confirmed the absence of hepatotoxicity, cardiotoxicity, and nephrotoxicity (16, 64). In a complementary study, Das et al. explored PEGylated estradiol-conjugated doxorubicin-loaded MWCNTs, leveraging estrogen receptor (ER)-mediated pathways for targeted nuclear delivery, resulting in 1.5-to 1.6-fold lower IC50 values in A549 and MCF-7 cells than in non-targeted systems. The imaging aspect of this system was particularly significant, as fluorescence-based intracellular tracking via confocal laser scanning microscopy (CLSM) confirmed the nuclear localization of MWCNTs within ER-positive A549 and MCF-7 cells. Rhodamine fluorescence labeling provides visual evidence of enhanced cellular uptake and intranuclear drug delivery, which is a crucial advancement for improving therapeutic efficacy. *In vivo* optical imaging and biodistribution studies further demonstrated higher tumor retention, with tumor accumulation 18 times higher than that of the saline control, underscoring the precision of this targeted approach. This targeted nuclear delivery mechanism, supported by robust imaging techniques, highlights the potential of CNT-based platforms for precision oncology, ensuring both effective drug delivery and the real-time monitoring of therapeutic outcomes (64, 66).

As research has progressed, CNTs have also established a foundation for biosensors and diagnostic platforms. One such application was discussed by Choudhary et al., who developed a CNT-based hybrid electrode for detecting human telomerase reverse transcriptase (hTERT) biomarkers, which enabled the early detection of lung cancer (67, 68). Li et al. introduced a CNT field-effect transistor biosensor for carcinoembryonic antigen detection. The sensor achieved ultra-high sensitivity (detection limit: 72 ag/mL) and a wide linear range (1 fg/mL to 1 ng/mL) in both phosphate-buffered saline and serum samples, demonstrating its utility in early lung cancer diagnosis (69). As previously discussed for PTT, the potential of CNTs for imaging and drug delivery has been well described in a study conducted by Zhang et al. They utilized magneto-fluorescent Fe_3O_4-carbon quantum dot–coated SWCNTs for tumor imaging. Using doxorubicin as a model drug, the SWCNT-PEG-Fe_3O_4-CQDs-doxorubicin-Apt construct enabled high-contrast magnetic resonance imaging (MRI) in tumor-bearing mice (60). In another study, MWCNT-conjugated CQDs containing gadolinium, doped with hetero-elements (nitrogen), exhibited red emission in confocal imaging, demonstrating selective uptake in A549 cells and robust imaging capabilities without nuclear infiltration, thus demonstrating the efficacy of CNTs for future imaging technologies (63). CNT-based imaging and theranostic platforms represent transformative advancements in lung cancer management. By integrating diagnostic and therapeutic functionalities, CNTs enable real-time tumor imaging, precise drug delivery, and treatment monitoring. Their versatility across imaging modalities and targeted therapy systems underscores their potential to address gaps in early detection and personalized treatment, paving the way for their integration into clinical practice.

7.7 BIOCOMPATIBILITY AND TOXICITY OF CARBON NANOTUBES

The toxicity of CNTs is closely linked to their structure, length, surface area, aggregation tendency, functional groups, and manufacturing methods. Adverse effects include oxidative stress, membrane damage, mitochondrial dysfunction, and DNA damage, all of which can lead to apoptosis or necrosis. However, these toxicological outcomes remain controversial, with studies reporting varied responses depending on dose, concentration, and cellular context. Advances in surface modifications and controlled delivery methods are crucial for balancing the therapeutic benefits of CNTs with their potential toxicity, particularly for respiratory applications, such as lung cancer therapy.

7.8 UNDERSTANDING TOXICITY OF CARBON NANOTUBES

CNTs present significant biocompatibility and toxicity challenges while offering promising applications in areas such as lung cancer therapy. These challenges arise owing to the inherent properties of CNTs, including their fibrous structure, high surface area, and chemical reactivity, which contribute to adverse effects at the pulmonary, systemic, and cellular levels. Among the different types of CNTs, MWCNTs have drawn particular attention because of their structural similarity to asbestos, raising questions regarding their long-term biopersistence and associated health risks. A thorough understanding of CNT toxicity is essential for optimizing their use in cancer therapy while minimizing their adverse effects (70).

Pulmonary toxicity remains one of the most significant challenges associated with CNTs, particularly when they are inhaled. Several studies have demonstrated that exposure to CNTs leads to oxidative stress, inflammation, and fibrotic lesions in lungs (71). Mercer et al. observed that inhaled MWCNTs in C57BL/6J mice resulted in alveolar macrophage infiltration, pulmonary tissue thickening, and granuloma formation over 56 days. The fibrous nature of CNTs complicates their clearance from lung tissue, leading to chronic inflammation and persistent fibrosis in a dose-dependent manner (70). Additionally, CNT exposure generates ROS, disrupts cellular homeostasis, and triggers inflammatory pathways. Elevated levels of pro-inflammatory cytokines, such as tumor necrosis factor-alpha (TNF-α) and interleukin-6 (IL-6), have been detected in the bronchoalveolar lavage fluid (BALF) of CNT-exposed animals, reinforcing the notion that CNT-induced inflammation is a critical factor in pulmonary toxicity (71, 72).

Systemic toxicity is another significant concern, particularly when CNTs enter the circulation either through intravenous administration or through secondary systemic distribution following inhalation. Research has shown that MWCNTs accumulate primarily in the liver, spleen, and kidneys following intravenous injection. Deng et al. reported that 14C-labeled MWCNTs exhibited partial clearance for over 90 days; however, some residual particles remained, highlighting their persistence in biological systems. In contrast, Tang et al. demonstrated that functionalized MWCNTs induced minimal changes in blood parameters and histopathology, underscoring the role of surface modifications in mitigating systemic adverse effects. Although systemic toxicity may not be as immediate as pulmonary effects, long-term accumulation in vital organs raises concerns regarding the delayed toxicological consequences (73).

At the cellular level, CNTs interact with various cells, including epithelial cells, macrophages, and fibroblasts, often resulting in oxidative stress and genotoxic effects. Cellular uptake of CNTs occurs mainly through endocytosis, with subsequent localization to lysosomes and mitochondria. Studies conducted on BEAS-2B human bronchial epithelial cells revealed dose-dependent suppression of cell viability following exposure to MWCNTs. TEM imaging confirmed CNT accumulation in lysosomes, which triggered mitochondrial dysfunction and DNA damage. Hence, functionalization plays a pivotal role in modulating these interactions (31, 74).

CNTs demonstrate dose-dependent toxicity, with higher doses resulting in severe fibrotic and inflammatory responses. Ground CNTs, which have reduced aggregation, are distributed more evenly within lung tissues, leading to milder fibrotic responses than intact CNTs. Despite efforts to mitigate toxicity, the biopersistence of CNTs continues to be a major concern, necessitating stringent safety evaluations and functionalization strategies to improve their biocompatibility.

7.9 STRATEGIES TO IMPROVE BIOCOMPATIBILITY OF CARBON NANOTUBES

Several strategies have been developed to enhance the biocompatibility of CNTs and mitigate their toxic effects, including surface engineering and functionalization, dosage optimization, and controlled release mechanisms. Surface modification is among the most effective approaches for improving the compatibility of CNTs with biological systems. Functionalization involves the attachment of chemical groups, polymers, or biomolecules to the CNT surface, which enhances dispersion and reduces cytotoxicity (75). Techniques such as carboxylation and PEGylation have been widely explored to increase water solubility, minimize aggregation, and decrease immune response. Coating CNTs with biopolymers such as chitosan or proteins further enhances their stability and reduces oxidative stress by limiting ROS production. These modifications help CNTs integrate more effectively into biological environments, reducing their inflammatory potential (31, 76).

Another important approach to improve biocompatibility is dosage optimization, as the toxicity of CNTs is largely dose dependent. Studies have demonstrated that high doses of CNTs can trigger severe inflammatory responses, fibrosis, and systemic toxicity. Therefore, optimizing the dosage is crucial for ensuring the safe application of CNT-based therapies. Preclinical studies play a key role in determining the maximum tolerated dose (MTD) and establishing a safe therapeutic window. Research has suggested that functionalized CNTs administered at lower doses elicit fewer adverse immune responses, highlighting the importance of dose control in biomedical applications. Additionally, precise control over dosage using nanocarrier formulations has been explored to ensure targeted delivery while minimizing exposure to non-targeted tissues (31, 75).

Controlled release mechanisms represent another promising strategy for enhancing the safety and efficacy of CNT-based therapies. Encapsulating CNTs in biocompatible carriers such as hydrogels, liposomes, or polymeric matrices makes it possible to regulate drug release and prevent burst effects that may lead to toxicity. These systems allow for the gradual and sustained release of therapeutic agents, improved pharmacokinetics, and reduced systemic exposure. Furthermore, stimuli-responsive release mechanisms, such as pH-sensitive coatings or thermally responsive polymers, enable CNTs to selectively release their payload in specific environments such as acidic tumor microenvironments. This targeted approach further minimizes adverse effects while enhancing therapeutic outcomes, making CNT-based therapies more viable for clinical applications (70, 77, 78).

Combining these strategies allows CNTs to be optimized for biomedical applications including lung cancer therapy. Surface engineering and functionalization help mitigate cytotoxicity, dosage optimization minimizes inflammatory responses, and controlled release mechanisms enhance targeted drug delivery. These advancements collectively contribute to safer and more practical applications of CNTs in medical treatment, paving the way for their successful translation into clinical settings.

7.10 REGULATORY CHALLENGES

The regulatory landscape for CNTs in biomedical applications presents several challenges, particularly in the context of their unique properties and lack of standardized testing protocols. The variability in the physical, chemical, and biological characteristics of CNTs, such as size, surface functionalization, and aggregation state, complicates safety assessments. To ensure the safe use of CNTs in cancer therapy and other applications, regulatory bodies must develop standardized methods to assess their biocompatibility, biodistribution, and long-term toxicity.

Although regulatory agencies, such as the FDA and EMA, have provided initial guidelines for evaluating nanomaterials, these frameworks often do not account for the specific risks posed by CNTs. Existing guidelines focus on aspects such as purity, sterility, and manufacturing consistency but fail to address the challenges associated with the fibrous structure and biopersistence of CNTs in the body. To bridge this gap, regulatory agencies must develop more specific protocols that consider the unique properties of CNTs (79). Additionally, advances in real-time analytical techniques could enable better tracking of CNT behavior *in vivo*, providing regulators with the tools required to monitor their interactions within biological systems.

Ethical concerns surrounding CNT-based therapies are also a significant component of regulatory discussion. The potential risks to patient health, including long-term inflammation, fibrosis, and carcinogenicity, necessitate extensive preclinical and clinical testing (80). Furthermore, equitable access to CNT-based therapies, particularly in underserved populations, raises critical ethical concerns. Although promising, these therapies may not be accessible to all patients, especially those in low-resource settings. Regulatory frameworks must ensure that these treatments are distributed considering the needs of diverse populations (81, 82).

In addition to patient safety, the environmental impact of CNT production and disposal is an important ethical issue. Nanoparticles, including CNTs, can pose risks to ecosystems if not properly managed. The release of CNTs into the environment can lead to contamination, thereby affecting both public health and biodiversity (83). Therefore, regulatory bodies must establish guidelines to ensure that CNT production and disposal are environmentally sustainable, mitigating the risks posed by these materials while promoting their beneficial use in medicine.

7.11 COMPARISON BETWEEN CARBON NANOTUBE AND EXISTING THERAPIES FOR LUNG CANCER

CNTs have been established as a promising platform for lung cancer treatment because of their advantages in drug delivery and targeting capabilities compared with other conventional therapies (Table 7.1).

7.12 PRECLINICAL AND CLINICAL STUDIES OF CARBON NANOTUBES IN LUNG CANCER

The distinctive capability of CNTs to be precisely functionalized with targeting ligands in conjunction with their capacity to carry high drug payloads has elicited considerable interest in their potential to enhance current lung cancer treatments. Preclinical investigations have revealed that CNTs can be utilized in diverse therapeutic strategies including targeted drug delivery and PTT. These advancements are particularly relevant in cancer pharmacotherapy because insufficient selectivity for cancer cells, systemic adverse effects, and drug resistance hinder traditional therapeutic approaches (14).

Recent preclinical investigations have revealed unprecedented therapeutic efficacy of CNTs in lung cancer treatment. Among the innovative nanostructures, these demonstrate particular excellence in targeted drug delivery, as they enhance the delivery of chemotherapeutics and siRNA to lung cancer cells while minimizing systemic toxicity and improving overall efficacy (86). In studies utilizing

TABLE 7.1
Comparison Between Carbon Nanotubes and Existing Lung Cancer Therapies

Criteria	Carbon Nanotubes	Existing Therapy	References
Targeting specificity	Due to the functionalization of CNTs, high specificity in drug delivery can be achieved.	Moderate specificity: can affect the subsidiary healthy tissues.	10, 11, 18, 27, 30
Therapeutic efficacy	Better efficacy because of improved drug delivery system and synergistic effect.	It is effective but with the possibility of drug resistance and non-specific targeting.	11, 27, 84, 85
Side effects	Lower due to targeted delivery of the drug.	High systemic toxicity.	16, 24, 30, 32
Innovation in mechanism	Includes novel mechanisms like photothermal therapy and gene therapy.	Depends on a conventional form of mechanism of action such as cell cycle arrest, and DNA damage.	10, 11, 18
Ease of administration	Includes advanced techniques (e.g. injectables and inhalable nanocarriers).	Includes conventional methods like oral, intravenous, or radiation procedures.	11, 16, 18, 24, 30
Cost	Expensive	Relatively less expensive.	10, 11, 18, 27, 30, 32
Safety and biocompatibility	Undocumented	Safety conditions and biocompatibility have been well documented.	10, 11, 16, 18, 24
Resistance to treatment	CNT can overcome drug resistance through precisely targeted approaches and combination therapies.	It remains a challenge.	10, 11, 18, 39–44
Regulatory challenges	It is still in process due to a lack of standardization.	Well established.	79–83

MWCNT as vectors of siRNA or drug carriers, significant rates of lung cancer xenograft eradication have been demonstrated, indicating their efficacy for gene therapy and chemotherapy (29, 87). This success in targeted delivery has prompted researchers to investigate their combined use with conventional treatments, revealing that CNTs exhibit significant synergistic potential when paired with traditional CT agents. For instance, subsequent encapsulation of cisplatin in CNTs has not only enhanced its effectiveness in targeting lung tumors but also markedly reduced its impact on the healthy cells surrounding a tumor (85). Researchers have developed numerous functionalization techniques to further enhance their clinical viability, including biocompatible polymer coating layers, which have successfully mitigated many of their inherent toxicities (Table 7.2).

MWCNT, multiwalled carbon nanotube; SWCNT, single-walled carbon nanotube.

Despite these promising discoveries, significant safety concerns have been reported during preclinical evaluations, particularly regarding toxicity profiles. Consequently, these challenges have led to the development and implementation of novel techniques for designing and modifying CNTs. Consequently, researchers are developing modified CNTs—shorter, curved, or functionalized—to maximize therapeutic benefits and minimize risks, ultimately progressing toward clinical trials (102). Most studies on CNTs are in the preclinical phase. However, a few clinical studies have recently investigated the potential of CNTs in lung cancer treatment, with applications spanning multiple areas of cancer care. In targeted drug delivery, chemotherapeutic efficacy is enhanced by functionalizing CNTs to selectively deliver drugs to lung cancer cells while reducing systemic toxicity and

TABLE 7.2
Recent Research Outcomes in Carbon Nanotubes for Lung Cancer Treatment

Type	Functionalized with	Drug/Protein	Reference
MWCNT	Cation	siRNA	88
MWCNT	Hyaluronic acid	Doxorubicin	17
SWCNT	Graphene oxide	Paclitaxel	15
MWCNT	Just carboxylate	Dexamethasone	89
MWCNT	Fluorochrome	Methotrexate	16
MWCNT	. . .	Platinum	90
MWCNT	Carboxylate	Betulinic acid	91
SWCNT	Pyrene butyric acid *N*-hydroxysuccinimide ester, and Polyethylene glycol	TRAIL or Apo2L	92
MWCNT	Cation	siRNA	84
SWCNT	Chitosan	Paclitaxel	93
SWCNT	Polyvinylpyrrolidone	Doxorubicin	94
SWCNT	Polyethylene glycol	Gemcitabine	29
MWCNT	α-Tocopheryl polyethylene glycol 1000 succinate	Docetaxel/coumarin-6	95
MWCNT	Polyethylene glycol	–	24
MWCNT	Graphene oxide	–	96
MWCNT	Chitosan	Docetaxel/coumarin-6	97
MWCNT	Polyethylene glycol	VP-16 and Bcl-2-targeted antisense	98
MWCNT	Polyethyleneimine	iRGD	99
SWCNT	Chitosan	Curcumin	100
MWCNT	Polyethylene glycol	–	101
SWCNT	Polyethylenimine with betaine	*Survivin* siRNA, and Doxorubicin	87
MWCNT	Chitosan	Methotrexate	78
MWCNT	Naringenin	Naringenin	45
MWCNT	Bromocriptine	Bromocriptine	10

associated side effects. However, safety concerns persist regarding the biocompatibility of CNTs. *In vivo* studies have demonstrated that MWCNTs can induce inflammation, fibrosis, and potential carcinogenicity (103). Researchers are designing various functionalization strategies to address these challenges and improve the CNT safety profiles. Furthermore, the transition from experimental to clinical settings encounters several additional challenges, including systematic assessment of long-term effects, manufacturing process quality control, and regulatory considerations.

7.13 FUTURE DIRECTIONS AND PROSPECTS

CNTs for the treatment of lung cancer create a complex terrain rich in potential that consists of both positive and negative components. Therefore, it is important to effectively address these problems and simultaneously seek ways to further develop the field and increase patient outcomes. CNTs show promise in various fields owing to their exceptional mechanical, electrical, and thermal properties. The results of using CNTs in lung cancer treatment are promising. Potential transformative advances in therapeutic efficacy, reduced systemic toxicities, and increased tailoring of the therapeutic approach to CNTs hold promise for improved patient outcomes. As these technologies evolve and become more robust, they are well positioned to play a critical role in revolutionizing the lung cancer therapeutic landscape and offering much-needed, effective, and targeted treatment options (104). Advanced innovations in CNT-based therapy are currently being developed rapidly in emerging technologies and synergistic approaches that promise to change cancer treatment, including lung cancer. Therapeutic advances involving the recent properties of CNTs, including their extensive surface area, chemical flexibility, and ability to cross biological barriers, have been employed.

Nanotechnology and materials science are expected to lead to the development of sophisticated CNT-based platforms for drug delivery, imaging, and therapy. The integration of CNTs with other nanomaterials such as gold nanoparticles or quantum dots can further enhance their functionality and enable multimodal approaches for lung cancer treatment (105). Their applications include medicine, gene and drug delivery, and environmental remediation. Challenges remain in improving the fabrication methods, functionalization, and characterization techniques to enhance CNT performance and sensitivity. Overcoming the current obstacles and focusing on application-driven development are crucial for realizing the full potential of CNT functional materials (106, 107).

Recent innovations in CNT-based therapies have been directed toward their application as anticancer agents in drug delivery systems. The ability of CNTs to encapsulate drugs and specifically target cancer cells provides this treatment with the highest possible efficacy, while reducing side effects (108). For instance, CNTs can be functionalized with targeting ligands that bind specifically to receptors on the surface of cancer cells, thus providing a more selective delivery of anticancer therapeutic agents directly at the tumor site. Such targeting becomes extremely important in lung cancer treatment, where conventional therapies produce systemic toxicity in their treatment and limited efficacy, in addition to providing artificial intelligence (AI)-influenced revolution in drug design. AI-driven methodologies aim to optimize the design of CNTs for targeted therapeutic applications by predicting how modifications could enhance drug loading or their release profiles. Algorithms and computing power will speed up the development of CNT-based therapies and flesh them out with personalized treatment strategies based on individual patient profiles. Therefore, the interface between AI and CNTs facilitated more effective and safer lung cancer treatments. AI can also optimize the functionalization of nanotubes to enhance their targeting capabilities and biocompatibility. For example, AI algorithms can predict the most practical combination of targeting ligands (e.g., antibodies and peptides) and surface modifications to ensure that CNTs selectively bind to cancer cells, while minimizing off-target effects. Additionally, AI can help design CNT-based drug delivery systems that release therapeutic agents in a controlled and sustained manner, maximizing their efficacy while reducing systemic toxicity (74, 109, 110).

Despite their potential in cancer therapy, several translational challenges remain. Significant barriers include biocompatibility, because CNTs can induce inflammatory responses and cytotoxic effects in biological systems. Current research focuses on modifying the surfaces of CNTs to improve biocompatibility and reduce toxicity, which are essential for safe clinical applications in the future (111, 112). Standardizing the manufacturing processes for CNTs has become imperative to ensure stable quality in clinical applications. Additional efforts are needed to define pathways whereby CNT-based therapies could first be entered into clinical development and followed on the market on a commercial scale while forming a robust regulatory framework, answering the questions of implementation and commercialization. Because CNTs are classified as nanomaterials, they face several restrictions and hurdles during their informal and formal approval processes, and such approval may prevent CNTs from reaching the market for a long time. Developing clear guidance on the assessment of CNT therapies will be crucial for them to gain acceptance in the clinic. Collaboration among academia, industries, and regulators is required to ease the shift from laboratory to clinical research (109, 113).

For multidrug delivery and chemosensitivity in cancers, some offer to improve combination therapies that appear as the best crossover to meet treatment measures for the treatment of cancers. Interdisciplinary collaboration is critical for the advancement of CNT-based therapies. Assembling a team that includes materials science, oncology, pharmacology, and computational biology allows for novel approaches to tackle insurmountable hurdles in cancer treatment. Such cooperative research collaborations have yielded breakthroughs in how CNTs function, their applications, and other biological system interactions, leading to improved therapeutic implications (114).

Another innovative advancement in CNT-based cancer treatment is gene therapy. Owing to their high carrying capacity, CNTs are being investigated for drug delivery applications, particularly for delivering therapeutic genes to cancer cells, as they represent a potential safe alternative to common viral vectors. In comparison, CNTs can be functionalized with nucleic acid payloads, such as siRNA, miRNA, or plasmid DNA, and transported and targeted to tumor cells, where gene expression can be modulated or apoptosis induced (115).

Ensuring the biocompatibility of CNTs remains a major challenge. CNTs hold great promise in preclinical models, but concerns about human health, including possible toxicity and long-term effects, must be factored into clinical use. When CNTs accumulate in vital organs such as the lungs, liver, and spleen, they can initiate inflammation, oxidative stress, and fibrosis. To resolve these issues, researchers have supported surface-functionalization approaches to protect the biocompatibility of CNTs. Additionally, the development of biodegradable CNTs or CNT-like materials can mitigate long-term toxicity (14, 116).

CNTs offer innovative approaches that extend beyond conventional therapies, leading to custom medical strategies driven by nanotechnology. Its unique properties, such as high surface area and conductivity, allow the development of multifunctional drug administration systems that can improve the effectiveness of chemotherapeutics while simultaneously reducing systemic toxicity. Continuous innovation in the design and application of CNTs is essential to unlock their therapeutic potential. The advent of functionalized CNTs has led to interesting advances in anticancer activities, showing their ability to adapt to specific carcinogenic cell directions and simultaneously improve therapeutic results (117). Collaborative efforts are underway to develop standardized protocols and regulatory compliance for clinical applications. Apart from lung cancer, CNT applications are expanding to other lung-related conditions, such as asthma, COPD, pulmonary fibrosis, acute respiratory distress syndrome, and infectious diseases, such as tuberculosis. These efforts have highlighted the utility of CNTs as carriers of drugs, biosensors, and regenerative therapies. Continued interdisciplinary research and collaboration across disciplines are essential to unlock the full potential of CNTs in respiratory therapy.

7.14 CONCLUSION

CNTs have emerged as revolutionary in lung cancer therapy, including their potential as drug delivery systems, applications in PTT and PDT, and dual-functional capabilities in imaging and treatment. Furthermore, their integration into theranostic platforms represents a promising approach for combining diagnosis and therapy to improve patient outcome. However, the translation of CNT-based treatments from laboratory research to clinical applications remains challenging. Biocompatibility, toxicity, and regulatory concerns must be addressed to ensure their safe and effective use. Surface engineering, functionalization, and dosage optimization are promising approaches for mitigating these concerns; however, further studies are required to advance these approaches. As research has progressed, the integration of CNTs into clinical practice could revolutionize cancer therapy, paving the way for more personalized and effective treatments. In conclusion, while challenges remain, the future of CNTs in lung cancer therapy appears promising, with the potential to significantly impact global oncology and patient management.

REFERENCES

1. Dai H. Carbon nanotubes: opportunities and challenges. *Surface Science*. 2002 Mar 10;500(1–3):218–41.
2. Chen J, Wei S, Xie H. A brief introduction of carbon nanotubes: history, synthesis, and properties. *Journal of Physics: Conference Series*. 2021 Jun 1; 1948(1):012184.

3. Tang L, Xiao Q, Mei Y, He S, Zhang Z, Wang R, Wang W. Insights on functionalized carbon nanotubes for cancer theranostics. *Journal of Nanobiotechnology*. 2021 Dec;19:1–28.
4. Zhang Y, Bai Y, Yan B. Functionalized carbon nanotubes for potential medicinal applications. *Drug Discovery Today*. 2010 Jun 1;15(11–12):428–35.
5. Aqel A, Abou El-Nour KM, Ammar RA, Al-Warthan A. Carbon nanotubes, science and technology part (I) structure, synthesis and characterisation. *Arabian Journal of Chemistry*. 2012 Jan 1;5(1):1–23.
6. Zare H, Ahmadi S, Ghasemi A, Ghanbari M, Rabiee N, Bagherzadeh M, Karimi M, Webster TJ, Hamblin MR, Mostafavi E. Carbon nanotubes: smart drug/gene delivery carriers. *International Journal of Nanomedicine*. 2021 Mar 1:1681–706.
7. Huang Z, Li X, Sun Y, Song Z, Wang L, Yu L, Wang Z, Liu J. Enhancing the electrochemical properties and processing techniques of carbon nanotubes for bio pharmacological applications: a research perspective. *Highlights in Science, Engineering and Technology*. 2023 Jul 12;58:297–303.
8. Popov VN. Carbon nanotubes: properties and application. *Materials Science and Engineering: R: Reports*. 2004 Jan 15;43(3):61–102.
9. Abdallah B, Elhissi AM, Ahmed W, Najlah M. Carbon nanotubes drug delivery system for cancer treatment. In *Advances in Medical and Surgical Engineering* 2020 Jan 1 (pp. 313–32). Academic Press.
10. Kamazani FM, Sotoodehnejad Nematalahi F, Siadat SD, Pornour M, Sheikhpour M. A success targeted nano delivery to lung cancer cells with multiwalled carbon nanotubes conjugated to bromocriptine. *Scientific Reports*. 2021 Dec 24;11(1):24419.
11. Sharma M, Alessandro P, Cheriyamundath S, Lopus M. Therapeutic and diagnostic applications of carbon nanotubes in cancer: recent advances and challenges. *Journal of Drug Targeting*. 2024 Mar 15;32(3):287–99.
12. Naief MF, Mohammed SN, Mohammed AM. Carbon nanotubes: a review on synthesis and drug delivery for cancer treatment. *Inorganic Chemistry Communications*. 2024 Jan 1;159:111694.
13. Sheikhpour M, Golbabaie A, Kasaeian A. Carbon nanotubes: a review of novel strategies for cancer diagnosis and treatment. *Materials Science and Engineering: C*. 2017 Jul 1;76:1289–304.
14. Sheikhpour M, Naghinejad M, Kasaeian A, Lohrasbi A, Shahraeini SS, Zomorodbakhsh S. The applications of carbon nanotubes in the diagnosis and treatment of lung cancer: a critical review. *International Journal of Nanomedicine*. 2020 Sep 24:7063–78.
15. Arya N, Arora A, Vasu KS, Sood AK, Katti DS. Combination of single walled carbon nanotubes/graphene oxide with paclitaxel: a reactive oxygen species mediated synergism for treatment of lung cancer. *Nanoscale*. 2013;5(7):2818–29.
16. Das M, Datir SR, Singh RP, Jain S. Augmented anticancer activity of a targeted, intracellularly activatable, theranostic nanomedicine based on fluorescent and radiolabeled, methotrexate-folic acid-multiwalled carbon nanotube conjugate. *Molecular Pharmaceutics*. 2013 Jul 1;10(7):2543–57.
17. Datir SR, Das M, Singh RP, Jain S. Hyaluronate tethered, "smart" multiwalled carbon nanotubes for tumor-targeted delivery of doxorubicin. *Bioconjugate Chemistry*. 2012 Nov 21;23(11):2201–13.
18. Fabbro C, Ali-Boucetta H, Da Ros T, Kostarelos K, Bianco A, Prato M. Targeting carbon nanotubes against cancer. *Chemical Communications*. 2012;48(33):3911–26.
19. Moore TL, Pitzer JE, Podila R, Wang X, Lewis RL, Grimes SW, Wilson JR, Skjervold E, Brown JM, Rao A, Alexis F. Multifunctional polymer-coated carbon nanotubes for safe drug delivery. *Particle & Particle Systems Characterization*. 2013 Apr;30(4):365–73.
20. Razak SAA, Mohd Gazzali A, Fisol FA, Abdulbaqi IM, Parumasivam T, Mohtar N, Wahab H. Advances in nanocarriers for effective delivery of docetaxel in the treatment of lung cancer: an overview. *Cancers*. 2021 Jan;13(3):400.
21. Singh RP, Sharma G, Singh S, Patne SC, Pandey BL, Koch B, Muthu MS. Effects of transferrin conjugated multiwalled carbon nanotubes in lung cancer delivery. *Materials Science and Engineering: C*. 2016 Oct 1;67:313–25.
22. Li B, Zhang XX, Huang HY, Chen LQ, Cui JH, Liu Y, Jin H, Lee BJ, Cao QR. Effective deactivation of A549 tumor cells in vitro and in vivo by RGD-decorated chitosan-functionalized single-walled carbon nanotube loading docetaxel. *International Journal of Pharmaceutics*. 2018 May 30;543(1–2):8–20.
23. Park JY, Lee GH, Yoo KH, Khang D. Overcoming multidrug-resistant lung cancer by mitochondrial-associated ATP inhibition using nanodrugs. *Journal of Nanobiotechnology*. 2023 Jan 12;21(1):12.
24. Kim SW, Lee YK, Lee JY, Hong JH, Khang D. PEGylated anticancer-carbon nanotubes complex targeting mitochondria of lung cancer cells. *Nanotechnology*. 2017 Oct 20;28(46):465102.
25. Eskandari M, Hosseini SH, Adeli M, Pourjavadi A. Polymer-functionalized carbon nanotubes in cancer therapy: a review. *Iranian Polymer Journal*. 2014 May;23:387–403.
26. Feng J, Zhang P, Wang D, Li Y, Tan J. New strategies for lung cancer diagnosis and treatment: applications and advances in nanotechnology. *Biomarker Research*. 2024 Nov 13;12(1):136.

27. Halwai K, Khanna S, Gupta G, Wahab S, Khalid M, Kesharwani P. Folate-conjugated carbon nanotubes as a promising therapeutic approach for targeted cancer therapy. *Journal of Drug Targeting*. 2024 Aug 14:1–6.
28. Wu L, Man C, Wang H, Lu X, Ma Q, Cai Y, Ma W. PEGylated multiwalled carbon nanotubes for encapsulation and sustained release of oxaliplatin. *Pharmaceutical Research*. 2013 Feb;30:412–23.
29. Razzazan A, Atyabi F, Kazemi B, Dinarvand R. In vivo drug delivery of gemcitabine with PEGylated single-walled carbon nanotubes. *Materials Science and Engineering: C*. 2016 May 1;62:614–25.
30. Shojaei S, Pourmadadi M, Homayoonfal M, Behnamrad P, Fathi-Karkan S, Rahdar A, Gerayli S, Pandey S. Revolutionizing lung cancer treatment: nanotechnology-driven advances in targeted drug delivery and novel therapeutic strategies. *Journal of Drug Delivery Science and Technology*. 2024 Sep 14:106186.
31. Mahajan S, Patharkar A, Kuche K, Maheshwari R, Deb PK, Kalia K, Tekade RK. Functionalized carbon nanotubes as emerging delivery system for the treatment of cancer. *International Journal of Pharmaceutics*. 2018 Sep 5;548(1):540–58.
32. Liu D, Pan X, Zhao X, Zhang H. Preparation and in vitro and in vivo evaluation of lung cancer-targeted carbon nanotube ultrasound contrast agents. *Journal of Biomedical Nanotechnology*. 2022 Jun 1;18(6):1684–91.
33. Qi Y, Yang W, Liu S, Han F, Wang H, Zhao Y, et al. Cisplatin loaded multiwalled carbon nanotubes reverse drug resistance in NSCLC by inhibiting EMT. *Cancer Cell International*. 2021 Dec;21(1):74.
34. Salas-Treviño D, Saucedo-Cárdenas O, Loera-Arias MD, Rodríguez-Rocha H, García-García A, Montes-de-Oca-Luna R, Piña-Mendoza EI, Contreras-Torres FF, García-Rivas G, Soto-Domínguez A. Hyaluronate functionalized multi-wall carbon nanotubes filled with carboplatin as a novel drug nanocarrier against murine lung cancer cells. *Nanomaterials*. 2019 Nov 6;9(11):1572.
35. Pu Z, Wei Y, Sun Y, Wang Y, Zhu S. Carbon nanotubes as carriers in drug delivery for non-small cell lung cancer, mechanistic analysis of their carcinogenic potential, safety profiling and identification of biomarkers. *International Journal of Nanomedicine*. 2022;17:6157.
36. Mitra J, Srilekha GK, Wagh N, Lakkakula J. Carbon nanotubes for drug delivery applications. In *Carbon Nanotubes* 2022 Oct 6 (pp. 161–207). Apple Academic Press.
37. Jiang P, Liang B, Zhang Z, Fan B, Zeng L, Zhou Z, Mao Z, Xu Q, Yao W, Shen Q. New insights into nanosystems for non-small-cell lung cancer: diagnosis and treatment. *RSC Advances*. 2023;13(28):19540–64.
38. Farooq MA, Aquib M, Farooq A, Haleem Khan D, Joelle Maviah MB, Sied Filli M, Kesse S, Boakye-Yiadom KO, Mavlyanova R, Parveen A, Wang B. Recent progress in nanotechnology-based novel drug delivery systems in designing of cisplatin for cancer therapy: an overview. *Artificial Cells, Nanomedicine, and Biotechnology*. 2019 Dec 4;47(1):1674–92.
39. Ashrafizadeh M, Zarrabi A, Hushmandi K, Hashemi F, Moghadam ER, Owrang M, Hashemi F, Makvandi P, Goharrizi MA, Najafi M, Khan H. Lung cancer cells and their sensitivity/resistance to cisplatin chemotherapy: role of microRNAs and upstream mediators. *Cellular Signalling*. 2021 Feb 1;78:109871.
40. Chen KB, Yang W, Xuan Y, Lin AJ. miR-526b-3p inhibits lung cancer cisplatin-resistance and metastasis by inhibiting STAT3-promoted PD-L1. *Cell Death & Disease*. 2021 Jul 28;12(8):748.
41. Chen Y, Zhou H, Yang S, Su D. Increased ABCC2 expression predicts cisplatin resistance in non-small cell lung cancer. *Cell Biochemistry and Function*. 2021 Mar;39(2):277–86.
42. Kryczka J, Kryczka J, Czarnecka-Chrebelska KH, Brzeziańska-Lasota E. Molecular mechanisms of chemoresistance induced by cisplatin in NSCLC cancer therapy. *International Journal of Molecular Sciences*. 2021 Aug 18;22(16):8885.
43. Ray R, Al Khashali H, Haddad B, Wareham J, Coleman KL, Alomari D, Ranzenberger R, Guthrie J, Heyl D, Evans HG. Regulation of cisplatin resistance in lung cancer cells by nicotine, BDNF, and a β-adrenergic receptor blocker. *International Journal of Molecular Sciences*. 2022 Oct 24;23(21):12829.
44. Tang H, Chen J, Han X, Feng Y, Wang F. Upregulation of SPP1 is a marker for poor lung cancer prognosis and contributes to cancer progression and cisplatin resistance. *Frontiers in Cell and Developmental Biology*. 2021 Apr 29;9:646390.
45. Morais RP, Novais GB, Sangenito S, et al. Naringenin-functionalized multiwalled carbon nanotubes: a potential approach for site-specific remote-controlled anticancer delivery for the treatment of lung cancer cells. *International Journal of Molecular Sciences*. 2020;21(12):4557. https://doi.org/10.3390/ijms21124557.
46. Li J, Zhu L, Kwok HF. Nanotechnology-based approaches overcome lung cancer drug resistance through diagnosis and treatment. *Drug Resistance Updates*. 2023 Jan 1;66:100904.

47. Curcio M, Farfalla A, Saletta F, Valli E, Pantuso E, Nicoletta FP, Iemma F, Vittorio O, Cirillo G. Functionalized carbon nanostructures versus drug resistance: promising scenarios in cancer treatment. *Molecules*. 2020 Apr 30;25(9):2102.
48. Umar H, Wahab HA, Attiq A, Amjad MW, Bukhari SN, Ahmad W. Platinum-based targeted chemotherapies and reversal of cisplatin resistance in non-small cell lung cancer (NSCLC). *Mutation Research/Fundamental and Molecular Mechanisms of Mutagenesis*. 2024 Mar 19:111856.
49. Naief MF, Mohammed SN, Mayouf HJ, Mohammed AM. A review of the role of carbon nanotubes for cancer treatment based on photothermal and photodynamic therapy techniques. *Journal of Organometallic Chemistry*. 2023 Jul 20:122819.
50. Sundaram P, Abrahamse H. Phototherapy combined with carbon nanomaterials (1D and 2D) and their applications in cancer therapy. *Materials*. 2020 Oct 28;13(21):4830.
51. Dias LD, Buzzá HH, Stringasci MD, Bagnato VS. Recent advances in combined photothermal and photodynamic therapies against cancer using carbon nanomaterial platforms for in vivo studies. *Photochem*. 2021 Oct 21;1(3):434–47.
52. Lagos KJ, Buzzá HH, Bagnato VS, Romero MP. Carbon-based materials in photodynamic and photothermal therapies applied to tumor destruction. *International Journal of Molecular Sciences*. 2021 Dec 21;23(1):22.
53. Qi K, Sun B, Liu SY, Zhang M. Research progress on carbon materials in tumor photothermal therapy. *Biomedicine & Pharmacotherapy*. 2023 Sep 1;165:115070.
54. Nasseri B, Alizadeh E, Bani F, Davaran S, Akbarzadeh A, Rabiee N, Bahadori A, Ziaei M, Bagherzadeh M, Saeb MR, Mozafari M. Nanomaterials for photothermal and photodynamic cancer therapy. *Applied Physics Reviews*. 2022 Mar 1;9(1).
55. Chen J, Zeng Z, Huang L, Luo S, Dong J, Zhou FH, Zhou K, Wang L, Kang L. Photothermal therapy technology of metastatic colorectal cancer. *American Journal of Translational Research*. 2020;12(7):3089.
56. Sundaram P. *The Use of Biopolymer Functionalized Single Walled Carbon Nanotubes for Effective Targeted Photodynamic Therapy of Colon Cancer Stem Cells* 2020. University of Johannesburg (South Africa).
57. Mei C, Wang N, Zhu X, Wong KH, Chen T. Photothermal-controlled nanotubes with surface charge flipping ability for precise synergistic therapy of triple-negative breast cancer. *Advanced Functional Materials*. 2018 Nov;28(45):1805225.
58. Sun X, Li D, Lv Y, Zhang M, Qiao D, Zhang Z, Ren H, Zhang Y, Yang Z, Gao J. Nanomaterials for the diagnosis and treatment of triple-negative breast cancer. *Wiley Interdisciplinary Reviews: Nanomedicine and Nanobiotechnology*. 2024 Nov;16(6):e2019.
59. Nag S, Mitra O, Tripathi G, Adur I, Mohanto S, Nama M, Samanta S, Gowda BJ, Subramaniyan V, Sundararajan V, Kumarasamy V. Nanomaterials-assisted photothermal therapy for breast cancer: state-of-the advances and future perspectives. *Photodiagnosis and Photodynamic Therapy*. 2024 Jan 14:103959.
60. Zhang M, Wang W, Cui Y, Chu X, Sun B, Zhou N, Shen J. Magnetofluorescent Fe3O4/carbon quantum dots coated single-walled carbon nanotubes as dual-modal targeted imaging and chemo/photodynamic/photothermal triple-modal therapeutic agents. *Chemical Engineering Journal*. 2018 Apr 15;338:526–38.
61. Wang S, Zhang Q, Yang P, Yu X, Huang LY, Shen S, Cai S. Manganese oxide-coated carbon nanotubes as dual-modality lymph mapping agents for photothermal therapy of tumor metastasis. *ACS Applied Materials & Interfaces*. 2016 Feb 17;8(6):3736–43.
62. Zhao Y, Zhao T, Cao Y, Sun J, Zhou Q, Chen H, Guo S, Wang Y, Zhen Y, Liang XJ, Zhang S. Temperature-sensitive lipid-coated carbon nanotubes for synergistic photothermal therapy and gene therapy. *ACS Nano*. 2021 Apr 27;15(4):6517–29.
63. Zhang M, Wang W, Wu F, Yuan P, Chi C, Zhou N. Magnetic and fluorescent carbon nanotubes for dual modal imaging and photothermal and chemo-therapy of cancer cells in living mice. *Carbon*. 2017 Oct;123:70–83.
64. Singh RP, Sonali, Mehata AK, Agrawal P, Singh S, Muthu MS. Carbon nanotubes: super-fabric tiny materials as lung cancer theranostics. *The Pharmstudent*. 2016;27:55–78.
65. Minati L, Antonini V, Dalla Serra M, Speranza G. Multifunctional branched gold-carbon nanotube hybrid for cell imaging and drug delivery. *Langmuir*. 2012 Nov 13;28(45):15900–6.
66. Das M, Singh RP, Datir SR, Jain S. Intranuclear drug delivery and effective in vivo cancer therapy via estradiol PEG-appended multiwalled carbon nanotubes. *Molecular Pharmaceutics*. 2013;10(9):3404–16.
67. Singh R, Deshmukh R. Carbon nanotube as an emerging theranostic tool for oncology. *Journal of Drug Delivery Science and Technology*. 2022 Aug;74:103586.

68. Wan Q, Zhang X, Gui Y. Theoretical study on Pt-doped carbon nanotubes used to detect typical exhaled gases of lung cancer. *Journal of Computational and Theoretical Nanoscience*. 2015;12(10):3412–17.
69. Li L, Liu X, Wei T, Wang K, Zhao Z, Cao J, et al. Carbon nanotube field-effect transistor biosensor with an enlarged gate area for ultra-sensitive detection of a lung cancer biomarker. *ACS Applied Materials & Interfaces*. 2023 Jun 7;15(22):27299–306.
70. Mercer RR, Hubbs AF, Scabilloni JF, Wang L, Battelli LA, Schwegler-Berry D, Castranova V, Porter DW. Distribution and persistence of pleural penetrations by multiwalled carbon nanotubes. *Particle and Fibre Toxicology*. 2010 Dec;7:1.
71. Kayat J, Gajbhiye V, Tekade RK, Jain NK. Pulmonary toxicity of carbon nanotubes: a systematic report. *Nanomedicine: Nanotechnology, Biology and Medicine*. 2011 Feb 1;7(1):40–9.
72. Shvedova AA, Kisin ER, Porter D, Schulte P, Kagan VE, Fadeel B, Castranova V. Mechanisms of pulmonary toxicity and medical applications of carbon nanotubes: two faces of Janus? *Pharmacology & Therapeutics*. 2009 Feb 1;121(2):192–204.
73. Tang S, Tang Y, Zhong L, Murat K, Asan G, Yu J, Jian R, Wang C, Zhou P. Short-and long-term toxicities of multi-walled carbon nanotubes in vivo and in vitro. *Journal of Applied Toxicology*. 2012 Nov;32(11):900–12.
74. Costa PM, Bourgognon M, Wang JT, Al-Jamal KT. Functionalised carbon nanotubes: from intracellular uptake and cell-related toxicity to systemic brain delivery. *Journal of Controlled Release*. 2016 Nov 10;241:200–19.
75. Maghimaa M, Sagadevan S, Boojhana E, Fatimah I, Lett JA, Moharana S, Garg S, Al-Anber MA. Enhancing biocompatibility and functionality: carbon nanotube-polymer nanocomposites for improved biomedical applications. *Journal of Drug Delivery Science and Technology*. 2024 Jul 9:105958.
76. Vardharajula S, Ali SZ, Tiwari PM, Eroğlu E, Vig K, Dennis VA, Singh SR. Functionalized carbon nanotubes: biomedical applications. *International Journal of Nanomedicine*. 2012 Oct 9:5361–74.
77. Zhou L, Forman HJ, Ge Y, Lunec J. Multiwalled carbon nanotubes: a cytotoxicity study in relation to functionalization, dose and dispersion. *Toxicology In Vitro*. 2017 Aug 1;42:292–8.
78. Cirillo G, Vittorio O, Kunhardt D, Valli E, Voli F, Farfalla A, Curcio M, Spizzirri UG, Hampel S. Combining carbon nanotubes and chitosan for the vectorization of methotrexate to lung cancer cells. *Materials*. 2019 Sep 6;12(18):2889.
79. Bianco A, Kostarelos K, Prato M. Making carbon nanotubes biocompatible and biodegradable. *Chemical Communications*. 2011;47(37):10182–8.
80. Meredith JR, Jin C, Narayan RJ, Aggarwal R. Biomedical applications of carbon-nanotube composites. *Frontiers in Bioscience*. 2013 Jan 1;5:610–21.
81. Nyström AM, Fadeel B. Safety assessment of nanomaterials: implications for nanomedicine. *Journal of Controlled Release*. 2012 Jul 20;161(2):403–8.
82. Al Faraj A, Fauvelle F, Luciani N, Lacroix G, Levy M, Crémillieux Y, Canet-Soulas E. In vivo biodistribution and biological impact of injected carbon nanotubes using magnetic resonance techniques. *International Journal of Nanomedicine*. 2011 Feb 15:351–61.
83. Vivekanandhan S, Schreiber M, Muthuramkumar S, Misra M, Mohanty AK. Carbon nanotubes from renewable feedstocks: a move toward sustainable nanofabrication. *Journal of Applied Polymer Science*. 2017 Jan 20;134(4).
84. Guo C, Al-Jamal WT, Toma FM, Bianco A, Prato M, Al-Jamal KT, Kostarelos K. Design of cationic multiwalled carbon nanotubes as efficient siRNA vectors for lung cancer xenograft eradication. *Bioconjugate Chemistry*. 2015;26(7):1370–9. https://doi.org/10.1021/acs.bioconjchem.5b00249.
85. Guven A, Rusakova IA, Lewis MT, Wilson LJ. Cisplatin@US-tube carbon nanocapsules for enhanced chemotherapeutic delivery. *Biomaterials [Internet]*. 2012 Feb 1;33(5):1455–61. https://www.sciencedirect.com/science/article/pii/S0142961211012762.
86. Murjani BO, Kadu PS, Bansod M, Vaidya SS, Yadav MD. Carbon nanotubes in biomedical applications: current status, promises, and challenges. *Carbon Letters*. 2022;32:1207–26. https://doi.org/10.1007/s42823-022-00364-4.
87. Cao Y, Huang HY, Chen LQ, Du HH, Cui JH, Zhang LW, Lee BJ, Cao QR. Enhanced lysosomal escape of pH-responsive polyethylenimine-betaine functionalized carbon nanotube for the codelivery of survivin small interfering RNA and doxorubicin. *ACS Applied Materials & Interfaces*. 2019;11(10):9763–76. https://doi.org/10.1021/acsami.8b20810.
88. Podesta JE, Al-Jamal KT, Herrero MA, Tian B, Ali-Boucetta H, Hegde V, Bianco A, Prato M, Kostarelos K. Antitumor activity and prolonged survival by carbon-nanotube-mediated therapeutic siRNA silencing in a human lung xenograft model. *Small*. 2009;5(10):1176–85. https://doi.org/10.1002/smll.200801572.

89. Lodhi N, Mehra NK, Jain NK. Development and characterization of dexamethasone mesylate anchored on multiwalled carbon nanotubes. *Journal of Drug Targeting*. 2013;21(1):67–76. https://doi.org/10.3109/1061186X.2012.729213.
90. Li J, Pant A, Chin CF, Ang WH, Ménard-Moyon C, Nayak TR, Gibson D, Ramaprabhu S, Panczyk T, Bianco A, Pastorin G. In vivo biodistribution of platinum-based drugs encapsulated into multiwalled carbon nanotubes. *Nanomedicine*. 2014;10(7):1465–75. https://doi.org/10.1016/j.nano.2014.01.004.
91. Tan JM, Karthivashan G, Arulselvan P, Fakurazi S, Hussein MZ. Characterization and in vitro studies of the anticancer effect of oxidized carbon nanotubes functionalized with betulinic acid. *Drug Design, Development and Therapy*. 2014;8:2333–43. https://doi.org/10.2147/DDDT.S70650.
92. Zakaria AB, Picaud F, Rattier T, Pudlo M, Saviot L, Chassagnon R, Lherminier J, Gharbi T, Micheau O, Herlem G. Nanovectorization of TRAIL with single wall carbon nanotubes enhances tumor cell killing. [published correction appears in *Nano Letters*. 2015 Aug 12;15(8):5664. https://doi.org/10.1021/acs.nanolett.5b02056. Dufour, Florent [Added]]. *Nano Letters*. 2015;15(2):891–5. https://doi.org/10.1021/nl503565t.
93. Yu B, Tan L, Zheng R, Tan H, Zheng L. Targeted delivery and controlled release of paclitaxel for the treatment of lung cancer using single-walled carbon nanotubes. *Materials Science & Engineering. C, Materials for Biological Applications*. 2016;68:579–84. https://doi.org/10.1016/j.msec.2016.06.025.
94. Faraj A, Shaik A, Halwani R, Alfuraih A. Magnetic targeting and delivery of drug-loaded SWCNTs theranostic nanoprobes to lung metastasis in breast cancer animal model: noninvasive monitoring using magnetic resonance imaging. *Molecular Imaging and Biology*. 2015 Oct 20:1–10.
95. Singh RP, Sharma G, Sonali, Singh S, Kumar M, Pandey BL, Koch B, Muthu MS. Vitamin E TPGS conjugated carbon nanotubes improved efficacy of docetaxel with safety for lung cancer treatment. *Colloids and Surfaces B: Biointerfaces*. 2016;141:429–42. https://doi.org/10.1016/j.colsurfb.2016.02.011.
96. Attri P, Gaur J, Choi S, Kim M, Bhatia R, Kumar N, Park JH, Cho AE, Choi EH, Lee W. Interaction studies of carbon nanomaterials and plasma activated carbon nanomaterials solution with telomere binding protein. *Scientific Reports* 2017 Jun 1;7(1):2636. https://doi.org/10.1038/s41598-017-02690-4.
97. Singh RP, Sharma G, Sonali, Singh S, Bharti S, Pandey BL, Koch B, Muthu MS. Chitosan-folate decorated carbon nanotubes for site specific lung cancer delivery. *Materials Science and Engineering: C*. 2017;77:446–58. https://doi.org/10.1016/j.msec.2017.03.225.
98. Heger Z, Polanska H, Krizkova S, Balvan J, Raudenska M, Dostalova S, Moulick A, Masarik M, Adam V. Codelivery of VP-16 and Bcl-2-targeted antisense on PEG-grafted oMWCNTs for synergistic in vitro anti-cancer effects in non-small and small cell lung cancer [published correction appears in *Colloids and Surfaces B-Biointerfaces*. 2022 Apr;212:112361. https://doi.org/10.1016/j.colsurfb.2022.112361]. *Colloids and Surfaces B-Biointerfaces*. 2017;150:131–40. https://doi.org/10.1016/j.colsurfb.2016.11.023.
99. Su Y, Hu Y, Wang Y, Xu X, Yuan Y, Li Y, Wang Z, Chen K, Zhang F, Ding X, Li M. A precision-guided MWNT mediated reawakening the sunk synergy in RAS for anti-angiogenesis lung cancer therapy. *Biomaterials*. 2017;139:75–90. https://doi.org/10.1016/j.biomaterials.2017.05.046.
100. Singh N, Sachdev A, Gopinath P. Polysaccharide functionalized single walled carbon nanotubes as nanocarriers for delivery of curcumin in lung cancer cells. *Journal of Nanoscience and Nanotechnology*. 2018;18(3):1534–41. https://doi.org/10.1166/jnn.2018.14222.
101. Suo X, Eldridge BN, Zhang H, Mao C, Min Y, Sun Y, Singh R, Ming X. P-glycoprotein-targeted photothermal therapy of drug-resistant cancer cells using antibody-conjugated carbon nanotubes. *ACS Applied Materials & Interfaces*. 2018;10(39):33464–73. https://doi.org/10.1021/acsami.8b11974.
102. Saito N, Haniu H, Usui Y, Aoki K, Hara K, Takanashi S, Shimizu M, Narita N, Okamoto M, Kobayashi S, Nomura H. Safe Clinical use of carbon nanotubes as innovative biomaterials. *Chemical Reviews [Internet]*. 2014 Jun 11;114(11):6040–79. https://doi.org/10.1021/cr400341h.
103. Hojo M, Maeno A, Sakamoto Y, Ohnuki A, Tada Y, Yamamoto Y, Ikushima K, Inaba R, Suzuki J, Taquahashi Y, Yokota S. Two-year intermittent exposure of a multiwalled carbon nanotube by intratracheal instillation induces lung tumors and pleural mesotheliomas in F344 rats. *Particle and Fibre Toxicology [Internet]*. 2022 May 19;19(1):38. https://doi.org/10.1186/s12989-022-00478-7.
104. Kumar M, Hilles AR, Almurisi SH, Bhatia A, Mahmood S. Micro and nanocarriers-based pulmonary drug delivery system: their current updates, challenges, and limitations–a review. *JCIS Open*. 2023 Dec 1;12:100095.
105. Sim S, Wong NK. Nanotechnology and its use in imaging and drug delivery. *Biomedical Reports*. 2021 May;14(5):42.
106. de Almeida Barcelos K, Garg J, Soares DC, de Barros AL, Zhao Y, Alisaraie L. Recent advances in the applications of CNT-based nanomaterials in pharmaceutical nanotechnology and biomedical engineering. *Journal of Drug Delivery Science and Technology*. 2023 Aug 7:104834.

107. Ijaz H, Mahmood A, Abdel-Daim MM, Sarfraz RM, Zaman M, Zafar N, Alshehery S, Salem-Bekhit MM, Ali MA, Eltayeb LB, Benguerba Y. Review on carbon nanotubes (CNTs) and their chemical and physical characteristics, with particular emphasis on potential applications in biomedicine. *Inorganic Chemistry Communications*. 2023 Jul 7:111020.
108. Garg P, Singhal G, Kulkarni P, Horne D, Salgia R, Singhal SS. Artificial intelligence–driven computational approaches in the development of anticancer drugs. *Cancers*. 2024 Nov 20;16(22):3884.
109. Awasthi S, Srivastava A, Kumar D, Pandey SK, Mubarak NM, Dehghani MH, Ansari K. An insight into the toxicological impacts of carbon nanotubes (CNTs) on human health: a review. *Environmental Advances*. 2024 Nov 13:100601.
110. Alowais SA, Alghamdi SS, Alsuhebany N, Alqahtani T, Alshaya AI, Almohareb SN, Aldairem A, Alrashed M, Bin Saleh K, Badreldin HA, Al Yami MS. Revolutionizing healthcare: the role of artificial intelligence in clinical practice. *BMC Medical Education*. 2023 Sep 22;23(1):689.
111. Prajapati SK, Malaiya A, Kesharwani P, Soni D, Jain A. Biomedical applications and toxicities of carbon nanotubes. *Drug and Chemical Toxicology*. 2022 Jan 2;45(1):435–50.
112. Choi C, Yun TG, Hwang B. Dispersion stability of carbon nanotubes and their impact on energy storage devices. *Inorganics*. 2023 Sep 25;11(10):383.
113. Kim M, Goerzen D, Jena PV, Zeng E, Pasquali M, Meidl RA, Heller DA. Human and environmental safety of carbon nanotubes across their life cycle. *Nature Reviews Materials*. 2024 Jan;9(1):63–81.
114. Gurunathan S, Kang MH, Qasim M, Kim JH. Nanoparticle-mediated combination therapy: two-in-one approach for cancer. *International Journal of Molecular Sciences*. 2018 Oct 20;19(10):3264.
115. Gao S, Xu B, Sun J, Zhang Z. Nanotechnological advances in cancer: therapy a comprehensive review of carbon nanotube applications. *Frontiers in Bioengineering and Biotechnology*. 2024 Mar 6;12:1351787.
116. Aoki K, Saito N. Biocompatibility and carcinogenicity of carbon nanotubes as biomaterials. *Nanomaterials*. 2020 Feb 4;10(2):264.
117. Ramesh GB, Singh P, Biswas K. Potentialities of bio-functionalized carbon nanotubes for different anti-cancerous activities. *Journal of Inorganic and Organometallic Polymers and Materials*. 2024 Mar 6:1–26.

8 Liposomes for the Management of Lung Cancer

Lokesh Nagar[1], *Annu Saini*[1], *Tarif Hussain*[1], *Nisha Gulati*[1], *Kamal Dua*[2] *and Harish Dureja*[1,2,3*]

[1] Department of Pharmaceutical Sciences, Maharshi Dayanand University, Rohtak 124001, India

[2] Faculty of Health, Australian Research Centre in Complementary and Integrative Medicine, University of Technology, Sydney, Ultimo NSW 2007, Australia

[3] Centre for Research Impact & Outcome, Chitkara College of Pharmacy, Chitkara University, Rajpura, Punjab 140401, India

*Corresponding author: Harish Dureja, harishdureja@mdurohtak.ac.in.

8.1 INTRODUCTION

Lung cancer develops in the lung tissues, typically starting in the cells that line the airways (Sabbula & Anjum, 2021). Lung cancer is of two types: small cell lung cancer (SCLC) and non–small cell lung cancer (NSCLC) (Markman, 2022). Lung cancer remains the primary cause of cancer-related fatalities worldwide, significantly impacting the global cancer healthcare burden. In 2020, approximately 1.8 million people lost their lives and an estimated 2.2 million new cases were diagnosed, as reported by the Global Cancer Observatory (Zhou et al., 2024). The occurrence of lung cancer varies across regions, with higher rates found in countries where tobacco consumption is widespread, industrial pollution is a major concern, and the population is aging (Barta et al., 2019). Although smoking is the leading risk factor, genetic susceptibility and environmental exposures also play a role in disease development (CDC, 2010). Despite progress in early diagnosis and treatment, survival rates remain low, especially in low- and middle-income nations, where healthcare services and screening programs are less accessible (Shah et al., 2019). This highlights the critical need for effective prevention strategies and equitable healthcare access to lessen the global burden of lung cancer. Lung cancer treatment continues to encounter significant obstacles despite advancements in medical research. A major challenge is the late detection of the disease, as symptoms often emerge only in advanced stages, limiting the effectiveness of curative interventions (Vicidomini, 2023). The complexity of lung cancer, characterized by diverse genetic mutations and tumor subtypes, necessitates personalized treatment strategies that may not be widely available, particularly in resource constrained regions (Testa et al., 2018). Resistance to therapies, including targeted treatments and immunotherapy, frequently leads to disease progression, posing another major concern (Xiang et al., 2024). Additionally, managing treatment-related side effects while preserving patients' quality of life remains a persistent issue (Molassiotis et al., 2021). Healthcare disparities, high costs of treatment, and restricted access to advanced diagnostic tools further hinder effective disease management, especially in underserved communities (Kurzrock et al., 2024). Moreover, enhancing screening programs and identifying reliable biomarkers for early detection are crucial for improving patient outcomes (Xie et al., 2021). Addressing these interconnected challenges requires continued research, innovation, and equitable access to healthcare services to enhance lung cancer treatment and survival rates.

DOI: 10.1201/9781003595830-8

Liposomal drug delivery systems are crucial for treating lung cancer, as they improve the effectiveness and safety of anticancer medications (Hamad et al., 2024). By encapsulating drugs within lipid bilayers, these systems can enhance drug stability, control release, and target cancer cells more precisely, reducing side effects and improving treatment outcomes (Yingchoncharoen et al., 2016). Liposomes are spherical vesicles made of phospholipid bilayers, providing a flexible platform for encapsulating both hydrophilic and hydrophobic drugs (Nsairat et al., 2022). Their compatibility with biological systems, ability to enhance drug solubility, and potential for targeted drug delivery offer significant benefits for treating the lung cancer (Carrasco-Esteban et al., 2021). By altering the surface of liposomes with specific ligands or antibodies, these systems can be directed more precisely toward cancer cells (Deshpande et al., 2013). This approach helps to reduce unintended targeting, thereby lowering off target toxicity and minimizing the adverse effects commonly associated with traditional chemotherapy (Chehelgerdi et al., 2023). Furthermore, liposomes can facilitate the controlled and prolonged release of drugs, enhancing therapeutic effectiveness and minimizing the need for frequent dosing (Liu et al., 2022). They also demonstrate improved permeability and retention effects, which allow for the concentration of drugs within tumor tissues, while minimizing impact on healthy lung tissue (Wu, 2021). Advancements in liposome technology, including the creation of PEGylated (stealth) liposomes, have significantly increased their circulation time in the bloodstream, thereby improving the efficiency of drug delivery (Taher et al., 2023). Thus, liposome-based drug delivery systems represent a promising strategy in the fight against lung cancer, offering improved efficacy, reduced toxicity, and better patient outcomes (Gu et al., 2020).

Drugs like paclitaxel have anticancer activity but have limited water solubility, which limits their use for pulmonary drug delivery (Gala et al., 2020). The pulmonary epithelium layer is very thin in nature, which provides a shorter residence period to drug; with this the drugs are not capable to target the lungs (Wang et al., 2024; Fulton & Najahi-Missaoui, 2023). The drug-loaded liposome has potential to reduce these types of issues and have potential to target the lung cancer (Golombek et al., 2018). The liposomal inhalation delivery may localize the effect of drug in lung cancer (Deng et al., 2021; Zhang et al., 2024; Araghi et al., 2023). This chapter highlights liposomal drug delivery and its advancement in treatment of lung cancer.

8.2 LIPOSOMES: AN OVERVIEW

Liposomes are spherical entities composed of one or more phospholipid layers surrounding an aqueous centre (Nakhaei et al., 2021). These vesicles form when phospholipids, such as phosphatidylcholine, interact with water (Thompson, 1990). The amphipathic properties of phospholipids, characterized by their hydrophilic head groups and hydrophobic fatty acid tails, promote their self-organization into bilayers (Jiang et al., 2023). These bilayers then bend to create closed, spherical vesicles (Lombardo & Kiselev, 2022). Liposomes vary in size from nanometres to micrometres and are categorized by their structural composition (Table 8.1). Unilamellar liposomes have a single phospholipid bilayer, whereas multilamellar liposomes have several concentric bilayers, with aqueous layers separating each bilayer (Akbarzadeh et al., 2013). Liposomes are commonly utilized in drug delivery, cosmetics, and biotechnology because of their biocompatibility and capacity to encapsulate both hydrophilic and hydrophobic compounds (Rudzińska et al., 2024). Liposomes present numerous benefits in drug delivery for lung cancer, which positions them as a valuable option in oncology (Sercombe et al., 2015). Due to their biocompatibility and capacity to avoid immune detection, these particles can remain in circulation for extended periods, enhancing the delivery of drugs to tumour locations more efficiently (Yao et al., 2020). Liposomal formulations safeguard drugs from early degradation, allowing for the delivery of therapeutic agents that would otherwise be unstable (Tomsen-Melero et al., 2022). These characteristics enhance therapeutic effectiveness, minimize side effects, and open possibilities for combination therapies, helping us to tackle the intricate challenges associated with treating lung cancer (La'ah & Chiou, 2024).

TABLE 8.1
Classification of Different Liposomes According to Their Sizes

Liposome	Size (μm)
Giant	10.00–1000.00
Oligolamellar	0.10–10.00
Multivesicular	2.00–40.00
Multilamellar liposomes	0.02–0.50
Large unilamellar	>0.06
Small unilamellar	0.02–0.05

Eugster & Luciani, 2024; Dhillon et al., 2024.

8.3 RATIONALE FOR LIPOSOMAL DRUG DELIVERY IN LUNG CANCER

Liposomal drug delivery in lung cancer offers several advantages, particularly in addressing challenges such as bioavailability, stability, and multidrug resistance (Chen et al., 2024). One key benefit is the improved bioavailability and stability of the encapsulated drugs. Liposomes, which are lipid-based nanoparticles, enhance the solubility and stability of drugs with poor water solubility, allowing for more effective drug delivery (Lee, 2020). This increased stability prevents rapid degradation of the drug in the bloodstream, ensuring that the therapeutic agents remain intact until they reach the target site in the lungs, which is critical for achieving optimal therapeutic outcomes (Labiris & Dolovich, 2003). Furthermore, liposomal drug delivery allows for targeted delivery to lung tissues, a critical factor in lung cancer treatment (Rudokas et al., 2016). The lipid bilayer of liposomes can be engineered to interact with specific receptors or structures on the surface of lung cancer cells, enabling precise drug delivery directly to the tumour site (Wang et al., 2023). In addition, liposomal drug delivery systems have shown promise in overcoming multidrug resistance (MDR), a significant hurdle in the treatment of lung cancer (Omidian et al., 2024). MDR occurs when cancer cells develop resistance to chemotherapy drugs, often due to efflux pumps or other mechanisms that prevent the drugs from reaching their intracellular targets (Duan et al., 2023). Liposomes can bypass these resistance mechanisms by delivering drugs directly into the cancer cells, avoiding the efflux pumps that would normally pump out the drug (Garg et al., 2024). This can restore the efficacy of chemotherapeutic agents that may have otherwise been rendered ineffective in conventional formulations (Anand et al., 2022).

8.4 MECHANISMS OF LIPOSOMAL DRUG DELIVERY IN LUNG CANCER

Liposomal drug delivery systems utilize various mechanisms for delivering chemotherapeutic agents effectively (Allahou et al., 2021). One critical mechanism is passive targeting via enhanced permeability and retention (EPR) effect. Tumours possess leaky vasculature and deficient lymphatic drainage, allowing liposomes to accumulate preferentially in tumour tissues (Subhan et al., 2021). This phenomenon facilitates the selective delivery of drugs to the tumour site. Passive targeting is particularly advantageous for lung cancer, where systemic delivery often leads to toxicity (Li et al., 2023). In addition to passive targeting, active targeting using ligands and antibodies enhances the specificity of liposomal drug delivery (Riaz et al., 2018). Functionalizing the surface of liposomes with targeting moieties, such as folic acid, transferrin, peptides, or monoclonal antibodies, allows them to bind selectively to overexpressed receptors on lung cancer cells (Nel et al., 2023). For instance, ligands targeting epidermal growth factor receptor (EGFR) or integrins enable precise delivery, improving therapeutic outcomes by directly interacting with tumour-specific markers (Y. Zhou et al., 2024). Finally, the effectiveness of liposomal delivery relies on intracellular drug release mechanisms. Once the liposomes are internalized via receptor-mediated endocytosis or other pathways, they must release their payload efficiently within cancer cells (Amin et al., 2022). Several stimuli-responsive liposome regimens have been developed to enhance intracellular drug release (Ashrafizadeh et al., 2022). These include pH-sensitive liposomes that destabilize in the acidic tumour microenvironment or within endosomes and enzyme-sensitive liposomes that degrade upon exposure to tumour-specific enzymes. These mechanisms ensure the release of active drugs at the site of action (Nikolova et al., 2022).

8.5 APPLICATIONS OF LIPOSOMAL DRUG DELIVERY SYSTEM FOR LUNG CANCER

Research indicates that liposomal formulations, particularly for inhalation, could greatly enhance the efficacy of chemotherapy in lung cancer patients (Gupta et al., 2022). Several studies have investigated the potential of liposomal drug delivery for treating lung cancer (Lin et al., 2017). Koshkina and team researched on liposomal-based drug delivery of paclitaxel for its therapeutic effectiveness in lung cancer. The results revealed that paclitaxel gave longer drug residence time and increased drug concentration in the lungs during inhalation delivery as compared to intravenous delivery (Koshkina et al., 2001). Garbuzenko et al. (2009) compared the efficacy of liposomal doxorubicin (DOX) delivered via intratracheal administration to intravenous delivery in mice. The study found that intratracheal administration was significantly more effective in controlling growth of lung cancer and also caused less side effects to healthy organs (Garbuzenko et al., 2008). In another study, Hitzman et al. (2006) prepared a liposome-based drug delivery of 5-Fluorouracil (5-FU) and compared the release patterns of liposomal formulations for aerosol and intracellular deliver. They observed the similar patterns of 5-FU release from liposomal carriers, suggesting that liposomal aerosol delivery effectively reaches the lower respiratory tract (Hitzman et al., 2006). Cisplatin, a commonly used chemotherapy drug for lung cancer, is known for its dose-limiting toxicity, including nephrotoxicity and peripheral neuropathy. To mitigate these side effects, liposomal cisplatin formulations have been explored. Phase I studies of liposomal cisplatin for treating primary or metastatic lung carcinoma showed minimal side effects, with no dose-limiting toxicity in most patients (Stathopoulos et al., 2011). However, the study identified low drug decomposition in the lung region, possibly due to the absence of a 5% CO_2 concentration in the nebulizer. Furthermore, the drug cisplatin liposome (SLIT Cisplatin), developed by Transave Inc. (later acquired by Insmed), showed promising results in clinical research (Arriagada et al., 2004).

Camptothecin (9-NC) and its analogues face challenges related to solubility and dose-limited toxicity, limiting their potential as effective cancer treatments. Animal studies on liposomal 9-NC for lung cancer treatment showed significant reductions in tumour size with fewer side effects. A phase I clinical trial with aerosolized liposomal 9-NC reported partial remission in one patient and tumour stabilization in three others. The treatment was well tolerated with mild side effects, such as pharyngitis, nausea, fatigue, and coughing (Zamboni et al., 2004).

Several examples of research studies on liposomal drug delivery and their outcomes in treatment of lung cancer are shown in Table 8.2.

8.6 CLINICAL TRIALS ON LIPOSOMAL-BASED DRUG DELIVERY IN LUNG CANCER

Table 8.3 lists some clinical trials of liposomal-based drug delivery for lung cancer treatment.

TABLE 8.2
Liposomal-Based Drug Delivery Systems in the Treatment of Lung Cancer

Serial Number	Drug	Aim of Study	Drug's Limitation	Particle Size	Cancer Cell Line/ Animal Model	Outcome	References
1	Everolimus	To formulate liposomal drug delivery of everolimus	Low solubility and bioavailability	91.40 ± 23.90 nm	NCI-H446 cell lines	Prevents tumour growth	Iwase & Maitani, 2012
2	Gefitinib	To formulate liposomal gefitinib DPI for treating lung cancer	High distribution in body limits its efficacy	69.81 ± 2.61 nm	Lung cancer rat model	Instantly absorbed in lungs intratracheally than powder gefitinib orally	Zhang et al., 2020
3	Gemcitabine-HCL	To prepare liposome as DPI to reduce toxicity and avoid metabolism	High toxicity	325.00 nm	A-549 cell lines	Innovative approach to manage lung cancer	Gandhi et al., 2015
4	Pirfenidone (PFD)	To formulate inhalable PFD liposomes for NSCLC	Low efficacy	211.80–243.20 nm	Spheroid model	Better efficacy	Parvathaneni et al., 2020
5	Afatinib and Cetuximab	To increase retention and permeability in tumour	Low efficacy	109.50 ± 3.41 nm	NSCLC xenografted model	Liposomes protects afatinib from binding to blood cells and minimize risk of idiosyncratic drug toxic reaction	Lu et al., 2019
6	Osimertinib	To formulate liposomal vesicles using osimertinib	Low safety profile and potency	141.70 ± 7.60 nm	–	Remarkable results in patients with advanced NSCLC	Skupin-Mrugalska & Minko, 2020
7	Doxorubicin	To formulate liposomal doxorubicin drug delivery for lung cancer	Hydrophobicity, immunogenicity. And cytotoxicity	141.70 ± 7.60 nm	A-549 cell lines and lung carcinoma model in mice	Liposomal drugs persist longer in lung tissue without triggering an immune response	Ogawara et al., 2008
8	Vinorelbine	To decrease serious side effects	Granulocytopenia	180.50 ± 35.20 nm	–	Optimal size for passive targeting with minimal side effects	Bahadori et al., 2014
9	Fisetin	To facilitate *in vivo* administration	Low water solubility	162.00 ± 3.00 nm	A549 cell lines	Suitable for *in vivo* use and effectively inhibits tumour growth in a dose- and time-dependent manner	Wang & Huang, 2017
10	Paclitaxel	To decrease toxicity	Dose-limited toxicities and low water solubilities	501.00 ± 15.43 nm	–	Enhance therapeutic effectiveness while minimizing harmful toxicity	Zhao et al., 2010

(*Continued*)

TABLE 8.2 (*Continued*)
Liposomal-Based Drug Delivery Systems in the Treatment of Lung Cancer

Serial Number	Drug	Aim of Study	Drug's Limitation	Particle Size	Cancer Cell Line/ Animal Model	Outcome	References
11	Ginsenoside CK+ parthenolide	To improve the potency of lung cancer treatment	It lacks selective targeting between healthy and cancerous cells and may cause toxicity, affecting quality of life	188.00–155.00 nm	A549 cell lines	The combined use of natural products enhances anti-tumour effects while minimizing toxicity	Jin et al., 2018
12	Betulinic acid	To decrease rigidity and successfully drug incorporation	Toxicity and low bioavailability	1.00–1.50 μm	Foxn1 mice model	No systemic toxicity with a 50% reduction in tumour growth	Wu et al., 2019
13	Curcumin	To formulate lower-cost anticancer treatment options	Low solubility and bioavailability	101.00 ± 6.70 nm	A549 cell lines	Enhanced drug absorption and highly effective anticancer treatment with superior potency	Zhang et al., 2018
14	Berberine-HCL	Active targeting of drug	Low bioavailability, stability, and hydrophobicity	2.20–3.50 μm	A549 cell lines and lung cancer model in mice	Active targeting is often more effective than passive targeting	Sailor et al., 2015
15	Selaginella doederleinii	To increase solubility, increase bioavailability, and reduce toxicity	Low solubility and bioavailability	62.50 nm	A549 and PC-9 cell lines	Greater tumour inhibition potential with no systemic toxicity	Syaefudin et al., 2016
16	Telmisartan + Docetaxel	To formulate a combination of telmisartan and docetaxel for targeting lung cancer stem cells	Low cytotoxicity and efficacy	133.20 ± 11.70 nm	H460 cell lines	Demonstrated increased cytotoxicity and reduced cancer stem cell activity	Arthur et al., 2020
17	Cisplatin + phenethyl isothiocyanate	To formulate liposomal combination therapy for enhancing anticancer effect for NSCLC	Less potency and toxic	127.00 ± 47.00 nm	A549, H596 NSCLC cell lines	The liposomal formulation increased the toxicity of the drug combination in NCI-H596 NSCLC cells, resulting in a 30.2 ± 6.2% cell survival rate, compared to 50.9 ± 3.2% with the free drug treatment	Yang et al., 2014
18	Honokiol + cisplatin	To achieve better therapeutic efficacy by combining liposomal Honokiol with cisplatin	Less efficacy and toxicity	–	A-549 cell lines	Enhances the anti-tumour effects against human lung cancer	Jiang et al., 2008

TABLE 8.3
Clinical Trials on Liposome for Lung Cancer Treatment

Serial Number	NCT Number	Title	Phases	Status	Diseases	Sources
1	EUCTR2013-004659-19-NL	Ciprofloxacin Dry Powder for Inhalation (DPI) in Non-Cystic Fibrosis Bronchiectasis	III	Passed	Bronchiectasis	https://www.clinicaltrialsregister.eu/ctr-search/search?query=eudract_number%3A2013-004659-19
2	NCT00409188	Cancer Vaccine Study for Unresectable Stage III Non–Small Cell Lung Cancer	III	Passed	Non–small cell lung cancer	https://clinicaltrials.gov/study/NCT00409188
3	NCT00080418	Liposome Entrapped Paclitaxel Easy to Use (LEP-ETU) in Patients with Advanced Cancer	I	Passed	Neoplasm	https://clinicaltrials.gov/study/NCT00080418
4	NCT01515007	Phase 3 Study with Ciprofloxacin Dispersion for Inhalation in Non-CF Bronchiectasis (ORBIT-3)	III	Passed	Non-cystic fibrosis bronchiectasis	https://clinicaltrials.gov/study/NCT01515007
5	NCT03038178	Liposomal Amikacin for Inhalation (LAI) in the Treatment of Mycobacterium Abscessus Lung Disease	II	Passed	Mycobacterium infections, nontuberculous mycobacteria, atypical	https://clinicaltrials.gov/study/NCT03038178
6	NCT01254708	Safety and Efficacy Study of Inhaled Ambisome for Prevention of Aspergillus Colonization in Lung Transplant Recipients	II	Passed	Lung transplant recipient	https://clinicaltrials.gov/study/NCT01254708
7	NCT01650545	Aerosol Liposomal Cyclosporine for Chronic Rejection in Lung Transplant Recipients	I and II	Passed	Lung transplantation and bronchiolitis obliterans	https://clinicaltrials.gov/study/NCT01650545
8	ISRCTN66069132	Aerosolised liposomal cyclosporin A (L-CsA) versus aerosolised placebo in the prevention of bronchiolitis obliterans (BO) in lung transplant (LT) patients	III	Passed	Bronchiolitis obliterans	https://www.isrctn.com/ISRCTN66069132
9	NCT00102531	Inhalation SLIT Cisplatin (Liposomal) for the Treatment of Osteosarcoma Metastatic to the Lung	I and II	Passed	Osteosarcoma metastatic	https://clinicaltrials.gov/study/NCT00102531
10	NCT00777296	Multidose Safety and Tolerability Study of Dose Escalation of Liposomal Amikacin for Inhalation (ARIKACE™)	I and II	Passed	Cystic fibrosis	https://clinicaltrials.gov/study/NCT00777296

8.7 CHALLENGES AND FUTURE PROSPECT

Despite remarkable progress in utilizing liposomal drug delivery for lung cancer therapy, several obstacles persist. A significant challenge is drug resistance, where cancer cells develop adaptive mechanisms that diminish the effectiveness of chemotherapy. Additionally, the heterogeneous nature of lung tumours makes it difficult to create a single liposomal formulation suitable for all patients. Another critical issue is the high cost of liposome production and development, which limits its accessibility, especially in low- and middle-income nations. Furthermore, ensuring drug stability, proper storage, and large-scale manufacturing remains a concern, necessitating further research to maintain consistent drug efficacy.

Looking ahead, liposomal drug delivery holds great potential for advancing lung cancer treatment. The introduction of stimuli-responsive liposomes, which release their drug payload in response to tumour-specific factors like pH, temperature, or enzyme activity, could significantly improve precision and treatment efficiency. Innovations in nanotechnology and surface engineering may further enhance liposomal formulations, helping to overcome drug resistance. Additionally, integrating liposomal drug delivery with immunotherapy and other advanced treatment approaches could lead to better therapeutic outcomes. Moreover, the expansion of inhalation-based liposomal therapies offers a promising strategy for localized drug delivery to the lungs, maximizing treatment benefits while reducing systemic toxicity.

8.8 CONCLUSION

Liposomal drug delivery is a ground-breaking strategy in lung cancer treatment, enhancing drug stability, precision targeting, and minimizing systemic toxicity. By utilizing both passive and active targeting mechanisms, liposomal formulations improve the effectiveness of traditional chemotherapy. However, challenges such as drug resistance, high manufacturing costs, and tumour variability remain obstacles. Ongoing research and technological advancements, including stimuli-responsive liposomes, nanotechnology, and personalized medicine, offer promising solutions to these issues. Further development of inhalation-based therapies and improved accessibility will be crucial in ensuring more effective and equitable lung cancer treatment for all patients.

REFERENCES

Akbarzadeh, A., Rezaei-Sadabady, R., Davaran, S., Joo, S. W., Zarghami, N., Hanifehpour, Y., Samiei, M., Kouhi, M., & Nejati-Koshki, K. (2013). Liposome: Classification, preparation, and applications. *Nanoscale Research Letters*, *8*(1). https://doi.org/10.1186/1556-276x-8-102

Aksamit, T., De Soyza, A., Bandel, T.-J., Criollo, M., Elborn, J. S., Operschall, E., Polverino, E., Roth, K., Winthrop, K. L., & Wilson, R. (2018). RESPIRE 2: A phase III placebo-controlled randomised trial of ciprofloxacin dry powder for inhalation in non-cystic fibrosis bronchiectasis. *European Respiratory Journal*, *51*(1), 1702053. https://doi.org/10.1183/13993003.02053-2017

Allahou, L. W., Madani, S. Y., & Seifalian, A. (2021). Investigating the application of liposomes as drug delivery systems for the diagnosis and treatment of cancer. *International Journal of Biomaterials*, *2021*, e3041969. https://doi.org/10.1155/2021/3041969

Amin, M., Seynhaeve, A. L. B., Sharifi, M., Falahati, M., & ten Hagen, T. L. M. (2022). Liposomal drug delivery systems for cancer therapy: The Rotterdam experience. *Pharmaceutics*, *14*(10), 2165. https://doi.org/10.3390/pharmaceutics14102165

Anand, U., Dey, A., Chandel, A. K. S., Sanyal, R., Mishra, A., Pandey, D. K., De Falco, V., Upadhyay, A., Kandimalla, R., Chaudhary, A., Dhanjal, J. K., Dewanjee, S., Vallamkondu, J., & Pérez de la Lastra, J. M. (2022). Cancer chemotherapy and beyond: Current status, drug candidates, associated risks and progress in targeted therapeutics. *Genes & Diseases*, *10*(4), 1367–1401. https://doi.org/10.1016/j.gendis.2022.02.007

Araghi, M., Mannani, R., Heidarnejad Maleki, A., Hamidi, A., Rostami, S., Safa, S. H., Faramarzi, F., Khorasani, S., Alimohammadi, M., Tahmasebi, S., & Akhavan-Sigari, R. (2023). Recent advances in non-small cell lung cancer targeted therapy; an update review. *Cancer Cell International*, *23*(1). https://doi.org/10.1186/s12935-023-02990-y

Arriagada, R., Bergman, B., Dunant, A., Le Chevalier, T., Pignon, J.-P., Vansteenkiste, J., & International Adjuvant Lung Cancer Trial Collaborative Group. (2004). Cisplatin-based adjuvant chemotherapy in patients with completely resected non-small-cell lung cancer. *The New England Journal of Medicine*, *350*(4), 351–360. https://doi.org/10.1056/NEJMoa031644

Arthur, P., Patel, N., Surapaneni, S. K., Mondal, A., Gebeyehu, A., Bagde, A., Kutlehria, S., Nottingham, E., & Singh, M. (2020). Targeting lung cancer stem cells using combination of Tel and Docetaxel liposomes in 3D cultures and tumor xenografts. *Toxicology and Applied Pharmacology*, *401*(15), 115112. https://doi.org/10.1016/j.taap.2020.115112

Ashrafizadeh, M., Delfi, M., Zarrabi, A., Bigham, A., Sharifi, E., Rabiee, N., Paiva-Santos, A. C., Kumar, A. P., Tan, S. C., Hushmandi, K., Ren, J., Zare, E. N., & Makvandi, P. (2022). Stimuli-responsive liposomal nanoformulations in cancer therapy: Pre-clinical & clinical approaches. *Journal of Controlled Release*, *351*(2022), 50–80. https://doi.org/10.1016/j.jconrel.2022.08.001

Bahadori, F., Topçu, G., Eroğlu, M. S., & Önyüksel, H. (2014). A new lipid-based nano formulation of vinorelbine. *AAPS PharmSciTech*, *15*(5), 1138–1148. https://doi.org/10.1208/s12249-014-0146-3

Barta, J. A., Powell, C. A., & Wisnivesky, J. P. (2019). Global epidemiology of lung cancer. *Annals of Global Health*, *85*(1). https://doi.org/10.5334/aogh.2419

Carrasco-Esteban, E., Domínguez-Rullán, J. A., Barrionuevo-Castillo, P., Pelari-Mici, L., Leaman, O., Sastre-Gallego, S., & López-Campos, F. (2021). Current role of nanoparticles in the treatment of lung cancer. *Journal of Clinical and Translational Research*, *7*(2), 140. https://pmc.ncbi.nlm.nih.gov/articles/PMC8177846/

CDC. (2010). *Pulmonary Diseases*. Nih.gov; Centers for Disease Control and Prevention (US). https://www.ncbi.nlm.nih.gov/books/NBK53021/

Chehelgerdi, M., Chehelgerdi, M., Allela, O.Q.B., Pecho, R.D.C., Jayasankar, N., Rao, D.P., Thamaraikani, T., Vasanthan, M., Viktor, P., Lakshmaiya, N., Saadh, M. J., Amajd, A., Abo-Zaid, M. A., Castillo-Acobo, R. Y., Ismail, A. H., Amin, A. H., & Akhavan-Sigari, R. (2023). Progressing nanotechnology to improve targeted cancer treatment: Overcoming hurdles in its clinical implementation. *Molecular Cancer*, *22*(1). https://doi.org/10.1186/s12943-023-01865-0

Chen, J., Hu, S., Sun, M., Shi, J., Zhang, H., Yu, H., & Yang, Z. (2024). Recent advances and clinical translation of liposomal delivery systems in cancer therapy. *European Journal of Pharmaceutical Sciences*, *193*(2024), 106688. https://doi.org/10.1016/j.ejps.2023.106688

Deng, Z., Kalin, G. T., Shi, D., & Kalinichenko, V. V. (2021). Nanoparticle delivery systems with cell-specific targeting for pulmonary diseases. *American Journal of Respiratory Cell and Molecular Biology*, *64*(3), 292–307. https://doi.org/10.1165/rcmb.2020-0306tr

Deshpande, P. P., Biswas, S., & Torchilin, V. P. (2013). Current trends in the use of liposomes for tumor targeting. *Nanomedicine (London, England)*, *8*(9). https://doi.org/10.2217/nnm.13.118

Dhillon, A., Singh, R., & Senwar, K. R. (2024). An extensive review on novel liposomes: Classification, methodology, characterization, current formulations. *International Journal of Drug Delivery Technology*, *14*(03), 1842–1852. https://doi.org/10.25258/ijddt.14.3.83

Duan, C., Yu, M., Xu, J., Li, B.-Y., Zhao, Y., & Kankala, R. K. (2023). Overcoming Cancer multi-drug resistance (MDR): Reasons, mechanisms, nanotherapeutic solutions, and challenges. *Biomedicine & Pharmacotherapy*, *162*(2023), 114643. https://doi.org/10.1016/j.biopha.2023.114643

Eugster, R., & Luciani, P. (2024). Liposomes: Bridging the gap from lab to pharmaceuticals. *Current Opinion in Colloid & Interface Science*, *75*(2025), 101875. https://doi.org/10.1016/j.cocis.2024.101875

Fulton, M. D., & Najahi-Missaoui, W. (2023). Liposomes in cancer therapy: How did we start and where are we now. *International Journal of Molecular Sciences*, *24*(7), 6615. https://doi.org/10.3390/ijms24076615

Gala, U. H., Miller, D. A., & Williams, R. O. (2020). Harnessing the therapeutic potential of anticancer drugs through amorphous solid dispersions. *Biochimica et Biophysica Acta (BBA)—Reviews on Cancer*, *1873*(1), 188319. https://doi.org/10.1016/j.bbcan.2019.188319

Gandhi, M., Pandya, T., Gandhi, R., Patel, S., Mashru, R., Misra, A., & Tandel, H. (2015). Inhalable liposomal dry powder of gemcitabine-HCl: Formulation, in vitro characterization and in vivo studies. *International Journal of Pharmaceutics*, *496*(2), 886–895. https://doi.org/10.1016/j.ijpharm.2015.10.020

Garbuzenko, O. B., Saad, M., Betigeri, S., Zhang, M., Vetcher, A. A., Soldatenkov, V. A., Reimer, D. C., Pozharov, V. P., & Minko, T. (2008). Intratracheal versus intravenous liposomal delivery of siRNA, antisense oligonucleotides and anticancer drug. *Pharmaceutical Research*, *26*(2), 382–394. https://doi.org/10.1007/s11095-008-9755-4

Garg, P., Malhotra, J., Kulkarni, P., Horne, D., Salgia, R., & Singhal, S. S. (2024). Emerging therapeutic strategies to overcome drug resistance in cancer cells. *Cancers*, *16*(13), 2478. https://doi.org/10.3390/cancers16132478

Golombek, S. K., May, J.-N., Theek, B., Appold, L., Drude, N., Kiessling, F., & Lammers, T. (2018). Tumor targeting via EPR: Strategies to enhance patient responses. *Advanced Drug Delivery Reviews*, *130*(2018), 17–38. https://doi.org/10.1016/j.addr.2018.07.007

Gu, Z., Da Silva, C. G., Van der Maaden, K., Ossendorp, F., & Cruz, L. J. (2020). Liposome-based drug delivery systems in cancer immunotherapy. *Pharmaceutics*, *12*(11). https://doi.org/10.3390/pharmaceutics12111054

Gupta, C., Jaipuria, A., & Gupta, N. (2022). Inhalable formulations to treat non-small cell lung cancer (NSCLC): Recent therapies and developments. *Pharmaceutics*, *15*(1), 139. https://doi.org/10.3390/pharmaceutics15010139

Hamad, I., Harb, A. A., & Bustanji, Y. (2024). Liposome-based drug delivery systems in cancer research: An analysis of global landscape efforts and achievements. *Pharmaceutics*, *16*(3), 400. https://doi.org/10.3390/pharmaceutics16030400

Hitzman, C. J., Elmquist, W. F., Wattenberg, L. W., & Wiedmann, T. S. (2006). Development of a respirable, sustained release microcarrier for 5-fluorouracil I: In vitro assessment of liposomes, microspheres, and lipid coated nanoparticles. *Journal of Pharmaceutical Sciences*, *95*(5), 1114–1126. https://doi.org/10.1002/jps.20591

Iwase, Y., & Maitani, Y. (2012). Preparation and in vivo evaluation of liposomal everolimus for lung carcinoma and thyroid carcinoma. *Biological and Pharmaceutical Bulletin*, *35*(6), 975–979. https://doi.org/10.1248/bpb.35.975

Jiang, Q., Fan, L., Yang, G., Guo, W.-H., Hou, W., Chen, L., & Wei, Y. (2008). Improved therapeutic effectiveness by combining liposomal honokiol with cisplatin in lung cancer model. *BMC Cancer*, *8*(1). https://doi.org/10.1186/1471-2407-8-242

Jiang, Y., Li, W., Wang, Z., & Lu, J. (2023). Lipid-based nanotechnology: Liposome. *Pharmaceutics*, *16*(1), 34. https://doi.org/10.3390/pharmaceutics16010034

Jin, X., Zhou, J., Zhang, Z., & Lv, H. (2018). The combined administration of parthenolide and ginsenoside CK in long circulation liposomes with targeted tLyp-1 ligand induce mitochondria-mediated lung cancer apoptosis. *Artificial Cells, Nanomedicine, and Biotechnology*, *46*(Suppl 3), 931–942. https://doi.org/10.1080/21691401.2018.1518913

Koshkina, N. V., Waldrep, J. C., Roberts, L. E., Golunski, E., Melton, S., & Knight, V. (2001). Paclitaxel liposome aerosol treatment induces inhibition of pulmonary metastases in murine renal carcinoma model. *Clinical Cancer Research: An Official Journal of the American Association for Cancer Research*, *7*(10), 3258–3262. https://pubmed.ncbi.nlm.nih.gov/11595722/

Kurzrock, R., Chaudhuri, A. A., Feller-Kopman, D., Florez, N., Gorden, J., & Wistuba, I. I. (2024). Healthcare disparities, screening, and molecular testing in the changing landscape of non-small cell lung cancer in the United States: A review. *Cancer and Metastasis Reviews*, *4*(43). https://doi.org/10.1007/s10555-024-10187-6

La'ah, A. S., & Chiou, S.-H. (2024). Cutting-edge therapies for lung cancer. *Cells*, *13*(5), 436. https://doi.org/10.3390/cells13050436

Labiris, N. R., & Dolovich, M. B. (2003). Pulmonary drug delivery. Part I: Physiological factors affecting therapeutic effectiveness of aerosolized medications. *British Journal of Clinical Pharmacology*, *56*(6), 588–599. https://doi.org/10.1046/j.1365-2125.2003.01892.x

Lee, M.-K. (2020). Liposomes for enhanced bioavailability of water-insoluble drugs: In vivo evidence and recent approaches. *Pharmaceutics*, *12*(3), 264. https://doi.org/10.3390/pharmaceutics12030264

Lee, Y., & Thompson, D. H. (2017). Stimuli-responsive liposomes for drug delivery. *Wiley Interdisciplinary Reviews: Nanomedicine and Nanobiotechnology*, *9*(5). https://doi.org/10.1002/wnan.1450

Li, J., Wang, Q., Xia, G., Adilijiang, N., Li, Y., Hou, Z., Fan, Z., & Li, J. (2023). Recent advances in targeted drug delivery strategy for enhancing oncotherapy. *Pharmaceutics*, *15*(9), 2233. https://doi.org/10.3390/pharmaceutics15092233

Lin, C., Wong, B. C. K., Chen, H., Bian, Z., Zhang, G., Zhang, X., Kashif Riaz, M., Tyagi, D., Lin, G., Zhang, Y., Wang, J., Lu, A., & Yang, Z. (2017). Pulmonary delivery of triptolide-loaded liposomes decorated with anti-carbonic anhydrase IX antibody for lung cancer therapy. *Scientific Reports*, *7*(1). https://doi.org/10.1038/s41598-017-00957-4

Liu, P., Chen, G., & Zhang, J. (2022). A review of liposomes as a drug delivery system: Current status of approved products, regulatory environments, and future perspectives. *Molecules*, *27*(4). https://doi.org/10.3390/molecules27041372

Lombardo, D., & Kiselev, M. A. (2022). Methods of liposomes preparation: Formation and control factors of versatile nanocarriers for biomedical and nanomedicine application. *Pharmaceutics*, *14*(3), 543. https://doi.org/10.3390/pharmaceutics14030543

Lu, X., Liu, S., Han, M., Yang, X., Sun, K., Wang, H., Mu, H., Du, Y., Wang, A., Ni, L., & Zhang, C. (2019). Afatinib-loaded immunoliposomes functionalized with cetuximab: A novel strategy targeting the epidermal growth factor receptor for treatment of non-small-cell lung cancer. *International Journal of Pharmaceutics*, *560*(2019), 126–135. https://doi.org/10.1016/j.ijpharm.2019.02.001

Markman, M. (2022, May 6). *Lung Cancer*. Cancer Treatment Centers of America. https://www.cancercenter.com/cancer-types/lung-cancer/types

Mitchell, P., Thatcher, N., Socinski, M. A., Wasilewska-Tesluk, E., Horwood, K., Szczesna, A., Martín, C., Ragulin, Y., Zukin, M., Helwig, C., Falk, M., Butts, C., & Shepherd, F. A. (2015). Tecemotide in unresectable stage III non-small-cell lung cancer in the phase III START study: Updated overall survival and biomarker analyses. *Annals of Oncology: Official Journal of the European Society for Medical Oncology*, *26*(6), 1134–1142. https://doi.org/10.1093/annonc/mdv104

Molassiotis, A., Yates, P., & Yorke, J. (2021). Editorial: Quality of life and side effects management in lung cancer treatment. *Frontiers in Oncology*, *11*(2021). https://doi.org/10.3389/fonc.2021.651797

Nakhaei, P., Margiana, R., Bokov, D. O., Abdelbasset, W. K., Jadidi Kouhbanani, M. A., Varma, R. S., Marofi, F., Jarahian, M., & Beheshtkhoo, N. (2021). Liposomes: Structure, biomedical applications, and stability parameters with emphasis on cholesterol. *Frontiers in Bioengineering and Biotechnology*, *9*(2021), 705886. https://doi.org/10.3389/fbioe.2021.705886

Nel, J., Elkhoury, K., Velot, É., Bianchi, A., Acherar, S., Francius, G., Tamayol, A., Grandemange, S., & Arab-Tehrany, E. (2023). Functionalized liposomes for targeted breast cancer drug delivery. *Bioactive Materials*, *24*(2023), 401–437. https://doi.org/10.1016/j.bioactmat.2022.12.027

Nikolova, M. P., Kumar, E. M., & Chavali, M. S. (2022). Updates on responsive drug delivery based on liposome vehicles for cancer treatment. *Pharmaceutics*, *14*(10), 2195. https://doi.org/10.3390/pharmaceutics14102195

Nsairat, H., Khater, D., Sayed, U., Odeh, F., Al Bawab, A., & Alshaer, W. (2022). Liposomes: Structure, composition, types, and clinical applications. *Heliyon*, *8*(5), e09394. https://doi.org/10.1016/j.heliyon.2022.e09394

Ogawara, K., Un, K., Minato, K., Tanaka, K., Higaki, K., & Kimura, T. (2008). Determinants for in vivo anti-tumor effects of PEG liposomal doxorubicin: Importance of vascular permeability within tumors. *International Journal of Pharmaceutics*, *359*(1–2), 234–240. https://doi.org/10.1016/j.ijpharm.2008.03.025

Omidian, H., Gill, E. J., & Cubeddu, L. X. (2024). Lipid nanoparticles in lung cancer therapy. *Pharmaceutics*, *16*(5), 644. https://doi.org/10.3390/pharmaceutics16050644

Parvathaneni, V., Kulkarni, N. S., Shukla, S. K., Farrales, P. T., Kunda, N. K., Muth, A., & Gupta, V. (2020). Systematic development and optimization of inhalable pirfenidone liposomes for non-small cell lung cancer treatment. *Pharmaceutics*, *12*(3). https://doi.org/10.3390/pharmaceutics12030206

Riaz, M. K., Riaz, M. A., Zhang, X., Lin, C., Wong, K. H., Chen, X., Zhang, G., Lu, A., & Yang, Z. (2018). Surface functionalization and targeting strategies of liposomes in solid tumor therapy: A review. *International Journal of Molecular Sciences*, *19*(1). https://doi.org/10.3390/ijms19010195

Rudokas, M., Najlah, M., Alhnan, M. A., & Elhissi, A. (2016). Liposome delivery systems for inhalation: A critical review highlighting formulation issues and anticancer applications. *Medical Principles and Practice*, *25*(2), 60–72. https://doi.org/10.1159/000445116

Rudzińska, M., Grygier, A., Knight, G., & Kmiecik, D. (2024). Liposomes as carriers of bioactive compounds in human nutrition. *Foods*, *13*(12), 1814. https://doi.org/10.3390/foods13121814

Sabbula, B. R., & Anjum, F. (2021). *Squamous Cell Lung Cancer*. PubMed; StatPearls Publishing. https://www.ncbi.nlm.nih.gov/books/NBK564510/

Sailor, G., Seth, A., Parmar, G., Chauhan, S., & Javia, A. (2015). Formulation and in vitro evaluation of berberine containing liposome optimized by 32 full factorial designs. *Journal of Applied Pharmaceutical Science*, *5*(7), 23–28. https://doi.org/10.7324/japs.2015.50704

Sercombe, L., Veerati, T., Moheimani, F., Wu, S. Y., Sood, A. K., & Hua, S. (2015). Advances and challenges of liposome assisted drug delivery. *Frontiers in Pharmacology*, *6*(286). https://doi.org/10.3389/fphar.2015.00286

Shah, S. C., Kayamba, V., Peek, R. M., & Heimburger, D. (2019). Cancer control in low- and middle-income countries: Is it time to consider screening? *Journal of Global Oncology*, *5*(5), 1–8. https://doi.org/10.1200/jgo.18.00200

Skupin-Mrugalska, P., & Minko, T. (2020). Development of liposomal vesicles for osimertinib delivery to EGFR mutation—positive lung cancer cells. *Pharmaceutics*, *12*(10), 939. https://doi.org/10.3390/pharmaceutics12100939

Stathopoulos, G. P., Antoniou, D., Dimitroulis, J., Stathopoulos, J., Marosis, K., & Michalopoulou, P. (2011). Comparison of liposomal cisplatin versus cisplatin in non-squamous cell non-small-cell lung cancer. *Cancer Chemotherapy and Pharmacology*, *68*(4), 945–950. https://doi.org/10.1007/s00280-011-1572-5

Subhan, M. A., Yalamarty, S. S. K., Filipczak, N., Parveen, F., & Torchilin, V. P. (2021). Recent advances in tumor targeting via EPR effect for cancer treatment. *Journal of Personalized Medicine*, *11*(6), 571. https://doi.org/10.3390/jpm11060571

Syaefudin, N., A Juniarti, L Rosiyana, A Setyani, & S Khodijah. (2016). Nanoparticles of *Selaginella doederleinii* leaf extract inhibit human lung cancer cells A549. *IOP Conference Series Earth and Environmental Science*, *31*(2016), 012029. https://doi.org/10.1088/1755-1315/31/1/012029

Taher, M., Susanti, D., Haris, M. S., Rushdan, A. A., Widodo, R. T., Syukri, Y., & Khotib, J. (2023). PEGylated liposomes enhance the effect of cytotoxic drug: A review. *Heliyon*, *9*(3), e13823. https://doi.org/10.1016/j.heliyon.2023.e13823

Testa, U., Castelli, G., & Pelosi, E. (2018). Lung cancers: Molecular characterization, clonal heterogeneity and evolution, and cancer stem cells. *Cancers*, *10*(8), 248. https://doi.org/10.3390/cancers10080248

Thompson, T. E. (1990). Phosphatidylcholine vesicles: Structure and formation. *Hepatology (Baltimore, Md.)*, *12*(3 Pt 2), 51S52S; discussion 53S–55S. https://pubmed.ncbi.nlm.nih.gov/2210657/

Tomsen-Melero, J., Merlo-Mas, J., Carreño, A., Sala, S., Córdoba, A., Veciana, J., González-Mira, E., & Ventosa, N. (2022). Liposomal formulations for treating lysosomal storage disorders. *Advanced Drug Delivery Reviews*, *190*(2022), 114531. https://doi.org/10.1016/j.addr.2022.114531

Vicidomini, G. (2023). Current challenges and future advances in lung cancer: Genetics, instrumental diagnosis and treatment. *Cancers*, *15*(14), 3710. https://doi.org/10.3390/cancers15143710

Wang, B., Wang, L., Yang, Q., Zhang, Y., Tang Qinglai, Yang, X., Xiao, Z., Lei, L., & Li, S. (2024). Pulmonary inhalation for disease treatment: Basic research and clinical translations. *Materials Today Bio*, *25*(2024), 100966. https://doi.org/10.1016/j.mtbio.2024.100966

Wang, J., & Huang, S. (2017). Fisetin inhibits the growth and migration in the A549 human lung cancer cell line via the ERK1/2 pathway. *Experimental and Therapeutic Medicine*, *3*(15). https://doi.org/10.3892/etm.2017.5666

Wang, S., Chen, Y., Guo, J., & Huang, Q. (2023). Liposomes for tumor targeted therapy: A review. *International Journal of Molecular Sciences*, *24*(3), 2643. https://doi.org/10.3390/ijms24032643

Wu, J. (2021). The enhanced permeability and retention (EPR) effect: The significance of the concept and methods to enhance its application. *Journal of Personalized Medicine*, *11*(8), 771. https://doi.org/10.3390/jpm11080771

Wu, J., Niu, Y., Jiao, Y., & Chen, Q. (2019). Fungal chitosan from Agaricus bisporus (Lange) Sing. Chaidam increased the stability and antioxidant activity of liposomes modified with biosurfactants and loading betulinic acid. *International Journal of Biological Macromolecules*, *123*(15), 291–299. https://doi.org/10.1016/j.ijbiomac.2018.11.062

Xiang, Y., Liu, X., Wang, Y., Zheng, D., Meng, Q., Jiang, L., Yang, S., Zhang, S., Zhang, X., Liu, Y., & Wang, B. (2024). Mechanisms of resistance to targeted therapy and immunotherapy in non-small cell lung cancer: Promising strategies to overcoming challenges. *Frontiers in Immunology*, *15*(2024). https://doi.org/10.3389/fimmu.2024.1366260

Xie, Y., Meng, W.-Y., Li, R., Wang, Y., Qian, X., Chan, C.-C., Yu, Z., Fan, X.-X., Pan, H., Xie, C., Wu, Q., Yan, P., Liu, L., Tang, Y., Yao, X.-S., Wang, M., & Leung, E. L.-H. (2021). Early lung cancer diagnostic biomarker discovery by machine learning methods. *Translational Oncology*, *14*(1), 100907. https://doi.org/10.1016/j.tranon.2020.100907

Yang, Y.-T., Shi, Y., Jay, M., & Di, A. J. (2014). Enhanced toxicity of cisplatin with chemosensitizer phenethyl isothiocyanate toward non-small cell lung cancer cells when delivered in liposomal nanoparticles. *Chemical Research in Toxicology*, *27*(6), 946–948. https://doi.org/10.1021/tx5001128

Yao, Y., Zhou, Y., Liu, L., Xu, Y., Chen, Q., Wang, Y., Wu, S., Deng, Y., Zhang, J., & Shao, A. (2020). Nanoparticle-based drug delivery in cancer therapy and its role in overcoming drug resistance. *Frontiers in Molecular Biosciences*, *7*(193). https://doi.org/10.3389/fmolb.2020.00193

Yingchoncharoen, P., Kalinowski, D. S., & Richardson, D. R. (2016). Lipid-based drug delivery systems in cancer therapy: What is available and what is yet to come. *Pharmacological Reviews*, *68*(3), 701–787. https://doi.org/10.1124/pr.115.012070

Zamboni, W. C., Jung, L. L., Egorin, M. J., Potter, D. M., Friedland, D. M., Belani, C. P., Agarwala, S. S., Wong, M. M., Fakih, M., Trump, D. L., Jin, R., Strychor, S., Vozniak, M., Troetschel, M., & Ramanathan, R. K. (2004). Phase I and pharmacologic study of intermittently administered 9-nitrocamptothecin in patients with advanced solid tumors. *Clinical Cancer Research*, *10*(15), 5058–5064. https://doi.org/10.1158/1078-0432.ccr-03-0288

Zhang, S., Li, R., Jiang, T., Gao, Y., Zhong, K., Cheng, H., Chen, X., & Li, S. (2024). Inhalable nanomedicine for lung cancer treatment. *Smart Materials in Medicine*, *5*(2024). https://doi.org/10.1016/j.smaim.2024.04.001

Zhang, T., Chen, Y., Ge, Y., Hu, Y., Li, M., & Jin, Y. (2018). Inhalation treatment of primary lung cancer using liposomal curcumin dry powder inhalers. *Acta Pharmaceutica Sinica B*, *8*(3), 440–448. https://doi.org/10.1016/j.apsb.2018.03.004

Zhang, T., Wang, R., Li, M., Bao, J., Chen, Y., Ge, Y., & Jin, Y. (2020). Comparative study of intratracheal and oral gefitinib for the treatment of primary lung cancer. *European Journal of Pharmaceutical Sciences*, *149*(2020), 105352. https://doi.org/10.1016/j.ejps.2020.105352

Zhao, L., Ye, Y., Li, J., & Wei, Y. (2010). Preparation and the in-vivo evaluation of paclitaxel liposomes for lung targeting delivery in dogs. *Journal of Pharmacy and Pharmacology*, *63*(1), 80–86. https://doi.org/10.1111/j.2042-7158.2010.01184.x

Zhou, J., Xu, Y., Liu, J., Feng, L., Yu, J., & Chen, D. (2024). Global burden of lung cancer in 2022 and projections to 2050: Incidence and mortality estimates from GLOBOCAN. *Cancer Epidemiology*, *93*(2024), 102693. https://doi.org/10.1016/j.canep.2024.102693

Zhou, Y., Tao, L., Qiu, J., Xu, J., Yang, X., Zhang, Y., Tian, X., Guan, X., Cen, X., & Zhao, Y. (2024). Tumor biomarkers for diagnosis, prognosis and targeted therapy. *Signal Transduction and Targeted Therapy*, *9*(1), 1–86. https://doi.org/10.1038/s41392-024-01823-2

https://clinicaltrials.gov/study/NCT00080418
https://clinicaltrials.gov/study/NCT00102531
https://clinicaltrials.gov/study/NCT00409188
https://clinicaltrials.gov/study/NCT00777296
https://clinicaltrials.gov/study/NCT01254708
https://clinicaltrials.gov/study/NCT01515007
https://clinicaltrials.gov/study/NCT01650545
https://clinicaltrials.gov/study/NCT03038178
https://www.clinicaltrialsregister.eu/ctr-search/search?query=eudract_number%3A2013-004659-19
https://www.isrctn.com/ISRCTN66069132

9 Polymeric Nanoparticles for the Management of Lung Cancer

Nisha Gulati[1], *Raman Jangra*[1], *Shobha Yadav*[1], *Kamal Dua*[1,2,4], *and Harish Dureja*[1,2,3*]

[1] Department of Pharmaceutical Sciences, Maharshi Dayanand University, Rohtak 124001, India

[2] Faculty of Health, Australian Research Centre in Complementary and Integrative Medicine, University of Technology Sydney, Ultimo, Australia

[3] Centre for Research Impact & Outcome, Chitkara College of Pharmacy, Chitkara University, Rajpura 140401, Punjab, India

[4] Discipline of Pharmacy, Graduate School of Health, University of Technology Sydney, Ultimo NSW 2007, Australia

*Corresponding author: Harish Dureja, harishdureja@mdurohtak.ac.in, harishdureja@gmail.com

9.1 INTRODUCTION

Lung cancer continues to be a serious worldwide health issue with its high incidence rates and poor prognosis. As of 2025, the U.S. expects 226,650 new lung cancer cases and 124,730 deaths. In India, lung cancer is a leading cause of cancer-related deaths, particularly among men. Projections for 2025 estimate over 161,000 new lung cancer cases, comprising approximately 81,219 cases in men and 30,109 in women. Even with developments in diagnosis techniques and treatment approaches, the prognosis for those with lung cancer, especially those discovered at advanced stages, is still poor. Lack of selectivity in traditional treatment approaches affects both healthy and malignant tissues and leads to treatment failure. Targeted cancer treatment now is more feasible due to recent advancements in the development of nanotechnology. Polymeric nanoparticles are one of several nanocarrier systems that provide a flexible and effective means of delivering genes, immunomodulators, and chemotherapeutic medications straight to tumor locations [1–3]. Lung cancer primarily consists of two major types: non–small cell lung cancer (NSCLC) and small cell lung cancer (SCLC). NSCLC accounts for more than 84% of all lung cancers and is a significant contributor to global cancer death rates [4]. The major problem associated with lung cancer is the lack of symptoms at the early stage of cancer; the disease is diagnosed at an advanced stage, which reduces the efficacy of treatment measures. Targeted therapy faces both difficulties and potential due to the genetic heterogeneity of lung cancers and the existence of oncogenic drivers such epidermal growth factor receptor mutations and anaplastic lymphoma kinase rearrangements. Off-target effects and drug resistance development are still major problems even with the availability of targeted drug delivery and immunotherapies, highlighting the need for more accurate delivery methods [5–6].

 DOI: 10.1201/9781003595830-9

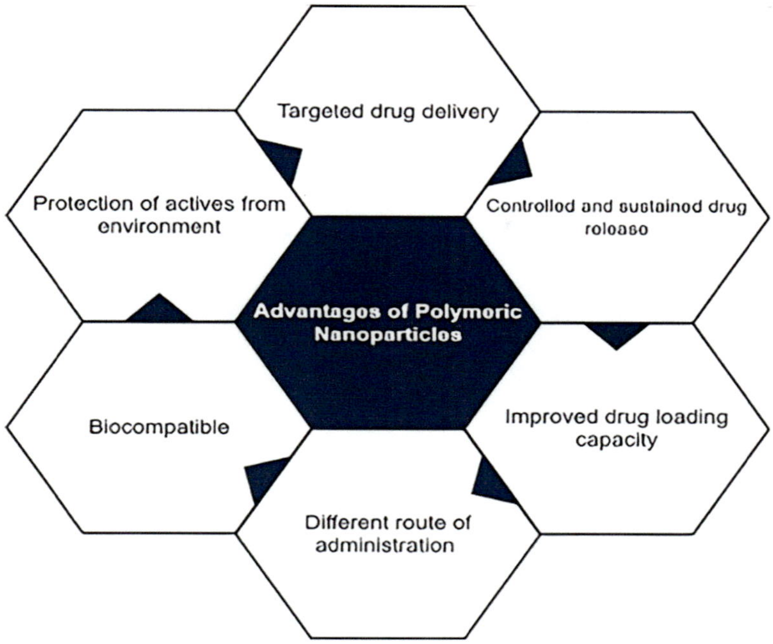

FIGURE 9.1 Advantages of polymeric nanoparticles for lung cancer

9.2 POLYMERIC NANOPARTICLES

Polymeric nanoparticles (PNPs) are colloidal carriers composed of natural or synthetic polymers, typically ranging in size from 10 to 1000 nm. These nanostructures can encapsulate, adsorb, or conjugate therapeutic agents, offering controlled release, improved solubility, and enhanced bioavailability. The other advantages of polymeric nanoparticles include improving the drug permeability, promoting site-specific drug delivery in tumors, and improving the solubility and bioavailability of poorly soluble anticancer medications. By avoiding efflux processes and offering continuous drug release, polymeric nanoparticles can overcome multidrug resistance. They are appropriate for clinical usage due to their biocompatibility, biodegradability, and possibility for non-invasive delivery. Furthermore, polymeric nanoparticles facilitate theranostics and combination therapies, providing personalized treatment [7–8]. Figure 9.1 depicts the advantages of various polymeric nanoparticles for lung cancer treatment.

9.3 MECHANISM OF ACTION OF POLYMERIC NANOPARTICLES FOR LUNG CANCER TREATMENT

PNPs hold great potential for lung cancer treatment, providing a targeted mechanism of drug delivery to tumor cells with minimal adverse effects to healthy tissue. This is accomplished when utilized correctly, due to favorable physical properties and surface modifications that allow for accumulation, release, and specificity of the drug [9]. The different mechanisms of actions of polymeric nanoparticles are given in Figure 9.2, whereas the advantages and disadvantages are given in Table 9.1.

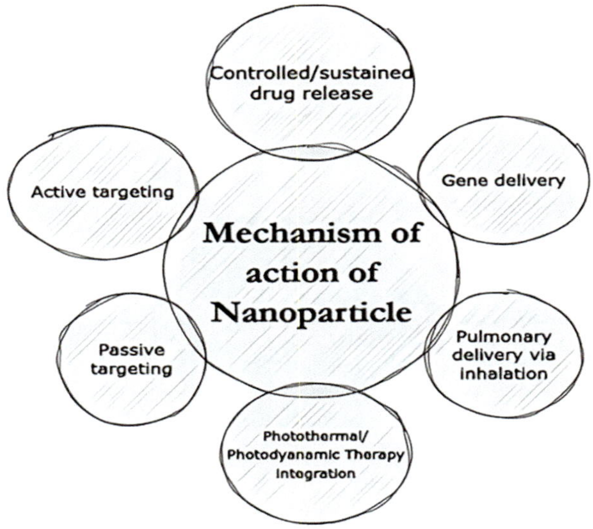

FIGURE 9.2 Mechanism of action of polymeric nanoparticles

TABLE 9.1
Mechanism of Action and Drug Delivery Using Polymeric Nanoparticles for Lung Cancer Treatment

Mechanism of Action	Definition	Advantage	Disadvantage	References
Passive targeting	Passive targeting promotes the placement of nanovectors inside the tumor microenvironment due to unique properties of the tumor microenvironment that do not typically exist in healthy tissue.	• Improved bioavailability • Selective accumulation • Reduced side effects • Ease of manufacture • Improved distribution	• Non-specific distribution • Dependency on RPE (inter-individual variability)	10
Active targeting	The active targeting nanoparticles conjugate target ligands on their surface, enhancing cellular absorption by receptor-mediated endocytosis and, consequently, drug accumulation in cancer cells. The interaction between tumor ligands conjugated on nanoparticle surfaces and cell-surface receptors or antigens on cancer cell surfaces is essential for this mechanism.	• High specificity • Increased delivery efficiency • Adaptable control possibility of customization	• Difficult to prepare • Costly • Risk of off-target effects induces variability of target organ response	11
Controlled/ sustained drug release	Sustained release refers to the release of a predetermined amount of drug at a programmed rate to provide drug delivery over a prolonged period of time. Sustained drug release is typically used in the case of drugs that have rapid metabolism and excretion from the body.	• Keeps the concentration of drugs at constant levels in blood or target tissues • Helps in reducing the medication frequency • Improves patient compliance	• Dose dumping • Invasive procedure to implant or remove the system • Uptake by RES that reduces its efficacy • Poor IVIVC • Costly	12

TABLE 9.1 *(Continued)*

Mechanism of Action and Drug Delivery Using Polymeric Nanoparticles for Lung Cancer Treatment

Mechanism of Action	Definition	Advantage	Disadvantage	References
Gene delivery	Gene therapy typically involves using oligonucleotides that provide a targeted approach to specifically regulate abnormal genetic expressions associated with cancer development within cancer cells.	• Provides higher efficiency as compared to conventional treatments • Helps in reducing toxicity	• Including limited targeting efficiency • Off-target effects • Development of resistance in tumor cells • Can cause toxicity to normal cells	13
Pulmonary delivery via inhalation	Pulmonary drug delivery system refers to both the direct administration of some medications to the lungs and the use of the lungs as a conduit for patients to actively or passively inhale medications in order to treat local or systemic illnesses. For the treatment of pulmonary disorders, this technique is the most direct and efficient method of drug delivery.	• Rapid • Efficient • Well targeted • Fewer side effects • Good patient compliance • Non-invasive	• Particle size and aerodynamic behavior • Regulatory and formulation requirements • Physiologically tolerated components • pH stability and enzyme degradation	14
Photothermal/ photodynamic therapy integration	These therapies are called photodynamic therapies as they depend on light and endogenous or exogenous absorbers to convert energy into cytotoxic reactive oxygen species (ROS) or increased local temperatures.	• Reduces chemotherapy resistance and compensatory signaling pathways • Enhances tumor drug delivery • Reduces off target toxicity • Decreases the risk of developing secondary cancers	• Less effective in oxygen deficient tumor environment • Potential off-target damage	15

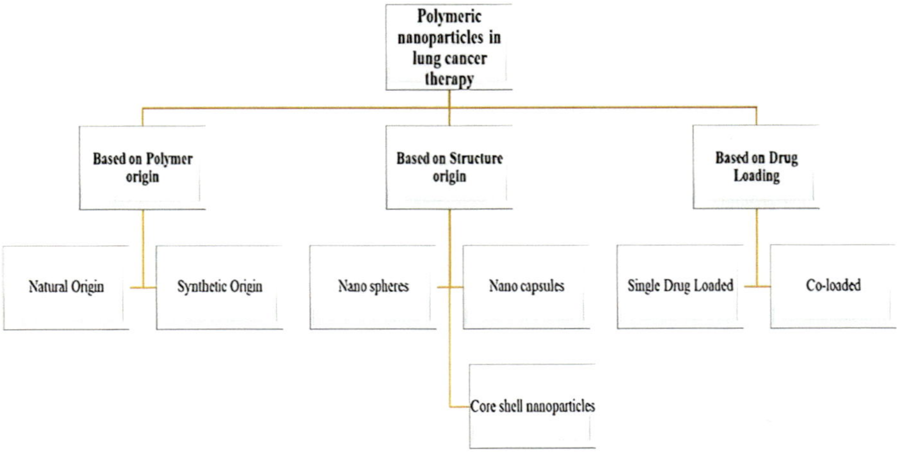

FIGURE 9.3 Classification of polymeric nanoparticles in lung cancer therapy

9.4 CLASSIFICATION OF POLYMERIC NANOPARTICLES IN LUNG CANCER THERAPY

9.4.1 Based on Polymer Origin

9.4.1.1 Natural Polymers

Natural polymers are perfect for the management of lung cancer because they have many advantages like low immunogenicity, biocompatibility, and biodegradability. Their environmental responsiveness enables targeted release, and their mucoadhesive qualities improve drug retention time at the tumor site [16]. They also allow for controlled and prolonged drug release. Polymers can be designed into gels or matrices that allow the drug's controlled release at the tumor site. Additionally, their chemical makeup makes surface modification simple, allowing targeted ligands like folic acid or antibodies to attach. They are environmentally friendly, made from renewable resources, support prolonged drug delivery, and can be functionalized for targeting [17].

Following are the various natural polymers used in the treatment of lung cancer.

- **Chitosan:** Chitosan is the most common natural polymer used in lung cancer treatment agent due to its mucoadhesive, biodegradable, and biocompatible properties. It enhances the solubility and the targeted delivery of anticancer drugs. Chitosan is additionally valued for its target and controlled drug delivery. The therapeutic efficacy of the drugs can be increased while adverse effects are reduced with the chitosan based drug delivery [18]. Various chitosan derivatives have anticancer properties; it also functions as a possible vector for immunotherapy and gene delivery. However, the major problems associated with its use as a natural polymer include its low solubility at physiological pH and formulation variability are still being studied [19].

  Khayati et al. formulated zoledronic-acid-loaded chitosan-based nanoparticles, via the microfluidics technique for the targeted cancer therapy. The formulations have controlled size, high drug entrapment, sustained release, and enhanced cytotoxicity compared to bulk methods and free zoledronic acid. Chitosan's biocompatibility and ability to form nanoparticles without crosslinkers further support its use as an efficient carrier for zoledronic acid in cancer treatment [20].
- **Alginate:** Alginate, a natural polymer, is most commonly used for the preparation of nanoparticles to release chemotherapeutic drugs such as doxorubicin, paclitaxel, and cisplatin in controlled and sustained manners, decreasing systemic toxicity. Nanoparticles based on alginate have exhibited tremendous potential in treating lung cancer because of their biocompatibility, biodegradable characteristics, and capability of hydrogel formation [21]. The nanoparticles support controlled and sustained release of a drug, with less systemic toxicity and greater therapeutic efficacy. Their mucoadhesive properties are ideal for inhalable delivery, enhancing drug retention within the lungs. Also, alginate nanoparticles are being surface modified for targeted delivery and employed as combined therapies, such as gene delivery, providing a potential route for lung cancer therapy [22].

  Ata, T. E et al. developed alginate–chitosan nanoparticles to enhance the delivery of Alectinib, a poorly soluble drug used in the treatment of NSCLC. The nanoparticles were developed by the solid–oil–water emulsification technique and ionotropic gelation method. The drug has good encapsulation efficiency, 161 nm particle size, and +21 mV zeta potential. The use of alginate and chitosan as a combined polymer enabled sustained release of Alectinib, improved cytotoxicity, and enhanced oral bioavailability in rats. The alginate and chitosan also improve the Alectinib's solubility and therapeutic efficacy, highlighting the potential of alginate chitosan nanocarriers for effective NSCLC treatment and overcoming limitations of conventional Alectinib formulations [23].

- **Gelatin:** Gelatin is a naturally occurring polymer that is used in the preparation of nanoparticles. These nanoparticles improve drug accumulation at the tumor site and reduce systemic toxicity by delivering regulated and sustained drug release. Their functional groups enable surface modification for lung cancer cell active targeting. Gelatin nanoparticles can also be made to be inhalable, which would improve lung deposition and treatment effectiveness. They can also be used for gene delivery and combination therapy [24].

  Chen et al. (2020) developed gelatin-loaded nanoparticles that carried both epigallocatechin gallate and cisplatin to treat lung cancer. They acted as a biodegradable carrier that helped in the stable loading of epigallocatechin gallate and cisplatin. The size of the gelatin-based nanoparticles were about 75 nm with positive surface charge. These particles were easily taken up by lung cancer cells, which increased the amount of cisplatin inside the cells. The combination of both drugs in gelatin nanoparticles showed greater cancer cell killing at lower doses than cisplatin alone, helping to overcome drug resistance effectively [25].
- **Hyaluronic acid:** Hyaluronic acid (HA) has shown great promise in the treatment of lung cancer, especially when it comes to targeted drug delivery. A HA receptor called CD44 is overexpressed in many lung cancer cells, making it possible to target specific cells with HA-functionalized nanoparticles. These systems decrease systemic toxicity, increase drug accumulation at tumor sites, and boost therapeutic efficacy [26]. Drug penetration may be further improved by the tumor microenvironment's enzymatic breakdown of HA. Clinical validation is crucial, even though preclinical results are encouraging. A promising approach to effective and individualized lung cancer treatment is the use of HA-based techniques [27].

  Wen et al., prepared hyaluronic acid nanoplatform, which carried cisplatin for lung cancer treatment. By attaching to receptors that are frequently overexpressed on the surface of tumor cells, hyaluronic acid nanoparticles improve cisplatin accumulation at the tumor site. This focused strategy improved treatment efficacy and decreased side effects. Hyaluronic acid's significance in targeted cancer therapy was highlighted by the system's combination of chemotherapy and chemodynamic therapy, which increased cancer cell death and decreased tumor growth [28].
- **Cellulose and derivatives:** Cellulose and its derivatives have emerged as unique biomaterials for lung cancer patients, mainly as carriers to enable targeted drug delivery. Derivatives such as carboxymethyl cellulose (CMC), hydroxypropyl cellulose, and cellulose acetate have been utilized to encapsulate certain chemotherapeutics to help increase drug solubility, stability, and controlled release. These systems can be engineered to target tumor-specific receptors, improving treatment efficacy and minimizing side effects. Additionally, cellulose-based scaffolds are being explored for lung tissue engineering and localized therapy [29].

  Cellulose, in the form of carboxymethyl cellulose, has played a vital role in modifying the surface of a zinc–benzenetricarboxylic acid metal–organic framework to create CMC–Zn@BTC nanoparticles for cancer therapy. Cellulose controlled the surface morphology, reduced the surface area and pore volume, and stabilized the nanocomposite structure. This cellulose-coated system enhanced anticancer effects by promoting apoptosis, causing DNA damage, and suppressing cell growth in lung and colon cancer cells. It also reduced cell migration and invasion by targeting proteins related to metastasis and proliferation [30].
- **Dextran:** Dextran-based nanoparticles offer a versatile and effective platform for lung cancer treatment as they can be engineered into nanocarriers for delivering chemotherapeutic agents, genes, or imaging probes directly to lung tumors. Functionalization with ligands targeting receptors overexpressed in lung cancer cells (e.g., folate, transferrin, or antibodies) enhances selective tumor accumulation and reduces systemic toxicity [31].

These nanoparticles provide controlled and sustained drug release, improving therapeutic efficacy. Moreover, dextran can be combined with stimuli-responsive elements (pH, redox) for smart drug release in the tumor microenvironment. Dextran-based systems show promise in advancing targeted lung cancer therapy [32].

Dong et al. covalently coupled two synergistic drugs docetaxel and docosahexaenoic acid through separate linkers to dextran (100 kDa), which led to a new dual-drug conjugate dextran-DHA-DTX. The dual-drug conjugate dextran–DHA–DTX self-assembled into nanoparticles with a diameter of 102.3 ± 8.3 nm and presented water solubility with improved pharmacokinetic profiles. Cellular uptake studies showed that the dual–drug conjugate penetrated tissues more efficiently than the parent DTX, by measuring the intracellular DTX content using HPLC/MS analysis and assessing the fluorescent intensity of dextran-Cy7.5 and dextran–DHA–Cy7.5. Notably, the dual-drug dextran–DHA–DTX drug conjugate reduced DTX concentrations in normal tissues by accumulating in the tissues. The dual-drug conjugate completely eradicated all the MCF-7 xenograft tumors without any side effects and demonstrated better antitumor activity than parent DTX, single-drug dextran–DTX, and dextran–DHA conjugates. Immuno-histochemistry studies demonstrated that DHA enhanced the antitumor activity of dextran–DTX drug conjugate both in vitro and in vivo [33].

- **Pectin:** Pectin-based nanoparticles are gaining attention in lung cancer therapy, as pectin is rich in galactose residues, which can bind to galectin-3, a protein overexpressed in many lung cancer cells, enabling active targeting. Pectin nanoparticles can encapsulate chemotherapeutic agents, improving drug solubility, stability, and controlled release [34]. They also enhance drug accumulation in tumors while minimizing systemic side effects. Additionally, pectin's pH-sensitive properties allow for selective drug release in the acidic tumor microenvironment. These features make pectin-based nanoparticles a promising approach for safe and effective lung cancer treatment [35].

  Chen et al. developed the DOX grafted pec-AD with new herbal medicine extract limonin for the targeting of lung cancer cells. The P(NIPAM195-co-AH54) cross-linking hydrogel was prepared and the hydrazone bond cross-linked hydrogel demonstrated good mechanics and self-healing properties. DOX and loaded limonin covalent coupling with the hydrogel lowered the in vivo toxicity and the hydrogel degraded with timeframe. This DOX grafted self-healing hydrogel that was biodegradable and highly effective in lung cancer treatment; it would have great use potential and application in cancer treatment [36].

9.4.1.2 Synthetic Polymers

- **Poly(lactic-co-glycolic acid) (PLGA):** PLGA-based nanoparticles are widely used polymers for treatment of lung cancer due to their advantages, for example, being FDA-approved and having good biocompatibility and biodegradability. Various chemotherapeutic drugs can be encapsulated by these nanoparticles, improving their stability, solubility, and controlled release. PLGA nanoparticles provide prolonged delivery of drugs directly to lung cancers while shielding them from early breakdown. PLGA nanoparticles are easily surface modified with targeting ligands (such as peptides or antibodies) so that cancer cells selectively absorb them. They can also be designed to react to stimuli such as pH or enzymes present in the tumor microenvironment.

  Gupta et. Al. created PLGA-based nanoparticles embedded with docetaxel (PLGA-Dtx) and studied their anticancer and immunomodulatory effects. The nanoparticles (~150 nm, 70% EE) exhibited significantly greater cytotoxicity in A549 lung cancer cells than free docetaxel, with limited toxicity in normal bronchial epithelial cells and human PBMCs. Furthermore, PLGA-Dtx modified cytokine profiles, suggesting possible immune activation. Overall, the results demonstrate the selective anticancer and immunostimulatory properties of PLGA-Dtx nanoparticles [37].

- **Polyethylene glycol (PEG):** PEG plays a crucial role in lung cancer therapy, primarily as a surface modifier in nanoparticle-based drug delivery systems. PEGylation enhances the stability, solubility, and circulation time of nanoparticles by reducing opsonization and clearance by the reticuloendothelial system. This "stealth" property allows PEG-coated nanoparticles to accumulate more effectively in lung tumors via the enhanced permeability and retention (EPR) effect. PEG can also be conjugated with chemotherapeutic agents or targeting ligands for improved specificity and efficacy.

  Jiang et al. developed polyethylene glycol 1000 succinate-based nanoparticles demonstrating significant advantages as a drug delivery carrier for crizotinib in lung cancer therapy. The copolymer's ability to self-assemble into stable, biocompatible micelles with sustained drug release contributes to improved therapeutic performance. The polyethylene glycol 1000 succinate enhanced cytotoxicity against NCI-H3122 lung cancer cells, facilitated efficient cellular uptake via endocytosis, and significantly increased apoptotic activity compared to free drug. These findings highlight the potential of vitamin E TPGS-modified PLA nanoparticles to overcome limitations of conventional therapy by improving drug solubility, stability, and anticancer efficacy [38].
- **Polycaprolactone (PCL):** PCL is a biodegradable, biocompatible polymer widely used for lung cancer therapy due to its ability to provide sustained and targeted drug delivery. Its slow degradation enables prolonged release of anticancer drugs like paclitaxel and doxorubicin, minimizing systemic toxicity. PCL can be formulated into nanoparticles, nanofibers, and microspheres, with surface modifications enhancing tumor specificity through ligands targeting EGFR or folate receptors. Hybrid systems incorporating siRNA or imaging agents offer potential for combined therapy and diagnostics. Despite challenges like limited hydrophilicity and slow release, PCL remains a promising material for innovative lung cancer treatment strategies and nanomedicine development.

  Patel et al. developed SB-loaded PCL/Pluronic F68 nanoparticles for pulmonary administration in lung cancer treatment. The Poly Caprolactone (PCL)/Pluronic F68 NPs showed a sustained release effect for up to 48 hours with emitted mass median aerodynamic diameter and geometric size distribution of 4.235 ± 0.124 and 1.958 ± 1.23, respectively. In a pharmacokinetics study, the PCL/Pluronic F68 NPs significantly enhanced SB bioavailability, with an increase in AUC of greater than four-fold compared to IV delivery. These findings suggest that SB-loaded PCL/Pluronic F68 nanoparticles could be a viable method of providing lung cancer treatment [39].
- **Polylactic acid (PLA):** PLA is a biodegradable, biocompatible polymer mostly used for lung cancer therapy because of its ability of controlled release and targeting drug delivery. PLA nanoparticles can encapsulate the different anticancer drugs, including chemotherapy agents like paclitaxel and docetaxel. Lung cancer is treated by formulating them into different forms like nanoparticles, microparticles, and nanofibers to deliver the drugs directly into the lungs. These particles are engineered to be small enough to be inhaled and to deposit within the lungs, but simultaneously large enough to prevent them from getting clogged in the airway [40].

  Nie et al. created polymeric micelles based on PEG-b-PLA to enhance quercetin's (QUR) physicochemical and chemotherapeutic effectiveness. QUR-micelles (QUR-PM) demonstrated a high distribution index and a nanoscale particle size. When compared to free QUR, QUR-PM had a greater cytotoxic effect on A549 cancer cells. After 24 hours of incubation, the IC50 values for QUR and QUR-PM were determined to be 42.3 and 30 μg/mL, respectively. After 48 hours of incubation, the IC50 values for QUR and QUR-PM were 14.5 and 7.2 μg/mL, respectively. Consequently, QUR-PM revealed a spectacular apoptosis of cancer cells. Together, these findings indicated that the micellar formulations created in this study have a high potential for killing lung cancer cells and may be regarded as an effective drug delivery method for insoluble medications like QUR [41].

- **Poly(ethylene imine) (PEI):** PEI is referred to as a non-biodegradable polymer. PEI, being a cationic polyamine, can bind or interact with anionic residues in DNA templates and polymerases through electrostatic interactions and thus significantly increase their transfection efficiency. Further, PEI's high positive surface charge is also found to be associated with strong cytotoxicity effects on cells due to the availability of several amine groups. Hence, neutralization of PEI's surface charge by chemical or physical modifications can greatly reduce its cytotoxicity while increasing biocompatibility.

  Jin et al. introduced a new delivery technique for microRNA-942–5p-sponges non-coding RNA delivery using fluorine-modified PEI 1.8 kDa. Compared to PEI 25 kDa, this new gene delivery system displayed a nearly sixfold increase in endocytosis and a relative increase in cell survivability. Note from in vivo studies that the positively charged PEI and hydrophobic and oleophobic properties of the fluorine modified group were significant for favorable biosafety and anti-tumor effects. This study provides an effective gene delivery method for the treatment of NSCLC [42].
- **Poly(2-hydroxyethyl methacrylate):** Poly(2-hydroxyethyl methacrylate) is a hydrophilic polymer that has received a lot of attention in materials science research. Most commonly, it is used as hydrogels, which can retain large amounts of water while still holding structure. PHEMA has biocompatibility and permeability properties that will be of use in the development studies of biomedical devices. As well, the polymer's controlled swelling is of interest in controlled release systems, where it has been studied for release kinetics of an encapsulated substance. The polymer has been studied to understand the impact on surface modification as a means of enhancing the interface between synthetic material and biological tissue [43].

  You and August created pH-sensitive poly(DMAEMA/HEMA) nanoparticles capable of controlled paclitaxel release in tumor environments. These nanoparticles processed alterations in pH and released 25% of the drug at pH 6.8 and 9% at pH 7.4 in 2 hours. They were created using emulsion and photopolymerization, well characterized, and non-cytotoxic, making them suitable for responsive and safe cancer drug delivery [44].
- **Polystyrene-co-maleic anhydride (SMA):** SMA is a copolymer of hydrophobic styrene and hydrophilic maleic anhydride units. The amphiphilic character of SMA permits its self-assembly into micelles in water. As a copolymer, it permits the drug encapsulation and their delivery directly to the tumor site, particularly through the use of the EPR effect. The capability of targeting specifically lung tumors and delivering therapeutic agents will result in treatments that are more effective and less toxic [45].

  Angelova and Yordanov manufactured poly(styrene-co-maleic acid) drug-loaded pirubicin nanoparticles by a nanoprecipitation approach in a mixed (water/acetone) solvent system; the drug was loaded by sorption during manufacturing. By controlling polymer concentration and the acetone–water volume ratio, they were able to produce nanoparticles with size ranging from 100 to 320 nm in diameter. Moreover, all nanoparticles showed negative zeta-potentials in physiological-like conditions; human serum albumin decreased the absolute value of the zeta-potential of both the drug-free and drug-loaded nanoparticles, but both remained negative, which implies that albumin adsorbed to the particle surface. The nanoparticles were stable in phosphate-buffered saline over the pH range of 4.5 to 9.3 (there was precipitation at lower pH and disintegration at higher pH). From an in vitro release study epirubicin released from the nanoparticles appeared to be associated strongly with the polymer; about 10% of loaded drug was released over 7 hours [46].
- **Polyurethane (PU):** Pus have been studied as a drug delivery system for lung cancer treatment based on their flexibility and biocompatibility. Pus can be formulated to be biodegradable or non-biodegradable, providing versatile benefits in drug release control and targeting cancer cells. In particular, polyurethane nanoparticles (PUNs) are under development to deliver drugs specifically to lung cancer treatment. Pus have vast potential as a

material to devise drug delivery systems with their flexibility, strength, and biocompatibility. They can be tailored to entrap drugs and deliver them in a controlled form at the tumor site, optimizing therapeutic efficiency and minimizing side effect

Iyer et al. developed GSH-responsive polyurethane nanoparticles (CGPUs) for targeted cisplatin delivery in lung cancer. These nanoparticles released cisplatin in response to elevated GSH levels in tumor cells, showing greater cytotoxicity in A549 cells than free cisplatin. In vivo studies demonstrated tumor-specific accumulation and significantly reduced tumor growth compared to saline and free drug, indicating CGPUs' potential for controlled and effective lung cancer therapy [47].

- **Poly(methyl methacrylate)** (PMMA)**:** PMMA is a clear, inexpensive, and strong synthetic polymer. Such nanoparticles can be utilized for encapsulation of drugs such as curcumin or photosensitizers, which are delivered to the cancer cells and in certain instances activated by light to kill cells. PMMA nanoparticles are utilized to deliver light-activatable photosensitizers to destroy cells through PDT.
- Ma and Fu created PMMA-curcumin nanoparticles. Characterization procedures were used to establish their diameters, and flow cytometry and a cell proliferation experiment were used to identify their impacts on lung cancer cells A549. Western blotting was used to identify the cells' expression of proteins linked to apoptosis. The findings demonstrated that the particle size of polyacrylic acid (PAA)-Cur-PMMA-NPs was 215.00 ± 6.00 nm. Compared to free Cur, the drug loading and encapsulation rates of nanospheres were significantly greater ($p < 0.05$). Following the administration of PAA-Cur-PMMA-NPs to the cells, the apoptotic rate and the expression of Bax, Caspase-3, and Caspase-9 increased ($p < 0.05$), whereas cell proliferation and Bcl-2 expression decreased. Therefore, it is anticipated that Cur-PMMA-NPs will be a successful intervention strategy to enhance the therapeutic effect on lung cancer in the future as they have the ability to stop the proliferation of lung cancer cells and trigger their apoptosis [48].

9.4.2 Based on Structure

- **Nanospheres:** Nanospheres are tiny particles, typically 100 nm or smaller in diameter, used in lung cancer therapy to deliver pharmaceuticals or as adjuncts to other therapies. Nanospheres have a number of advantages, such as targeted drug delivery, increased efficacy, and reduced toxicity compared to traditional therapies. Nanospheres can be engineered to deliver drugs to only cancer cells without killing the normal tissue, thus increasing efficacy in therapy. They can be engineered to deliver therapeutic genes to lung cancer cells, leading to long-term cancer control. Nanospheres can also be utilized to encapsulate chemotherapy drugs, thus improving their stability, targeting, and release profiles

  Zhang et al. created the gene circuit apYr-ads-8–5HRE-cfosp-iNOS-IFNG, which can express IFNγ. They then employed this plasmid in conjunction with C225 (cetuximab) to create gene-loaded immunomagnetic albumin nanospheres (IMANS). Additionally, we looked at the in vitro and in vivo therapeutic effects of gene-loaded IMANS in conjunction with radiation therapy on human lung cancer. The findings demonstrated that this gene circuit was built piece by piece and verified that the gene circuit was responsible for the rise in INFγ expression. Radiation therapy in conjunction with gene-loaded IMANS showed better outcomes both in vivo and in vitro. To sum up, gene-loaded IMANS improved combination therapy effectiveness, resolved gene transfer issues, and selectively targeted lung cancer cells [49].
- **Nanocapsules:** Nanocapsules are nanoparticles that are hollow and have a solid shell with a liquid space in the center of it. They are vesicular systems made up of polymeric membrane and therefore encapsulate an inner liquid core. They are new and promising systems utilized in drug delivery and the self-healing of objects (materials). The encapsulation

method has considerable advantages related to protection of the encapsulated substances from adverse effects of the environment and precision targeting of and controlled release of active ingredient(s). The can be functionalized by embedding into the shell of the nanocapsules molecules that have specific properties, and therefore, the contents of the nanocapsules are released upon an interaction with a specific biomolecule that aids targeting drug-delivery

Gündel et al. produced etoposide-loaded lipid-core nanocapsules (ETO NC) retaining a unimodal particle size of ≈150 nm, a drug concentration of 0.5 mg/mL, and an encapsulation efficiency of ≈99%. After 28 days, the formulation was stable at 5°C. The release experiment had 70% of the drug dialyzed from ETO NC in 72 hours, with the first phase having a biexponential curve demonstrating a sustained phase (33%, t½ = 173.2 hours) after the burst phase (60%, t½ = 1.3 hours). After 48 hours, the cell viability of A549 lung adenocarcinoma cells exposed to 10 µmol/L [50].

- **Core–shell nanoparticle:** Core–shell nanoparticles are a two-part nanostructure system made up of a core material in the middle and a shell coating on the outside. By designing them in this way, the beneficial properties of each material can be combined, optimizing stability, functionality, and targeted delivery in biomedical, catalytic, and environmental fields. Core is the material in the middle that can be metal (e.g., gold, iron oxide), polymer, silica, or drug-loaded material. Shell is a protective or modifying coating layer made up of silica, polymers (e.g., PEG, PLGA), lipids, or other metals. They can also be designed to be stimuli-responsive, releasing drugs when exposed to specific stimuli such as pH, radiation, or glutathione, enabling targeted drug delivery

  Menon et al. developed a multifunctional dual drug-loaded nanoparticle (MDNP) as a platform for the targeting of folate receptors, with a carboxymethyl chitosan shell encapsulating a poly(N-isopropylacrylamide) shell over a PLGA core, which was designed for the induction of localized chemo-radiotherapy for efficient treatment of lung cancers. The formulation allowed for the controlled release of the drugs entrapped within, NU7441—a strong radiosensitizer—and the FDA-approved chemotherapeutic agent gemcitabine, employed for lung cancer chemo-radiotherapy. The nanoparticles were also found to be highly stable, with remarkable hemocompatibility, superior in vitro cytocompatibility with alveolar Type I cells, and dose-dependent caveolae-mediated in vitro cellular uptake by lung cancer cells. Moreover, they could be co-loaded with superparamagnetic iron oxide (SPIO) nanoparticles and visualized by MRI in vivo. Initial in vivo results showed a low toxicity of the particles and their efficacy in chemo-radiotherapy for the effective inhibition of lung tumors [51].

9.4.3 Based on Drug Loading

- **Single-drug-loaded nanoparticles:** Single-drug-loaded nanoparticles are nanocarriers usually 1–100 nm in diameter and are engineered to encapsulate and deliver one drug molecule. Single-drug-loaded nanoparticles have benefits such as controlled drug release, improved bioavailability, and decreased side effects compared to conventional drug delivery. They are also applied in different applications such as targeted drug delivery to a specific tissue or cell and controlled drug release. Single drug-loaded nanoparticles may be prepared by different techniques such as nanoprecipitation, solvent evaporation, and emulsion–solvent evaporation processes. In the therapy of lung cancer, particularly in NSCLC, single drug-loaded nanoparticles aid in overcoming challenges such as multidrug resistance, limited drug penetration in tumors, and high side effects due to systemic chemotherapy.

Fonseca et al. developed Paclitaxel-loaded PLGA nanoparticles (Ptx-PLGA-Nps) as an injectable alternative to Taxol, with the purpose of improving efficacy and removing the toxic effects of Cremophor EL. The nanoparticles demonstrated biphasic drug release and showed greater cell cytotoxicity against NCI-H69 lung cancer cells than Taxol, with longer exposure times indicating additional therapeutic impact [52].

- **Co-loaded nanoparticles**: Co-loaded nanoparticles are advanced drug delivery systems that are engineered to co-deliver and release two or more drugs within a single nano-sized vehicle. The co-delivery mode is advantageous in the treatment of complex diseases like cancer, where the drug combination synergistically results in an increased therapeutic effect and reduced drug resistance. By loading pharmacologically different agents, for example, a chemotherapeutic agent with a gene therapy agent, within a single nanoparticle, it enables concerted delivery of the agents to the target location, which results in synergistic interactions and reduced toxicity to the surrounding healthy tissues. The nanoparticles can be engineered from a wide range of materials, including lipids (liposomes), polymers (e.g., PLGA and PEG), dendrimers, or metals (e.g., gold or iron oxide), depending on the intended use. Controlled and site-specific release is enabled using responsive materials that respond to appropriate stimuli, for example, pH, temperature, or enzymes within the microenvironment of the disease. Furthermore, co-loaded nanoparticles are being explored in theranostics, where they co-deliver therapeutic agents and diagnostic agents, thereby enabling real-time monitoring of treatment. Their development is a significant step toward more effective and individualized healthcare solutions.
- Li et al. developed paclitaxel and curcumin co-loaded solid lipid nanoparticles (PAC-CUR-SLNs) for lung cancer applications. These nanoparticles (~190 nm) had high encapsulation efficiency (>94%) and caused synergistic cytotoxicity in A549 cells in vitro. When tested in BALB/c mice, the PAC-CUR-SLN group had a greater than 12-fold reduction in tumor volume and a high tumor inhibition rate of 82.7% without significant body weight loss, suggesting improved efficacy with relation to reduced toxicity compared to monotherapies. This shows that PAC-CUR-SLNs have potential for focused lung cancer therapies [53].

9.4.4 Design and Functionalization of Polymeric Nanoparticles for Lung Cancer Treatment

Table 9.2 lists the different methods of preparation for polymeric nanoparticles with their principles, advantages, and limitations.

9.5 CHALLENGES AND LIMITATIONS

Polymeric nanoparticles have many potential advantages in lung cancer therapy, yet they still have many limitations. Thus, many challenges in the treatment of lung cancer stem from barriers to drug delivery (e.g. physiological barriers such as mucociliary clearance and alveolar macrophages). Also, some limitations are due to tumor heterogeneity, and some of limitations are due to polymeric nanoparticles' lack of ability to penetrate deeply into tissue, therefore leading to inefficient targeting. Other limitations arise from formulation shortcomings, such as reproducibility, batch-to-batch variations, and solvent toxicity when scaling up for manufacture. Biocompatibility limitations, such as potential toxicity due to degradation products or prolonged accumulation in the body, have been raised as concerns for using polymeric nanoparticles in vivo. Furthermore, stability in storage and clinical limitations to delivery strategies, particularly for inhalation, would still impose major barriers to moving toward clinical trials [66–68].

TABLE 9.2
Design and Functionalization of Polymeric Nanoparticles for Lung Cancer Treatment

Method	Principle	Suitable Polymers	Advantages	Limitations	References
Nanoprecipitation (solvent displacement)	Polymer and drug dissolved in organic solvent, added to aqueous phase, leading to nanoparticle formation.	PLGA, PCL, PLA, PEG, PVA, Poloxomer, etc.	Easy to use, reproducible, versatility, reduced toxicity	Unsuitable for hydrophilic compounds, challenges with multiple polymer system, usage of organic solvent	3
Emulsion–solvent evaporation	Formation of oil-in-water (O/W) emulsion; solvent removed by evaporation.	DMF, *NN*-methyl-2-pyrrolidone, Polytetrafluoroethylene, etc.	Easily accessible, versatility, enhanced solubility and bioavailability, improved penetration	Slow process, less entrapment efficiency, large size nanoparticle, heterogenous particle size, high toxicity	54
Double emulsion (W/O/W)	A double emulsion or an emulsion of an emulsion is an emulsion in which the first emulsion is distributed in additional liquid (continuous phase). Double emulsions (W/O/W) are complex systems consisting of an inner (W/O) emulsion distributed in a second aqueous continuous phase.	Polyglycerol ester of polyricinoleic acid (PGPR), CMCC	Improved taste of drug, reduce side effects, drug targeting, improved absorption of drug	Thermodynamic instability, large droplet size, environment sensitivity. limited range of emulsifiers	55
Spray drying	Spray drying is the rapid evaporation of a liquid feed with a polymer into a fine spray and subsequent exposure to hot drying gas, usually air, resulting in the rapid evaporation of the solvent and formation of dried particles.	PLGA, PVA, chitosan, alginate, PLA	Scalability and efficiency, uniform particle size, controlled drug release, rapid drying, cost effectiveness	Thermal and mechanical stress, low yield, particle aggregation	56
Ionic gelation	Ionic gelation, or ionotropic gelation, is a synthesis of polymeric nanoparticles by electrostatic attraction of oppositely charged ions, typically located in a polymeric network.	Polyacrylates, polycaprolactone, alginate, chitosan, gelatin	Environmental friendly, simple and low cost, natural polymer compatibility	pH sensitivity, greater polydispersity and particle size, aggregation potential, and limited size and distribution control	57
Supercritical fluid technology	In SFT, a polymer solution and other constituents are dissolved in a supercritical fluid, for example, CO_2. The rapid expansion of this solution (most often by pressure reduction) leads to precipitation of the polymer from solution into nanoparticles.	Alginate, chitosan, dextran, PLGA	Controlled particle size and shape, better encapsulation efficiency, scalable production, improved stability	High equipment costs, potential for drug degradation, potential for residue, need for optimization	58
Salting out	This involves the addition of the aqueous phase and the salting-out agent triggers the organic solvent's diffusion into the aqueous phase, later reducing the solubility of the polymer and drug, and hence leading to their precipitation and nanoparticle formation.	PLGA, chitosan, gelatin, PLA, alginate	No heat required, good loading efficiency, scalability, encapsulation of protein and nucleic acids	Potential incompatibility, limited for lipophilic drugs, larger particle sizes, less precise control	59

TABLE 9.3

Clinical Trial Studies on Polymeric Nanoparticles for the Management of Lung Cancer

Serial Number	NCT Number	Polymer	Drug	Phases	Status	Comments	Reference
1	NCT01620190	Albumin	Paclitaxel	Phase 2	Completed	Paclitaxel albumin-stabilized nanoparticle formulation (nab-paclitaxel) for patients who had been previously treated for advanced non–small cell lung cancer represents real advances in improving drug delivery and has less toxic side effects when compared with drug formulations that are not stabilized. Part of what the design uses is albumin to enhance tumor targeting and the administering of the drug without toxic solvents or steroid premedication which might be useful for patients who had previously been heavily pretreated.	60
2	NCT00729612	Albumin	Paclitaxel + carboplatin	Phase 2	Completed	The combination of the paclitaxel albumin-stabilized nanoparticle formulation and carboplatin is a well-established and effective first-line treatment option for patients with advanced (stage IIIB/IV) or recurrent NSCLC. Nab-paclitaxel enhances the delivery of paclitaxel and reduces solvent-related toxicities, while carboplatin possesses synergistic cytotoxic effects. The regimen has been shown to improve response rates and a manageable safety profile compared to standard chemotherapy, thus serving as an effective therapeutic option in the management of advanced NSCLC.	61
3	NCT04033354	Albumin	HLX10, paclitaxel, carboplatin	Phase 3	Ongoing	This phase III randomized, double-blind, multicenter study aims to evaluate the efficacy and safety of HLX10, a recombinant anti-PD-1 humanized monoclonal antibody, in combination with chemotherapy (carboplatin plus nab-paclitaxel) compared with chemotherapy alone in treatment-naive patients with locally advanced or metastatic squamous NSCLC. By combining immunotherapy with the established carboplatin–nab-paclitaxel regimen, this study aims to improve overall survival and response rates through its ability to maximize anti-tumor immune activation.	62
4	NCT06199895		Paclitaxel	Phase 2	Ongoing	This research uses paclitaxel polymeric micelles as a potential treatment for taxane-resistant tumors focusing on improved drug delivery and efficacy while also limiting toxicity and side effects. Based on the data and analysis of paclitaxel polymeric micelles in several resistant tumor types, it is reasonable to consider this to be an effective option for patients who have limited treatment options and considers an important hurdle in cancer treatment.	63
5	NCT05782426		Paclitaxel, platinum combined with sindilizumab	Phase 2	Ongoing	This research is a promising first-line treatment using a multimodal, paclitaxel polymeric micelle, platinum chemotherapy, and sindilizumab for treating advanced non-squamous NSCLC. This study takes advantage of improved drug delivery through poly-clustered micelles to increase the efficacy of immunotherapy, while providing an alternative treatment with a better safety profile and less adverse events for our patients.	64
6	NCT00748163	Albumin	Paclitaxel albumin-STABILIZED nanoparticle formulation and sunitinib malate	Phase 2	Withdrawn	Utilizing the paclitaxel albumin-delivering nanoparticle signing (nab-paclitaxel) alongside sunitinib as first-line therapy in patients with stage IV NSCLC is an approach that uses effective chemotherapy and anti-angiogenic targeted therapy. Such an approach intends to elicit a tumor response by attacking cancer cells and cutting off their blood supply at the same time, which has the potential to improve outcomes in NSCLC patients with advanced disease.	65

9.6 FUTURE PERSPECTIVES

Polymeric nanoparticles present a remarkable opportunity for future lung cancer therapy. Surface modification and ligand-based targeting should enhance tumor specificity and reduce off-target effects. Further advances in stimuli-responsive and smart delivery systems will offer tighter control and more site-specific drug release. The additional use of imaging agents with the nanoparticle delivery system may allow theranostics and real-time monitoring. Personalizing the nanoparticle formulation to specifically target the tumor's genetic and microenvironments could yield better outcomes. As regulatory processes improve, formulation processes advance, and safety and efficiency evolve, polymeric nanoparticles have the potential of exciting possibilities in more effective and safer treatments of lung cancer [69–70].

9.7 CONCLUSION

Polymeric nanoparticles are a promising mechanism for targeted and controlled release anticancer drugs for the treatment of lung cancer. Polymer nanoparticles can increase drug solubility, decrease systemic toxicity, and improve therapeutic response, which could provide a valuable strategy useful in modern oncology. However, individual biological barriers, formulation challenges, and regulatory approval need to be better addressed in order for polymeric nanoparticles to successfully translate to being used in clinical practice. In future, polymeric nanoparticles have the potential to substantially improve the efficacy and precision of treatment for lung cancer through continual research and further advancements in technology.

REFERENCES

1. Li, Y., Yan, B., & He, S. (2023). Advances and challenges in the treatment of lung cancer. *Biomedicine & Pharmacotherapy*, *169*, 115891. https://doi.org/10.1016/j.biopha.2023.115891
2. Lahiri, A., Maji, A., Potdar, P. D., Singh, N., Parikh, P., Bisht, B., . . . & Paul, M. K. (2023). Lung cancer immunotherapy: Progress, pitfalls, and promises. *Molecular Cancer*, *22*(1), 40. https://doi.org/10.1186/s12943-023-01740-y
3. https://www.cancer.org/content/dam/cancer-org/research/cancer-facts-and-statistics/annual-cancer-facts-and-figures/2025/2025-cancer-facts-and-figures-acs.pdf?utm
4. Nooreldeen, R., & Bach, H. (2021). Current and future development in lung cancer diagnosis. *International Journal of Molecular Sciences*, *22*(16), 8661. https://doi.org/10.3390/ijms22168661
5. Polanski, J., Jankowska-Polanska, B., Rosinczuk, J., Chabowski, M., & Szymanska-Chabowska, A. (2016). Quality of life of patients with lung cancer. *OncoTargets and Therapy*, 1023–1028. https://doi.org/10.2147/OTT.S100685
6. Marchevsky, A. M. (2006). Problems in pathologic staging of lung cancer. *Archives of Pathology & Laboratory Medicine*, *130*(3), 292–302. https://doi.org/10.5858/2006-130-292-PIPSOL
7. Beach, M. A., Nayanathara, U., Gao, Y., Zhang, C., Xiong, Y., Wang, Y., & Such, G. K. (2024). Polymeric nanoparticles for drug delivery. *Chemical Reviews*, *124*(9), 5505–5616.
8. Madej, M., Kurowska, N., & Strzalka-Mrozik, B. (2022). Polymeric nanoparticles—tools in a drug delivery system in selected cancer therapies. *Applied Sciences*, *12*(19), 9479. https://doi.org/10.3390/app12199479
9. Xiao, X., Teng, F., Shi, C., Chen, J., Wu, S., Wang, B., . . . & Li, W. (2022). Polymeric nanoparticles—promising carriers for cancer therapy. *Frontiers in Bioengineering and Biotechnology*, *10*. https://doi.org/10.3389/fbioe.2022.1024143
10. Bazak, R., Houri, M., Achy, S. E., Hussein, W., & Refaat, T. (2014). Passive targeting of nanoparticles to cancer: A comprehensive review of the literature. *Molecular and Clinical Oncology*, *2*(6), 904–908.
11. Muhamad, N., Plengsuriyakarn, T., & Na-Bangchang, K. (2018). Application of active targeting nanoparticle delivery system for chemotherapeutic drugs and traditional/herbal medicines in cancer therapy: A systematic review. *International Journal of Nanomedicine*, *13*, 3921–3935.
12. Liu, S., Li, L., Zhang, X., & Meng, Q. (2023). Nanotherapies from an oncologist doctor's view. *Smart Materials in Medicine*, *4*, 183–198.

13. Lee, H.-Y., Mohammed, K. A., & Nasreen, N. (2016). Nanoparticle-based targeted gene therapy for lung cancer. *American Journal of Cancer Research*, *6*(5), 1118–1134.
14. Wang, B., Wang, L., Yang, Q., Zhang, Y., Qinglai, T., Yang, X., . . . & Li, S. (2024). Pulmonary inhalation for disease treatment: Basic research and clinical translations. *Materials Today Bio*, *25*, 100966.
15. Overchuk, M., Weersink, R. A., Wilson, B. C., & Zheng, G. (2023). Photodynamic and photothermal therapies: Synergy opportunities for nanomedicine. *ACS Nano*, *17*(9), 7979–8003.
16. Wong, K. H., Lu, A., Chen, X., & Yang, Z. (2020). Natural ingredient-based polymeric nanoparticles for cancer treatment. *Molecules*, *25*(16), 3620. https://doi.org/10.3390/molecules25163620
17. Jin, X., Heidari, G., Hua, Z., Lei, Y., Huang, J., Wu, Z., . . . & Xu, Y. (2024). Nanoengineered polymers and other organic materials in lung cancer treatment: Bridging the gap between research and clinical applications. *European Polymer Journal/European Polymer Journal*, 112891. https://doi.org/10.1016/j.eurpolymj.2024.112891
18. Kazmi, I., Shaikh, M. A. J., Afzal, O., Alfawaz Altamimi, A. S., Almalki, W. H., Alzarea, S. I., . . . & Gupta, G. (2023). Chitosan-based nano drug delivery system for lung cancer. *Journal of Drug Delivery Science and Technology*, *81*, 104196. https://doi.org/10.1016/j.jddst.2023.104196
19. Silva, A. C., Costa, M. P., Zacaron, T. M., Ferreira, K. C. B., Braz, W. R., Fabri, R. L., . . . & Tavares, G. D. (2024). The role of inhaled chitosan-based nanoparticles in lung cancer therapy. *Pharmaceutics*, *16*(8), 969. https://doi.org/10.3390/pharmaceutics16080969
20. Khayati, M., Safari, H., Bagheri, F., & Kheiri, H. (2025). Evaluation of the cytotoxic effect of chitosan-zoledronic acid nanoparticles and free zoledronic acid against a549 lung cancer cells using microfluidic technology. *Microfluidics and Nanofluidics*, *29*(4), 1–12.
21. He, L., Shang, Z., Liu, H., & Yuan, Z. (2020). Alginate-based platforms for cancer-targeted drug delivery. *BioMed Research International*, *2020*, 1–17. https://doi.org/10.1155/2020/1487259
22. Iravani, S., & Varma, R. S. (2022). Alginate-based micro- and nanosystems for targeted cancer therapy. *Marine Drugs*, *20*(10), 598. https://doi.org/10.3390/md20100598
23. Ata, T. E., Al-Ani, I., Karameh, N., Atta, M. R., & Dayyih, W. A. (2025). Alectinib-loaded chitosan-alginate nanoparticles: A novel synthesis method with in vitro and in vivo evaluations. *Pharmaceutics*, *17*(4), 492.
24. Jiang, X., Du, Z., Zhang, X., Zaman, F., Song, Z., Guan, Y., . . . & Huang, Y. (2023). Gelatin-based anticancer drug delivery nanosystems: A mini review. *Frontiers in Bioengineering and Biotechnology*, *11*. https://doi.org/10.3389/fbioe.2023.1158749
25. Chen, Y. J., Wang, Z. W., Lu, T. L., Gomez, C. B., Fang, H. W., Wei, Y., & Tseng, C. L. (2020). The synergistic anticancer effect of dual drug-(cisplatin/epigallocatechin gallate) loaded gelatin nanoparticles for lung cancer treatment. *Journal of Nanomaterials*, *2020*(1), 9181549.
26. Alsaikhan, F. (2023). Hyaluronic acid-empowered nanotheranostics in breast and lung cancers therapy. *Environmental Research*, *237*(Pt 1), 116951. https://doi.org/10.1016/j.envres.2023.116951
27. Xue, G., Tang, L., Pan, X., Li, S., & Zhao, J. (2024). Hyaluronic acid-targeted topotecan liposomes improve therapeutic efficacy against lung cancer in animals. *Frontiers in Oncology*, *14*. https://doi.org/10.3389/fonc.2024.1520274
28. Wen, Q., Li, J., Deng, H., Wang, B., Huang, J., Dai, J., . . . & Fu, S. (2025). Hyaluronic acid modified metal-organic frameworks loading cisplatin achieve combined chemodynamic therapy and chemotherapy for lung cancer. *International Journal of Biological Macromolecules*, 140238.
29. Li, Y., Liu, W., Wang, Y., & Lu, S. (2025). Cellulose based nano-scaffolds for targeted cancer therapies: Current status and future perspective. *International Journal of Nanomedicine*, *20*, 199–213. https://doi.org/10.2147/ijn.s500261
30. Suares, D., Shetty, S., & Shetty, S. (2023). *Cellulose-Based Drug Delivery Systems in Lung Disorders*. Springer Nature, pp. 83–102. https://doi.org/10.1007/978-981-19-7656-8_4
31. Kumari, M., Sharma, S., Kanwar, N., Naman, S., & Baldi, A. (2024). Dextran-based drug delivery approaches for lung diseases: A review. *Current Drug Delivery*, *21*(11), 1474–1496. https://doi.org/10.2174/0115672018267737231116100812
32. Prasher, P., Sharma, M., Singh, S. K., Haghi, M., MacLoughlin, R., Chellappan, D. K., . . . & Dua, K. (2022). Advances and applications of dextran-based nanomaterials targeting inflammatory respiratory diseases. *Journal of Drug Delivery Science and Technology*, *74*, 103598. https://doi.org/10.1016/j.jddst.2022.103598
33. Dong, P., Lv, H., Jia, W., Liu, J., Wang, S., Li, X., . . . & Shi, Y. (2022). Polysaccharide dextran-based conjugate for selective co-delivery of two synergistic drugs docetaxel and docosahexaenoic acid to tumor cells. *Drug Delivery*, *30*(1), 40–50.

34. Bhat, A. A., Gupta, G., Alharbi, K. S., Afzal, O., Altamimi, A. S. A., Almalki, W. H., . . . & Dua, K. (2022). Polysaccharide-based nanomedicines targeting lung cancer. *Pharmaceutics*, *14*(12), 2788. https://doi.org/10.3390/pharmaceutics14122788
35. Carene, M., & Neergheen, V. S. (2023). Pectin a multifaceted biopolymer in the management of cancer: A review. *Heliyon*, e22236. https://doi.org/10.1016/j.heliyon.2023.e22236
36. Chen, D., Ma, X., Zhu, J., Wang, Y., Guo, S., & Qin, J. (2023). Pectin based hydrogel with covalent coupled doxorubicin and limonin loading for lung tumor therapy. *Colloids and Surfaces B Biointerfaces*, *234*, 113670.
37. Gupta, P., Singh, A., Verma, A., Kant, S., Pandey, A., Khare, P., & Prakash, V. (2022). The anti-tumor and immunomodulatory effects of PLGA-based docetaxel nanoparticles in lung cancer: The potential involvement of necroptotic cell death through reactive oxygen species and calcium build-up. *Vaccines*, *10*(11), 1801.
38. Jiang, Z. M., Dai, S. P., Xu, Y. Q., Li, T., Xie, J., Li, C., & Zhang, Z. H. (2015). Crizotinib-loaded polymeric nanoparticles in lung cancer chemotherapy. *Medical Oncology*, *32*, 1–8.
39. Patel, P., Raval, M., Manvar, A., Airao, V., Bhatt, V., & Shah, P. (2022). Lung cancer targeting efficiency of silibinin loaded poly caprolactone/pluronic F68 inhalable nanoparticles: In vitro and In vivo study. *PLOS One*, *17*(5), e0267257.
40. García-Rodríguez, A., Gutiérrez, J., Villacorta, A., Arribas Arranz, J., Romero-Andrada, I., Lacoma, A., . . . & Rubio, L. (2024). Polylactic acid nanoplastics (PLA-NPLs) induce adverse effects on an in vitro model of the human lung epithelium: The Calu-3 air-liquid interface (ALI) barrier. *Journal of Hazardous Materials*, *475*, 134900. https://doi.org/10.1016/j.jhazmat.2024.134900
41. Nie, Z.-Y., Zhang, M.-Z., Zheng, D.-H., Zhang, M.-X., Jia, H.-Q., Zhang, X.-D., Yue, G.-C., Li, X.-M., Liu, P., Zhang, Y.-F., & Liang, K. (2017). Enhanced antitumoral activity of quercetin against lung cancer cells using biodegradable poly(lactic acid)-based polymeric nanoparticles. *Journal of Biomaterials and Tissue Engineering*, *7*(4), 269–274.
42. Jin, Y., Yu, W., Zhang, W., Wang, C., Liu, Y., Yuan, W.-E., & Feng, Y. (2023). A novel fluorinated polyethyleneimine with microRNA-942–5p-sponges polyplex gene delivery system for non-small-cell lung cancer therapy. *Journal of Colloid and Interface Science*, *648*, 287–298. https://doi.org/10.1016/j.jcis.2023.05.153
43. Liang, X.-J., Wang, N., Jin, N., Wu, N., & Liang, X.-J. (2012). Construction of paclitaxel-loaded poly (2-hydroxyethyl methacrylate)-g-poly (lactide)- 1,2-dipalmitoyl-sn-glycero-3-phosphoethanolamine copolymer nanoparticle delivery system and evaluation of its anticancer activity. *International Journal of Nanomedicine*, 1313. https://doi.org/10.2147/ijn.s29371
44. You, J.-O., & Auguste, D. T. (2008). Feedback-regulated paclitaxel delivery based on poly(N,N-dimethylaminoethyl methacrylate-co-2-hydroxyethyl methacrylate) nanoparticles. *Biomaterials*, *29*(12), 1950–1957. https://doi.org/10.1016/j.biomaterials.2007.12.041
45. Larson, N., Greish, K., Bauer, H., Maeda, H., & Ghandehari, H. (2011). Synthesis and evaluation of poly(styrene-co-maleic acid) micellar nanocarriers for the delivery of tanespimycin. *International Journal of Pharmaceutics*, *420*(1), 111–117. https://doi.org/10.1016/j.ijpharm.2011.08.011
46. Angelova, N., & Yordanov, G. (2014). Nanoparticles of poly(styrene-co-maleic acid) as colloidal carriers for the anticancer drug epirubicin. *Colloids and Surfaces A: Physicochemical and Engineering Aspects*, *452*, 73–81.
47. Iyer, R., Nguyen, T., Padanilam, D., Xu, C., Saha, D., Nguyen, K. T., & Hong, Y. (2020). Glutathione-responsive biodegradable polyurethane nanoparticles for lung cancer treatment. *Journal of Controlled Release*, *321*, 363–371.
48. Ma, X., & Fu, F. (2020). Effect of curcumin coated with polymethyl methacrylate nanoparticles on lung cancer cells. *Nanoscience and Nanotechnology Letters*, *12*(8), 1015–1021. https://doi.org/10.1166/nnl.2020.3205
49. Zhang, H., Liang, C., Hou, X., Wang, L., & Zhang, D. (2016). Study of the combined treatment of lung cancer using gene-loaded immunomagnetic albumin nanospheres in vitro and in vivo. *International Journal of Nanomedicine*, 1039.
50. Gündel, S., Dallemole, D. R., Alves, A. D. C. S., da Rosa, S. A., Oliveira, F. D. P., Ourique, A. F., . . . & Pohlmann, A. R. (2024). Nanoencapsulation of etoposide promotes increased long-term DNA damage, greater induction of senescence and reduces population regrowth of lung cancer cells. *Journal of Drug Delivery Science and Technology*, *100*, 106091.
51. Menon, J. U., Kuriakose, A., Iyer, R., Hernandez, E., Gandee, L., Zhang, S., . . . & Nguyen, K. T. (2017). Dual-drug containing core-shell nanoparticles for lung cancer therapy. *Scientific Reports*, *7*(1).

52. Fonseca, C., Simões, S., & Gaspar, R. (2002). Paclitaxel-loaded PLGA nanoparticles: Preparation, physicochemical characterization and in vitro anti-tumoral activity. *Journal of Controlled Release*, *83*(2), 273–286.
53. Li, M., Fang, G., Zahid, F., Saleem, R., Ishrat, G., Ali, Z., . . . & Din, F. U. (2024). Co-delivery of paclitaxel and curcumin loaded solid lipid nanoparticles for improved targeting of lung cancer: In vitro and in vivo investigation. *Heliyon*, *10*(9), e30290.
54. Shepherd, S. J., Issadore, D., & Mitchell, M. J. (2021). Microfluidic formulation of nanoparticles for biomedical applications. *Biomaterials*, 120826.
55. Mudrić, J., Šavikin, K., Ibrić, S., & Đuriš, J. (2019). Double emulsions (W/O/W emulsions): Encapsulation of plant bioactives. *Lekovite Sirovine*, *39*, 76–83.
56. Arpagaus, C. (2019). PLA/PLGA nanoparticles prepared by nano spray drying. *Journal of Pharmaceutical Investigation*, *49*(4), 405–426.
57. Hoang, N. H., Le Thanh, T., Sangpueak, R., Treekoon, J., Saengchan, C., Thepbandit, W., . . . & Buensanteai, N. (2022). Chitosan nanoparticles-based ionic gelation method: A promising candidate for plant disease management. *Polymers*, *14*(4), 662.
58. Dhanda, D. S., Tyagi, P., Mirvish, S. S., & Kompella, U. B. (2013). Supercritical fluid technology based large porous celecoxib-PLGA microparticles do not induce pulmonary fibrosis and sustain drug delivery and efficacy for several weeks following a single dose. *Journal of Controlled Release*, *168*(3), 239–250.
59. Sarcan, E. T., Silindir-Gunay, M., & Ozer, A. Y. (2018). Theranostic polymeric nanoparticles for NIR imaging and photodynamic therapy. *International Journal of Pharmaceutics*, *551*(1–2), 329–338.
60. https://clinicaltrials.gov/search?term=NCT01620190. Accessed dated 03-06-2025
61. https://clinicaltrials.gov/search?term=NCT00729612. Accessed dated 03-06-2025
62. https://clinicaltrials.gov/search?term=NCT04033354. Accessed dated 03-06-2025
63. https://clinicaltrials.gov/search?term=NCT06199895. Accessed dated 03-06-2025
64. https://clinicaltrials.gov/search?term=NCT05782426. Accessed dated 03-06-2025
65. https://clinicaltrials.gov/study/NCT00748163. Accessed dated 03-06-2025
66. Gavas, S., Quazi, S., & Karpiński, T. M. (2021). Nanoparticles for cancer therapy: Current progress and challenges. *Nanoscale Research Letters*, *16*, 173. https://doi.org/10.1186/s11671-021-03628-6
67. Sun, L., Wang, X., He, Y., Chen, B., Shan, B., Yang, J., . . . & Liang, R. (2023). Polyurethane scaffold-based 3D lung cancer model recapitulates *in vivo* tumor biological behavior for nanoparticulate drug screening. *Regenerative Biomaterials*, *10*, rbad091. https://doi.org/10.1093/rb/rbad091
68. Priya, S., Murali, A., Mohan, S., Lakshminarayanan, A., Sekar, S., Ramesh, R., Devendiran, M., & Han, S. S. (2023). *In vitro* anti-prostate adenocarcinoma and lung cancer studies of phenoxyaniline-*block*-poly(methyl methacrylate) based nanocomposites *via* controlled radical polymerization. *Nanoscale Advances*, *5*(21), 5870–5879. https://doi.org/10.1039/d3na00644a
69. Nagaich, U. (2018). Polymeric nanocapsules: An emerging drug delivery system. *Journal of Advanced Pharmaceutical Technology & Research*, *9*(3), 65. https://doi.org/10.4103/japtr.japtr_325_18
70. Yang, X., & Xie, Y. (2021). Recent advances in polymeric core-shell nanocarriers for targeted delivery of chemotherapeutic drugs. *International Journal of Pharmaceutics*, *608*, 121094. https://doi.org/10.1016/j.ijpharm.2021.121094

10 Solid Lipid Nanoparticles for the Management of Lung Cancer

Anjana Sharma[1], Pradhi Srivastava[2], Chandresh Patel[2], Maitrayee Ghosh[2], and Nitin Sharma[2]*

[1] Lloyd Institute of Management and Technology, Knowledge Park II, Greater Noida 201306, India

[2] Department of Pharmaceutics, Amity Institute of Pharmacy, Amity University Uttar Pradesh, Noida, Uttar Pradesh 201301, India

*Corresponding author: Nitin Sharma, nitinsharmaa.0202@gmail.com; sharma.nitin0909@gmail.com.

10.1 INTRODUCTION

Globally, cancer of the lung is the primary reason for cancer-related mortality with a five-year survival rate (1, 2). The mortality rate of lung cancer is much higher than that of the cancers of the pancreas, colon, and breast (3). In 2022, lung cancer was expected to be responsible for 1.8 million deaths and 2.5 million new cases globally (4). Males are more susceptible than females, and the incidence rate rises with age (5). Cigarette smoking remains the most common risk factor for lung cancer, but environmental exposures to substances like biomass fuels, asbestos, and arsenic also contribute significantly. The levels of these exposures differ greatly across different regions worldwide (6). Lung cancer arises from a combination of genetic mutations and significant epigenetic modifications. These alterations include changes in histone structure, remodeling of nucleosomes, and variations in DNA methylation patterns (7). Small cell lung cancer (SCLC), which is one of the primary types of lung cancer, contributes 10–15% of total cases of lung cancer. Additionally, it is distinguished by the existence of neuroendocrine cells. The second main type of lung cancer is non–small cell lung cancer (NSCLC), which contributes to 85–90% of total cases of lung cancer (8).

Currently, the three primary therapeutic options for lung cancer are surgery, radiation therapy, and chemotherapy, depending on the various stages and subtypes. Surgery and radiotherapy are considered localized treatments, aiming specifically at the primary cancer and nearby affected regions. Yet, it often faces challenges in accurately identifying malignant cancers that exhibit indistinct borders, such as SCLC. This difficulty arises because the blurred edges of these lesions complicate the differentiation between cancerous and non-cancerous tissues, leading to potential misdiagnosis and treatment complications (9). Radiation therapy uses high-energy X-rays to reduce cancer size; thus, radiation is usually given in combination with surgery or chemotherapy, or as neoadjuvant therapy to reduce the cancer size before surgical intervention. Yet, radiological treatment alone is often less effective for treating locally or regionally advanced lung cancer due to cancer resistance to radiation, systemic progression, and the development of local or distant metastases. Genomic research has facilitated the more effective use of anticancer drugs. Still, many chemotherapy agents for lung cancer have become less effective in patients with a positive mutation of the p53 cancer marker (10). Chemotherapeutic agents are designed to target and disrupt the rapid proliferation of cancer cells. However, these toxic compounds also affect the fast-growing cells in hair follicles, bone marrow,

DOI: 10.1201/9781003595830-10

and the gastrointestinal tract, resulting in the unwanted side effects commonly associated with cancer treatment (11). Chemotherapy is used for metastatic cancers; however, issues such as systemic toxicity and the development of drug resistance are frequently encountered. Additionally, small molecule drugs tend to have a brief peak plasma concentration, while large carriers struggle to deliver precise targeting. Therefore, analysis and categorization of lung cancer gene mutations have led to the development of targeted therapies, which have demonstrated strong therapeutic benefits (9). Over the past few years nanoparticles have found great applications in targeted drug delivery, gene delivery, and diagnostics. The toxicity concern of the nano-carriers must also be addressed adequately before their administration. Nanotoxicity can be prevented via several methods like coating, grafting, and ligation (12).

Solid lipid nanoparticles (SLNPs) are considered to be the first lipid-based nanocarriers, composed of lipids which are solid at 37°C and emulsifiers stabilizes them (13, 14). SLNPs typically have submicron sizes, with diameters between 50 to 1000 nm (15). The safety profile of SLNPs is largely due to their use of biocompatible lipids, which are well-tolerated by both the lungs and the body. As a result, SLNPs are strongly recommended for pulmonary drug delivery, whether administered as a suspension or in dry powder form, without causing inflammation. SLNPs have been also used for intranasal delivery of nalbuphine for the management of pain. Gamma scintigraphy study has shown that the prepared SLNPs of nalbuphine were able to reach the target and were retained in the brain (16). These nanoparticles offer several benefits, including easy transformation, suitability with lipophilic drugs, delivering active agents at the desired site, improved drug stability, minimized toxicity, and ability to bypass the first-pass effect, distinguishing them from other colloidal carriers (17). SLNPs are essential for improving drug delivery at the target site through passive and active targeting mechanisms. Their small particle size facilitates passive targeting by enhancing permeation and retention (EPR), which is particularly effective due to the unique characteristics of cancer cells, such as hypervascularization and inadequate lymphatic drainage. The surface of SLNPs can be readily modified and enhanced with specific targeting ligands, facilitating active targeting to receptors that are overexpressed on malignant cancer cells through receptor-mediated endocytosis (18). Active targeting of therapeutic substances using SLNPs can be achieved by decorating the nanocarriers using several ligands including the epidermal growth factor receptor (EGFR), transferrin receptor, folate receptor, cluster of differentiation 44 (CD44), other receptors such as $\alpha_v\beta_3$ integrin receptor, luteinizing hormone-releasing hormone receptor, or tyrosine kinase Axl receptor (19).

In this chapter, we discuss the applications of SLNPs for the effective treatment of lung cancer. It highlights the specific applications of SLNPs for the delivery of several categories of therapeutic substances, which include chemo, gene, immuno, radio, photothermal, and photodynamic therapeutics. Moreover, these applications are not limited to the mentioned areas and can be expanded to other related areas.

10.2 LUNG CANCER

The pathophysiology of lung cancer is very complicated and yet to be completely understood. It has been hypothesized that persistent exposure to carcinogens like smoking cigarettes causes abnormality in growth and maturation of lung epithelium. Regular smoking or exposure to carcinogens triggers the gene mutation process, thereby affecting protein synthesis by disrupting the cell cycle and promoting carcinogenesis (20). Numerous molecular abnormalities are being investigated as possible targets for new molecularly focused treatments and as prognostic indicators. As lung cancer develops, several growth factors and receptors are encoded by proto-oncogenes that are triggered by different methods. Oncogenes like mutated KRAS and c-myc along with EGFR, cyclin D1, and BCLII proteins found overexpressed in cancer (21). The downregulation and hypermethylation of tumor suppressor genes (TSGs) at the promoter sites of FHIT, RASFF1A, and SEMA3B were found to be the cause of lung cancer (22). Less often engaged TSGs include PTEN, hOGG1

(DNA repair), and BAP1 (ubiquitination), whose irregular inactivation suggests participation in their related cell regulatory processes. Many genes (*APC*, *CDH13*, *RARβ*, *FHIT*, *RASSF1A*, *TIMP3*, *p16*, *MGMT*, *SEMA3B*, and *DAPK*) have their expression inactivated primarily by tumor-acquired promoter hypermethylation (23, 24). The Rb/p16/cyclin D1, p53/MDM2/p19ARF, wnt/APC, EGFR/Ras, PP2a, and telomerase pathways, in addition to several genes involved in different DNA repair processes, are among the cell regulatory pathways that are dysregulated in lung cancer. Deregulation of angiogenesis, autocrine/paracrine growth factor circuits, and apoptosis is also seen. The genome-wide allelotyping changes indicating allele loss (genetic alterations) and methylation modifications (epigenetic alterations) are beginning to offer hypothetical estimates for the fewest occurrences (20–30) that result in the development of lung cancer (25).

10.3 SOLID LIPID NANOPARTICLES

SLNPs are colloidal carriers that are composed of biocompatible and biodegradable lipids including triglycerides, cetyl palmitate, and glyceryl monostearate. Since they can enhance the solubility, stability, and bioavailability of drugs that are not especially soluble in water, SLNPs are commonly used in drug delivery systems. SLNPs are stabilized with emulsifiers or surfactants and typically range from 50 to 1000 nm in size and have advantages such as high loading of therapeutic substances and easy alteration in their structure for targeting purposes. Encapsulated around this core is a stabilizing surfactant layer that protects the particles from agglomeration and ensures colloidal stability. The lipid matrix shields the entrapped therapeutic substances from environmental degradation and allows their controlled release (26). SLNPs are useful because they are non-toxic, can contain both hydrophilic and lipophilic drugs, and can be administered orally, topically, or intravenously. They may also be able to target specific areas by changing their surface with ligands or antibodies (27).

SLNPs comprise of several essential components that work together to ensure functionality, stability, and efficacy in drug delivery systems. There is the basic constituent that constitutes the core of the nanoparticles: the lipid matrix. In this matrix, solid lipids retain a solid state at body temperatures, ensuring long-term drug release. Some of the most commonly used lipids include fatty acids such as stearic acid and palmitic acid, triglycerides such as tristearin and tripalmitin, waxes such as cetyl palmitate, and glycerides such as glyceryl monostearate and glyceryl behenate. Several factors, including the solubility and compatibility of the drug with the lipid matrix, determine the choice of lipid. These lipids are necessary for encapsulating the drug and protecting it from degradation. The second critical constituent is surfactants, acting as stabilizers. Surfactants ensure that SLNPs disperse uniformly in solution while preventing particle agglomeration by reducing the surface tension. Examples of surfactants are ionic, such as polysorbates (e.g. Tween 80), Poloxamers, sodium dodecyl sulfate, and sodium cholate, and non-ionic surfactants (28).

The size, surface characteristics, and stability of the SLNPs are highly influenced by the selection and concentration of the surfactant. Co-surfactants are added in some cases to improve particle stability and effectiveness of encapsulation in the SLNPs. Bile salts and ethanol can also be considered as excellent examples of co-surfactants. These compounds result in improved overall stability with functionality along with the main surfactants. Further, to prevent the SLNPs from opsonization, the hydrophilic material such as polyethylene glycol is being mixed. Pegylation prevents the uptake of SLNPs from the reticuloendothelial system and prolongs systemic circulation time. The types of various components and quantity in the formulation govern several physiological characteristics (shape, size, encapsulation efficiency, drug release, and zeta potential) and their biological performance (permeation, binding with various proteins, extravasation and cellular uptake) (29).

10.4 SOLID-LIPID-NANOPARTICLE-MEDIATED TARGETING APPROACHES TO LUNG CANCER TREATMENT

SLNPs have shown their therapeutic potential for the effective targeting of active substances. Several reposts and literature have suggested that SLNPs have effectively loaded the therapeutically active substances and delivered them to the desired site. This approach not only increased the therapeutic potential but also reduced the associated toxicity to the peripheral tissues. Physical targeting ensures the EPR of therapeutic substance–loaded SNLs to cancer vicinity whereas active targeting can be ensured by ligating the several targeting moieties on the surface of SLNPs, which help the delivery system to navigate the cancer site. These targeting approaches are discussed here in more detail (30).

10.4.1 Passive Targeting

The microenvironmental characteristics of lung cancer allow targeting of the cancer through the EPR effect. This could be due to vascular leakage and breakdown of basal epithelia of cancer microenvironment. Enhanced penetration of drug-loaded SLNPs across the interstitial space and the surroundings of the cancer vicinity is due to enhanced angiogenesis and basal epithelial dysregulation. Due to the enhanced retention of the cancer vicinity, the drug-loaded SLNPs remain in place for a long time after penetration through the epithelia. The mechanisms that contribute to the improved retention of SLNPs in the cancer vicinity include cellular crowding, altered interstitial fluid dynamics, and increased extracellular matrix deposition. Included among these three associated ways in which the EPR effect impacts molecular distribution are nanoparticle extravasation from blood vessels, nanoparticle diffusion into cancer tissue, and nanoparticle engagement with intracellular or extracellular targets in the cancer vicinity (29).

10.4.2 Active Targeting

The active targeting approach facilitates SLNPs to increase the circulation time and ensure the availability of therapeutic substances at the cancer site. This involves the ligation of several targeting moieties with the delivery system, which ensures their uptake by the receptors overexpressed on the cancer cells. At the same time, active targeting ensures non-exposure to chemotherapeutic substances by the healthy cell, thus reducing the side effects (31). Folate receptors are one of them, which are overexpressed on the NSCLC, especially in the adenocarcinoma, which makes this receptor a promising targeting moiety for lung cancer targeting. A folate receptor is a glycosylphosphatidylinositol-anchored membrane protein that is responsible for the transportation of folic acid (vitamin for DNA synthesis and cell growth). In research by Patel *et al.*, folate-receptor-mediated silibinin-loaded nano carriers were developed for the effective treatment of lung cancer. The author reported improved deposition of developed nanocarriers in lung tissue (32). Similarly, Rosiere *et al.*, also developed folate grafted paclitaxel loaded inhalable SLNPs for lung cancer treatment. The outcome of the study concludes that the developed formulation possesses improved therapeutic value in *in vivo* model (33).

Another receptor in this category is the EGFR, which is a transmembrane tyrosine kinase receptor that overexpresses cancer cells including mutated lung cancer. Similar to the folate receptor, the EGFR also plays an important role in cell proliferation and survival. Nan *et al.* developed lipid polymeric nanoparticles loaded with co-delivered cisplatin and doxorubicin for EGFR-mediated targeting of lung carcinomas. The author demonstrated the better therapeutic value of developed lipid polymeric nanoparticles with a cancer inhibition ratio up to 74.5% (34). In a similar context, another receptor is the transferrin receptor, which is a membrane-bound glycoprotein overexpressed on the cancer cell and responsible for the iron transportation which is an essential mechanism for DNA synthesis, cell proliferation, and survival. Guo *et al.* developed transferrin-conjugated, doxorubicin-loaded, lipid-coated nanoparticles for the effective management of lung cancer. The outcome of the study demonstrated that the developed nanoparticles offered better antiproliferation activity in lung cells as compared to the non-ligated nanoparticles, which confirmed the application of transferrin ligation on the nanoparticles for better therapeutic value (35, 36). Figure 10.1 primarily depicts active and passive accumulation (EPR effect) of SLNPs. This involves functionalizing the surface of the SLN with specific targeting moieties (such as antibodies, peptides, or small molecules) that recognize and bind to receptors overexpressed on the surface of tumour cells or tumour-associated cells.

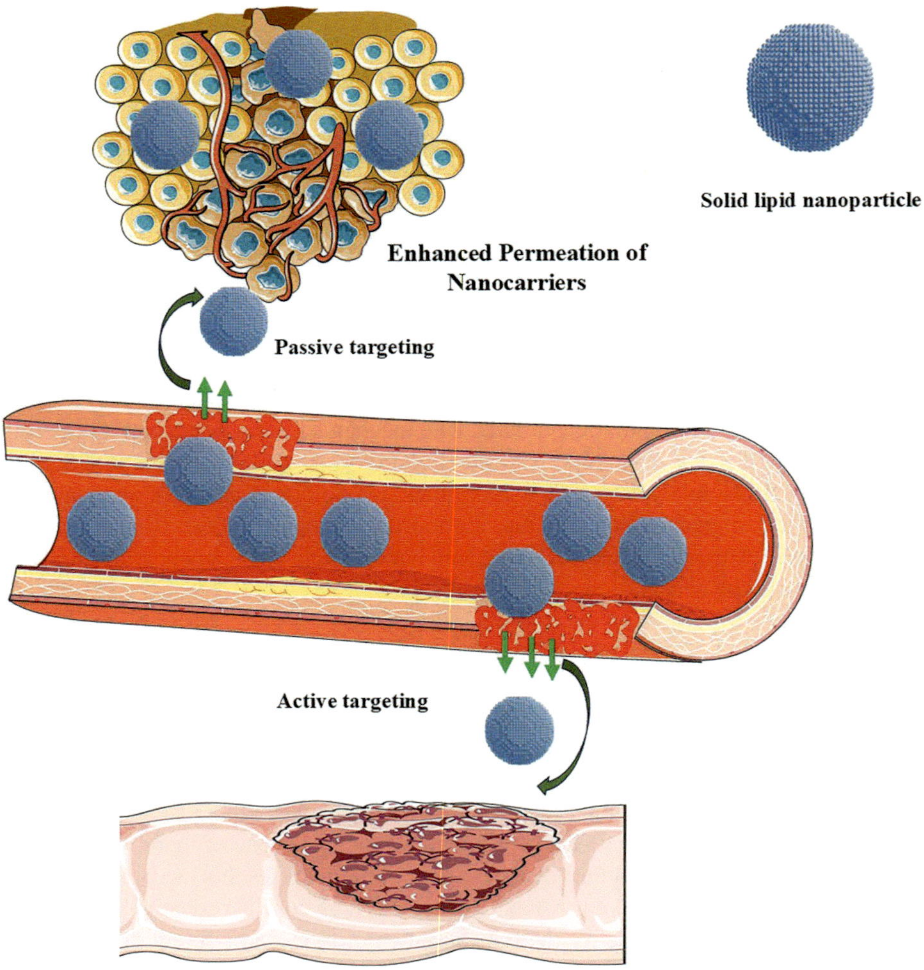

FIGURE 10.1 Active and passive targeting into a tumour tissue using SLNPs

10.5 THERAPEUTICS APPLICATIONS OF SOLID LIPID NANOPARTICLES IN LUNG CANCER

Lung cancer continues to be among the leading causes of cancer-related mortality worldwide, which prioritizes the urgent need for the development of novel therapeutic strategies. In this respect, lipid-based nanoparticles, especially SLNPs, have appeared as a versatile and efficient platform due to their biocompatibility, stability, and ability to host a variety of drugs. Here we discuss the application of SLNPs in several therapies, including chemotherapy, gene therapy, immunotherapy, radiotherapy, phototherapy, and photodynamic therapy. Few applications of SLNPs are also represented in Table 10.1.

10.5.1 SLNPs in Chemotherapy

SLNPs are one of the most effective platforms for the delivery of chemotherapeutic agents in the therapy of lung cancer having characteristics like small size, high stability, biocompatibility, and the ability to encapsulate hydrophilic and hydrophobic drugs and overcome some key limitations of traditional chemotherapy like systemic toxicity, poor solubility, and lack of cancer specificity (37). These nanocarriers improve the therapeutic index of drug, while protecting them from enzymatic degradation, and also provide controlled drug release, reducing systemic side effects and the dosing frequency (38, 39). Due to the EPR effect, SLNPs tend to accumulate passively in cancer tissues by taking advantage of leaky vasculature and poor lymphatic drainage characteristics of cancer vicinity. Functionalization with ligands such as folic acid and transferrin allows for active targeting; after binding to the corresponding overexpressed receptors on the surface of cancer cells, these nanocarriers ensure targeted drug delivery to avoid damage to healthy tissues (40).

SLNPs have also been used to encapsulate various chemotherapeutic substances such as paclitaxel, docetaxel, doxorubicin, and cisplatin. In 2022, Pi *et al.* prepared SLNPs of paclitaxel and curcumin to improve their therapeutic potential and evaluated by *in vitro* and *in vivo* models. The authors evaluated the developed formulation in different cancer cell lines; among them A549 cells (lung cancer) showed the highest inhibitory effect. IC_{50} values decreased 6.08-, 66.16-, and 7.32-fold when compared with paclitaxel and 2.58-, 1.47-, and 1.29-fold when compared to free paclitaxel+curcumin after 24, 48, and 72 hours respectively. Also, an *in vivo* study showed that cancer growth was controlled by their SLNP formulation, and its efficacy was increased (37). In 2013 Mosallaei *et al.* prepared docetaxel-loaded SLNPs and evaluated for their *in vitro* cellular uptake in cancer cells by HPLC method, and they found that docetaxel-loaded SLNPs could significantly deliver and accumulate more therapeutic substance in cancer cells after 4, 12, and 24 hours when compared with the commercial formulation. They also evaluated the prepared formulation in C-26 tumour-bearing mice model and found it decreased the death rates in the mice model. From this study, the authors concluded that their prepared docetaxel SLNPs had good efficacy in MTT assay, higher cellular uptake, and better cancer regression compared with commercial formulations (41). Also, Wang *et al.* focused on the development of Phyllanthi-Tannin-loaded SLNPs for lung cancer therapy. The authors developed SLNPs using the thin film hydration method and characterized their spherical shape with an average particle size of approximately 50 nm and favourable physicochemical properties. Evaluation by *in vitro* studies on Lewis lung cancer cells revealed that Phyllanthi-Tannin-loaded SLNPs exhibited significantly higher cytotoxicity compared to free Phyllanthi Tannin, with IC_{50} values of 20.74 μg/mL for Phyllanthi-Tannin-loaded SLNPs versus 67.43 μg/mL for free Phyllanthi Tannin. Their *in vivo* evaluation involving cancer-bearing mice showed that SLNPs achieved cancer inhibition rate of 64.55% at the dosage of 0.4 g/kg while the inhibition rate was observed 59.97% with a higher dosage (2 g/kg) of free Phyllanthi Tannin. Additionally, safety assessments indicated that SLNPs avoided some of the organ damage associated with Phyllanthi Tannin treatment, suggesting an improved therapeutic profile (42).

10.5.2 SLNPs in Gene Therapy

SLNPs appear as a promising carrier for delivering genetic material into cells in lung cancer therapy. SLNPs provide an effective, non-viral method of delivering nucleic acids, including small interfering RNA (siRNA), microRNA (miRNA), and plasmid DNA, which overcome the major problems of poor stability, enzymatic degradation, and poor cellular uptake linked with conventional gene therapy for lung cancer. SLNPs are well suited for gene therapy due to their biocompatibility, protection of nucleic acids from degradation, and facile functionalization, allowing for targeting delivery. Inside the lung cells, SLNPs release the gene of interest into the cytoplasm so that the genetic material made available to exert the therapeutic effect (43).

The research of Choi *et al.* focused on the use of cationic SLNPs for gene therapy to target the p53 TSG in NSCLC. Using the melt homogenization method, the authors developed a novel SLNP formulation with tricaprin and other lipids at different weight ratios. They transfected H1299 lung cancer cells with plasmid DNA expressing p53, assessed transfection efficiency by flow cytometry, and showed gene expression by RT-PCR and Western blot analysis—a substantial restoration of wild-type p53 mRNA and protein. Such results indicate that the transfection efficiency of SLNPs formulation was higher compared to the commercially available Lipofectin. Effective induction of apoptosis and significant growth inhibition in such cells can be achieved by *p53* gene expression. The data from this study suggest that SLNP-mediated delivery of the *p53* gene may have the potential for nonviral gene therapy strategies toward lung cancer, where the apoptotic pathways may be re-established and cancer growth can be hindered (44). More studies prove the potential efficacy of SLNPs in gene therapy with combination with chemotherapy. For example, the study of Han *et al.* focused on developing and characterizing surface-modified SLNPs for the co-delivery of doxorubicin and plasmid DNA in the treatment of lung cancer. The study exploited transferrin as a targeting ligand to enhance the delivery of these nanoparticles exactly to the lung cancer cells. The *in vitro* cytotoxicity tests proved that transferrin-modified SLNPs were well tolerated with low toxicity; cell viability was above 80–100% in A549 human lung adenocarcinoma cells. This formulation was also evaluated on mice with A549 cancer and the results showed better therapeutic effect compared to not modified SLNPs. Thus, chemotherapy and gene therapy have successfully been combined within one single drug delivery system, namely the multifunctional system (45).

10.5.3 SLNPs in Immunotherapy

SLNPs have a very important role in delivering immune checkpoint inhibitors, including anti-PD-1 and anti-PD-L1 antibodies, which block pathways that cancer cells use to evade immune detection (46). Encapsulation enhances the stability of these agents, improves their pharmacokinetics, and enables targeted delivery to the cancer microenvironment. It ensures a higher therapeutic effect while minimizing off-target immune activation and consequently side effects. In the clinical setting, blocking the PD-1/PD-L1 interaction with anti-PD-1 or anti-PD-L1 monoclonal antibodies has been successful in several cancer, including NSCLC. Nafee *et al.* investigated the efficacy of Myricetin-loaded SLNPs in treating lung cancer, using both *in vitro* and *in vivo* models. Their *in-vitro* study demonstrated that the SLNPs significantly enhanced cellular uptake and antitumour activity, leading to a threefold reduction in IC_{50} values compared to free Myricetin. The authors also performed confocal imaging to confirm enhanced fluorescence intensity, which indicated improved cellular internalization. In *in vivo* evaluation, SLNPs showed over 80% drug release within 8 hours and effective deposition in the targeted region, suggesting potential for enhanced therapeutic efficacy against lung carcinoma. Overall, the findings highlight the promising role of SLNPs in lung cancer therapy, showcasing improved drug delivery and antitumor effects (47).

SLNPs are also being investigated for delivering immunostimulatory cytokines, which enhance the body's immune response against cancer. In this approach, Banerjee *et al.* in 2019 addressed the immunomodulatory effects of a peptide-modified SLNP formulation of paclitaxel and its potential as an immunotherapy for lung cancer, using a murine melanoma model. They used B16F10 melanoma cells to compare paclitaxel with dacarbazine, a standard chemotherapeutic agent. Their study design included *in vitro* assays such as apoptosis, invasion, and calreticulin exposure and *in-vivo* models of cancer growth and immune responses. Their results showed that paclitaxel significantly enhanced the infiltration of CD8+ T cells into cancer and induced calreticulin exposure on cancer cells, which are markers of immunogenic cell death. Moreover, treatment with paclitaxel resulted in a remarkable decrease in both cancer volume and the formation of metastases in the lungs in comparison with dacarbazine. This result indicates that paclitaxel entrapped in SLNPs outperformed traditional chemotherapy and elicited a favourable systemic immune response without observable toxicity (48).

10.6 RADIOTHERAPY

Radiotherapy is the sole treatment option for all disease stages and patient performance level. Several studies show that 77% of the affected patients have been provided radiotherapy but still it is in dormancy in several parts of globe. One study has revealed that use of radiotherapy alone resulted in

TABLE 10.1
Application of Solid Lipid Nanoparticles in the Treatment of Lung Cancer

Serial Number	Drug Incorporated	Preparation Method	Models Used	Inference	Reference
1	Gefitinib	Ultrasonic melt-emulsification	A549 cell line	IC_{50} was reduced by 1.79 and 1.94 fold by SLNPs when compared to pure gefitinib	49)
2	Bedaquiline	Hot micro emulsification and solvent diffusion	A549 cells	SLNPs increased the therapeutic value of the drug and decreased the associated toxicity.	50)
			Wistar rat	Cancer volume was reduced and the accumulation of bedaquiline was increased in the lung tissue.	
3	Curcumin	Emulsification-ultrasonication	A549 cells	After 48 hours, SLNPs reduced cytotoxicity (IC_{50}) and exhibited high cellular uptake compared to plain curcumin.	51
4	Docetaxel	High energy	BALB/C mice	Lung metastasis was prevented spontaneously in 4T1 tumour-bearing mice.	52
5	Mangiferin	Emulsification-solvent evaporation method	A549 cells	Transferrin-modified SLNPs showed higher cellular uptake and antiproliferative effect when compared to plain mangiferin.	53
			Mice	Transferrin-modified SLNPs showed a decrease in cancer size and exhibited higher biodistribution.	
6	Allicin	High-pressure homogenization	A549 and HFF cells	After 48 hours, viability of A549 cells was reduced while HFF cells remained unaffected.	54
7	Paclitaxel and curcumin	High-pressure homogenization	A549 cells	Cell uptake of SLNPs was found to increase and cancer cell viability was reduced.	55
			BALB/C mice	SLNPs increased anticancer efficacy, which reduced cancer volume without affecting the body weight.	

a 4% survival and attainment of 8.3% in five-year locoregional control. The incorporation of SLNPs in cancer therapy resulted in improved outcomes in lung cancer treatment (56).

Joshi HA *et al.* conducted a study aimed at developing a non-toxic and effective radiation countermeasure by enhancing the bioavailability of baicalein through the SLNPs. The research focused on evaluating the radioprotective and radio-sensitizing effects of these nanoparticles. To address the issue of poor bioavailability associated with lipophilicity, SLNPs were fabricated and characterized to ensure effective drug entrapment and release kinetics. The study assessed the radioprotective effects of SLNPs across various experimental models. In murine splenic lymphocytes, SLNPs demonstrated significant protection against apoptosis induced by 4 Gy ionizing radiation. Furthermore, in a whole-body irradiation model in mice, oral administration of SLNPs resulted in approximately 70% protection against mortality caused by 7.5 Gy whole body irradiation. The bioavailability assessment revealed that the oral relative bioavailability of baicalein was enhanced by over 300% with SLNPs compared to free baicalein. Additionally, oral dosing of SLNPs led to a transient increase in neutrophil abundance in peripheral blood. In human lung cancer cells, treatment with SLNPs showed radio-sensitization, evidenced by decreased survival and clonogenic potential, as well as an increase in cellular reactive oxygen species (ROS) levels. This increase is ROS likely attributed to higher uptake and inhibition of TrxR. The study concluded that SLNPs significantly improved the bioavailability of baicalein, offering enhanced radioprotection to normal cells while sensitizing cancer cells to radiation-induced killing. These findings suggest that SLNPs could serve as a promising adjuvant in cancer radiotherapy, providing protection to normal cells while enhancing therapeutic efficacy against cancer cells (57).

Bromma K. *et al.* conducted a study to explore the uptake and radiation dose enhancement of gold nanoparticles (GNPs) in synchronized versus unsynchronized cancer cell populations. The researchers utilized the application of SLNPs system to deliver smaller GNPs, specifically 5 nm in diameter, at a low concentration of 0.2 nm into human breast cancer cells (MDA-MB-231, a triple-negative cell line). The investigation aimed to compare GNP uptake and survival fractions between synchronized and unsynchronized cancer cell populations, measure the uptake of the 5 nm GNP delivered via SLNPs, and assess the radiation dose enhancement using 2 Gy of 6 MeV photon radiation, while also supporting experimental findings with model predictions based on Monte Carlo calculations. The results revealed a remarkable 39-fold enhancement in GNPs uptake in unsynchronized cancer cell populations and a 73-fold enhancement in synchronized populations when using SLNPs delivery. This increased nanoparticle uptake per cell led to a 27% enhancement in cancer cell death in synchronized populations following radiation treatment. Monte Carlo modelling corroborated these findings, predicting significant improvements in cell killing with a clinical treatment plan involving 30 fractions of 2 Gy radiation doses. Notably, synchronizing cells and employing SLNPs–GNPs resulted in an extraordinary 1000-fold enhancement in cell killing. The study concludes that synchronizing cancer cells and utilizing the SLNPs–GNP system can significantly improve radiation therapy outcomes, presenting a novel and cost-effective strategy for enhancing GNPs-mediated cancer therapy (58).

10.7 PHOTOTHERMAL THERAPY

Photothermal therapy, a promising treatment strategy, that uses the near-infrared (NIR) laser photoabsorbers to create heat for thermal ablation of cancer cells after exposure NIR laser irradiation. Compared with conventional therapeutic modalities, photothermal therapy exhibits unique advantages in cancer therapy, including high specificity, minimal invasiveness precise spatial-temporal selectivity, and so forth. In addition to treating cancer cells in the primary malignancy or local metastases in adjacent lymph nodes, photothermal treatment can be used in conjunction with existing therapeutic methods to treat cancer cells in the metastatic areas (59).

Xue S *et al.* conducted a study to design lipid-decorated bismuth-based nanoflowers intended to function as both a radiosensitizer and a photothermal therapy agent, with the goal of improving treatment efficiency for lung metastatic breast cancer treatment while addressing the challenges of resistance and side effects associated with traditional radiotherapy. The therapeutic efficacy of combining radiotherapy with photothermal therapy was assessed through Western blot assays aimed at determining cancer apoptosis. The results indicated that the optimized lipid-decorated bismuth-based nanoflowers exhibited significant potential as a novel inhalable integrated microsphere, enhancing the therapeutic efficiency of radiotherapy when combined with photothermal therapy for lung metastatic breast cancer. This study suggests that lipid-decorated bismuth-based nanoflowers could provide a promising new approach for thermo-radiotherapy, potentially leading to improved treatment outcomes for patients suffering from lung metastatic breast cancer (60).

10.8 PHOTODYNAMIC THERAPY

Photodynamic therapy is a type of non-invasive form of contemporary non-ionizing radiation. The basic principle of this treatment is the local or systemic administration of a photosensitive substance that remains intact by malignant cells and tissue through both active and passive accumulation coupled with single wavelength activation light. The photosensitive chemical absorbs the right wavelength of light, which starts the activation stages of ROS formation and causes cancerous cells to be destroyed selectively. Several photosensitive compounds have been discovered so far and targeted to the cancer site using SLNPs (61).

Jiang S *et al.* performed a study aimed to encapsulate curcumin into SLNPs to improve its therapeutic activity in photodynamic therapy by addressing its poor solubility and hydrolytic instability. Curcumin-loaded SLNPs were prepared using an emulsification and low-temperature solidification method. The functions of curcumin-loaded SLNPs were evaluated on A549 cells for photodynamic therapy. The study measured the phototoxicity of curcumin-loaded SLNPs under light excitation at a concentration of 15 μM and analysed cell cycle distribution, ROS production, mitochondrial membrane potential disruption, and protein expression related to apoptosis using Western blot analysis. The results demonstrated that curcumin-loaded SLNPs induced approximately 2.27-fold higher toxicity than free curcumin at a low concentration of 15 μM under light excitation, leading to increased cell cycle arrest at the G2/M phase (62).

Yeo S *et al.* conducted a study aimed to enhancing the bioavailability and effectiveness of photodynamic therapy for cancer treatment by synthesizing purpurin-18-N-propylimide methyl ester as photosensitive compound and delivered through SLNPs. The efficacy of photodynamic therapy using these SLNPs was evaluated through the 1,3-diphenylisobenzofuran assay and photocytotoxicity tests on HeLa (human cervical carcinoma) and A549 (human lung carcinoma) cell lines. Photocytotoxicity assays indicated that while the formulations exhibited toxicity under light irradiation, they showed no toxicity in the dark, confirming their photodynamic activity. Notably, the smallest-sized SLNP formulation displayed the most effective photodynamic activity, suggesting that P18 N PI ME-loaded SLNPs could represent a promising strategy for improving photodynamic therapy in cancer. These findings highlight the potential of utilizing SLNPs to enhance the delivery and efficacy of photosensitizers in treating cancer (63).

10.9 DIAGNOSTIC APPLICATIONS OF SOLID LIPID NANOPARTICLES IN LUNG CANCER THERAPY

Clinically, lung cancer can be diagnosed by either imaging or pathology. The imaging method (non-invasive) is mainly low-dose computed tomography but their high degree of false positive results of about 96.4% and radiation exposure are found to be the significant obstacles in the early diagnosis of lung cancer in many patients (64–66). The invasive methods like surgery, biopsy are considered to be as the gold standard for lung cancer diagnosis. Significant obstacles for those at risk of developing cancer are posed by the invasiveness, high expenses, and risk of surgical complications, particularly in developing and underdeveloped nations (67). Therefore, the combined effects of advanced diagnostic techniques like SECT, MRI, and PET with lipid nanoparticles exemplify the promising treatment in lung cancer. The effect may be attributed to their small size, ease of functionalization, improved drug loading capacity, enhanced penetration, and improved retention time in the targeted tissue thereby, preventing the healthy tissues (68). The cost-effective diagnostic imaging technique should offer the most possible information about the cancer or other damaged organs (69). Anatomical parts of a cancer and surrounding tissues can be located and spatial (topographic) data can be obtained using MRI and CT imaging techniques. These approaches cannot detect cancer in its early stages because their precision in recognizing cellular abnormalities at the molecular level is not accurate. Several studies have been conducted in molecular biology, nanotechnology, and other fields to increase the sensitivity of image contrast agents and had a significant impact on early diagnosis of cancer.

For *in vivo* optical imaging, enhanced optical characteristics and multiplexing ability make them superior options. Despite several advantages, optical imaging faces the limitations of tissue penetration, auto fluorescence, and qualitative nature. Therefore, a combination of imaging based on both fluorescence and radionuclides has synergistic benefits over either technique alone (70). Several near-infrared fluorescence imaging dyes have been developed and applied in various preclinical and clinical cancer diagnoses. Researchers have studied the synthesis of indocyanine green (ICG) derivatives for the early detection of cancer in humans as it was the first dye that was US-FDA approved. The major limitations of ICG are its non-specificity to tumour and rapid clearance, short half-life (2–4 minutes), and high rate of plasma clearance (~18.5% per minute) in humans with fluorescence quenching (71). Therefore, it is essential to tag targeting agents, including proteins, peptides, antibodies, and their fragments. Potential applications of ICG have received a lot of attention for the early identification of superficial cancers from organs like the breast and the use of photodynamic treatment to treat cancers that arise from other organs (72). It is essential to achieve cancer-targeted ICG administration and subsequent accumulation and retention in the cancer in order to diagnose and treat using PDT (73). The most popular imaging technique is fluorescent probe-based optical imaging, in which incident radiation in the invisible or near-infrared region excites the probes, causing them to produce energy at a lower level. Optical imaging can provide detailed information on individual cells and intracellular molecules such as RNA, DNA, proteins, peptides, receptors, biomarkers, and epitope distributions, but it cannot explain anatomical information (74). Upon comparison to molecule-based contrast agents, nanoparticles tagged with contrast agents have proven to enhance disease identification with target explicitly via improved circulation. The contrast agent can be encapsulated to target the sites and minimize the side effects (75). Researchers have prepared multifunctional SLNPs called Gd-FITC-SLNPs, which is loaded by octadecylamine fluorescein isothiocyanate and gadolinium diethylenetriaminepenta acetic acid. These SLNPs serve as an excellent contrast agent for MRI with unique tumour-absorbable property. They are non-toxic to human colon cancer cells (HT29) and mice colon cancer cells (CT26), and they exhibit remarkable T1 relaxivity. Cancer cells were shown to absorb these contrast compounds *in vitro* in a concentration-dependent manner. When transrectally administered, these SLNPs showed markedly improved colorectal imaging *in vivo* MRI tumours in contrast to conventional infusion of Gd-DTPA. The concentrated presence of Gd-FITC-SLNPs within the tumour was validated by confocal fluorescence imaging. This method demonstrates how SLNPs can improve optical fluorescence imaging in colon cancerous tumours, providing crucial data for the detection and treatment of lung cancer (76).

10.10 CONCLUSION

SLNPs have emerged as promising nano-delivery systems in lung cancer therapy through several modes. They have several advantages over other carriers because they are biocompatible/biodegradable, drug targeting, and non-immunogenic. Their ability to encapsulate chemotherapeutic substances ensures the safety to the peripheral cells while ensuring the chemotherapeutic substances to the cancer site. Despite these advantages, problems such as nanotoxicity, scalability, stability, and regulatory aspects should be properly addressed to ensure successful clinical translation. Continued study into improving SLNPs formulations, understanding their pharmacokinetics, and assuring biocompatibility will be critical to expanding their clinical uses.

REFERENCES

1. Leiter A, Veluswamy RR, Wisnivesky JP. The global burden of lung cancer: current status and future trends. *Nature Reviews Clinical Oncology*. 2023;20(9):624–39.
2. Smeltzer MP, Wynes MW, Lantuejoul S, Soo R, Ramalingam SS, Varella-Garcia M, et al. The international association for the study of lung cancer global survey on molecular testing in lung cancer. *Journal of Thoracic Oncology*. 2020;15(9):1434–48.
3. Zappa C, Mousa SA. Non-small cell lung cancer: current treatment and future advances. *Translational Lung Cancer Research*. 2016;5(3):288.
4. Bray F, Laversanne M, Sung H, Ferlay J, Siegel RL, Soerjomataram I, et al. Global cancer statistics 2022: GLOBOCAN estimates of incidence and mortality worldwide for 36 cancers in 185 countries. *CA: A Cancer Journal for Clinicians*. 2024;74(3):229–63.
5. Porro C, La Torre ME, Tartaglia N, Benameur T, Santini M, Ambrosi A, et al. The potential role of nutrition in lung cancer establishment and progression. *Life*. 2022;12(2):270.
6. Zhou J, Xu Y, Liu J, Feng L, Yu J, Chen D. Global burden of lung cancer in 2022 and projections to 2050: incidence and mortality estimates from GLOBOCAN. *Cancer Epidemiology*. 2024;93:102693.
7. Liu Y, Mao C, Wang M, Liu N, Ouyang L, Liu S, et al. Cancer progression is mediated by proline catabolism in non-small cell lung cancer. *Oncogene*. 2020;39(11):2358–76.
8. Gilad S, Lithwick-Yanai G, Barshack I, Benjamin S, Krivitsky I, Edmonston TB, et al. Classification of the four main types of lung cancer using a microRNA-based diagnostic assay. *The Journal of Molecular Diagnostics*. 2012;14(5):510–17.
9. Li S, Xu S, Liang X, Xue Y, Mei J, Ma Y, et al. Nanotechnology: breaking the current treatment limits of lung cancer. *Advanced Healthcare Materials*. 2021;10(12):2100078.
10. Huang C-Y, Ju D-T, Chang C-F, Reddy PM, Velmurugan BK. A review on the effects of current chemotherapy drugs and natural agents in treating non–small cell lung cancer. *Biomedicine*. 2017;7(4).
11. Pérez-Herrero E, Fernández-Medarde A. Advanced targeted therapies in cancer: drug nanocarriers, the future of chemotherapy. *European Journal of Pharmaceutics and Biopharmaceutics*. 2015;93:52–79.
12. Sharma N, Kurmi BD, Singh D, Mehan S, Khanna K, Karwasra R, et al. Nanoparticles toxicity: an overview of its mechanism and plausible mitigation strategies. *Journal of Drug Targeting*. 2024;32(5):457–69.
13. Badrzadeh F, Rahmati-Yamchi M, Badrzadeh K, Valizadeh A, Zarghami N, Farkhani SM, et al. Drug delivery and nanodetection in lung cancer. *Artificial Cells, Nanomedicine, and Biotechnology*. 2016;44(2):618–34.
14. Ghasemiyeh P, Mohammadi-Samani S. Solid lipid nanoparticles and nanostructured lipid carriers as novel drug delivery systems: applications, advantages and disadvantages. *Research in Pharmaceutical Sciences*. 2018;13(4):288–303.
15. Bayón-Cordero L, Alkorta I, Arana L. Application of solid lipid nanoparticles to improve the efficiency of anticancer drugs. *Nanomaterials*. 2019;9(3):474.
16. Khanna K, Sharma N, Rawat S, Khan N, Karwasra R, Hasan N, et al. Intranasal solid lipid nanoparticles for management of pain: a full factorial design approach, characterization & gamma scintigraphy. *Chemistry and Physics of Lipids*. 2021;236:105060.
17. Sharma P, Mehta M, Dhanjal DS, Kaur S, Gupta G, Singh H, et al. Emerging trends in the novel drug delivery approaches for the treatment of lung cancer. *Chemico-Biological Interactions*. 2019;309:108720.
18. Abdelaziz HM, Freag MS, Elzoghby AO. Solid lipid nanoparticle-based drug delivery for lung cancer. *Nanotechnology-Based Targeted Drug Delivery Systems For Lung Cancer*: (Elsevier); 2019. p. 95–121.
19. Apostolou M, Fatokun AA, Assi S, Khan I. Targeted lipid-based drug delivery systems for lung cancer therapy. *Applied Sciences*. 2024;14(15):6759.
20. Cagle PT, Allen TC, Olsen RJ. Lung cancer biomarkers: present status and future developments. *Archives of Pathology and Laboratory Medicine*. 2013;137(9):1191–8.
21. Fong KM, Sekido Y, Minna JD. The molecular basis of lung carcinogenesis. *The Molecular Basis of Human Cancer*: Springer; 2002. p. 379–405.
22. Ito M, Ito G, Kondo M, Uchiyama M, Fukui T, Mori S, et al. Frequent inactivation of RASSF1A, BLU, and SEMA3B on 3p21. 3 by promoter hypermethylation and allele loss in non-small cell lung cancer. *Cancer Letters*. 2005;225(1):131–9.

23. Belinsky SA, Nikula KJ, Palmisano WA, Michels R, Saccomanno G, Gabrielson E, et al. Aberrant methylation of p16 INK4a is an early event in lung cancer and a potential biomarker for early diagnosis. *Proceedings of the National Academy of Sciences*. 1998;95(20):11891–6.
24. Zöchbauer-Müller S, Fong KM, Virmani AK, Geradts J, Gazdar AF, Minna JD. Aberrant promoter methylation of multiple genes in non-small cell lung cancers. *Cancer Research*. 2001;61(1):249–55.
25. Girard L, Zochbauer-Muller S, Virmani AK, Gazdar AF, Minna JD. Genome-wide allelotyping of lung cancer identifies new regions of allelic loss, differences between small cell lung cancer and non-small cell lung cancer, and loci clustering. *Cancer Research*. 2000;60(17):4894–906.
26. Harish V, Mohd S, Tewari D, Pandey NK, Vishwas S, Babu MR, et al. Unravelling the role of solid lipid nanoparticles in drug delivery: journey from laboratory to clinical trial. *Journal of Drug Delivery Science and Technology*. 2023;85:104616.
27. Ahmed S, Mahmood S, Ansari MD, Gull A, Sharma N, Sultana Y. Nanostructured lipid carrier to overcome stratum corneum barrier for the delivery of agomelatine in rat brain; formula optimization, characterization and brain distribution study. *International Journal of Pharmaceutics*. 2021;607:121006.
28. Mainuddin, Kumar A, Ratnesh RK, Singh J, Dumoga S, Sharma N, et al. Physical characterization and bioavailability assessment of 5-fluorouracil-based nanostructured lipid carrier (NLC): in vitro drug release, hemolysis, and permeability modulation. *Medical Oncology*. 2024;41(5):95.
29. German-Cortés J, Vilar-Hernández M, Rafael D, Abasolo I, Andrade F. Solid lipid nanoparticles: multitasking nano-carriers for cancer treatment. *Pharmaceutics*. 2023;15(3):831.
30. Sun T, Zhang YS, Pang B, Hyun DC, Yang M, Xia Y. Engineered nanoparticles for drug delivery in cancer therapy. *Angewandte Chemie International Edition*. 2014;53:12320–64.
31. Ulbrich K, Hola K, Subr V, Bakandritsos A, Tucek J, Zboril R. Targeted drug delivery with polymers and magnetic nanoparticles: covalent and noncovalent approaches, release control, and clinical studies. *Chemical Reviews*. 2016;116(9):5338–431.
32. Patel P, Raval M, Airao V, Ali N, Shazly GA, Khan R, et al. Formulation of folate receptor-targeted silibinin-loaded inhalable chitosan nanoparticles by the QbD approach for lung cancer targeted delivery. *ACS Omega*. 2024;9(9):10353–70.
33. Rosiere R, Van Woensel M, Gelbcke M, Mathieu V, Hecq J, Mathivet T, et al. New folate-grafted chitosan derivative to improve delivery of paclitaxel-loaded solid lipid nanoparticles for lung tumor therapy by inhalation. *Molecular Pharmaceutics*. 2018;15(3):899–910.
34. Nan Y. Lung carcinoma therapy using epidermal growth factor receptor-targeted lipid polymeric nanoparticles co-loaded with cisplatin and doxorubicin. *Oncology Reports*. 2019;42(5):2087–96.
35. Guo Y, Wang L, Lv P, Zhang P. Transferrin-conjugated doxorubicin-loaded lipid-coated nanoparticles for the targeting and therapy of lung cancer. *Oncology Letters*. 2015;9(3):1065–72.
36. Bayón-Cordero L, Alkorta I, Arana L. Application of solid lipid nanoparticles to improve the efficiency of anticancer drugs. *Nanomaterials*. 2019;9(3):474.
37. Pi C, Zhao W, Zeng M, Yuan J, Shen H, Li K, et al. Anti-lung cancer effect of paclitaxel solid lipid nanoparticles delivery system with curcumin as co-loading partner in vitro and in vivo. *Drug Delivery*. 2022;29(1):1878–91.
38. Gupta T, Singh J, Kaur S, Sandhu S, Singh G, Kaur IP. Enhancing bioavailability and stability of curcumin using solid lipid nanoparticles (CLEN): a covenant for its effectiveness. *Frontiers in Bioengineering and Biotechnology*. 2020;8:879.
39. Üner M, Yener G. Importance of solid lipid nanoparticles (SLN) in various administration routes and future perspectives. *International Journal of Nanomedicine*. 2007;2(3):289–300.
40. Pawar H, Surapaneni SK, Tikoo K, Singh C, Burman R, Gill MS, et al. Folic acid functionalized long-circulating co-encapsulated docetaxel and curcumin solid lipid nanoparticles: in vitro evaluation, pharmacokinetic and biodistribution in rats. *Drug Delivery*. 2016;23(4):1453–68.
41. Mosallaei N, Jaafari MR, Hanafi-Bojd MY, Golmohammadzadeh S, Malaekeh-Nikouei B. Docetaxel-loaded solid lipid nanoparticles: preparation, characterization, in vitro, and in vivo evaluations. *Journal of Pharmaceutical Sciences*. 2013;102(6):1994–2004.
42. Wang B, Wu K, Liu R, Huang Y, Chang Z, Gao Y, et al. Phyllanthi tannin loaded solid lipid nanoparticles for lung cancer therapy: preparation, characterization, pharmacodynamics and safety evaluation. *Molecules*. 2023;28(21):7399.
43. del Pozo-Rodriguez A, Solinís MÁ, Rodríguez-Gascón A. Applications of lipid nanoparticles in gene therapy. *European Journal of Pharmaceutics and Biopharmaceutics*. 2016;109:184–93.
44. Choi SH, Jin S-E, Lee M-K, Lim S-J, Park J-S, Kim B-G, et al. Novel cationic solid lipid nanoparticles enhanced p53 gene transfer to lung cancer cells. *European Journal of Pharmaceutics and Biopharmaceutics*. 2008;68(3):545–54.

45. Han Y, Zhang P, Chen Y, Sun J, Kong F. Co-delivery of plasmid DNA and doxorubicin by solid lipid nanoparticles for lung cancer therapy. *International Journal of Molecular Medicine*. 2014;34(1):191–6.
46. Yu X, Fang C, Zhang K, Su C. Recent advances in nanoparticles-based platforms targeting the PD-1/PD-L1 pathway for cancer treatment. *Pharmaceutics*. 2022;14(8):1581.
47. Nafee N, Gaber DM, Elzoghby AO, Helmy MW, Abdallah OY. Promoted antitumor activity of myricetin against lung carcinoma via nanoencapsulated phospholipid complex in respirable microparticles. *Pharmaceutical Research*. 2020;37:1–24.
48. Banerjee I, De M, Dey G, Bharti R, Chattopadhyay S, Ali N, et al. A peptide-modified solid lipid nanoparticle formulation of paclitaxel modulates immunity and outperforms dacarbazine in a murine melanoma model. *Biomaterials Science*. 2019;7(3):1161–78.
49. Sherif AY, Harisa GI, Alanazi FK, Nasr FA, Alqahtani AS. PEGylated SLN as a promising approach for lymphatic delivery of gefitinib to lung cancer. *International Journal of Nanomedicine*. 2022;17:3287.
50. Najib Ullah SNM, Afzal O, Altamimi ASA, Alossaimi MA, Almalki WH, Alzahrani A, et al. Bedaquiline-loaded solid lipid nanoparticles drug delivery in the management of non-small-cell lung cancer (NSCLC). *Pharmaceuticals*. 2023;16(9):1309.
51. Rahman MA, Ali A, Rahamathulla M, Salam S, Hani U, Wahab S, et al. Fabrication of sustained release curcumin-loaded solid lipid nanoparticles (cur-SLNs) as a potential drug delivery system for the treatment of lung cancer: optimization of formulation and in vitro biological evaluation. *Polymers*. 2023;15(3):542.
52. da Rocha MCO, da Silva PB, Radicchi MA, Andrade BYG, de Oliveira JV, Venus T, et al. Docetaxel-loaded solid lipid nanoparticles prevent tumor growth and lung metastasis of 4T1 murine mammary carcinoma cells. *Journal of Nanobiotechnology*. 2020;18:1–20.
53. Zhou Q, Hou K, Fu Z. Transferrin-modified mangiferin-loaded SLNs: preparation, characterization, and application in A549 lung cancer cell. *Drug Design, Development and Therapy*. 2023:1767–78.
54. Kharazmi F, Neamati A, Tabrizi MH. Evaluation of the anticancer effects of allicin-loaded solid lipid nanoparticles in the A549 lung cancer cell line. *BioNanoScience*. 2025;15(1):1–9.
55. Li M, Fang G, Zahid F, Saleem R, Ishrat G, Ali Z, et al. Co-delivery of paclitaxel and curcumin loaded solid lipid nanoparticles for improved targeting of lung cancer: in vitro and in vivo investigation. *Heliyon*. 2024;10(9).
56. Vinod SK, Hau E. Radiotherapy treatment for lung cancer: current status and future directions. *Respirology*. 2020;25:61–71.
57. Joshi HA, Patwardhan RS, Sharma D, Sandur SK, Devarajan PV. Pre-clinical evaluation of an innovative oral nano-formulation of baicalein for modulation of radiation responses. *International Journal of Pharmaceutics*. 2021;595:120181.
58. Bromma K, Rieck K, Kulkarni J, O'Sullivan C, Sung W, Cullis P, et al. Use of a lipid nanoparticle system as a Trojan horse in delivery of gold nanoparticles to human breast cancer cells for improved outcomes in radiation therapy. *Cancer Nanotechnology*. 2019;10:1–17.
59. Zou L, Wang H, He B, Zeng L, Tan T, Cao H, et al. Current approaches of photothermal therapy in treating cancer metastasis with nanotherapeutics. *Theranostics*. 2016;6(6):762.
60. Xue S, Jiao J, Miao S, Wang L, Liu Y, Zhang Q, et al. Lipid-coated bismuth nanoflower as the thermos-radio sensiti for therapy of lung metastatic breast cancer: preparation, optimisation, and characterisation. *IET Nanobiotechnology*. 2022;16(9):305–15.
61. Crous A, Abrahamse H. Photodynamic therapy of lung cancer, where are we? *Frontiers in Pharmacology*. 2022;13:932098.
62. Jiang S, Zhu R, He X, Wang J, Wang M, Qian Y, et al. Enhanced photocytotoxicity of curcumin delivered by solid lipid nanoparticles. *International Journal of Nanomedicine*. 2017:167–78.
63. Yeo S, Wu H, Yoon I, Kim H-S, Song YK, Lee WK. Enhanced photodynamic therapy efficacy through solid lipid nanoparticle of purpurin-18-N-propylimide methyl ester for cancer treatment. *International Journal of Molecular Sciences*. 2024;25(19):10382.
64. Zarinshenas R, Amini A, Mambetsariev I, Abuali T, Fricke J, Ladbury C, Salgia R. Assessment of barriers and challenges to screening, diagnosis, and biomarker testing in early-stage lung cancer. *Cancers*. 2023;15(5):1595.
65. Park H, Sholl LM, Hatabu H, Awad MM, Nishino M. Imaging of precision therapy for lung cancer: current state of the art. *Radiology*. 2019;293(1):15–29.
66. Succony L, Rassl D, Barker A, McCaughan F, Rintoul R. Adenocarcinoma spectrum lesions of the lung: detection, pathology and treatment strategies. *Cancer Treatment Reviews*. 2021;99:102237.
67. de Koning HJ, van Der Aalst CM, de Jong PA, Scholten ET, Nackaerts K, Heuvelmans MA, et al. Reduced lung-cancer mortality with volume CT screening in a randomized trial. *New England Journal of Medicine*. 2020;382(6):503–13.

68. Mukherjee S, Patra CR. Therapeutic application of anti-angiogenic nanomaterials in cancers. *Nanoscale*. 2016;8(25):12444–70.
69. Zhang R, Xiong C, Huang M, Zhou M, Huang Q, Wen X, et al. Peptide-conjugated polymeric micellar nanoparticles for dual SPECT and optical imaging of EphB4 receptors in prostate cancer xenografts. *Biomaterials*. 2011;32(25):5872–9.
70. Goel S, England CG, Chen F, Cai W. Positron emission tomography and nanotechnology: a dynamic duo for cancer theranostics. *Advanced Drug Delivery Reviews*. 2017;113:157–76.
71. Srikar R, Upendran A, Kannan R. Polymeric nanoparticles for molecular imaging. *Wiley Interdisciplinary Reviews: Nanomedicine and Nanobiotechnology*. 2014;6(3):245–67.
72. Larush L, Magdassi S. Formation of near-infrared fluorescent nanoparticles for medical imaging. *Nanomedicine*. 2011;6(2):233–40.
73. Kim TH, Chen Y, Mount CW, Gombotz WR, Li X, Pun SH. Evaluation of temperature-sensitive, indocyanine green-encapsulating micelles for noninvasive near-infrared tumor imaging. *Pharmaceutical Research*. 2010;27:1900–13.
74. Hahn MA, Singh AK, Sharma P, Brown SC, Moudgil BM. Nanoparticles as contrast agents for in-vivo bioimaging: current status and future perspectives. *Analytical and Bioanalytical Chemistry*. 2011;399:3–27.
75. Doiron AL, Homan KA, Emelianov S, Brannon-Peppas L. Poly (lactic-co-glycolic) acid as a carrier for imaging contrast agents. *Pharmaceutical Research*. 2009;26:674–82.
76. Sun J, Zhang S, Jiang S, Bai W, Liu F, Yuan H, et al. Gadolinium-loaded solid lipid nanoparticles as a tumor-absorbable contrast agent for early diagnosis of colorectal tumors using magnetic resonance colonography. *Journal of Biomedical Nanotechnology*. 2016;12(9):1709–23.

11 Dendrimers for the Management of Lung Cancer

Annu Saini[1], *Lokesh Nagar*[1], *Tarif Hussain*[1], *Prabhakar Kumar Verma*[1], *Kamal Dua*[2], *and Harish Dureja*[1,2,3]*

[1] Department of Pharmaceutical Sciences, Maharshi Dayanand University, Rohtak 124001, India

[2] Faculty of Health, Australian Research Centre in Complementary and Integrative Medicine, University of Technology, Sydney, Ultimo NSW 2007, Australia

[3] Centre for Research Impact & Outcome, Chitkara College of Pharmacy, Chitkara University, Rajpura, Punjab 140401, India

*Corresponding author: Harish Dureja, harishdureja@mdurohtak.ac.in.

11.1 INTRODUCTION

Lung cancer (LC) is one of the most common life-threatening cancers with high morbidity and mortality all over the world (Alexander et al., 2014). About 1.8 million people died of lung cancer and about 2.5 million people were diagnosed across the globe in 2022, according to Global Cancer Observatory (GLOBOCAN) statistics. This represents about 12.4% of all cancer cases and 18.7% of cancer fatalities (GLOBOCAN, 2022). LC is broadly classified from the histology of the cancer cells as small cell lung cancer (SCLC) and non–small cell lung cancer (NSCLC) (Siddique et al., 2024). Although less frequent, SCLC is characterized by its quick development, early body metastases, significant vascularization, genomic instability, and several mitotic events (Gazdar et al., 2017). On the other hand, NSCLC is a highly heterogeneous disease and is divided into squamous cell carcinoma (SCC), adenocarcinoma (ADC), and large cell carcinoma (LCC) (Bajbouj et al., 2021; Chen et al., 2014). Surgical removal of malignant tumors, radiation, and chemotherapy are the current strategies for managing lung cancer (Howington et al., 2013). The most effective therapeutic approach is surgical tumor resection, most frequently used in stage I and stage II NSCLC (Deboever et al., 2022). However, tumor resectability is often limited in advanced stages and when lung cancer has metastasized (stage III and VI NSCLC; and SCLC when the disease is not detected at very early stages) (Chiu et al., 2021). Due to the lack of early-stage identification, most lung cancers are discovered in advanced stages, which involve local tumor invasion or distant metastases, which are not treatable with surgeries and require treatment with chemotherapies and radiotherapies (Nooreldeen & Bach, 2021). Radiotherapy and chemotherapy are important therapeutic modalities that can be used as adjuvant treatment in patients who have undergone surgical resection or are not candidates for resection or as neoadjuvant therapy to improve the resectability of the tumor (Provencio et al., 2022; Aguado et al., 2022). Chemotherapeutic agents act by inhibiting rapidly dividing cells and affect normal cells with high proliferation rates (Mollaei et al., 2021). Platinum-based chemotherapy (cisplatin or carboplatin) and taxanes (paclitaxel or docetaxel) or gemcitabine are commonly used as standard treatment for advanced NSCLC (Cosaert & Quoix, 2002). Similarly, platinum-based chemotherapy plus etoposides is used as standard chemotherapy for SCLC. These therapies are said to be efficacious depending on many factors, such as the severity of the disease

DOI: 10.1201/9781003595830-11

and the response to therapy in each organism (Cappuzzo, 2009; Norouzi & Hardy, 2021). However, a significant issue that patients receiving these therapies face is that these therapies are unable to differentiate between tumor cells and healthy cells in the body, which can lead to severe adverse effects (Cortinovis et al., 2021). These effects can be avoided by using targeted therapies where drugs or mechanisms are used to take out tumor cells that are specifically targeted to the tumors (Shuel, 2022). Targeted treatment, a new kind of chemotherapy, is directed at specific proteins or genes related to a specific kind of cancer or tumor vasculature necessary to develop the cancer (Garg et al., 2024). Scientists are continuously investigating new and innovative methods to enhance the administration of chemotherapeutic drugs with the fewest possible negative effects (Pandey et al., 2024). With the help of nanotechnology, it could potentially be a way that solves current medical issues in an innovative and paradigm-changing manner (Freitas et al., 2021). These days, a certain class of nanoparticles (NPs) known as dendrimers is widely used for oncological applications and diagnosis (Crintea et al., 2023). Dendrimers are a class of nanomolecules widely utilized as systems for delivering drugs. Due to their unique properties, they have become one of the most promising drug delivery systems and have drawn much interest from researchers (Mittal et al., 2021).

11.2 DENDRIMERS: STRUCTURE AND PROPERTIES

A dendrimer is a synthetic polymeric macromolecule of the nanometer size range (between 1 and 100 nm in diameter), with a high degree of external versatility that is distinguished by its branched three-dimensional structure that can be used as a nanodrug delivery system (Ambekar et al., 2020; Santos et al., 2019). The word "dendrimer" is derived from the Greek term "dendron," which means tree or branch, and "meros," which means part. Buhleier et al. (1997) synthesized the first dendrimer having a cascade-like repetitive structure comprising mono and diamine at the periphery of the central core (Huang et al., 2023). Similarly, Tomalia et al. (1985) synthesized a new class of polymer called "stardust macromolecule" (Arora et al., 2022). In 1985, a new category of macromolecules was introduced with the discovery of what we now call "nano cascade spheres" and "starburst dendritic macromolecules." These macromolecules became known as dendrimers (Patel et al., 2022). Three different bodies make up dendrimers: a central core, tree-like branching structures linked to the core, and terminal functional groups that are connected to the branches' periphery (Ferreiro et al., 2023). The branches are arranged around the central core in a layer-by-layer fashion. There are two main ways to synthesize dendrimers: convergent and divergent approaches (Filipczak et al., 2021). The divergent approach involves growing successive generations of dendrons by adding monomers to a reactive core, increasing the number of functional groups on the surface with each generation and increasing the dendrimer's capacity to encapsulate different therapeutic agents (Abbasi et al., 2014). In contrast, the convergent approach involves building dendrons independently before joining them to a central core (Hsu et al., 2016). Although more control over the finished construction is possible, this is frequently more complicated (Barrett et al., 2009).

11.3 DENDRIMERS IN DRUG DELIVERY

The structural properties and functional versatility of dendrimers have so far been attractive for pharmaceutical research due to their unique capability as targeted drug delivery carriers and as diagnostics imaging tools for lung cancer treatment (Rastogi et al., 2022). These highly branched macromolecules provide effective carriers for greatly improving the solubility, stability, and bioavailability of the therapeutic agent while the reduction of systemic toxicity (Chehelgerdi et al., 2023). As their surface functionality is modifiable, their monodisperse size, aqueous solubility, and aqueous internal cavity would make them promising drug delivery systems, and their structure allows efficient encapsulation of therapeutic agents (Sherje et al., 2018). Due to their unique ability to be designed for specific therapeutic needs, various types of dendrimers, such as polyamidoamine (PAMAM), poly (propylene imine) (PPI), and poly-L-lysine (PLL) dendrimers, have been explored and investigated (Singh et al., 2016). The number of generations of these dendrimers determines their effectiveness in drug delivery, as the number of branching layers in the molecule (Wang et al., 2022). For instance, higher-generation dendrimers have larger diametric sizes and high molecular weight. Dendrimers from higher generations, such as G5, usually have greater dimensions (5–8 nm) and are better suited for therapeutic uses (Malik et al., 2000). G5 dendrimers can effectively cross biological barriers that can accumulate in the tumor tissues due to the improved permeability and retention effect, which promotes passive targeting of cancerous cells (Lee et al., 2005). However, lower-generation dendrimers can be rapidly removed from circulation due to their smaller size and may not be able to reach the desired therapeutic efficacy (Somani & Dufès, 2014). Recent literature reveals that dendrimers offer several advantages that overcome the limitations of traditional chemotherapeutic agents. Dendrimer-based drug delivery demonstrates better pharmacokinetic behavior and distribution patterns than standard free drugs (Mukherjee et al., 2022; Sathe & Bharatam, 2022). Dendrimer modification with specific ligands recognizes cancer cell receptors for active targeting mode, whereas enhancing permeation and retention effect is used for passive targeting mode. For instance, modification of PAMAM dendrimer can improve therapeutic agent specificity for lung cancer cells, thereby enhancing its therapeutic effect and minimizing off-target effect (Kesharwani & Iyer, 2015; Kaur et al., 2016).

11.4 APPLICATION OF DENDRIMERS FOR LUNG CANCER MANAGEMENT

Dendrimers have many advantages compared to other nano-based drug systems. These agents are tremendous in the management/treatment of multiple diseases, including lung cancers by the entrapment of chemotherapeutics and other NPs (metal) inside their 3D hollow design due to their unique properties such as surface functional groups and structure layout (Sharma et al., 2017; Fateh et al., 2024). Different dendrimers have made the process of drug delivery easier. Three types of dendrimers have been optimized and reported for their therapeutic efficacy in drug and gene–gene inhibitor delivery (PAMAM, PPI, and PL dendrimers) (Pandita et al., 2014).

11.4.1 Polyamidoamine Dendrimers

PAMAM dendrimers are a family of ultrabranch, ultrasmall polymeric macromolecules that have become immensely attractive for their potential clinical applications, particularly in lung cancer treatment (Asthana et al., 2005). They have excellent properties such as monodispersed tree-like dendrimers with multipurpose surfaces, hyper branching, and excellent biocompatibility, with a natural tendency to encapsulate imaging agents, genes, or other molecules targeting molecules (Gaballu et al., 2018; Surekha et al., 2021). Their distinct structure makes them ideal choices for delivering drugs, as it allows precise control over size and surface chemistry (Asthana et al., 2005). Different generations of PAMAM dendrimers can be produced; the more advanced generations have bigger diameters and better drug-loading capacities (Tarach & Janaszewska, 2021). Because of this characteristic, they can more successfully encapsulate hydrophobic medications and enhance the stability and solubility in biological environments (Kheraldine et al., 2021). To efficiently administer the anticancer medication paclitaxel to stage 1 non–small cell lung cancer (NSCLC), *Zhang et al.* developed a unique targeted drug delivery system using GE11-peptide conjugated paclitaxel (PTX) dendrimers (GE11-PAMAM-PTX) in their research. 293T cells were used to examine the cytotoxic effects of both formulations using the MTT assay, which revealed concentration-dependent anticancer activity. Receptor-mediated endocytosis via the over-expressed epidermal growth factor receptor (EGFR) allowed GE-PAMAM-PTX to have higher cellular uptake than PAMAM-PTX in both the 293T and L132 cell lines, according to the study. GE-PAMAM-PTX encapsulated paclitaxel with an encapsulation efficiency of over 95% and loading efficiency of 25%. The GE-PAMAM-PTX conjugates also demonstrated a sustained release profile of roughly 85% paclitaxel over 50 hours. Crucially, the GE11 ligand allowed for precise delivery to lung cancer cells since the

GE-PAMAM-PTX conjugates proved to be more cytotoxic than either PAMAM-PTX or pure paclitaxel was. By more effectively binding to EGFR on cancerous cells, the GE11 peptide increased therapeutic efficacy and may have lessened the negative effects of traditional chemotherapy (Zhang et al., 2017).

To create a targeted drug delivery system for NSCLCs, *Liu et al.* investigated a novel LC targeting peptide (LCTP) and attached FITC to a G4-acetylated PAMAM dendrimer (PAMAM-Ac-FITC-LCTP). In vitro tests demonstrated that PAMAM-Ac-FITC-LCTP demonstrated considerably increased cellular uptake in NCI-H460 cancer cells compared to PAMAM-Ac-FITC and 293T cells, confirming the specificity of LCTP for NSCLC. Further validation demonstrated that free LCTP reduced the uptake of PAMAM-Ac-FITC-LCTP in NCI-H460 cells, but not in 293T cells. Ex vivo imaging demonstrated that PAMAM-Ac-FITC-LCTP had a greater tumor distribution than PAMAM-Ac-FITC, but a similar distribution in other organs. This suggests that the LCTP-conjugated dendrimer targets lung cancer cells specifically. With increased selectivity and uptake in lung cancer cells but low uptake in non-cancerous cells, this study shows that LCTP-conjugated PAMAM dendrimers are intriguing candidates for targeted drug administration and diagnostic applications in NSCLC (Liu et al., 2010).

Delivering therapeutic molecules to tumor cells with good efficiency and selectivity while reducing harm to healthy cells is the main problem in RNA interference (RNAi)-based therapeutics (Kara et al., 2022). The study of Ayatollahi *et al.* aims to develop a targeted delivery system to increase the transfection efficiency of small hairpin RNA (shRNA) against Bcl-xL, an anti-apoptotic protein overexpressed in lung cancer and to increase apoptosis in cancer cells. To improve the transfection efficiency of the dendrimer, the authors modified polyamidoamine (PAMAM) dendrimers with 10-bromodecanoic acid (10C) and 10C-PEG. The AS1411 aptamer, a nucleolin-targeting ligand with excellent affinity and selectivity, was used to target lung cancer cells (A549) precisely. To silence Bcl-xL expression and increase apoptosis, Bcl-xL shRNA plasmids were designed to be delivered into the cells by the system. In A549 cells, the aptamer-modified PAMAM dendrimer dramatically increased the transfection efficiency when compared to non-targeted vectors, causing a 14% increase in late apoptosis and a 25% decrease in Bcl-xL expression. With their excellent selectivity, efficiency, and safety, this work presents a viable approach to targeted non-viral gene silencing in lung cancer cells using RNAi-based therapy (Ayatollahi et al., 2017).

PAMAM dendrimers have become well-known as efficient nanocarriers for targeted drug administration in lung cancer treatment because of their special structural characteristics and capacity to be functionalized for increased specificity (Patle & Dongre, 2024). To enhance the PTX delivery. *Bhatt et al.* developed PEGylated G4 PAMAM dendrimer coupled to cell-penetrating peptides, octa-arginine (R), and vitamin E succinate (VES) to increase the transport of paclitaxel (PTX) for lung cancer treatment. NMR, infrared, and gel permeation chromatography were used to characterize RVES-PD. The PTX-loaded RVES-PD nanoparticles' physicochemical characteristics, such as their size, drug loading, and release patterns, were assessed. According to the study, RVES-PD nanoparticles outperformed nontargeted VES-PD in terms of cellular absorption in human lung cancer A549 cells. Cytotoxicity assays demonstrated that RVES-PD-PTX was more cytotoxic than either nontargeted VES-PD-PTX or free PTX at concentrations between 0 and 50 μg/mL. Additionally, the Annexin V assay demonstrated that RVES-PD-PTX caused apoptosis in cancer cells and inhibited spheroid development more successfully than the other formulations (Bhatt et al., 2020).

For the targeted delivery of erlotinib (ERL), a tyrosine kinase inhibitor used to treat non–small cell lung cancer (NSCLC), *Fatani et al.* set out to create a non-invasive PAMAM dendrimer-based nanocomposite. The primary objective of this investigation was to improve the solubility and cellular uptake while reducing its systemic toxicity and off-target effects by formulating two different PAMAM dendrimers of G4 and G5 generations. Studies conducted on NSCLC cells in vitro revealed markedly increased cellular absorption, increased apoptosis, and suppression of tumor cell proliferation. Prolonged release from both generation dendrimers was observed using drug release studies. However, the G5-FITC released more drugs than the G4-FITC. In vitro cytotoxicity tests revealed that both drug-loaded and the blank dendrimers dramatically decreased the viability of cancer cells, indicating that dendrimers had inherent anticancer capabilities G4-FITC was more cytotoxic than G5-FITC, as G4-FITC has a negatively charged surface as opposed to the neutral charge of G5-FITC (Fatani et al., 2023).

For effective co-delivery of cis-diamminedichloroplatinum II (CDDP) and HuR siRNA to LC cells that overexpress folate receptor-α (FRA), Amreddy *et al.* produced G4 PAMAM dendrimers functionalized with folic acid (FA). This approach was designed to minimize toxicity to healthy cells while specifically targeting cancer cells. The nanoparticles (Den-PEI-CDDP-HuR) based on PAMAM dendrimer encapsulated CDDP (35.65% encapsulation efficiency) and HuR siRNA with good physicochemical properties, such as appropriate particle size and zeta potential. Using Western blot analysis, it was discovered that HuR siRNA dramatically reduced the expression of HuR in A549 and H1299 lung cancer cells. Furthermore, Den-PEI-CDDP-HuR nanoparticles were found to induce DNA damage in both cell lines. Controlled cytotoxicity was shown for the co-delivery system, with cancer cells having higher apoptosis levels (25% and 32% higher than free CDDP and Den-PEI-CDDP, respectively) and healthy cells spared. Furthermore, siRNA was able to penetrate the cytoplasm and carry out targeted gene silencing because to the nanoparticles' improved cellular absorption and effective endosomal escape. The findings suggest that the Den-PEI-CDDP-HuR system, with its focused drug delivery and less toxicity to healthy tissues, is a potential therapy approach for lung cancer (Amreddy et al., 2018).

In an investigation, Pooja *et al.* explored *N*-acetyl-D-glucosamine conjugated PAMAM dendrimer for target camptothecin delivery to the lung cells. *N*-acetyl-D-glucosamine was chosen for its ability to target glucose transporters and lecithin receptors overexpressed in A549 cells. NAG-labelled dendrimer (NAG-Dend) was developed to improve CPT delivery. The study showed that the CPT-loaded NAG-Dendimer (Nag-Dend-CPT) had improved cytotoxic effects since its cellular absorption was substantially faster and higher compared to that of the unlabeled dendrimer formulation. NAG-Dend-CPT was 4.5-fold more toxic to A549 than free camptothecin and exhibited time and dose-dependent anticancer activity. Moreover, Dend-CPT also enhanced reactive oxygen species (ROS) generation, promoted higher apoptosis, and showed better inhibition of A549 cell migration compared to Dend-CPT. These studies were confirmed in vivo in lung tumor-bearing mice, where NAG-Dend selectively accumulated in the lung tumor. The study shows that NAG-conjugated dendrimers are an effective nanocarrier technology for improving the effectiveness and delivery of anticancer drugs (Pooja et al., 2020).

11.4.2 Poly (Propylene Imine) PPI Dendrimers

PPI dendrimers are polar spherical vesicles with side chains of alkyl amines (Singh et al., 2021). In 1978, Vögtle and team developed "cascade" structures of low molecular weight amines; it is one of the oldest dendrimers to have ever been designed (Shcharbin et al., 2019). Later two groups (Worner/Mulhaupt and de Brabander-Van den Berg/Meijer) introduced the synthesis of PPI dendrimers on a larger scale using the modified Vögtle method (Tomalia et al., 2020; Li et al., 2023). Peripheral functional groups of the branching terminate as the branching does, that is as leaves of the branching (Rai et al., 2023). The main advantages of these dendrimers are ease of surface modification, which leads to high-generation dendrimers, and versatility of drug delivery and high functionality (Chis et al., 2020).

In recent years, nanomedicine has received significant interest in targeting and delivery of anticancer drugs. PPI dendrimers have been evaluated for their use in drug delivery, with the advantage of high loading capacity and easy modification to target the drug (Jarmila et al., 2024). Zhang *et al.* investigated the creation of a high-loading, targeted delivery system for drugs based on zwitterionic PPI dendrimers to enhance anti-tumor therapy. Through a thiol-ene click reaction, zwitterionic modification of PPI dendrimers was accomplished in this study. Paclitaxel and targeting peptides c(RGDfC) were then introduced. The developed amphiphilic dendrimers were self-assembled to form micelles, followed by encapsulating it with doxorubicin (DOX) to form PPIMPRC-DOX micelles. The purpose of these modifications was to enhance the delivery of paclitaxel and doxorubicin (DOX), a chemotherapy drug, to tumor sites. The study demonstrated good stability of these micelles in protein solutions and enhanced cytotoxicity in HeLa and A549 cells through MTT assay. Cellular uptake of PPIMPRC-DOX micelles is enhanced as the pH decreases due to charge reversal. It is likely because the micelles have a zwitterionic surface that resists nonspecific protein adsorption, has reduced cytotoxicity, and is modified with targeting peptides that recognize αvβ integrin. In vitro release experiments revealed that the release efficiency was enhanced in a mildly acidic environment (tumor conditions) (Zhang et al., 2024).

In an investigation, Bhargava *et al.* designed a mannosylated PPI dendrimer to study whether the transport of the antitumor drug gemcitabine could be used for treating lung cancer. The study was to take advantage of the mannose-specific endocytosis receptors that lung macrophages express to enhance targeted drug delivery. The researchers encapsulated mannose and gemcitabine into 5th-generation PPI dendrimers. MTT assays for cytotoxicity and hemolytic toxicity studies showed that the GmcH-MPPI dendrimers could induce the necessary toxicity in A549 lung adenocarcinoma cells. The study demonstrated that the dendrimers outperformed the free drug in terms of drug loading, targeting ability, bioavailability, and extended-release kinetics. Biodistribution and pharmacokinetic profiles were favorable in vivo in albino rats, and gemcitabine showed higher concentrations in convolved lungs when compared with free drugs. The study showed that mannose-conjugated PPI dendrimers loaded with gemcitabine could be a good drug delivery system for lung cancer chemotherapy with improved efficacy and reduced side effects by targeted delivery to mannose-specific receptors on lung macrophages (Bhargava et al., 2017).

11.4.3 Poly-L-Lysine Dendrimers

Poly-L-lysine dendrimers or PLL dendrimers are amino-acid-based polymers, which consist of asymmetrical shapes and are being investigated for their potential in cancer treatment (Zheng et al., 2021). Multiple studies suggest that dendrimers, including PLL dendrimers, possess inherent anti-cancer and anti-metastatic activities, even without therapeutic agent loading (Mendes et al., 2017; Bober et al., 2022). PLL dendrimers are among the three major types of dendrimers that are being investigated for lung cancer treatment, the other two being PAMAM and PPI dendrimers (Wu et al., 2013). PLL dendrimers consist of lysines as branching units and amines as the terminal groups (Kim et al., 2018). Synthetic production of PPL dendrimers has been carried out to produce a new dendritic polymer with reduced toxicity (Chis et al., 2020). It can condense with oligonucleotides, and thus it can help in transporting genetic matter (Chauhan, 2018). The application of PLL dendrimers to a tumor microenvironment is optimal to utilize the antiangiogenic and these promote apoptosis and destroy necrosis properties (Jiang et al., 2023).

188Re-ImDendrim was directly injected into lung tumors of five patients under CT guidance at an activity of 162 MBq/cc of tumor. The 188Re-ImDendrim was directly injected into lung tumors of five patients under CT guidance at 162 MBq/cc of tumor activity. The dose was divided and given in several injections over two weeks for larger tumors. SPECT controls were then performed following administration at various time points and treatment response was assessed by SUVs on PET/CT. All patients were successfully stereotactically administered 188Re-ImDendrim under local anesthesia and the radioactive product diffused homogeneously to the tumor volume and remained for 72 hours post-administration without significant leakage. As adverse events, one patient had transient hemoptysis1. At 12 weeks, all targeted tumors had two complete responses. Safe and well tolerated with promising first treatment results, it was concluded that percutaneous single and iterative administration of this 188Re-Imdendrim brachytherapy device into lung cancers is safe (Tahar et al., 2019).

A dendrimer-formulated drug delivery system's effectiveness depends critically on the NP's safety and target selectivity. Masuda *et al.* developed the G6 PLL-based dendrimer "KG6E" with

TABLE 11.1
Application of Dendrimers for Lung Cancer Management

Generation	Dendrimer Type	Drug/Gene	Conjugation	Cell Lines	Outcome	References
G4	Fucosylated PAMAM dendrimers	Chrysin	CRY-FUC-PAMAM	A549	CRY-FUC-PAMAM demonstrated improved cytotoxicity and IC50 value compared to free CRY; improved cellular uptake and apoptosis of CRY in the A549 cell line	Rai et al., 2024
G4	PAMAM dendrimer modified with folic acid (FA)	Cis-diaminedichloroplatinum 11 and HuR siRNA	Den-PEI-CDDP-HuR	H1299 and A549	Enhanced cytotoxicity against lung cancer cells; greater therapeutic effect of CDDP and HuR siRNA co-delivery than individual treatments	Amreddy et al., 2018
G4 and G5	PAMAM dendrimer nanocomposites	Erlotinib	ERL-G4-FITC and ERL-G5-FITC	A549	ERL dendrimers showed better proliferation inhibition against A549 lung cancer cells	Fatani et al., 2023
G4	PAMAM dendrimer	Doxorubicin	G4SA-GFLG-DOX	A549	G4SA-GFLG-DOX nanoDDS improved DOX penetration in 3D lung tumor spheroids by 3.1-fold and enhanced tumor specificity	Almuqbil et al., 2020
G3	PAMAM dendrimer	Paclitaxel	–	293T, L132	PTX PAMAM dendrimer demonstrated improved therapeutic efficacy of PTX in lung cancer	Zhang et al., 2017
G4	PAMAM dendrimer	Camptothecin	PAMAM-den-CPT	A549	PAMAM dendrimer/Camptothecin complex exhibited controlled drug release and improved anticancer activity against NSCLCs	Oledzka et al., 2023
G4	Alkyl modified-PAMAM with10-bromodecanoic acid (10C) and 10C-PEG	Bcl-xl shRNA	Aptamer (AS1411) conjugated dendrimers	A549	Significant gene silencing and cellular uptake and enhanced apoptosis in the A549 cell line	Ayatollahi et al., 2017

(*Continued*)

Table 11.1 (*Continued*)
Application of Dendrimers for Lung Cancer Management

Generation	Dendrimer Type	Drug/Gene	Conjugation	Cell Lines	Outcome	References
G 3.5	PAMAM dendrimer	Camptothecin	NAG-Dend-CPT	A549	NAG-Dend-CPT demonstrated enhanced cytotoxicity than unlabeled Dend-CPT; 4.5-fold more toxic to A549 than native CPT	Pooja et al., 2020
G 5	PAMAM dendrimer-entrapped gold nanoparticles (Au DEN)	Zwitterionic Gadolinium (III)	–	A549	Dendrimer demonstrated good cytocompatibility and enhanced antifouling capability	Liu et al., 2019
G4	Triazine-modified dendrimer	MDM2 siRNA	4,6-Diamino-1,3,5-triazine	PC9 cells	Effective delivery of MDM2 siRNA through triazine-modified dendrimer to NSCLCs cells; downregulated MDM2 and inhibited the tumor growth in a xenograft model	Huang et al., 2016
G4	Pegylated PAMAM dendrimer	Paclitaxel	G4-PTX-PEG-Biotin	A549	Improved cellular uptake and cytotoxicity	Rompicharla et al., 2019
G5	Mannosylated PPI dendrimer	Gemcitabine	GmcH MPPI	A549	High gemcitabine loading, sustained release, and improved lung targeting through receptor-mediated endocytosis	Bhargava et al., 2017
G6	Glutamic acid-modified PLL dendritic system (KG6E)	Doxorubicin	Anti-mucin 1 aptamer (anti-MUC1 apt) and KG6E	A549	Anti-mucin1 apt/KG6E nanocarrier successfully targeted the A549 lung cells	Masuda et al., 2016
G6	Cationic PLL dendrimer	Doxorubicin	PLL-DOX nanocomplex	-	Improved cytotoxicity and therapeutic efficacy of DOX in prostate and lung cancer models	Al-Jamal et al., 2013
G6	PEGylated PLL dendrimer	Doxorubicin	DOX-conjugated PAMAM	MAT13762 IIIB	Pulmonary instillation of PEGylated polylysine dendrimer-doxorubicin reduces lung tumors (>95%) with less toxicity than IV doxorubicin	Kaminskas et al., 2014
G5	PLL dendrimer	Diffusible imidazolic probe and 188-rhenium	188-Rhenium-ImDendrim	-	Brachytherapy with 188-Re-ImDendrim is safe, well tolerated and has promising efficacy when administered percutaneously	(Tahar et al., 2019)

glutamic acid alteration to achieve this goal. Electrostatic interactions result from a negatively charged surface, which makes cationic polymers hazardous. Thus, the anti-MUCI was designed to attach to the KG6E to avoid toxicity and remain safe. The conjugate "anti-MUC1-apt/KG6E" was created when this dendrimer and anti-mucin1 aptamer effectively formed a connection. It was found that the sequence 19-mer had the highest binding affinity out of all the sequences studied. It was determined that the conjugate (anti-MUC1-apt/KG6E) had a high affinity for the LC cell line A549 by testing its binding tendency with the cell line using flow cytometry. Scrambled apt/KG6E showed reduced affinity for the cells at all assessed concentrations compared to scrambled apt/KG6E. Anti-MUC1-apt/KG6E internalization in the A549 cells was evaluated by conjugate localization in the cell line using confocal microscopy. It was found that the anti-MUC1-apt/KG6E was transported into endosomal and lysosomal compartments, thus indicating that MUC1 recognized the anti-MUC1-apt/KG6E complex immediately (Masuda et al., 2016).

Examples of research studies on dendrimers and their outcomes in treatment of lung cancer are given in Table 11.1

11.5 CHALLENGES AND FUTURE PROSPECTS

Although dendrimers have considerable opportunities as nanocarrier systems for managing and treating lung cancer, their clinical application often has to overcome multiple hurdles. Cytotoxicity is a major challenge, especially for cationic dendrimers that tend to be cytotoxic to healthy cells at high coverages. Biocompatibility and regulatory approvals of nano-based drug delivery are other major challenges. The future prospective developments include optimization of dendrimer-based drug delivery, integration of targeting ligands, and complete clinical studies to ensure the safety of dendrimers. Over the past decades, dendrimers have constantly developed and may become a promising approach for lung cancer management.

11.6 CONCLUSION

With unique structural features, including a hollow core, multiple functional groups, and controlled branching, dendrimers have become promising nanotechnology-based drug delivery systems for lung cancer therapy. PAMAM has been widely studied among various dendrimers for delivering oncological drugs such as Paclitaxel and RNAi, with the aim of targeted drug delivery and positive outcomes. Non-specific toxicity of cationic dendrimers is, however, still a major concern. Surface functionalization is a promising strategy to reduce toxicity while maintaining circulation time and improving tumor targeting. The advancements place dendrimers as a superior platform for cancer nanomedicine. Although promising, more investigation is required to determine the safety and clinical effectiveness of dendrimers before they may be developed commercially. Dendrimer-based formulations could advance lung cancer treatment, which would allow for an improved and targeted therapeutic offer for the patient.

REFERENCES

Abbasi, E., Aval, S., Akbarzadeh, A., Milani, M., Nasrabadi, H., Joo, S., Hanifehpour, Y., Nejati-Koshki, K., & Pashaei-Asl, R. (2014). Dendrimers: Synthesis, applications, and properties. *Nanoscale Research Letters*, *9*(1), 247. https://doi.org/10.1186/1556-276x-9-247

Aguado, C., Chara, L., Antoñanzas, M., Matilla Gonzalez, J. M., Jiménez, U., Hernanz, R., Mielgo-Rubio, X., Trujillo-Reyes, J. C., & Couñago, F. (2022). Neoadjuvant treatment in non-small cell lung cancer: New perspectives with the incorporation of immunotherapy. *World Journal of Clinical Oncology*, *13*(5), 314–322. https://doi.org/10.5306/wjco.v13.i5.314

Alexander, M., Kirsa, S., Wolfe, R., MacManus, M., Ball, D., Solomon, B., & Burbury, K. (2014). Thromboembolism in lung cancer—an area of urgent unmet need. *Lung Cancer*, *84*(3), 275–280. https://doi.org/10.1016/j.lungcan.2014.02.009

Al-Jamal, K. T., Al-Jamal, W. T., Wang, J. T.-W., Rubio, N., Buddle, J., Gathercole, D., Zloh, M., & Kostarelos, K. (2013). Cationic poly-l-lysine dendrimer complexes doxorubicin and delays tumor growth in vitro and in vivo. *ACS Nano*, *7*(3), 1905–1917. https://doi.org/10.1021/nn305860k

Almuqbil, R. M., Heyder, R. S., Bielski, E. R., Durymanov, M., Reineke, J. J., & da Rocha, S. R. (2020). Dendrimer conjugation enhances tumor penetration and efficacy of doxorubicin in extracellular matrix-expressing 3D lung cancer models. *Molecular Pharmaceutics*, *17*(5), 1648–1662. https://doi.org/10.1021/acs.molpharmaceut.0c00083

Ambekar, R. S., Choudhary, M., & Kandasubramanian, B. (2020). Recent advances in dendrimer-based nanoplatform for cancer treatment: A review. *European Polymer Journal*, *126*, 109546. https://doi.org/10.1016/j.eurpolymj.2020.109546

Amreddy, N., Babu, A., Panneerselvam, J., Srivastava, A., Muralidharan, R., Chen, A., Zhao, Y. D., Munshi, A., & Ramesh, R. (2018). Chemo-biologic combinatorial drug delivery using folate receptor-targeted dendrimer nanoparticles for lung cancer treatment. *Nanomedicine: Nanotechnology, Biology and Medicine*, *14*(2), 373–384. https://doi.org/10.1016/j.nano.2017.11.010

Arora, V., Abourehab, M. A., Modi, G., & Kesharwani, P. (2022). Dendrimers as prospective nanocarrier for targeted delivery against lung cancer. *European Polymer Journal*, *180*, 111635. https://doi.org/10.1016/j.eurpolymj.2022.111635

Asthana, A., Chauhan, A. S., Diwan, P. V., & Jain, N. K. (2005). Poly(amidoamine) (PAMAM) dendritic nanostructures for controlled site specific delivery of acidic anti-inflammatory active ingredient. *AAPS PharmSciTech*, *6*(3), E536–E542. https://doi.org/10.1208/pt060367

Ayatollahi, S., Salmasi, Z., Hashemi, M., Askarian, S., Oskuee, R. K., Abnous, K., & Ramezani, M. (2017). Aptamer-targeted delivery of Bcl-xL shRNA using alkyl modified PAMAM dendrimers into lung cancer cells. *The International Journal of Biochemistry & Cell Biology*, *92*, 210–217. https://doi.org/10.1016/j.biocel.2017.10.005

Bajbouj, K., Al-Ali, A., Ramakrishnan, R. K., Saber-Ayad, M., & Hamid, Q. (2021). Histone modification in NSCLC: Molecular mechanisms and therapeutic targets. *International Journal of Molecular Sciences*, *22*(21), 11701. https://doi.org/10.3390/ijms222111701

Barrett, T., Ravizzini, G., Choyke, P., & Kobayashi, H. (2009). Dendrimers in medical nanotechnology. *IEEE Engineering in Medicine and Biology Magazine*, *28*(1), 12–22. https://doi.org/10.1109/memb.2008.931012

Bhargava, M., Bhargava, S., & Bhargava, V. (2017). P3.02c-002 mannosylated poly (propylene imine) dendrimer mediated lung delivery of anticancer bioactive. *Journal of Thoracic Oncology*, *12*(1), S1272. https://doi.org/10.1016/j.jtho.2016.11.1797

Bhatt, H., Ghosh, B., & Biswas, S. (2020). Cell-penetrating peptide and α-tocopherol-conjugated poly(amidoamine) dendrimers for improved delivery and anticancer activity of loaded paclitaxel. *ACS Applied Bio Materials*, *3*(5), 3157–3169. https://doi.org/10.1021/acsabm.0c00179

Bober, Z., Aebisher, D. B., & Aebisher, D. (2022). Application of dendrimers in anticancer diagnostics and therapy. *Molecules*, *27*(10), 3237. https://doi.org/10.3390/molecules27103237

Cappuzzo, F. (2009). Gemcitabine for the treatment of advanced nonsmall cell lung cancer. *OncoTargets and Therapy*, 209. https://doi.org/10.2147/ott.s4645

Chauhan, A. (2018). Dendrimers for drug delivery. *Molecules*, *23*(4), 938. https://doi.org/10.3390/molecules23040938

Chehelgerdi, M., Chehelgerdi, M., Al-lela, O. Q., Cosme, D., Jayasankar, N., Rao, D. P., Thamaraikani, T., Vasanthan, M., Viktor, P., Lakshmaiya, N., Saadh, M. J., Amajd, A., Abo-Zaid, M. A., Roxana Yolanda Castillo-Acobo, Ismail, A. H., Amin, A. H., & Sigari, R. A. (2023). Progressing nanotechnology to improve targeted cancer treatment: Overcoming hurdles in its clinical implementation. *Molecular Cancer*, *22*(1). https://doi.org/10.1186/s12943-023-01865-0

Chen, Z., Fillmore, C. M., Hammerman, P. S., Kim, C. F., & Wong, K.-K. (2014). Non-small-cell lung cancers: A heterogeneous set of diseases. *Nature Reviews Cancer*, *14*(8), 535–546. https://doi.org/10.1038/nrc3775

Chis, A. A., Dobrea, C., Morgovan, C., Arseniu, A. M., Rus, L. L., Butuca, A., Juncan, A. M., Totan, M., Vonica-Tincu, A. L., Cormos, G., Muntean, A. C., Muresan, M. L., Gligor, F. G., & Frum, A. (2020). Applications and limitations of dendrimers in biomedicine. *Molecules*, *25*(17), 3982. https://doi.org/10.3390/molecules25173982

Chiu, L.-C., Lin, S.-M., Lo, Y.-L., Kuo, S. C.-H., Yang, C.-T., & Hsu, P.-C. (2021). Immunotherapy and vaccination in surgically resectable non-small cell lung cancer (NSCLC). *Vaccines*, *9*(7), 689. https://doi.org/10.3390/vaccines9070689

Cortinovis, D., Bidoli, P., Canova, S., Colonese, F., Gemelli, M., Lavitrano, M. L., Banna, G. L., Liu, S. V., & Morabito, A. (2021). Novel cytotoxic chemotherapies in small cell lung carcinoma. *Cancers*, *13*(5), 1152. https://doi.org/10.3390/cancers13051152

Cosaert, J., & Quoix, E. (2002). Platinum drugs in the treatment of non-small-cell lung cancer. *British Journal of Cancer*, *87*(8), 825–833. https://doi.org/10.1038/sj.bjc.6600540

Crintea, A., Motofelea, A. C., Șovrea, A. S., Constantin, A.-M., Crivii, C.-B., Carpa, R., & Duțu, A. G. (2023). Dendrimers: Advancements and potential applications in cancer diagnosis and treatment—an overview. *Pharmaceutics*, *15*(5), 1406. https://doi.org/10.3390/pharmaceutics15051406

Deboever, N., Mitchell, K. G., Feldman, H. A., Cascone, T., & Sepesi, B. (2022). Current surgical indications for non-small-cell lung cancer. *Cancers*, *14*(5), 1263. https://doi.org/10.3390/cancers14051263

Fatani, W. K., Aleanizy, F. S., Alqahtani, F. Y., Alanazi, M. M., Aldossari, A. A., Shakeel, F., Haq, N., Abdelhady, H., Alkahtani, H. M., & Alsarra, I. A. (2023). Erlotinib-loaded dendrimer nanocomposites as a targeted lung cancer chemotherapy. *Molecules/Molecules Online/Molecules Annual*, *28*(9), 3974–3974. https://doi.org/10.3390/molecules28093974

Fateh, S. T., Aghaii, A. H., Aminzade, Z., Shahriari, E., Roohpour, N., Koosha, F., & Dezfuli, A. S. (2024). Inorganic nanoparticle-cored dendrimers for biomedical applications: A review. *Heliyon*, *10*(9), e29726. https://doi.org/10.1016/j.heliyon.2024.e29726

Ferreiro, M. P., Abelairas, A. M., Criado, A., Gómez, I. J., & Mosquera, J. (2023). Dendrimers: Exploring their wide structural variety and applications. *Polymers*, *15*(22), 4369. https://doi.org/10.3390/polym15224369

Filipczak, N., Yalamarty, S. S. K., Li, X., Parveen, F., & Torchilin, V. (2021). Developments in treatment methodologies using dendrimers for infectious diseases. *Molecules*, *26*(11). https://doi.org/10.3390/molecules26113304

Freitas, L. F., Ferreira, A. H., Thipe, V. C., Gustavo, Caroline, Jorge, Riello, F. N., Nogueira, K., Cruz, Giovanna, Rodrigues, A. S., Sousa, T. S., Alves, V. M., & Lugão, A. B. (2021). The state of the art of theranostic nanomaterials for lung, breast, and prostate cancers. *Nanomaterials*, *11*(10), 2579. https://doi.org/10.3390/nano11102579

Gaballu, F. A., Dehghan, G., Ghaffari, M., Yekta, R., Ravasjani, S. A., Baradaran, B., Dolatabadi, J. E. N., & Hamblin, M. R. (2018). PAMAM dendrimers as efficient drug and gene delivery nanosystems for cancer therapy. *Applied Materials Today*, *12*, 177–190. https://doi.org/10.1016/j.apmt.2018.05.002

Garg, P., Malhotra, J., Kulkarni, P., Horne, D., Salgia, R., & Singhal, S. S. (2024). Emerging therapeutic strategies to overcome drug resistance in cancer cells. *Cancers*, *16*(13), 2478–2478. https://doi.org/10.3390/cancers16132478

Gazdar, A. F., Bunn, P. A., & Minna, J. D. (2017). Small-cell lung cancer: What we know, what we need to know and the path forward. *Nature Reviews Cancer*, *17*(12), 725–737. https://doi.org/10.1038/nrc.2017.87

GLOBOCAN. (2022). *Statistics at a glance, 2022 top 5 most frequent cancers*. https://gco.iarc.who.int/media/globocan/factsheets/populations/900-world-fact-sheet.pdf

Howington, J. A., Blum, M. G., Chang, A. C., Balekian, A. A., & Murthy, S. C. (2013). Treatment of stage I and II non-small cell lung cancer: Diagnosis and management of lung cancer, 3rd ed: American college of chest physicians evidence-based clinical practice guidelines. *Chest*, *143*(5 Suppl), e278S–e313S. https://doi.org/10.1378/chest.12-2359

Hsu, H.-J., Bugno, J., Lee, S., & Hong, S. (2016). Dendrimer-based nanocarriers: A versatile platform for drug delivery. *Wiley Interdisciplinary Reviews: Nanomedicine and Nanobiotechnology*, *9*(1), e1409. https://doi.org/10.1002/wnan.1409

Huang, A. Y.-T., Kao, C.-L., Selvaraj, A., & Peng, L. (2023). Solid-phase dendrimer synthesis: A promising approach to transform dendrimer construction. *Materials Today Chemistry*, *27*, 101285. https://doi.org/10.1016/j.mtchem.2022.101285

Huang, Q., Li, L., Li, L., Chen, H., Dang, Y., Zhang, J., Shao, N., Chang, H., Zhou, Z., Liu, C., He, B., Wei, H., & Xiao, J. (2016). MDM2 knockdown mediated by a triazine-modified dendrimer in the treatment of non-small cell lung cancer. *Oncotarget*, *7*(28), 44013–44022. https://doi.org/10.18632/oncotarget.9768

Jarmila, P., Veronika, M., & Peter, M. (2024). Advances in the delivery of anticancer drugs by nanoparticles and chitosan-based nanoparticles. *International Journal of Pharmaceutics X*, 100281. https://doi.org/10.1016/j.ijpx.2024.100281

Jiang, Y., Lyu, Z., Ralahy, B., Liu, J., Roussel, T., Ding, L., Tang, J., Kosta, A., Giorgio, S., Tomasini, R., Liang, X.-J., Dusetti, N., Iovanna, J., & Peng, L. (2023). Dendrimer nanosystems for adaptive tumor-assisted drug delivery via extracellular vesicle hijacking. *Proceedings of the National Academy of Sciences*, *120*(7). https://doi.org/10.1073/pnas.2215308120

Kaminskas, L. M., McLeod, V. M., Ryan, G. M., Kelly, B. D., Haynes, J. M., Williamson, M., Thienthong, N., Owen, D. J., & Porter, C. J. H. (2014). Pulmonary administration of a doxorubicin-conjugated dendrimer enhances drug exposure to lung metastases and improves cancer therapy. *Journal of Controlled Release*, *183*, 18–26. https://doi.org/10.1016/j.jconrel.2014.03.012

Kara, G., Calin, G. A., & Ozpolat, B. (2022). RNAi-based therapeutics and tumor targeted delivery in cancer. *Advanced Drug Delivery Reviews*, *182*, 114113. https://doi.org/10.1016/j.addr.2022.114113

Kaur, D., Jain, K., Mehra, N. K., Kesharwani, P., & Jain, N. K. (2016). A review on comparative study of PPI and PAMAM dendrimers. *Journal of Nanoparticle Research*, *18*(6). https://doi.org/10.1007/s11051-016-3423-0

Kesharwani, P., & Iyer, A. K. (2015). Recent advances in dendrimer-based nanovectors for tumor-targeted drug and gene delivery. *Drug Discovery Today*, *20*(5), 536–547. https://doi.org/10.1016/j.drudis.2014.12.012

Kheraldine, H., Rachid, O., Habib, A. M., Al Moustafa, A.-E., Benter, I. F., & Akhtar, S. (2021). Emerging innate biological properties of nano-drug delivery systems: A focus on PAMAM dendrimers and their clinical potential. *Advanced Drug Delivery Reviews*, 113908. https://doi.org/10.1016/j.addr.2021.113908

Kim, Y., Park, E. C., & Na, D. G. (2018). Recent progress in dendrimer-based nanomedicine development. *Archives of Pharmacal Research*, *41*(6), 571–582. https://doi.org/10.1007/s12272-018-1008-4

Lee, C. C., MacKay, J. A., Fréchet, J. M. J., & Szoka, F. C. (2005). Designing dendrimers for biological applications. *Nature Biotechnology*, *23*(12), 1517–1526. https://doi.org/10.1038/nbt1171

Li, L., Deng, Y., Zeng, Y., Yan, B., Deng, Y., Zheng, Z., Li, S., Yang, Y., Hao, J., Xiao, X., & Wang, X. (2023). The application advances of dendrimers in biomedical field. *View*, *4*(6). https://doi.org/10.1002/viw.20230023

Liu, J., Liu, J., Chu, L., Wang, Y., Duan, Y., Feng, L., Wang, C., Yang, L., & Kong, D. (2010). Novel peptide-dendrimer conjugates as drug carriers for targeting nonsmall cell lung cancer. *International Journal of Nanomedicine*, *59*. https://doi.org/10.2147/ijn.s14601

Liu, J., Xiong, Z., Zhang, J., Peng, C., Maculewicz, B. K., Shen, M., & Shi, X. (2019). Zwitterionic gadolinium(III)-complexed dendrimer-entrapped gold nanoparticles for enhanced computed tomography/magnetic resonance imaging of lung cancer metastasis. *ACS Applied Materials & Interfaces*, *11*(17), 15212–15221. https://doi.org/10.1021/acsami.8b21679

Malik, N., Wiwattanapatapee, R., Klopsch, R., Lorenz, K., Frey, H., Weener, J. W., Meijer, E. W., Paulus, W., & Duncan, R. (2000). Dendrimers. *Journal of Controlled Release*, *65*(1–2), 133–148. https://doi.org/10.1016/s0168-3659(99)00246-1

Masuda, M., Kawakami, S., Wijagkanalan, W., Suga, T., Fuchigami, Y., Yamashita, F., & Hashida, M. (2016). Anti-MUC1 Aptamer/negatively charged amino acid dendrimer conjugates for targeted delivery to human lung adenocarcinoma A549 cells. *Biological & Pharmaceutical Bulletin*, *39*(10), 1734–1738. https://doi.org/10.1248/bpb.b16-00508

Mendes, L. P., Pan, J., & Torchilin, V. (2017). Dendrimers as nanocarriers for nucleic acid and drug delivery in cancer therapy. *Molecules*, *22*(9), 1401. https://doi.org/10.3390/molecules22091401

Mittal, P., Saharan, A., Verma, R., Altalbawy, F. M. A., Alfaidi, M. A., Batiha, G. E.-S., Akter, W., Gautam, R. K., Uddin, M. S., & Rahman, M. S. (2021). Dendrimers: A new race of pharmaceutical nanocarriers. *BioMed Research International*, *2021*, e8844030. https://doi.org/10.1155/2021/8844030

Mollaei, M., Hassan, Z. M., Khorshidi, F., & Langroudi, L. (2021). Chemotherapeutic drugs: Cell death- and resistance-related signaling pathways. Are they really as smart as the tumor cells? *Translational Oncology*, *14*(5), 101056. https://doi.org/10.1016/j.tranon.2021.101056

Mukherjee, S., Mukherjee, S., Abourehab, M. A. S., Sahebkar, A., & Kesharwani, P. (2022). Exploring dendrimer-based drug delivery systems and their potential applications in cancer immunotherapy. *European Polymer Journal*, *177*, 111471. https://doi.org/10.1016/j.eurpolymj.2022.111471

Nooreldeen, R., & Bach, H. (2021). Current and future development in lung cancer diagnosis. *International Journal of Molecular Sciences*, *22*(16), 8661. https://doi.org/10.3390/ijms22168661

Norouzi, M., & Hardy, P. (2021). Clinical applications of nanomedicines in lung cancer treatment. *Acta Biomaterialia*, *121*, 134–142. https://doi.org/10.1016/j.actbio.2020.12.009

Oledzka, E., Pa0nik, K., Izabela Domańska, I., Pisklak, M. Z., Piotrowska, U., Sobczak, M., Szeleszczuk, Ł., & Laskowska, A. (2023). Poly(amidoamine) dendrimer/camptothecin complex: From synthesis to in vitro cancer cell line studies. *Molecules/Molecules Online/Molecules Annual*, *28*(6), 2696. https://doi.org/10.3390/molecules28062696

Pandey, P., Garg, A., Singh, V., Rai, G., & Mishra, N. (2024). Clinical trials and future prospects of autophagy and ROS in cancer. *Cancer Drug Discovery and Development/Cancer Drug Discovery and Development*, 337–369. https://doi.org/10.1007/978-3-031-66421-2_15

Pandita, D., Poonia, N., Kumar, S., Lather, V., & Madaan, K. (2014). Dendrimers in drug delivery and targeting: Drug-dendrimer interactions and toxicity issues. *Journal of Pharmacy and Bioallied Sciences*, *6*(3), 139. https://doi.org/10.4103/0975-7406.130965

Patel, P., Patel, V., & Patel, P. M. (2022). Synthetic strategy of dendrimers: A review. *Journal of the Indian Chemical Society*, *99*(7), 100514. https://doi.org/10.1016/j.jics.2022.100514

Patle, R. Y., & Dongre, R. S. (2024). Recent advances in PAMAM mediated nano-vehicles for targeted drug delivery in cancer therapy. *Journal of Drug Targeting*, 1–21. https://doi.org/10.1080/1061186x.2024.2428966

Pooja, D., Reddy, T. S., Kulhari, H., Kadari, A., Adams, D. J., Bansal, V., & Sistla, R. (2020). N-acetyl-d-glucosamine-conjugated PAMAM dendrimers as dual receptor-targeting nanocarriers for anticancer drug delivery. *European Journal of Pharmaceutics and Biopharmaceutics*, *154*, 377–386. https://doi.org/10.1016/j.ejpb.2020.07.020

Provencio, M., Calvo, V., Romero, A., Spicer, J., & Cruz-Bermúdez, A. (2022). Treatment sequencing in resectable lung cancer: The good and the bad of adjuvant versus neoadjuvant therapy. *American Society of Clinical Oncology Educational Book*, *42*, 711–728. https://doi.org/10.1200/edbk_358995

Rai, D. B., Medicherla, K., Pooja, D., & Kulhari, H. (2023). Dendrimer-mediated delivery of anticancer drugs for colon cancer treatment. *Pharmaceutics*, *15*(3), 801. https://doi.org/10.3390/pharmaceutics15030801

Rai, D. B., Solanki, R., Patel, S., Pooja, D., & Kulhari, H. (2024). Designing of fucosylated dendrimers as a biocompatible carrier for the targeted delivery of chrysin to human lung cancer cells. *Next Materials*, *5*, 100257. https://doi.org/10.1016/j.nxmate.2024.100257

Rastogi, V., Yadav, P., Porwal, M., Sur, S., & Verma, A. (2022). Dendrimer as nanocarrier for drug delivery and drug targeting therapeutics: A fundamental to advanced systematic review. *International Journal of Polymeric Materials*, *73*(4), 310–332. https://doi.org/10.1080/00914037.2022.2158334

Rompicharla, S. V. K., Kumari, P., Bhatt, H., Ghosh, B., & Biswas, S. (2019). Biotin functionalized PEGylated poly(amidoamine) dendrimer conjugate for active targeting of paclitaxel in cancer. *International Journal of Pharmaceutics*, *557*, 329–341. https://doi.org/10.1016/j.ijpharm.2018.12.069

Santos, A., Veiga, F., & Figueiras, A. (2019). Dendrimers as pharmaceutical excipients: Synthesis, properties, toxicity and biomedical applications. *Materials*, *13*(1), 65. https://doi.org/10.3390/ma13010065

Sathe, R. Y., & Bharatam, P. V. (2022). Drug-dendrimer complexes and conjugates: Detailed furtherance through theory and experiments. *Advances in Colloid and Interface Science*, *303*, 102639. https://doi.org/10.1016/j.cis.2022.102639

Sharma, A. K., Gothwal, A., Kesharwani, P., Alsaab, H., Iyer, A. K., & Gupta, U. (2017). Dendrimer nanoarchitectures for cancer diagnosis and anticancer drug delivery. *Drug Discovery Today*, *22*(2), 314–326. https://doi.org/10.1016/j.drudis.2016.09.013

Shcharbin, D., Milowska, K., Dzmitruk, V., Ionov, M., Javier, & Bryszewska, M. (2019). Dendrimers and hyperbranched structures for biomedical applications. *European Polymer Journal*, *119*, 61–73. https://doi.org/10.1016/j.eurpolymj.2019.07.013

Sherje, A. P., Jadhav, M., Dravyakar, B. R., & Kadam, D. (2018). Dendrimers: A versatile nanocarrier for drug delivery and targeting. *International Journal of Pharmaceutics*, *548*(1), 707–720. https://doi.org/10.1016/j.ijpharm.2018.07.030

Shuel, S. L. (2022). Targeted cancer therapies. *Canadian Family Physician*, *68*(7), 515–518. https://doi.org/10.46747/cfp.6807515

Siddique, F., Shehata, M., Ghazal, M., Contractor, S., & El-Baz, A. (2024). Lung cancer subtyping: A short review. *Cancers*, *16*(15), 2643. https://doi.org/10.3390/cancers16152643

Singh, J., Jain, K., Mehra, N. K., & Jain, N. K. (2016). Dendrimers in anticancer drug delivery: Mechanism of interaction of drug and dendrimers. *Artificial Cells, Nanomedicine, and Biotechnology*, *44*(7), 1626–1634. https://doi.org/10.3109/21691401.2015.1129625

Singh, V., Sahebkar, A., & Kesharwani, P. (2021). Poly (propylene imine) dendrimer as an emerging polymeric nanocarrier for anticancer drug and gene delivery. *European Polymer Journal*, *158*, 110683. https://doi.org/10.1016/j.eurpolymj.2021.110683

Somani, S., & Dufès, C. (2014). Applications of dendrimers for brain delivery and cancer therapy. *Nanomedicine*, *9*(15), 2403–2414. https://doi.org/10.2217/nnm.14.130

Surekha, B., Kommana, N. S., Dubey, S. K., Kumar, A., Shukla, R., & Kesharwani, P. (2021). PAMAM dendrimer as a talented multifunctional biomimetic nanocarrier for cancer diagnosis and therapy. *Colloids and Surfaces B: Biointerfaces*, *204*, 111837. https://doi.org/10.1016/j.colsurfb.2021.111837

Tahar, H. B., Chen, J., Song, P., Zhao, J., Quan, M., Li, C., Gu, X., Yang, G., & Gao, Y. (2019). Novel CT-guided 188-rhenium brachytherapy device for local primary and secondary lung malignancies. *Journal of Global Oncology*, *5*(Suppl), 96. https://doi.org/10.1200/jgo.2019.5.suppl.96

Tarach, P., & Janaszewska, A. (2021). Recent advances in preclinical research using PAMAM dendrimers for cancer gene therapy. *International Journal of Molecular Sciences*, *22*(6), 2912. https://doi.org/10.3390/ijms22062912

Tomalia, D. A., Nixon, L. S., & Hedstrand, D. M. (2020). The role of branch cell symmetry and other critical nanoscale design parameters in the determination of dendrimer encapsulation properties. *Biomolecules*, *10*(4), 642. https://doi.org/10.3390/biom10040642

Wang, J., Li, B., Qiu, L., Qiao, X., & Yang, H. (2022). Dendrimer-based drug delivery systems: History, challenges, and latest developments. *Journal of Biological Engineering*, *16*(1). https://doi.org/10.1186/s13036-022-00298-5

Wu, J., Huang, W., & He, Z. (2013). Dendrimers as carriers for siRNA delivery and gene silencing: A review. *The Scientific World Journal*, *2013*(1). https://doi.org/10.1155/2013/630654

Zhang, T., Zhang, L., Shen, Y., Xiao, H., Han, Z., Wang, L., & Ni, S. (2024). Zwitterionic targeting nanodrugs prepared from poly(propyleneimine) dendrimers for enhanced anti-tumor therapy. *Colloids and Surfaces a Physicochemical and Engineering Aspects*, *694*, 134193. https://doi.org/10.1016/j.colsurfa.2024.134193

Zhang, W., Wang, Y., Kan, X., Wang, X., & Geng, D. (2017). Preparation and evaluation of peptide-dendrimer-paclitaxel conjugates for treatment of heterogeneous stage 1 nonsmall cell lung cancer in 293T and L132 cell lines. *Tropical Journal of Pharmaceutical Research*, *16*(4), 737. https://doi.org/10.4314/tjpr.v16i4.1

Zheng, M., Pan, M., Zhang, W., Lin, H., Wu, S., Lu, C., Tang, S., Liu, D., & Cai, J. (2021). Poly(α-l-lysine)-based nanomaterials for versatile biomedical applications: Current advances and perspectives. *Bioactive Materials*, *6*(7), 1878–1909. https://doi.org/10.1016/j.bioactmat.2020.12.001

12 Nanofibers for the Management of Lung Cancer

Rabab Fatima[1*], *Sadiq Umar*[2], *and Sahil Hussain*[3]
[1] School of Health Sciences & Technology, UPES, Dehradun 248007, Uttarakhand, India
[2] Department of Oral Biology, College of Dentistry, University of Illinois, Chicago
[3] College of Health and Life Sciences, Hamad Bin Khalifa University, Doha, Qatar
*Corresponding author: Rabab Fatima, rababtfatima@gmail.com.

12.1 INTRODUCTION

With an estimated 2.2 million new cases and 1.8 million deaths each year, lung cancer continues to be the largest cause of cancer-related mortality globally. With differing incidence rates in different geographic locations, the disease poses a serious threat to global health (de Groot et al., 2018). Lung cancer occurrences have somewhat decreased in the east as a result of falling smoking rates, but the incidence is still rising in developing countries as a result of increased tobacco use and environmental variables. With direct medical expenses reaching $130 billion yearly and indirect costs through lost productivity estimated at $140 billion globally, the financial burden of the disease is significant (Kratzer et al., 2024). Treatment strategies are evolving and are made more challenging by the diversity of lung cancer subtypes. About 85% of cases are non–small cell lung cancer (NSCLC), with the remaining 15% being small cell lung cancer (SCLC). Within NSCLC, adenocarcinoma (40%), squamous cell carcinoma (25–30%), and large cell carcinoma (10–15%) present distinct therapeutic challenges (Schabath & Cote, 2019). In order to achieve the best possible therapeutic results, traditional cancer treatments must overcome several challenges. About 60% of patients become resistant to first-line treatments within a year, making drug resistance a major problem. Numerous events during cancer progression, such as increased drug efflux, modified drug targets, and improved DNA repair mechanisms, contribute to the emergence of this resistance (Spiro & Silvestri, 2005). Up to 80% of patients experience severe side effects from the systemic toxicity of conventional chemotherapeutics, which has a substantial negative impact on their quality of life. Another major issue is the poor absorption of the intended chemotherapeutic. Low solubility and restricted membrane permeability are characteristics of many promising treatment candidates that lead to less-than-ideal drug concentrations at tumor locations where only 0.7–5% of the dosage given by current administration methods reaches the tumor tissue.

The advancement of nanotechnology, however, has further fueled the development and fabrication of innovative nanomaterials, including liposomes, nanosheets, nanoparticles, dendrimers, nanorods, and nanofibers, to revolutionize drug delivery strategies in several diseases including cancer. Nanomaterials with dimensions under 100 nm offer significant advantages, such as smaller sizes and larger surface area-to-volume ratios, making them highly effective for applications in drug delivery and biosensing (Singh et al., 2024). Challenges such as low drug encapsulation efficiency, rapid renal clearance, inconsistent dosage maintenance, and burst drug release in nanotechnology-based delivery systems have effectively been overcome through the advent and rise of nanomaterials

DOI: 10.1201/9781003595830-12

like nanofibers in this field (Pandey et al., 2024). Electrospun nanofibers in this regard have recently garnered significant attention as a fabrication material among various nanomaterials due to their exceptional properties such as high biocompatibility, excellent surface modification capabilities, adaptable surface conformations, increased surface area, high drug loading capacity, superior entrapment efficiency, a compatible microstructure, cost effectiveness, and the ability to integrate multiple therapeutic approaches (Rasekh et al., 2024). A highly sensitive, non-invasive, and label-free salivary sensor platform utilizing carbon nanofiber (CNF) surfaces anchored with bimetallic Nickel@silver core shell nanoparticles (Ni@Ag/CNFs) has been reported as an effective pre-diagnostic screening tool for lung cancer by detecting abnormal concentrations of potential biomarkers, including salivary imidazole compounds demonstrating its efficiency in analyzing real-time saliva samples, especially toward detection of salivary imidazole compounds such as urocanic acid, histidine, and histamine within the clinically relevant range in lung cancer diagnosis (Sunil et al., 2023). TiO_2 nanofibers were found to induce angiogenic, fibrotic, and epithelial-mesenchymal transition markers, along with enhanced genomic instability and reduced sensitivity to Cisplatin. Tumors derived from cells exposed to TiO_2 nanofibers also reported these markers, with increased dedifferentiation, and a notable presence of erythrocytes with subsequent loss of cisplatin sensitivity, indicating that TiO_2 nanofibers promote an aggressive tumor phenotype in lung epithelial cells (Medina-Reyes et al., 2019). A novel electrochemical aptasensor showed high sensitivity toward A549 cells with a detection limit of 1.2×10^3 cells/mL (Kivrak et al., 2020) Implantable nanofibers composed of poly(caprolactone) poly(ethylene glycol) (PCL/PEG) and loaded with 17-dimethylaminoethylamino-17-demethoxy geldanamycin (17-DMAG) demonstrated enhanced anticancer efficacy through the inhibition of HSP90 expression and telomerase activity in A549 lung cells. The designed nanofibers significantly reduced the HSP90 mRNA expression and telomerase activity, positioning these implantable scaffolds as a promising strategy to eradicate residual lung cancer tissues for prevention of local tumor recurrence (Mellatyar et al., 2018).

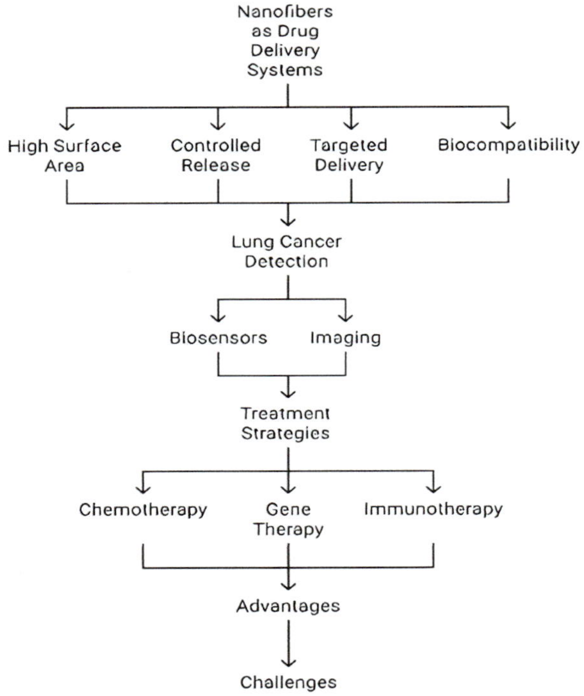

FIGURE 12.1 Nanofibers and their diverse applications in the biomedical field including lung cancer treatment

Traditional chemotherapy and radiation therapy often result in severe side effects because they affect both cancerous both healthy cells. Nanofibers help in minimize these effects by selectively targeting the cancer cells. With the use of biosensors specifically, nanofibers can help detect lung cancer at an early stage, significantly improving the chances of successful treatment (Tayebi-Khorrami et al., 2024). Nanofibers allow for efficient drug delivery, which can increase the drug's efficacy while reducing the required dosage and its side effects (AL-Attabi et al., 2024). Nanofibers, which usually have sizes between 50 and 500 nm, can be distinguished by their distinct physical and chemical characteristics (Figure 12.1). Nanofibers are highly attractive for drug delivery mainly due to their exceptionally high surface-area-to-volume ratio, further augmented by the presence of intrinsic pores within the fibers, facilitate enhanced encapsulation and therapeutic loading (Ahmadian et al., 2024). This increased surface area also mitigates diffusion limitations, enabling a higher percentage of drug release over the course of duration. The other important aspect to highlight about the use of nanofiber technology in drug delivery is the fiber diameter, porosity, and drug binding mechanisms, which can be precisely tailored by optimizing material selection and process parameters (Müller & Keck, 2004). Electrospun nanofibers hold significant promise as delivery carriers for drugs, cells, proteins, RNA, and DNA, offering controlled release directly at the target site. Beyond drug delivery, their distinctive properties contribute to cell proliferation, mass transport, cell adhesion, filtration, and sensing (Kuang et al., 2024). Moreover, they play a crucial role in *in vitro* tumor modeling, highlighting their vast potential in advancing lung cancer therapy applications.

12.2 METHODS OF NANOFIBER SYNTHESIS

The various methods for synthesizing nanofibers are discussed here; of them electrospinning is the most widely used and explored.

12.2.1 Electrospinning

Electrospinning is an innovative technique for producing nanofibers using high-voltage electric fields. It is the majorly reported technique for nanofiber production for synthesis of single, multi, and needleless forms, entailing meticulous control over the following variables: processing variables (voltage: 10–30 kV, flow rate: 0.5–2 mL/h), ambient conditions (temperature: 20–25°C, humidity: 30–50%), and solution parameters (concentration: 10–20 weight%, viscosity: 100–800 cP). This method is applicable to both melt- and solution-based polymers, with various parameters playing a crucial role in determining the fibers' morphology (Gao et al., 2024). Electrospun nanofibers offer unique characteristics, such as a high surface area, precisely controllable pore sizes, and uniform dimensions, which are challenging to achieve with traditional fiber spinning methods. These features make them highly suitable for applications in agriculture, antifouling filtration membranes, defense, tissue engineering, drug delivery, wound care, gas sensing, the pharmaceutical industry, catalysis, electronic devices, energy storage, environmental science, and biomedical fields related to cancer management (Ji et al., 2024). Owing to their significant advantages over conventional fibers, over 100 polymers have been successfully electrospun for diverse applications (Abid et al., 2019).

12.2.2 Coaxial Electrospinning

Coaxial electrospinning is a useful method to produce core shell structured nnaofibers often used in delayed diffusion and enhancing the protection of sensitive payloads (D. Han & Steckl, 2019). An attempt to determine how PCL fiber properties change based on varying solution parameters and processing conditions was studied. Fiber meshes formed from coaxial spinning were reinforced with acetylsalicylic acid (ASA) as a model agent for both inflammation and thrombotic prevention, and ASA release patterns of conventional blend electrospun fibers against coaxial electrospun fibers demonstrated robust evidences against the both the stated conditions (Repanas & Wolkers, 2015)

12.2.3 Bubbfil Electrospinning

This method is mainly known for the large-scale production of nanofiber material used in myriad biomedical conditions as it reduces the surface tension by using polymer or melt bubbles or membranes to form multiple jets for efficient and mass production of nanofibers. With jet ejection velocities reaching up to around 100 m/s, jets can reach the collector within 0.001 second, producing fibers with dimensions as fine as 5 nm. Bubbfil electrospinning offered a straightforward approach to efficiently fabricate nanofibers with a high surface area-to-volume ratio and enhanced porosity. This process, however, does not imply specific requirements on the solution part, enhancing its versatility further. The diameter of the nanofibers produced is heavily influenced by the size of the bubbles (R. Chen et al., 2014). The bubble size, in turn, can be controlled by adjusting factors such as the inlet pressure, the solution temperature, and the positioning of the air tube's top.

12.2.4 Solvent-Free Electrospinning

Solvent-free electrospinning is highly suitable for applications in tissue engineering and wound dressing materials, as the resulting fibrous structures are amenable to subsequent treatments. Despite its potential, this technique has received limited attention from researchers (Uddin et al., 2024). Solvent-free e-spinning methods, such as melt e-spinning, have received limited attention despite their advantages, including eliminating solvent emissions and achieving highly efficient precursor utilization (often exceeding 90 wt%). Importantly, solvent-free electrospun (e-spun) fibers, free from solvent residues, are particularly promising for applications like tissue engineering and wound dressings. Recent advances and future trends in solvent-free e-spinning, including melt e-spinning, supercritical CO_2-assisted e-spinning, anion-curing e-spinning, UV-curing e-spinning, and thermocuring e-spinning, highlight the potential for these techniques to evolve into efficient, controllable, and eco-friendly processes for producing ultrafine fibers (Ahmad et al., 2023). Defined as an e-spinning approach that avoids conventional solvents or ensures nearly complete conversion of the precursor solution into ultrafine fibers with minimal evaporation losses, solvent-free e-spinning achieves a precursor utilization ratio of over 90%. Compared to solution e-spinning, it offers numerous advantages such as eliminating solvent residue in fibers, preventing solvent evaporation into the atmosphere, and reducing the high costs associated with solvent recycling (W.-H. Han et al., 2022). These benefits make solvent-free e-spinning a promising technique for applications in biomedicine, tissue engineering, and cancer-related treatments (B. Zhang et al., 2017)

12.2.5 Magnetic-Field-Assisted Electrospinning

A needle-based electrospinning setup to fabricate magnetic nanofibers using polycarbonate-urethane as the polymer matrix filled with nickel nanoparticles was shown to modify the counter-electrode by incorporating permanent magnets to generate a constant magnetic field perpendicular to the electric field, enabling the production of aligned magnetic nanofibers between the magnetic poles (Iannotti et al., 2023). Similarly, another magnetic-field-assisted electrospinning process using electrospun PAN/FeCo nanofibers was studied, where fiber alignment improved with decreasing voltage and increasing FeCo content (Mei et al., 2019). Additionally, nickel ferrite–barium titanate core-shell nanofibers were synthesized through magnetic field-assisted electrospinning, enabling their assembly into disks or films within uniform or gradient magnetic fields (Sreenivasulu et al., 2017; Blachowicz & Ehrmann, 2020).

12.2.6 Self Assembly

These type of nanofibers have gained significant attention owing to their excellent biocompatibility, ease of modification, and versatile design using via bottom-up approach. These nanofibers have wide implications as well as applications in three-dimensional cell culture, tissue engineering, and regenerative medicine field. They have proven to be very effective as drug delivery systems in cancer treatment due to enhanced solubility of hydrophobic drug, improving accumulation at tumor sites, and reducing side effects to neighboring cells (Niece et al., 2003). Hydrophobic antitumor drugs such as Paclitaxel, Camptothecin, and Ellipticine have been explored and encapsulated within peptide nanofibers. They have demonstrated enhanced antitumor efficacy, suggesting that the two-dimensional structure of peptide nanofibers often outperforms the three-dimensional configuration of nanoparticles for drug delivery. A recent investigation in this context has signified that peptide-based nanofibers prepared from the self-assembly technique exhibit superior biocompatibility, better tumor targeting within smaller timeframes, and rapid systemic elimination compared to spherical nanomaterials such as poly(lactic-co-glycolic acid), gold, polystyrene, cadmium and selenium quantum dots, and carbon rods (Liu et al., 2013).

12.2.7 Centrifugal Spinning

Over the past decade, significant research has focused on the potential of nano- and microfiber scaffolds to support wound healing, drug delivery, tissue regeneration, and skin protection. Among the available techniques, centrifugal spinning stands out for its relatively simple mechanism and capacity to produce fibers in large quantities (Weitz et al., 2008). However, many polymeric materials with multifunctional properties suitable for tissue applications remain largely unexplored. Centrifugally spun polymeric fibers can offer ideal characteristics for tissue scaffolds, but an effective scaffold must balance several critical requirements such as degradation rate that matches the pace of tissue regrowth, appropriate layers and porous structures to support cellular proliferation, and a high surface-to-volume ratio with diverse surface compositions to enhance cellular attachment, mobility, growth and differentiation (Xu et al., 2023). By leveraging various methods, it is possible to create scaffold templates with biocompatible and biodegradable properties. Understanding the impact of fiber matrix morphology on bioactivity and controlled tissue development is therefore essential for advancing these applications. The typical configuration of centrifugal spinning equipment, often referred to as a cyclone machine, has been described in numerous studies. It features an electric rotor connected to a driveshaft, where a spinneret is mounted (X. Zhang & Lu, 2014). Fibers are produced by driving the material through the spinneret's precision nozzles using circular motion. The collection process is versatile, allowing fibers to be gathered either as a self-supporting mat or as a continuous mat. Key factors influencing the deposition of fibers on a platform or the formation of three-dimensional structures include the spinneret design, angular or rotational speed, and the aerodynamics within the enclosure (X. Zhang & Lu, 2014).

12.3 DRUG-LOADING MECHANISM IN NANOFIBERS

Therapeutics must first become immobile in the polymer matrix when using nanofibers as drug carriers for the release control mechanisms to function. Although the technique varies depending on the polymer and the treatment, immobilization can be broadly classified as either binding or entrapment (Ye et al., 2019). Drugs or proteins can be physically contained within a polymer matrix by entrapment (Xu et al., 2017). Examples of drug encapsulation in nanofiber drug carriers include entrapping the drug in an intermediate carrier that contains a polymer matrix that forms the nanofiber, or crosslinking the polymers (Mohammadian & Eatemadi, 2017). In its bulk form, alginate is a typical polymer used for entrapment when crosslinked with divalent ions such as calcium involving factors such as hydrophobicity, electrostatic contact, and hydrogen-bond formation for deal binding. For instance, molecular dynamic simulation showed that PCL can bind to Cucurbitacin B and Cucurbitacin I, via hydrogen bonding as well as hydrophobic interactions. However, through electrostatic interactions, the cationic polymer chitosan was reported to connect with negatively charged payloads. Direct mixing, which achieves loading efficiencies of 70–90%, chemical conjugation which modifies the surface with a coupling efficiency >85%, and sophisticated topologies such as core-shell structures are some of the methods for drug integration (Leung & Ko, 2011). Certain triggers in the tumor microenvironment, such as pH shifts (pH 6.5–6.8), increased enzyme levels, or outside stimuli like light or temperature, cause smart release systems to react. These devices provide sustained administration over a period of one to four weeks and achieve tailored release with a small burst effect (<20%).

12.4 DRUG RELEASE MECHANISMS IN NANOFIBER

This can primarily be attributed to *in vivo* fiber degradation, solid-state diffusion through the fiber matrix, and desorption from the fiber surface. Phosphate-buffered saline (PBS) is commonly employed in *in vitro* studies to evaluate drug release behavior from nanofiber-based carriers. Upon exposure to PBS, the solution surrounds the fiber and permeates the interstitial spaces between the individual fibers (Rasouli et al., 2019). Drugs or proteins thus adhered to the fiber surfaces are released as the aqueous phase hydrates and swells the matrix of nanofiber. Surface drug release occurs when the drugs desorb from the fiber surface and by rapid diffusion into the surrounding aqueous environment (Chou et al., 2015). This multi-stage release process highlights the intricate interaction between nanofiber structure, drug immobilization, and environmental conditions in controlled drug delivery systems. Drugs usually get tethered on the exterior of the nanopores within the nanofibers are also prone to the desorption mechanism, which is not restricted to the outer surface of the nanofibers (Figure 12.2). The surface area would be significantly greater than the fiber's outer surface area given the inner pores' nanometer size scale and the likelihood that they are largely correlated (M. Chen et al., 2014). Degradation of the nanofiber can potentially result in drug release. Drugs trapped within the polymer matrix may be released upon breakdown, which again is reported less important in the previously stated polymers and more important in fast-degrading ones like chitosan, alginate, and PVA (Rather et al., 2022). The BSA release from PVA fibers is an example of an event where the drug release rate was overly rapid owing to polymer degradation. A notable burst release was seen with PVA nanofibers, which degrade quickly in water. However, the release rate decreases significantly when the fibers are coated with parylene, which is less biodegradable.

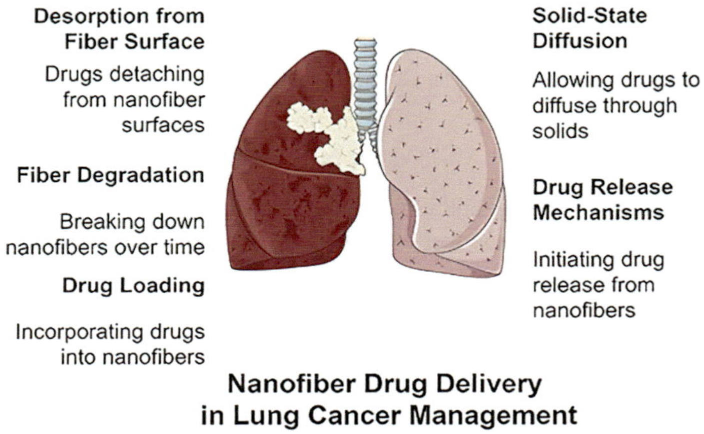

FIGURE 12.2 Mechanism of drug loading and drug release through nanofibers in lung cancer treatment

Since the drug release rates based on the two definitions of degradation can differ, particularly when comparing homogeneous and core-shell nanofibers, the idea of surface and volume degradation is also crucial for nanofiber drug carriers (Risdian et al., 2015).

Because the pharmaceuticals on the nanofiber surface are so close to the surrounding liquid, the desorption release mechanism is one of the fastest processes among those discussed and can therefore significantly aid in burst release. There are other ways to further regulate the rate of release, such as combining with other polymers to lessen the surface area that contains the drug (Y. Yang et al., 2024). The lengths that bulk liquid must travel to permeate the accessible surfaces of the core polymer and that the drug in the polymer must diffuse through are increased when the drug-bearing polymer is placed in the core and enclosed in the shell fiber. Additionally, the shell polymer appears as an additional layer of material that needs to break down in order to fully release the drug.

12.5 APPLICATIONS IN LUNG CANCER TREATMENT

The successful delivery of diverse anticancer agents benefits from nanofibers when these materials enhance drug loading and control delivery while preserving drug bioactivity and ensuring optimal encapsulation (Balaji et al., 2015). Gefitinib used for NSCLC has low awareness about oral intake because of its narrow water dissolution abilities. Drug delivery systems built using nanofiber technology show great potential for making cancer therapy more precise while reducing detrimental side effects. The combination of PVA and PEG nanofibers with collagen delivers gefitinib in an effective manner without causing substantial harm to lung cancer cells (Perumalsami et al., 2025). Detecting circulating tumor cells (CTC) through liquid biopsy represents an emerging method to diagnose early solid tumor progression. Laboratory research has produced a PLGA nanofiber (PN) NanoVelcro chip that functions as a novel device to rapidly collect CTCs from peripheral blood samples. Laboratory findings revealed that EGFR mutations detected in tumor areas also existed in CTCs leading to enhanced tumor progression diagnosis accuracy (Jiang et al., 2018). Both free silibinin (SLB) and SLB-loaded nanofibers were evaluated for their cytotoxic effects on lung cancer cells. While neither formulation had any significant effect on BEAS-2B cells, both SLB and SLB-loaded nanofibers demonstrated cytotoxicity against A549 lung cancer cells by increasing intracellular reactive oxygen species (ROS) levels. Both formulations triggered the expression of apoptotic proteins, with the SLB-loaded nanofibers showing more pronounced effect compared to free SLB, suggesting that SLB-loaded nanofibers are more efficient in targeting cancer cells (Wu et al., 2023). A slow and sustained release of DOX from core-shell nanofibers containing 5 wt% cobalt ferrite was observed for 300 and 600 hours at pH 5.5 and 7.4 respectively without using an alternating magnetic field (AMF), and over 168 and 360 hours under a prominent AMF. In the presence of core-shell nanofibers containing 100 μg DOX and 5% cobalt ferrite nanoparticles, more than 98.3% of A549 cancer cells were effectively killed when exposed to AMF. Flow cytometric analysis further assessed that only 19.1% of cancer cells remained viable under an external magnetic field and just 8.85% survived under AMF. Notably *in vivo* experiments demonstrated a massive reduction in tumor growth, a decrease in relative tumor volume to 0.5, when treated with magnetic core-shell nanofibers containing 100 μg DOX and 5% cobalt ferrite nanoparticles (Bahmani et al., 2024). An enhanced slow and sustained release rate of DOX, following an initial burst release, was achieved using electrospun chitosan/PLA/GO/TiO_2/DOX nanofibrous scaffolds with thicknesses of 30 and 50 μm over a two-week incubation period. The release rate of DOX from the nanofibers was comparatively faster at pH 5.3 than to pH 7.4. The Korsmeyer-Peppas kinetic model when applied determined the release mechanism of drug from the nanofibers. Cell viability assays further confirmed that the nanofibers exhibited a stronger inhibitory effect on the proliferation of lung cancer cells in the presence of AMF (Samadi et al., 2018).

Nanofibers synthesized by electrospinning hold significant clinical potential as a biomimetic tumor microenvironment model for drug screening, offering a controllable platform for localized, prolonged drug release in cancer therapy. They are highly sensitive diagnostic tools well designed for capturing and isolating CTCs from the systemic blood and for detecting an array of cancer-associated biomarkers (Erickson et al., 2022). An innovative core shell nanofiber delivery system successfully deciphered an extended release of DOX and siRNA (over 30 days) using DOX-Bcl-2 liposome-incorporated CS/PEO/PCL core-shell nanofibers against lung cancer management. The maximum downregulation of Bcl-2 siRNA, induction of apoptosis in cancer cells, and reduction in cell viability of A549 lung carcinoma cells were observed with the DOX-Bcl-2 liposome-incorporated CS/PEO/PCL core-shell nanofibers. *In vivo* results demonstrated the most significant suppression of tumor growth, with no change in the body weight of mice treated with the nanofibers (Kordbacheh et al., 2024). Gene expression analysis revealed notable changes in the expression levels of Bax and Bcl-2, as well as caspases-3 and -9, indicating that MET-loaded nanofibers induced apoptosis more effectively than free MET (Samadzadeh et al., 2021a). Samadzadeh et al. developed nanofibers incorporating metformin hydrochloride (MET) and investigated their mechanism of action. MET activates AMP-activated protein kinase (AMPK), which inhibits the mammalian target of rapamycin (mTOR) pathway, thereby suppressing protein synthesis and cancer cell proliferation. The nanofibers exhibited rapid drug release within the first day, sustained for over two weeks. Within 48 hours, MET-loaded nanofibers showed significant cytotoxicity against A549 lung cancer cells, effectively increasing intracellular ROS levels and inducing apoptosis. Additionally, these nanofibers caused notable changes in the expression of apoptosis-related proteins, including Bax, Bcl-2, caspase-3, and caspase-9, compared to free MET (Samadzadeh et al., 2021).

Biodegradable nanofiber meshes loaded with cisplatin were developed using a scalable electrospinning technique, employing two biocompatible polymers: polycaprolactone and poly(glycerol monostearate-co-caprolactone). The selected materials bestowed necessary mechanical properties suitable for dynamic lung tissue applications. The material showed superhydrophobic behavior because of its rough nanostructure alongside hydrophobic components within the polymer mixture. The non-wetting nature of these microparticles enabled cisplatin to maintain a linear release pattern for roughly 90 days until total elimination thus enabling further LLC cell anticancer assessment. *In vivo* studies further evaluated the efficacy of the cisplatin-loaded superhydrophobic meshes in preventing local cancer recurrence following surgical resection in a murine LLC model. Results showed a significant improvement ($p = 0.0006$) in median recurrence-free survival, extending to over 23 days, compared to standard intraperitoneal cisplatin therapy administered at the same dose (Kaplan et al., 2016). The release mechanisms of CMPT and PTX from core shell nanofibers were examined both *in vitro* and *in vivo*, and the results were compared to the direct administration of pristine CMPT and PTX. The mean fiber diameters for composite and core–shell nanofibers varied based on shell feeding rates, measuring approximately 225, 330, 520, and 640 nm for rates of 0.3, 0.5, and 0.7 mL/hour, respectively. *In vivo* release studies reported that the blood concentrations of CMPT and PTX administered core–shell nanofibers in moue models peaked at 26.8 ± 0.04 μg/mL and 26.5 ± 0.05 μg/mL at 36 hours and 24 hours, respectively, with gradual reductions observed over 84 hours for CMPT and 48 hours for PTX. The maximum cytotoxicity reported was 75% using PVA/κ-carrageenan/CMPT/Au/pegylated-PU/PTX core–shell nanofibers. *In vivo* antitumor studies suggest that the synergistic effects of Au nanoparticles combined with CMPT and PTX in significantly reducing tumor volume without affecting mouse body weight highlighting the therapeutic potential of PVA/κ-carrageenan/CMPT/Au/pegylated-PU/PTX core–shell nanofibers, highlighting their efficacy in simultaneously delivering CMPT, PTX, and Au nanoparticles for enhanced lung cancer treatment (Irani & Nodeh, 2022). Li et al. investigated the interaction between human lung epithelial A549 cells and electrospun nanofibers made from PVA and type I collagen by varying the diameter of the nanofibers. The A549 cells adhered to the nanofiber mats through filopodia attachment and adopted an extended morphology, triggering epithelial–mesenchymal transition (EMT) (Li et al., 2018). Pd-decorated WO_3 nanofibers with their highly porous structure led to better toluene detection alongside excellent H_2S molecule sensitivity. The Pd-decorated WO_3 nanofibers exhibited dynamic capabilities for detecting volatile organic compounds and showed special potential for detecting lung cancer VOCs. The sensors demonstrated their potential for particular

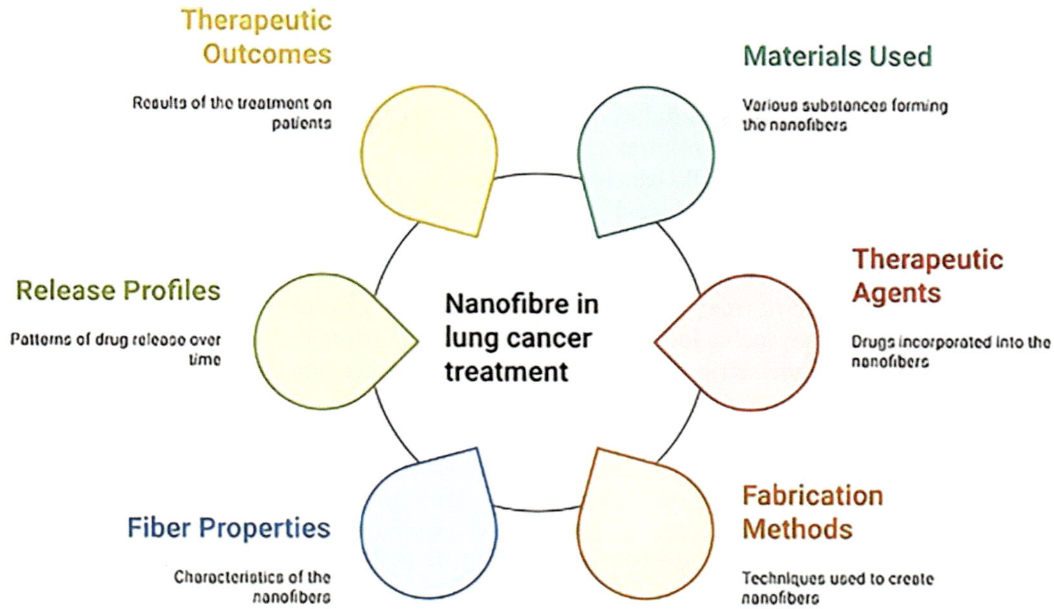

FIGURE 12.3 Implications of nanofibers in lung cancer management

and delicate detection methods by reaching a detection limit of 20 ppb during operation at 350°C (R_{air}/R_{gas} = 1.32) (Kim et al., 2014). Present-day drug delivery systems based on nanofibers enable optimized therapeutic outcomes for lung cancer patients because they combine drug loading capabilities with controlled release methods. The precise control of polymer selection alongside loading and release methods enables nanofibers to provide effective drug delivery with sustained release capabilities and targeted distribution, which increases drug effectiveness while decreasing systemic damage (Figure 12.3). The development of optimized treatment systems, which incorporate stimuli-responsive nanofibers together with combination therapies, demonstrates enormous potential to enhance lung cancer patient treatment effectiveness. Nanomaterial-based drug delivery for lung cancer shows advantages, which include precise drug delivery and well-regulated drug release and improved drug absorption for insoluble compounds alongside reduced negative drug effects. The promising future research possibilities become clear from these benefits discovered. The modern world of nanomedicines features five major classes of substances including lipid nanoparticles followed by polymer nanoparticles alongside gold nanoparticles together with magnetic nanoparticles which operate alongside mesoporous silica. A new drug-loaded Ptx-SA nanofiber was assessed as a self-assembled structure using a carrier-free design approach in lung cancer. Drug-loaded fibers displayed higher cellular absorption and demonstrated better tumor inhibition properties against free drug formulations when tested both in tissue culture and in animal models. The nanoparticle-mediated improvements in drug efficacy appear to result from specific protein expression regulation (Pan et al., 2021).

12.6 NANOFIBER-BASED SENSORS

The combination of tunable morphology and high surface area together with facile manufacturing processes and synergistic properties makes nanofibers ideal candidates for the next generation of predictive oncology diagnostic sensing devices especially when used with artificial intelligence and machine learning. These detection devices indicate boosted performance levels in terms of detection sensitivity as well as selectivity function and quick analytical response toward chemical and physical elements and biological substances Yang et al. (2018). Researchers developed a straightforward operation lung-on-a-chip device that incorporated electrospun nanofiber membranes made from poly(lactic-co-glycolic acid) (PLGA) for both chip substrate and cellular scaffold applications. Scientists introduced gefitinib directly into a 3-μm thick nanofiber membrane built to reproduce alveolar respiratory tissue. According to their research, A549 cells triggered apoptosis in endothelial cells which allowed tumors to invade. Study data indicated that IGF-1 released by HFL1 cells protected tumor cells through pathway activation of PI3K/Akt signaling, which rendered them less sensitive to chemotherapy agents after blocking the drug-induced EGFR-related signal cascade. By simulating alveolar biochemical factors within the tumor microenvironment *in vitro*, particularly in the respiratory membrane, this research offers promising insights for advancing localized treatment of lung tumors (X. Yang et al., 2018). Integration with optical, piezoelectric, and electrochemical transducers holds the potential to further enhance their functionality and may serve as pivotal tools for next generation sensing technologies for lung cancer detection based on nanofiber. Additionally, nanofiber-based sensors using PVDF are being explored for respiratory function monitoring and act as a noninvasive tool in lung cancer screening and furthering its apt management. Point-of-care devices are making positive and transformative impacts for accurate screening and testing without the need for a specialized laboratory infrastructure (Rezaei & Mahmoudifard, 2019). POC devices are rapid, are sensitive, and allow real time monitoring and better disease management. PVA and chitosan combined with gold nanofibers have enhanced the detection limit accuracy making them excellent point of care diagnostics (Iftikhar et al., 2023).

12.7 NATURAL-EXTRACT-LOADED NANOFIBERS

Natural phytoconstituent-based nanofibers exhibit a distinct approach to leverage the bioactive compounds derived from plants for therapeutic, diagnostic, and drug delivery applications in several diseases including cancer. The nanofibers combined with the benefits of natural phytoconstituents enhance their anti-inflammatory, antioxidant, and anticancer properties owing to mainly the high surface area, porosity, and controlled drug release of nanofibers and their stealth properties. The unique combination of biocompatibility, enhanced bioavailability, and controlled release makes them apt for advanced therapeutic applications such as cancer management and gene therapies as well. These innovative and natural nanofibers are known and have been extensively investigated to integrate bioactive plant-derived compounds, such as resveratrol, curcumin, and quercetin and so forth, with the integration of nanotechnology to improve stability, solubility, and the therapeutic efficacy of the system (Sridhar et al., 2014). The high surface area and porous structure further enable efficient drug loading and sustained release, targeting specific tissues and tumor microenvironment in cancer therapy (S. Chen et al., 2018). Moreover, the natural origin of phytoconstituents ensures minimal toxicity, with improved safety profile, while the modifiable aspect of nanofibers ensures targeted and controlled anticancer drug delivery, and in applications such as wound healing, and antimicrobial coatings as well. Drug-eluting medical implants are enormously utilized, particularly in cancer treatment. Regulatory agencies such as FDA have approved numerous nano-formulated devices that release active pharmaceutical ingredients, increasing the demand for further value-added innovations. In this context, curcumin-loaded polycaprolactone (PCL) nanofibers have been explored as drug-eluting stents or implants for anticancer activity against lung and breast cancers and exhibited 15% greater cytotoxicity compared to commercial 1% cisplatin-loaded PCL nanofibers (Sridhar et al., 2014). To improve water solubility and bioavailability, a straightforward approach using solvent evaporation and the amorphous solid dispersion method was reported, where poly(ethylene oxide) (PEO) was chosen as the polymeric solid dispersion matrix due to its favorable features. Additionally, PEO served as a template for *in situ* synthesis of silver nanoparticles (Ag NPs). Nanofibers co-delivering the anticancer drug Niclosamide and Ag NPs demonstrated superior anticancer activity *in vitro* compared to nanofibers containing either Niclosamide alone or Ag NPs (Dubey & Gopinath, 2016). Curcumin-based nanofibers acted by inhibiting various cell signaling pathways, including NF-KB, mTOR, complex I, cytokines, expression of p-p65, Ki67, and angiogenesis-associated genes. They also effectively induced apoptosis through ROS production and activation of caspase dependent pathways in alveolar cancer cells than free curcumin alone (Ataei et al., 2023).

12.8 FUTURE PERSPECTIVES AND CONCLUSION

Nanofibers hold transformative potential for early detection of lung cancer owing to sophisticated and controlled release of drug at the deliberated sites which in turn improve patient outcomes drastically. These versatile structures have transformed traditional diagnostic paradigms by enabling detection at molecular levels previously unattainable through conventional methods such as X-rays and CT scans in both early detection and therapeutic delivery. In the realm of diagnostics, nanofiber-based biosensors have emerged as powerful tools for identifying cancer-specific biomarkers. These sophisticated sensors utilize functionalized nanofibers that respond to the presence of proteins, genetic materials, and metabolites associated with lung cancer. Nanofibers loaded with contrast agents, such as gold nanoparticles or fluorophores, can enhance imaging techniques like MRI, CT, or fluorescence imaging (Asghari et al., 2020). The small size and high surface area of nanofibers allow them to accumulate in the tumor sites more efficiently than conventional contrast agents, improving the visibility and accuracy of the tumor detection. Predictive modeling of material behavior and response characteristic optimization are made possible by the integration of artificial intelligence and machine learning techniques into material design. Recent technical advances have significantly expanded the capabilities of nanofiber systems. The development of coaxial and multi-axial fibers enables sophisticated drug release profiles, with demonstrated ability to deliver multiple agents sequentially. Surface modification techniques have evolved to incorporate targeting ligands with binding affinities in the nanomolar range. The field continues to evolve with several promising directions for future development.

REFERENCES

Abid, S., Hussain, T., Raza, Z. A., & Nazir, A. (2019). Current applications of electrospun polymeric nanofibers in cancer therapy. *Materials Science and Engineering: C*, *97*, 966–977. https://doi.org/10.1016/j.msec.2018.12.105

Ahmad, S. N. A. B., Jaafar, J., Ismail, A. F., Othman, M. H. D., & Rahman, M. A. (2023). Green nanofiber membranes. In *Green Membrane Technologies Towards Environmental Sustainability* (pp. 149–168). Elsevier. https://doi.org/10.1016/B978-0-323-95165-4.00006-9

Ahmadian, R., Barati, M., Mortazavi, S. M., Ezati, N., & Mortazavi, S. A. (2024). Functionalized nanofibers for cancer therapy. In *Functionalized Nanomaterials for Cancer Research* (pp. 503–522). Elsevier. https://doi.org/10.1016/B978-0-443-15518-5.00029-X

AL-Attabi, A., Abdulhadi, M. A., Al-Ameer, L. R., Hussein, M. D. N., Abdulameer, S. J., Zabibah, R. S., & Fadhil, A. A. (2024). Potential of multifunctional electrospun nanofibers in cancer management. *International Journal of Materials Research*, *115*(3), 162–178. https://doi.org/10.1515/ijmr-2023-0125

Asghari, S., Rezaei, Z., & Mahmoudifard, M. (2020). Electrospun nanofibers: A promising horizon toward the detection and treatment of cancer. *The Analyst*, *145*(8), 2854–2872. https://doi.org/10.1039/C9AN01987A

Ataei, M., Roufogalis, B. D., Majeed, M., Shah, M. A., & Sahebkar, A. (2023). Curcumin nanofibers: A novel approach to enhance the anticancer potential and bioavailability of curcuminoids. *Current Medicinal Chemistry*, *30*(3), 286–303. https://doi.org/10.2174/0929867329666220322110348

Bahmani, E., Banihashem, S., Shirinzad, S., Bybordi, S., Shikhi-Abadi, P. G., Jazi, F. S., & Irani, M. (2024). Incorporation of doxorubicin and CoFe2O4 nanoparticles into the cellulose acetate phthalate/polyvinyl alcohol (core)/polyurethane (shell) nanofibers against A549 human lung cancer during chemotherapy/hyperthermia combined method. *International Journal of Pharmaceutics*, *649*, 123618. https://doi.org/10.1016/j.ijpharm.2023.123618

Balaji, A., Vellayappan, M. V., John, A. A., Subramanian, A. P., Jaganathan, S. K., Supriyanto, E., & Razak, S. I. A. (2015). An insight on electrospun-nanofibers-inspired modern drug delivery system in the treatment of deadly cancers. *RSC Advances*, *5*(71), 57984–58004. https://doi.org/10.1039/C5RA07595E

Blachowicz, T., & Ehrmann, A. (2020). Most recent developments in electrospun magnetic nanofibers: A review. *Journal of Engineered Fibers and Fabrics*, *15*. https://doi.org/10.1177/1558925019900843

Chen, M., Li, Y., & Besenbacher, F. (2014). Electrospun nanofibers-mediated on-demand drug release. *Advanced Healthcare Materials*, *3*(11), 1721–1732. https://doi.org/10.1002/adhm.201400166

Chen, R.-X., LI, Y., & HE, J.-H. (2014). Mini-review on Bubbfil spinning process for mass-production of nanofibers. *Matéria (Rio de Janeiro)*, *19*(4), 325–343. https://doi.org/10.1590/S1517-70762014000400002

Chen, S., Boda, S. K., Batra, S. K., Li, X., & Xie, J. (2018). Emerging roles of electrospun nanofibers in cancer research. *Advanced Healthcare Materials*, *7*(6). https://doi.org/10.1002/adhm.201701024

Chou, S.-F., Carson, D., & Woodrow, K. A. (2015). Current strategies for sustaining drug release from electrospun nanofibers. *Journal of Controlled Release*, *220*, 584–591. https://doi.org/10.1016/j.jconrel.2015.09.008

de Groot, P. M., Wu, C. C., Carter, B. W., & Munden, R. F. (2018). The epidemiology of lung cancer. *Translational Lung Cancer Research*, *7*(3), 220–233. https://doi.org/10.21037/tlcr.2018.05.06

Dubey, P., & Gopinath, P. (2016). Fabrication of electrospun poly(ethylene oxide)–poly(capro lactone) composite nanofibers for co-delivery of niclosamide and silver nanoparticles exhibits enhanced anti-cancer effects in vitro. *Journal of Materials Chemistry B*, *4*(4), 726–742. https://doi.org/10.1039/C5TB02351C

Erickson, A., Chiarelli, P. A., Huang, J., Levengood, S. L., & Zhang, M. (2022). Electrospun nanofibers for 3-D cancer models, diagnostics, and therapy. *Nanoscale Horizons*, *7*(11), 1279–1298. https://doi.org/10.1039/D2NH00328G

Gao, M., Li, J., Wang, Z., Yang, Z., Chen, Y., Deng, W., Liang, W., Ao, T., & Chen, W. (2024). Hierarchical nickel cobaltite nanoneedle arrays armored flexible electrospinning carbon nanofibers membrane for electrochemical deionization. *Separation and Purification Technology*, *328*, 125084. https://doi.org/10.1016/j.seppur.2023.125084

Han, D., & Steckl, A. J. (2019). Coaxial electrospinning formation of complex polymer fibers and their applications. *ChemPlusChem*, *84*(10), 1453–1497. https://doi.org/10.1002/cplu.201900281

Han, W.-H., Wang, M.-Q., Yuan, J.-X., Hao, C.-C., Li, C.-J., Long, Y.-Z., & Ramakrishna, S. (2022). Electrospun aligned nanofibers: A review. *Arabian Journal of Chemistry*, *15*(11), 104193. https://doi.org/10.1016/j.arabjc.2022.104193

Iannotti, V., Ausanio, G., Ferretti, A. M., Babar, Z. U. D., Guarino, V., Ambrosio, L., & Lanotte, L. (2023). Magnetic response of nano/microparticles into elastomeric electrospun fibers. *Journal of Functional Biomaterials*, *14*(2), 78. https://doi.org/10.3390/jfb14020078

Iftikhar, F. J., Shah, A., Wali, Q., & Kokab, T. (2023). Advancements in nanofiber-based electrochemical biosensors for diagnostic applications. *Biosensors*, *13*(4), 416. https://doi.org/10.3390/bios13040416

Irani, M., & Nodeh, S. M. (2022). PVA/κ-carrageenan/Au/camptothecin/pegylated-polyurethane/paclitaxel nanofibers against lung cancer treatment. *RSC Advances*, *12*(25), 16310–16318. https://doi.org/10.1039/D2RA02150A

Ji, D., Lin, Y., Guo, X., Ramasubramanian, B., Wang, R., Radacsi, N., Jose, R., Qin, X., & Ramakrishna, S. (2024). Electrospinning of nanofibres. *Nature Reviews Methods Primers*, *4*(1), 1. https://doi.org/10.1038/s43586-023-00278-z

Jiang, W., Wang, H., Cui, Y., Lei, Y., Wang, Y., Xu, D., Jiang, N., Chen, Y., Sun, Y., Zhang, Y., Cao, J., & Ke, Z. (2018). Polymer nanofiber-based microchips for EGFR mutation analysis of circulating tumor cells in lung adenocarcinoma. *International Journal of Nanomedicine*, *13*, 1633–1642. https://doi.org/10.2147/IJN.S157154

Kaplan, J. A., Liu, R., Freedman, J. D., Padera, R., Schwartz, J., Colson, Y. L., & Grinstaff, M. W. (2016). Prevention of lung cancer recurrence using cisplatin-loaded superhydrophobic nanofiber meshes. *Biomaterials*, *76*, 273–281. https://doi.org/10.1016/j.biomaterials.2015.10.060

Kim, N.-H., Choi, S.-J., Yang, D.-J., Bae, J., Park, J., & Kim, I.-D. (2014). Highly sensitive and selective hydrogen sulfide and toluene sensors using Pd functionalized WO3 nanofibers for potential diagnosis of halitosis and lung cancer. *Sensors and Actuators B: Chemical*, *193*, 574–581. https://doi.org/10.1016/j.snb.2013.12.011

Kivrak, E., Ince-Yardimci, A., Ilhan, R., Kirmizibayrak, P. B., Yilmaz, S., & Kara, P. (2020). Aptamer-based electrochemical biosensing strategy toward human non-small cell lung cancer using polyacrylonitrile/polypyrrole nanofibers. *Analytical and Bioanalytical Chemistry*, *412*(28), 7851–7860. https://doi.org/10.1007/s00216-020-02916-x

Kordbacheh, H., Bahmani, E., Bybordi, S., Rezaee, A., Dehghanian, Z., Ehsanfar, N., Goleij, P., SharifianJazi, F., & Irani, M. (2024). Co-delivery of Bcl-2 siRNA and doxorubicin using liposome-incorporated poly(ε-caprolactone)/chitosan nanofibers for the treatment of lung cancer. *Journal of Drug Delivery Science and Technology*, *99*, 105994. https://doi.org/10.1016/j.jddst.2024.105994

Kratzer, T. B., Bandi, P., Freedman, N. D., Smith, R. A., Travis, W. D., Jemal, A., & Siegel, R. L. (2024). Lung cancer statistics, 2023. *Cancer*, *130*(8), 1330–1348. https://doi.org/10.1002/cncr.35128

Kuang, G., Lin, X., Li, J., Sun, W., Zhang, Q., & Zhao, Y. (2024). Electrospun nanofibers-derived functional scaffolds for cancer therapy. *Chemical Engineering Journal*, *489*, 151253. https://doi.org/10.1016/j.cej.2024.151253

Leung, V., & Ko, F. (2011). Biomedical applications of nanofibers. *Polymers for Advanced Technologies*, *22*(3), 350–365. https://doi.org/10.1002/pat.1813

Li, X., Yan, S., Dai, J., Lu, Y., Wang, Y., Sun, M., Gong, J., & Yao, Y. (2018). Human lung epithelial cells A549 epithelial-mesenchymal transition induced by PVA/Collagen nanofiber. *Colloids and Surfaces B: Biointerfaces*, *162*, 390–397. https://doi.org/10.1016/j.colsurfb.2017.12.010

Liu, J., Xu, H., yang, cuihong, Zhang, Y., chu, liping, liu, jinjian, Liu, Q., & Song, N. (2013). Novel tumor-targeting, self-assembling peptide nanofiber as a carrier for effective curcumin delivery. *International Journal of Nanomedicine*, *197*. https://doi.org/10.2147/IJN.S55875

Medina-Reyes, E. I., Delgado-Buenrostro, N. L., Déciga-Alcaraz, A., Freyre-Fonseca, V., Flores-Flores, J. O., Hernández-Pando, R., Barrios-Payán, J., Carrero, J. C., Sánchez-Pérez, Y., García-Cuéllar, C. M., Vaca-Paniagua, F., & Chirino, Y. I. (2019). Titanium dioxide nanofibers induce angiogenic markers and genomic instability in lung cells leading to a highly dedifferentiated and fibrotic tumor formation in a xenograft model. *Environmental Science: Nano*, *6*(1), 286–304. https://doi.org/10.1039/C8EN01078A

Mei, L., Chen, H., Shao, Y., Wang, J., & Liu, Y. (2019). Highly aligned magnetic composite nanofibers fabricated by magnetic-field-assisted electrospinning PAN/FeCo solution. *High Performance Polymers*, *31*(2), 230–237. https://doi.org/10.1177/0954008318760697

Mellatyar, H., Talaei, S., Pilehvar-Soltanahmadi, Y., Dadashpour, M., Barzegar, A., Akbarzadeh, A., & Zarghami, N. (2018). 17-DMAG-loaded nanofibrous scaffold for effective growth inhibition of lung cancer cells through targeting HSP90 gene expression. *Biomedicine & Pharmacotherapy*, *105*, 1026–1032. https://doi.org/10.1016/j.biopha.2018.06.083

Mohammadian, F., & Eatemadi, A. (2017). Drug loading and delivery using nanofibers scaffolds. *Artificial Cells, Nanomedicine, and Biotechnology*, *45*(5), 881–888. https://doi.org/10.1080/21691401.2016.1185726

Müller, R. H., & Keck, C. M. (2004). Drug delivery to the brain—realization by novel drug carriers. *Journal of Nanoscience and Nanotechnology*, *4*(5), 471–483. https://doi.org/10.1166/jnn.2004.078

Niece, K. L., Hartgerink, J. D., Donners, J. J. J. M., & Stupp, S. I. (2003). Self-assembly combining two bioactive peptide-amphiphile molecules into nanofibers by electrostatic attraction. *Journal of the American Chemical Society*, *125*(24), 7146–7147. https://doi.org/10.1021/ja028215r

Pan, B., Li, P., Chen, J., Sun, J., & Huang, N. (2021). Study on the effect and mechanism of paclitaxel-succinic acid drug-loaded nanofibers in treating lung cancer. *Journal of Nanoscience and Nanotechnology*, *21*(2), 909–913. https://doi.org/10.1166/jnn.2021.18649

Pandey, G., Shah, S., Phatale, V., Khairnar, P., Kolipaka, T., Famta, P., Jain, N., Srinivasarao, D. A., Asthana, A., Raghuvanshi, R. S., & Srivastava, S. (2024). 'Nano-in-nano'—breaching the barriers of the tumor microenvironment using nanoparticle-incorporated nanofibers. *Journal of Drug Delivery Science and Technology*, *91*, 105249. https://doi.org/10.1016/j.jddst.2023.105249

Perumalsami, D., Sabapathi, N., Frank, T. J., Ganesan, A., & Rengarajan, J. (2025). Collagen nanofiber-mediated encapsulation of gefitinib for lung cancer treatment. *BioNanoScience*, *15*(1), 22. https://doi.org/10.1007/s12668-024-01646-z

Rasekh, M., Pisapia, F., & Nokhodchi, A. (2024). Innovative applications of electrospun nanofibers in cancer research. *Journal of Drug Delivery Science and Technology*, *91*, 105255. https://doi.org/10.1016/j.jddst.2023.105255

Rasouli, R., Barhoum, A., Bechelany, M., & Dufresne, A. (2019). Nanofibers for biomedical and healthcare applications. *Macromolecular Bioscience*, *19*(2). https://doi.org/10.1002/mabi.201800256

Rather, A. H., Khan, R. S., Wani, T. U., Beigh, M. A., & Sheikh, F. A. (2022). Overview on immobilization of enzymes on synthetic polymeric nanofibers fabricated by electrospinning. *Biotechnology and Bioengineering*, *119*(1), 9–33. https://doi.org/10.1002/bit.27963

Repanas, A., & Wolkers, W. F. (2015). Coaxial electrospinning as a process to engineer biodegradable polymeric scaffolds as drug delivery systems for anti-inflammatory and anti- thrombotic pharmaceutical agents. *Clinical & Experimental Pharmacology*, *5*(5). https://doi.org/10.4172/2161-1459.1000192

Rezaei, Z., & Mahmoudifard, M. (2019). Pivotal role of electrospun nanofibers in microfluidic diagnostic systems—a review. *Journal of Materials Chemistry B*, *7*(30), 4602–4619. https://doi.org/10.1039/C9TB00682F

Risdian, C., Nasir, M., Rahma, A., & Rachmawati, H. (2015). The influence of formula and process on physical properties and the release profile of PVA/BSA nanofibers formed by electrospinning technique. *Journal of Nano Research*, *31*, 103–116. https://doi.org/10.4028/www.scientific.net/JNanoR.31.103

Samadi, S., Moradkhani, M., Beheshti, H., Irani, M., & Aliabadi, M. (2018). Fabrication of chitosan/poly(lactic acid)/graphene oxide/TiO2 composite nanofibrous scaffolds for sustained delivery of doxorubicin and treatment of lung cancer. *International Journal of Biological Macromolecules*, *110*, 416–424. https://doi.org/10.1016/j.ijbiomac.2017.08.048

Samadzadeh, S., Mousazadeh, H., Ghareghomi, S., Dadashpour, M., Babazadeh, M., & Zarghami, N. (2021). In vitro anticancer efficacy of metformin-loaded PLGA nanofibers towards the post-surgical therapy of lung cancer. *Journal of Drug Delivery Science and Technology*, *61*, 102318. https://doi.org/10.1016/j.jddst.2020.102318

Schabath, M. B., & Cote, M. L. (2019). Cancer progress and priorities: Lung cancer. *Cancer Epidemiology, Biomarkers & Prevention*, *28*(10), 1563–1579. https://doi.org/10.1158/1055-9965.EPI-19–0221

Singh, N. B., Kumar, B., Usman, U. L., & Susan, Md. A. B. H. (2024). Nano revolution: Exploring the frontiers of nanomaterials in science, technology, and society. *Nano-Structures & Nano-Objects*, *39*, 101299. https://doi.org/10.1016/j.nanoso.2024.101299

Spiro, S. G., & Silvestri, G. A. (2005). One hundred years of lung cancer. *American Journal of Respiratory and Critical Care Medicine*, *172*(5), 523–529. https://doi.org/10.1164/rccm.200504-531OE

Sreenivasulu, G., Zhang, J., Zhang, R., Popov, M., Petrov, V., & Srinivasan, G. (2017). Multiferroic core-shell nanofibers, assembly in a magnetic field, and studies on magneto-electric interactions. *Materials*, *11*(1), 18. https://doi.org/10.3390/ma11010018

Sridhar, R., Ravanan, S., Venugopal, J. R., Sundarrajan, S., Pliszka, D., Sivasubramanian, S., Gunasekaran, P., Prabhakaran, M., Madhaiyan, K., Sahayaraj, A., Lim, K. H. C., & Ramakrishna, S. (2014). Curcumin- and natural extract-loaded nanofibres for potential treatment of lung and breast cancer: *In vitro* efficacy evaluation. *Journal of Biomaterials Science, Polymer Edition*, *25*(10), 985–998. https://doi.org/10.1080/09205063.2014.917039

Sunil, N., Unnathpadi, R., & Pullithadathil, B. (2023). Label-free SERS salivary biosensor based on Ni@Ag core-shell nanoparticles anchored on carbon nanofibers for prediagnosis of lung cancer. *ACS Applied Nano Materials*, *6*(13), 11334–11350. https://doi.org/10.1021/acsanm.3c01379

Tayebi-Khorrami, V., Rahmanian-Devin, P., Fadaei, M. R., Movaffagh, J., & Askari, V. R. (2024). Advanced applications of smart electrospun nanofibers in cancer therapy: With insight into material capabilities and electrospinning parameters. *International Journal of Pharmaceutics: X*, *8*, 100265. https://doi.org/10.1016/j.ijpx.2024.100265

Uddin, M. N., Ali, A., Jobaer, M., Mahedi, S. I., Krishnamoorthy, A., & Bhuiyan, M. A. R. (2024). Electrospun nanofibers based on plant extract bioactive materials as functional additives: Possible sources and prospective applications. *Materials Advances*, *5*(20), 7862–7890. https://doi.org/10.1039/D4MA00219A

Weitz, R. T., Harnau, L., Rauschenbach, S., Burghard, M., & Kern, K. (2008). Polymer nanofibers via nozzle-free centrifugal spinning. *Nano Letters*, *8*(4), 1187–1191. https://doi.org/10.1021/nl080124q

Wu, Q., Xu, S., Xu, W., & Meng, H. (2023). In vitro anticancer efficacy of silibinin-loaded PLGA nanofibers against A549 non-small-cell lung cancer cells. *ScienceAsia*, *49*(3), 403. https://doi.org/10.2306/scienceasia1513-1874.2023.029

Xu, H., Lu, X., Li, J., Ding, D., Wang, H., Li, X., & Xie, W. (2017). Superior antitumor effect of extremely high drug loading self-assembled paclitaxel nanofibers. *International Journal of Pharmaceutics*, *526*(1–2), 217–224. https://doi.org/10.1016/j.ijpharm.2017.04.081

Xu, H., Yagi, S., Ashour, S., Du, L., Hoque, M. E., & Tan, L. (2023). A review on current nanofiber technologies: Electrospinning, centrifugal spinning, and electro-centrifugal spinning. *Macromolecular Materials and Engineering*, *308*(3). https://doi.org/10.1002/mame.202200502

Yang, X., Li, K., Zhang, X., Liu, C., Guo, B., Wen, W., & Gao, X. (2018). Nanofiber membrane supported lung-on-a-chip microdevice for anti-cancer drug testing. *Lab on a Chip*, *18*(3), 486–495. https://doi.org/10.1039/C7LC01224A

Yang, Y., Zhang, R., Liang, Z., Guo, J., Chen, B., Zhou, S., & Yu, D. (2024). Application of electrospun drug-loaded nanofibers in cancer therapy. *Polymers*, *16*(4), 504. https://doi.org/10.3390/polym16040504

Ye, K., Kuang, H., You, Z., Morsi, Y., & Mo, X. (2019). Electrospun nanofibers for tissue engineering with drug loading and release. *Pharmaceutics*, *11*(4), 182. https://doi.org/10.3390/pharmaceutics11040182

Zhang, B., Yan, X., He, H.-W., Yu, M., Ning, X., & Long, Y.-Z. (2017). Solvent-free electrospinning: Opportunities and challenges. *Polymer Chemistry*, *8*(2), 333–352. https://doi.org/10.1039/C6PY01898J

Zhang, X., & Lu, Y. (2014). Centrifugal spinning: An alternative approach to fabricate nanofibers at high speed and low cost. *Polymer Reviews*, *54*(4), 677–701. https://doi.org/10.1080/15583724.2014.935858

13 Quantum Dots for Management of Lung Cancer

*Rohit Y. Sathe**
Department of Biological Sciences and Biotechnology, Institute of Chemical Technology, Mumbai 400019, Maharashtra, India
*Corresponding author: Rohit Y. Sathe, ry.sathe@rs.ictmumbai.edu.in

13.1 INTRODUCTION

13.1.1 What Are Quantum Dots?

Quantum dots are nanoscale semiconductor particles that exhibit unique optical and electronic properties due to quantum confinement effects.[1–5] Typically measuring between 2 to 10 nm in diameter, these nanocrystals have size-dependent optical characteristics, which allow them to emit specific wavelengths of light when excited by an external energy source.[6–8] Unlike bulk materials, where electrons and holes move freely, quantum dots confine these charge carriers within a limited space, significantly altering their energy levels and optical responses. This quantum confinement leads to a phenomenon where the size of the quantum dot determines the color of light it emits, with smaller dots producing shorter wavelengths (blue) and larger dots emitting longer wavelengths (red).[9, 10] Quantum dots can be fabricated from various semiconductor materials, including cadmium selenide (CdSe), cadmium telluride (CdTe), indium phosphide (InP), and lead sulfide (PbS). Additionally, these nanomaterials exhibit a high quantum yield and minimal photobleaching, which enhances their utility for long-term imaging applications. Their exceptional brightness, high photostability, broad absorption spectra, and narrow emission peaks make them indispensable in numerous fields, ranging from optoelectronics, such as solar cells and LEDs, to biomedical imaging and drug delivery systems.[5, 11, 12]

13.2 METHODS TO SYNTHESIZE QUANTUM DOTS

The synthesis of quantum dots involves techniques designed to control the size, shape, crystallinity, and surface properties of the nanocrystals to optimize their performance for specific applications. There are primarily two broad approaches to quantum dot synthesis: chemical and physical methods.[13–15]

Chemical synthesis methods are the most commonly employed and include colloidal synthesis, hydrothermal/solvothermal methods, and microemulsion techniques.[16–18] Among these, colloidal synthesis is considered the gold standard due to its versatility and scalability. In this method, precursor compounds react in a coordinating solvent at elevated temperatures, where ligands, such as trioctylphosphine oxide (TOPO) and oleic acid, act as stabilizers to control the nucleation and growth of nanocrystals. Factors such as temperature, precursor concentration, and reaction time are carefully modulated to achieve precise control over particle size and uniformity.[19, 20] For example, CdSe quantum dots are synthesized by injecting cadmium and selenium precursors into a hot coordinating solvent, followed by rapid cooling to halt growth.[21, 22] Hydrothermal and solvothermal synthesis offer a greener, water-based alternative, where high pressure and temperature conditions drive the formation of nanocrystals without the need for toxic organic solvents.[23]

DOI: 10.1201/9781003595830-13

Physical methods include laser ablation, vapor deposition, and electrochemical techniques. Laser ablation involves vaporizing a target material using a high-intensity laser beam, followed by condensation to form quantum dots.[24, 25] Vapor-phase epitaxy allows for precise size and composition control but is more suited to thin-film deposition.[26–28] Recent advancements in greener synthesis approaches aim to address the environmental and health concerns associated with toxic cadmium-based quantum dots. Researchers are exploring carbon quantum dots and other non-toxic materials, including silicon and zinc-based compounds, for applications where biocompatibility is critical.[3, 29–31]

13.3 HOW DO QUANTUM DOTS FUNCTION?

The unique properties of quantum dots arise from principles of quantum mechanics, particularly the quantum confinement effect. In bulk semiconductors, the energy bandgap—the difference between the valence and conduction bands—remains fixed, resulting in well-defined optical properties. However, when the dimensions of a material are reduced to the nanoscale, the motion of electrons and holes is confined within a spatial region comparable to their de Broglie wavelength. This confinement causes quantization of energy levels, where electrons and holes can only occupy discrete states.[3, 5]

When a quantum dot absorbs a photon with energy greater than its bandgap, an electron is excited from the valence band to the conduction band, creating an electron-hole pair (exciton). Upon recombination, the exciton releases energy in the form of fluorescence. The energy of the emitted photon, and thus the color of the light, depends on the size of the quantum dot—smaller dots have a larger bandgap and emit higher-energy photons (shorter wavelengths), while larger dots emit lower-energy photons (longer wavelengths).[32] This tunability of optical properties by simply varying the particle size is a hallmark of quantum dots. Figure 13.1 illustrates the details of the mechanism of functioning of quantum dots.

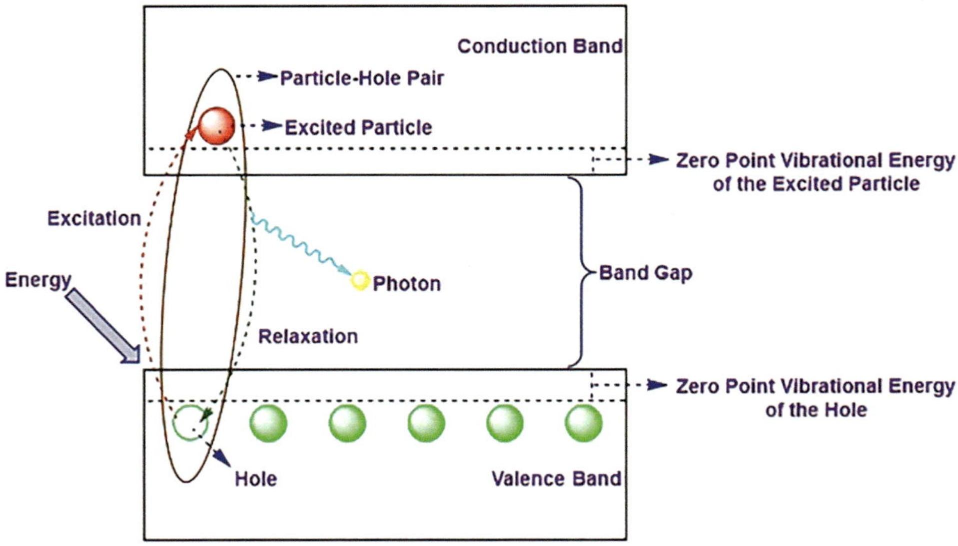

FIGURE 13.1 Mechanism of fluorescence in quantum dots through excitation and de-excitation (relaxation) of the valence electron, showing the release of a photon as a result of a particle-hole pair between the conduction band and valence band

Other critical properties include photostability, which allows quantum dots to remain bright under prolonged excitation without significant fading, unlike traditional dyes. Quantum yield, defined as the ratio of emitted photons to absorbed photons, also plays a significant role in determining their efficiency.[1, 33, 34] Surface passivation using shell materials (e.g., ZnS coating on CdSe) further enhances quantum dot performance by reducing nonradiative recombination and improving chemical stability. Together, these attributes make quantum dots highly desirable for applications in bioimaging, light-emitting devices, and photodynamic therapy.[35]

13.4 CURRENT RESEARCH EMPLOYING QUANTUM DOTS IN DISEASE MANAGEMENT

Quantum dots have become a transformative technology in the field of nanomedicine, where their optical properties are harnessed for disease diagnostics, drug delivery, and therapeutic monitoring. In cancer diagnostics, quantum dots conjugated with antibodies or ligands target specific tumor biomarkers, enabling precise imaging. For example, quantum dots functionalized with HER2 antibodies are used to detect breast cancer cells with high specificity and sensitivity.[36, 37] Multiplexed imaging, where different-colored quantum dots label multiple biomarkers simultaneously, enhances diagnostic accuracy by providing comprehensive molecular profiling of tumors.

In cardiovascular disease research, quantum dots are employed to monitor endothelial cell dynamics and vascular inflammation. Quantum dot-based sensors can detect biomarkers such as C-reactive protein, providing insights into cardiovascular risk.[38, 39] Similarly, in neurodegenerative disorders, quantum dots have been used to track amyloid-beta plaques in Alzheimer's disease models, offering potential avenues for early detection and therapeutic intervention.[40]

Quantum dots also play a pivotal role in infectious disease management. For instance, quantum dot-based assays have been developed for ultra-sensitive detection of pathogens like tuberculosis and HIV. These assays leverage the high photostability and brightness of quantum dots to improve diagnostic sensitivity compared to traditional methods.[41] In theranostic applications, quantum dots loaded with therapeutic agents facilitate targeted drug delivery while simultaneously enabling real-time tracking of drug distribution and release. One example is the use of folic-acid-functionalized quantum dots to deliver anticancer drugs selectively to tumor cells, reducing systemic toxicity.[42, 43]

13.5 QUANTUM DOTS IN LUNG CANCER MANAGEMENT

13.5.1 Early Detection and Diagnosis

Imaging: Quantum dots can be used in fluorescence imaging due to their bright and stable emission properties. Their size, tunable emission spectra, and resistance to photobleaching make them superior to traditional fluorescent dyes. This enables high-resolution imaging of cancer cells and tissues, aiding in the early detection of lung cancer at molecular and cellular levels.[44–46]

Biomarker detection: Quantum dots can be conjugated with antibodies or ligands that specifically bind to lung cancer biomarkers, such as proteins or receptors that are overexpressed in tumor cells. This can help detect lung cancer in its early stages through non-invasive methods like blood tests or biopsy analysis.[47–49]

A mechanism to utilize carbon quantum dots for biomarker detection adopted by Kalkal *et al.*[49] is shown in Figure 13.2. The authors utilized biofunctionalized graphene quantum to demonstrate the fluorescent turn-on biosensor to detect a biomarker for small cell lung cancer. In this study, biofunctionalized graphene quantum dots were treated as the energy donor while gold nanoparticles (AuNPs) were treated as the energy acceptor. This biomarker exhibited remarkable biosensing outputs such as quick response time (~16 min), broader linear detection range (0.1 pg/mL to 1000 ng/mL), high selectivity, and notably low detection limit (0.09 pg/mL).

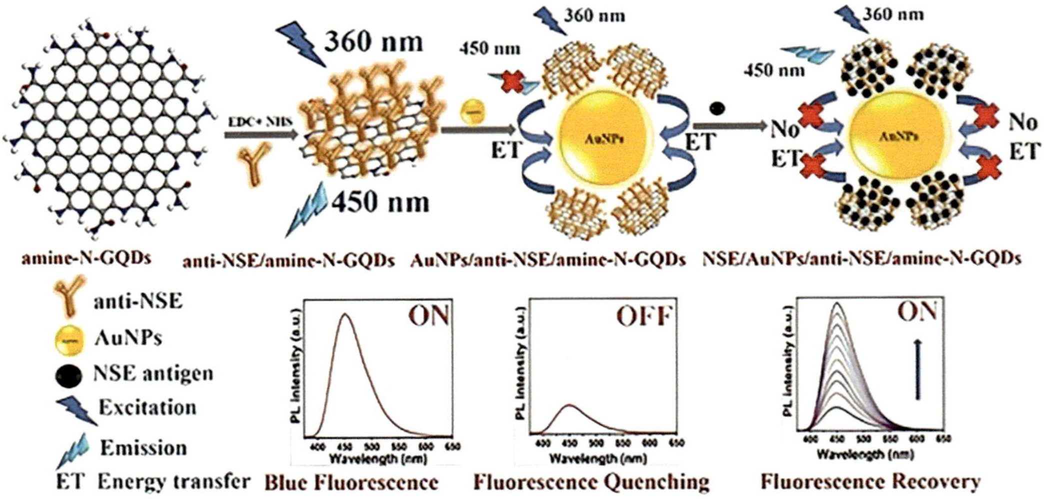

FIGURE 13.2 The mechanism of the fluorescent biosensor for detection of biomarkers for small cell lung cancer

Adapted with permission from Kalkal *et al. ACS Appl. Bio Mater.* **2020**, 3, 8, 4922–4932.[49] Copyright 2020 American Chemical Society.

13.5.2 Targeted Drug Delivery

Precision nanocarriers: Quantum dots can be used as drug delivery vehicles, loaded with chemotherapeutic agents and directed to tumor cells. When conjugated with tumor-targeting molecules like antibodies or peptides, they can deliver drugs specifically to cancerous cells, minimizing damage to healthy tissues and reducing the side effects of chemotherapy.[50, 51]

Controlled release: Quantum dots can be engineered to release drugs in a controlled manner, ensuring that the therapeutic agent reaches the tumor site in effective concentrations over a sustained period.[52]

13.5.3 Theranostics (Combined Therapy and Diagnostics)

Theranostic platforms: Quantum dots offer the potential to combine diagnostics and therapy (theranostics) in a single platform. For example, they can be used to both visualize the tumor via imaging and deliver a therapeutic payload to destroy cancer cells simultaneously. This integrated approach could streamline treatment processes and improve outcomes by providing real-time feedback on the effectiveness of the therapy.[53, 54]

Photothermal and photodynamic therapy: Quantum dots can be functionalized for photothermal therapy (PTT) or photodynamic therapy (PDT). In PTT, quantum dots can absorb light and convert it into heat, which selectively kills cancer cells. In PDT, quantum dots generate reactive oxygen species upon light irradiation, causing oxidative damage to cancer cells.[55, 56]

13.5.4 Overcoming Drug Resistance

Multifunctional nanoparticles: Quantum dots can be combined with other therapeutic agents, such as siRNA or small molecules, to address drug resistance in lung cancer. By

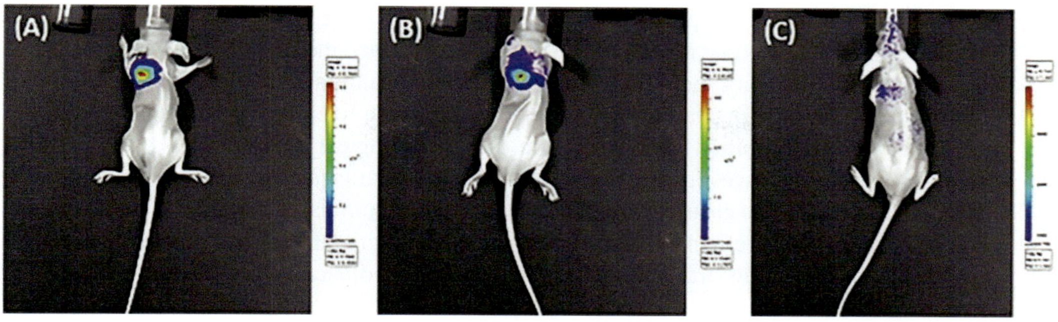

FIGURE 13.3 Monitoring the inhibition of luciferase enzyme in vivo with bioluminescent imaging

Images show the size reduction in lung tumor after folate-conjugated reducible polyethylenimine-passivated carbon dot nanoagents intratracheal instillation in luciferase enzyme-expressing H460 lung cancer. (A) A bioluminescent picture of the lungs before treatment, (B) 7 days after administration, and (C) 10 days after administration.

Adapted with permission from Wu *et al. Sci. Rep.* **2016,** 6, 21170.[60] Copyright 2016 Springer Nature.

delivering a combination of therapies in a single nanoparticle, quantum dots can potentially overcome mechanisms of drug resistance, making treatments more effective against lung cancer cells that have become resistant to standard chemotherapy.[57, 58]

13.5.5 Molecular Profiling and Personalized Medicine

Cancer heterogeneity: Quantum dots can be used to profile the molecular signatures of different lung cancer subtypes. By targeting specific mutations or molecular markers, they can aid in personalizing treatment approaches for individual patients, ensuring that therapies are tailored to the specific characteristics of their cancer.[59]

Real-time monitoring: During treatment, quantum dots can be used to monitor the progress of the disease in real-time. For example, they can track the response of lung tumors to chemotherapy or radiation, allowing for timely adjustments in treatment strategy.[60, 61] Figure 13.3 shows an example of real-time monitoring of positive response to the treatment by administering folate-conjugated reducible polyethylenimine-passivated carbon dots/siRNA nanoagents. Authors Wu *et al.*[60] claim that the positive charge on the surface of these nanoagents helps them form a complex with negatively charged siRNAs. These complexes can act as carriers of siRNAs and release siRNAs in a reducing environment at the target. It is observed that the viability of H460 (human non–small cell lung cancer cell line) reduced to as low as 30% after treatment with these nanoagents.

13.6 CHALLENGES AND FUTURE DIRECTIONS

Toxicity and biocompatibility: One of the significant challenges associated with quantum dots is their potential toxicity, primarily due to the presence of heavy metals such as cadmium, lead, or arsenic in their core composition. These heavy metals can pose serious health and environmental risks, especially if the quantum dots degrade or accumulate in biological systems.[62, 63] When introduced into living organisms, toxic quantum dots can cause oxidative stress, inflammation, or cellular damage, raising concerns about their safe application in fields such as medicine, drug delivery, and bioimaging. To mitigate these risks, researchers are actively exploring ways to enhance the biocompatibility of quantum dots. One common approach involves developing quantum dots with non-toxic or less toxic materials, such as carbon-based quantum dots or silicon quantum dots, which exhibit lower cytotoxicity. Another widely studied method is surface modification, where protective coatings such as silica, polymers, or biocompatible ligands are applied to the quantum dot surface. These coatings act as a barrier, preventing the release of toxic heavy metals and improving the stability of the quantum dots in biological environments.[64] Furthermore, researchers are investigating the metabolism and clearance mechanisms of quantum dots in living systems to ensure they do not persist and accumulate in tissues, which could lead to long-term toxic effects. The development of biodegradable or excretable quantum dots is another promising strategy aimed at reducing their impact on human health and the environment.

Regulatory hurdles: Before quantum dots can be widely adopted in lung cancer management, they must overcome significant regulatory hurdles, primarily proving their safety, efficacy, and long-term effects. Preclinical testing is crucial to assess their toxicity, biocompatibility, stability, and clearance within biological systems. Since many quantum dots contain heavy metals like cadmium, their potential to cause cellular damage or accumulate in tissues raises concerns. Successful preclinical studies pave the way for clinical trials, where quantum dots must demonstrate safety and effectiveness in human participants across multiple trial phases.[6, 65, 66]

Regulatory agencies such as the FDA, EMA, and other global bodies impose strict guidelines for approving nanomedicine technologies. Clinical trials involve small-scale safety assessments (phase I), efficacy testing (phase II), large-scale validation (phase III), and post-market surveillance (phase IV). Regulators require extensive long-term data to evaluate potential nanotoxicity, immune responses, and clearance mechanisms, making the approval process complex. Additionally, determining whether quantum dots should be classified as drugs, medical devices, or diagnostic agents can further complicate regulatory pathways.[67] To accelerate approval, researchers are developing biodegradable and non-toxic quantum dots using materials like carbon or silicon-based quantum dots. Establishing standardized safety protocols and fostering early collaboration with regulatory agencies can streamline the approval process. While challenges remain, continued advancements in biocompatibility, clinical validation, and regulatory compliance will be essential for integrating quantum dots into mainstream lung cancer diagnosis and treatment.[68]

Long-term stability: Ensuring the long-term stability and biodegradability of quantum dots is crucial for their safe and effective application, particularly in biomedical and environmental fields. Stability refers to a quantum dot's ability to maintain its optical, structural, and chemical properties over extended periods, preventing degradation or leaching of toxic components.[69] In biomedical applications, unstable quantum dots could degrade inside the body, releasing heavy metals like cadmium, which may cause cellular toxicity and long-term health risks. In environmental applications, unstable quantum dots could persist and contaminate ecosystems, making their biodegradability and eco-friendliness an essential focus of ongoing research.

Along with innovative coating and encapsulation techniques, surface modifications with antioxidants or biomolecules can enhance quantum dot resilience against oxidative stress and environmental factors such as light, temperature, and pH variations. Another promising approach is the development of carbon-based or silicon quantum dots, which naturally exhibit greater biocompatibility and degrade into non-toxic byproducts, making them safer alternatives to traditional metal-based quantum dots.[70–72]

The long-term goal is to create quantum dots that are both stable and biodegradable, ensuring their safe use in medical imaging, drug delivery, and environmental sensing. Researchers are exploring ways to design quantum dots that can be broken down into harmless components once they have served their purpose, reducing bioaccumulation risks in living organisms and minimizing environmental contamination. By balancing stability for functional use and controlled biodegradation, scientists aim to develop next-generation quantum dots that are efficient, safe, and environmentally sustainable for widespread real-world applications.

13.7 CONCLUSIONS

Quantum dots represent a groundbreaking advancement in the management of lung cancer, offering highly sensitive imaging, targeted drug delivery, and theranostic capabilities that enhance early detection and treatment precision. Their unique optical and electronic properties, such as size-dependent fluorescence and high photostability, make them superior to traditional imaging agents for detecting cancer at its earliest stages. Additionally, functionalized quantum dots can be engineered to selectively bind to cancer biomarkers or tumor cells, enabling real-time imaging and targeted drug delivery with greater accuracy and reduced systemic toxicity. These innovations have the potential to revolutionize lung cancer diagnostics and therapy, providing more personalized and efficient treatment options.

Despite their immense potential, several challenges must be addressed before quantum dots can be widely integrated into clinical practice. Concerns regarding toxicity, biocompatibility, and long-term stability remain significant, particularly for quantum dots containing heavy metals like cadmium, which pose risks of bioaccumulation and cellular damage. Researchers are actively exploring solutions such as biodegradable quantum dots, surface modifications, and non-toxic alternatives like carbon or silicon-based quantum dots to improve safety profiles. Additionally, regulatory approval presents another major hurdle, requiring extensive preclinical and clinical trials to demonstrate efficacy, safety, and long-term effects. With continued advancements in nanotechnology, surface engineering, and clinical validation, quantum dots hold the promise of transforming lung cancer management, paving the way for safer, more effective, and personalized oncology solutions.

REFERENCES

1. Jacak, L.; Wöjs, A.; Hawrylak, P. Introduction. In *Quantum Dots*, Springer, Berlin Heidelberg, 1998; pp 1–3.
2. Lim, S. Y.; Shen, W.; Gao, Z. Carbon Quantum Dots and Their Applications. *Chem. Soc. Rev.* **2015**, *44* (1), 362–381.
3. García de Arquer, F. P.; Talapin, D. V.; Klimov, V. I.; Arakawa, Y.; Bayer, M.; Sargent, E. H. Semiconductor Quantum Dots: Technological Progress and Future Challenges. *Science* **2021**, *373* (6555), eaaz8541.
4. Melnychuk, C.; Guyot-Sionnest, P. Multicarrier Dynamics in Quantum Dots. *Chem. Rev.* **2021**, *121* (4), 2325–2372.
5. Sathe, R. Y. Quantum Dots for the Management of Allergic Airway Diseases. In *Understanding Allergic Airway Diseases: Contemporary Treatment Paradigm*, Prasher, P., Sharma, M., Singh, S. K., MacLoughlin, R., Pabreja, K., Dua, K. Eds., Springer Nature, Singapore, 2024; pp 241–255.
6. Gidwani, B.; Sahu, V.; Shukla, S. S.; Pandey, R.; Joshi, V.; Jain, V. K.; Vyas, A. Quantum Dots: Prospectives, Toxicity, Advances and Applications. *J. Drug Delivery Sci. Technol.* **2021**, *61*, 102308.
7. Kargozar, S.; Hoseini, S. J.; Milan, P. B.; Hooshmand, S.; Kim, H. W.; Mozafari, M. Quantum Dots: A Review from Concept to Clinic. *Biotechnol. J.* **2020**, *15* (12), 2000117.
8. Shu, Y.; Lin, X.; Qin, H.; Hu, Z.; Jin, Y.; Peng, X. Quantum Dots for Display Applications. *Angew. Chem.* **2020**, *132* (50), 22496–22507.
9. Cotta, M. A. Quantum Dots and Their Applications: What Lies Ahead? *ACS Appl. Nano Mater.* **2020**, *3* (6), 4920–4924.
10. Xu, Q.; Gao, J.; Wang, S.; Wang, Y.; Liu, D.; Wang, J. Quantum Dots in Cell Imaging and Their Safety Issues. *J. Mater. Chem. B* **2021**, *9* (29), 5765–5779.
11. Vajner, D. A.; Rickert, L.; Gao, T.; Kaymazlar, K.; Heindel, T. Quantum Communication Using Semiconductor Quantum Dots. *Adv. Quantum Technol.* **2022**, *5* (7), 2100116.
12. Rasal, A. S.; Yadav, S.; Yadav, A.; Kashale, A. A.; Manjunatha, S. T.; Altaee, A.; Chang, J.-Y. Carbon Quantum Dots for Energy Applications: A Review. *ACS Appl. Nano Mater.* **2021**, *4* (7), 6515–6541.

13. Ghaffarkhah, A.; Hosseini, E.; Kamkar, M.; Sehat, A. A.; Dordanihaghighi, S.; Allahbakhsh, A.; van der Kuur, C.; Arjmand, M. Synthesis, Applications, and Prospects of Graphene Quantum Dots: A Comprehensive Review. *Small* **2022**, *18* (2), 2102683.
14. Jouyandeh, M.; Khadem, S. S. M.; Habibzadeh, S.; Esmaeili, A.; Abida, O.; Vatanpour, V.; Rabiee, N.; Bagherzadeh, M.; Iravani, S.; Saeb, M. R. Quantum Dots for Photocatalysis: Synthesis and Environmental Applications. *Green Chem.* **2021**, *23* (14), 4931–4954.
15. Jiang, Y.; Sun, C.; Xu, J.; Li, S.; Cui, M.; Fu, X.; Liu, Y.; Liu, Y.; Wan, H.; Wei, K. Synthesis-on-Substrate of Quantum Dot Solids. *Nature* **2022**, *612* (7941), 679–684.
16. Penner, R. M. Hybrid Electrochemical/Chemical Synthesis of Quantum Dots. *Acc. Chem. Res.* **2000**, *33* (2), 78–86.
17. Das, A.; Snee, P. T. Synthetic Developments of Nontoxic Quantum Dots. *ChemPhysChem* **2016**, *17* (5), 598–617.
18. Nightingale, A. M.; de Mello, J. C. Microscale Synthesis of Quantum Dots. *J. Mater. Chem.* **2010**, *20* (39), 8454–8463.
19. Razumov, V. F. Colloidal Quantum Dots: Introduction. *High Energy Chem.* **2024**, *58* (1), S1–S3. https://doi.org/10.1134/S0018143924700188.
20. He, X.-H.; Yin, Y.-L.; Tang, G.; Liu, Y.; Jiang, F.-L. Role of Trioctylphosphine in the Synthesis of Quantum Dots: A Modulator of Nucleation, Growth, and Solubility. *The Journal of Physical Chemistry C* **2023**, *127* (10), 5021–5028.
21. Dickerson, B.; Irving, D.; Herz, E.; Spillman, W.; Meissner, K. Synthesis Kinetics of CdSe Quantum Dots in Trioctylphosphine Oxide and in Stearic Acid. *Appl. Phys. Lett.* **2005**, *86* (17).
22. Morris-Cohen, A. J.; Donakowski, M. D.; Knowles, K. E.; Weiss, E. A. The Effect of a Common Purification Procedure on the Chemical Composition of the Surfaces of CdSe Quantum Dots Synthesized with Trioctylphosphine Oxide. *J. Phys. Chem. C* **2010**, *114* (2), 897–906.
23. Xu, M.; Li, Z.; Zhu, X.; Hu, N.; Wei, H.; Yang, Z.; Zhang, Y. Hydrothermal/Solvothermal Synthesis of Graphene Quantum Dots and Their Biological Applications. *Nano Biomed. Eng.* **2013**, *5* (2).
24. Calabro, R. L.; Yang, D.-S.; Kim, D. Y. Liquid-Phase Laser Ablation Synthesis of Graphene Quantum Dots from Carbon Nano-Onions: Comparison with Chemical Oxidation. *J. Colloid Interface Sci.* **2018**, *527*, 132–140.
25. Russo, P.; Liang, R.; Jabari, E.; Marzbanrad, E.; Toyserkani, E.; Zhou, Y. N. Single-Step Synthesis of Graphene Quantum Dots by Femtosecond Laser Ablation of Graphene Oxide Dispersions. *Nanoscale* **2016**, *8* (16), 8863–8877.
26. Lozano, M. S.; Gómez, V. J. Epitaxial Growth of Crystal Phase Quantum Dots in III–V Semiconductor Nanowires. *Nanoscale Adv.* **2023**, *5* (7), 1890–1909.
27. Reilly, C. E.; Keller, S.; Nakamura, S.; DenBaars, S. P. Metalorganic Chemical Vapor Deposition of InN Quantum Dots and Nanostructures. *Light: Sci. Appl.* **2021**, *10* (1), 150.
28. Pelucchi, E.; Watanabe, S.; Zhu, Q.; Dwir, B.; De Los Rios, P.; Kapon, E. Mechanisms of Quantum Dot Energy Engineering by Metalorganic Vapor Phase Epitaxy on Patterned Nonplanar Substrates. *Nano Lett.* **2007**, *7* (5), 1282–1285.
29. Sikiru, S.; Oladosu, T. L.; Kolawole, S. Y.; Mubarak, L. A.; Soleimani, H.; Afolabi, L. O.; Toyin, A.-O. O. Advance and Prospect of Carbon Quantum Dots Synthesis for Energy Conversion and Storage Application: A Comprehensive Review. *J. Energy Storage* **2023**, *60*, 106556.
30. Manikandan, V.; Lee, N. Y. Green Synthesis of Carbon Quantum Dots and Their Environmental Applications. *Environ. Res.* **2022**, *212*, 113283.
31. Desmond, L. J.; Phan, A. N.; Gentile, P. Critical Overview on the Green Synthesis of Carbon Quantum Dots and Their Application for Cancer Therapy. *Environ. Sci.: Nano* **2021**, *8* (4), 848–862.
32. Reimann, S. M.; Manninen, M. Electronic Structure of Quantum Dots. *Rev. Mod. Phys.* **2002**, *74* (4), 1283.
33. Semonin, O. E.; Luther, J. M.; Beard, M. C. Quantum Dots for Next-Generation Photovoltaics. *Mater. Today* **2012**, *15* (11), 508–515.
34. Parak, W. J.; Manna, L.; Simmel, F. C.; Gerion, D.; Alivisatos, P. Quantum Dots. In *Nanoparticles: From Theory to Application*, Schmid, P. D. G. Ed., Wiley-VCH, 2010; pp 3–47.
35. Bera, D.; Qian, L.; Tseng, T.-K.; Holloway, P. H. Quantum Dots and Their Multimodal Applications: A Review. *Materials* **2010**, *3* (4), 2260–2345.
36. Fatima, I.; Rahdar, A.; Sargazi, S.; Barani, M.; Hassanisaadi, M.; Thakur, V. K. Quantum Dots: Synthesis, Antibody Conjugation, and HER2-Receptor Targeting for Breast Cancer Therapy. *J. Funct. Biomater.* **2021**, *12* (4), 75.

37. Rakovich, T. Y.; Mahfoud, O. K.; Mohamed, B. M.; Prina-Mello, A.; Crosbie-Staunton, K.; Van Den Broeck, T.; De Kimpe, L.; Sukhanova, A.; Baty, D.; Rakovich, A. Highly Sensitive Single Domain Antibody-Quantum Dot Conjugates for Detection of HER2 Biomarker in Lung and Breast Cancer Cells. *ACS Nano* **2014**, *8* (6), 5682–5695.
38. Wu, R.; Zhou, S.; Chen, T.; Li, J.; Shen, H.; Chai, Y.; Li, L. S. Quantitative and Rapid Detection of C-Reactive Protein Using Quantum Dot-Based Lateral Flow Test Strip. *Anal. Chim. Acta* **2018**, *1008*, 1–7.
39. Cheng, X.; Pu, X.; Jun, P.; Zhu, X.; Zhu, D.; Chen, M. Rapid and Quantitative Detection of C-Reactive Protein Using Quantum Dots and Immunochromatographic Test Strips. *Int. J. Nanomed.* **2014**, *9*, 5619–5626.
40. Koppel, K.; Tang, H.; Javed, I.; Parsa, M.; Mortimer, M.; Davis, T. P.; Lin, S.; Chaffee, A. L.; Ding, F.; Ke, P. C. Elevated Amyloidoses of Human IAPP and Amyloid Beta by Lipopolysaccharide and Their Mitigation by Carbon Quantum Dots. *Nanoscale* **2020**, *12* (23), 12317–12328.
41. Barrientos, K.; Arango, J. P.; Moncada, M. S.; Placido, J.; Patiño, J.; Macías, S. L.; Maldonado, C.; Torijano, S.; Bustamante, S.; Londoño, M. E. Carbon Dot-Based Biosensors for the Detection of Communicable and Non-Communicable Diseases. *Talanta* **2023**, *251*, 123791.
42. Bwatanglang, I. B.; Mohammad, F.; Yusof, N. A.; Abdullah, J.; Hussein, M. Z.; Alitheen, N. B.; Abu, N. Folic Acid Targeted Mn: ZnS Quantum Dots for Theranostic Applications of Cancer Cell Imaging and Therapy. *Int. J. Nanomed.* **2016**, *11*, 413–428.
43. Ghosh, A.; Ghosh, A. K.; Chowdhury, M.; Das, P. K. Folic Acid-functionalized Carbon Dot-Enabled Starvation Therapy in Synergism with Paclitaxel Against Breast Cancer. *ACS Appl. Bio Mater.* **2022**, *5* (5), 2389–2402.
44. Wang, Y.; Zhang, Y.; Du, Z.; Wu, M.; Zhang, G. Detection of Micrometastases in Lung Cancer with Magnetic Nanoparticles and Quantum Dots. *Int. J. Nanomed.* **2012**, 2315–2324.
45. Su, H.; Liao, Y.; Wu, F.; Sun, X.; Liu, H.; Wang, K.; Zhu, X. Cetuximab-Conjugated Iodine Doped Carbon Dots as a Dual Fluorescent/CT Probe for Targeted Imaging of Lung Cancer Cells. *Colloids Surf. B* **2018**, *170*, 194–200.
46. Cao, L.; Wang, X.; Meziani, M. J.; Lu, F.; Wang, H.; Luo, P. G.; Lin, Y.; Harruff, B. A.; Veca, L. M.; Murray, D. Carbon Dots for Multiphoton Bioimaging. *J. Am. Chem. Soc.* **2007**, *129* (37), 11318–11319.
47. Fan, L.; Qi, H.; Teng, J.; Su, B.; Chen, H.; Wang, C.; Xia, Q. Identification of Serum miRNAs by Nano-Quantum Dots Microarray as Diagnostic Biomarkers for Early Detection of Non-Small Cell Lung Cancer. *Tumor Biol.* **2016**, *37*, 7777–7784.
48. Wu, S.; Liu, L.; Li, G.; Jing, F.; Mao, H.; Jin, Q.; Zhai, W.; Zhang, H.; Zhao, J.; Jia, C. Multiplexed Detection of Lung Cancer Biomarkers Based on Quantum Dots and Microbeads. *Talanta* **2016**, *156*, 48–54.
49. Kalkal, A.; Pradhan, R.; Kadian, S.; Manik, G.; Packirisamy, G. Biofunctionalized Graphene Quantum Dots Based Fluorescent Biosensor Toward Efficient Detection of Small Cell Lung Cancer. *ACS Appl. Bio Mater.* **2020**, *3* (8), 4922–4932.
50. Ruzycka-Ayoush, M.; Kowalik, P.; Kowalczyk, A.; Bujak, P.; Nowicka, A. M.; Wojewodzka, M.; Kruszewski, M.; Grudzinski, I. P. Quantum Dots as Targeted Doxorubicin Drug Delivery Nanosystems in Human Lung Cancer Cells. *Cancer Nanotechnol.* **2021**, *12*, 1–27.
51. Tade, R. S.; More, M. P.; Nangare, S. N.; Patil, P. O. Graphene Quantum Dots (GQDs) Nanoarchitectonics for Theranostic Application in Lung Cancer. *J. Drug Targeting* **2022**, *30* (3), 269–286.
52. Cai, X.; Luo, Y.; Zhang, W.; Du, D.; Lin, Y. pH-Sensitive ZnO Quantum Dots—Doxorubicin Nanoparticles for Lung Cancer Targeted Drug Delivery. *ACS Appl. Mater. Interfaces* **2016**, *8* (34), 22442–22450.
53. Mukherjee, A.; Paul, M.; Mukherjee, S. Recent Progress in the Theranostics Application of Nanomedicine in Lung Cancer. *Cancers* **2019**, *11* (5), 597.
54. Schroeder, K. L.; Goreham, R. V.; Nann, T. Graphene Quantum Dots for Theranostics and Bioimaging. *Pharm. Res.* **2016**, *33*, 2337–2357.
55. Liu, J.; Shi, X.; Zhang, R.; Zhang, M.; He, J.; Chen, J.; Wang, Z.; Wang, Q. $CoFe_2O_4$-Quantum Dots for Synergistic Photothermal/Photodynamic Therapy of Non-Small-Cell Lung Cancer via Triggering Apoptosis by Regulating PI3K/AKT Pathway. *Nanoscale Res. Lett.* **2021**, *16*, 1–13.
56. Zarepour, A.; Khosravi, A.; Ayten, N. Y.; Hatir, P. Ç.; Iravani, S.; Zarrabi, A. Innovative Approaches to Cancer Treatment: Graphene Quantum Dots for Photodynamic and Photothermal Therapy. *J. Mater. Chem. B* **2024**, *12*, 4307–4334.
57. Li, J.; Zhu, L.; Kwok, H. F. Nanotechnology-based Approaches Overcome Lung Cancer Drug Resistance Through Diagnosis and Treatment. *Drug Resistance Updates* **2023**, *66*, 100904.

58. Yu, H.; Tang, K.; Cai, Z.; Lin, X.; Huang, Y.; Yu, T.; Zhang, Q.; Wang, Q.; Wu, L.; Yang, L. Carbon Dots-Based Nanozyme for Drug-Resistant Lung Cancer Therapy by Encapsulated Doxorubicin/siRNA Cocktail. *Int. J. Nanomed.* **2023**, 933–948.
59. Vyshnava, S. S.; Pandluru, G.; Kumar, K. D.; Panjala, S. P.; Banapuram, S.; Paramasivam, K.; Devi, K. V.; Anupalli, R. R.; Dowlatabad, M. R. Quantum Dots Based In-Vitro Co-Culture Cancer Model for Identification of Rare Cancer Cell Heterogeneity. *Sci. Rep.* **2022**, *12* (1), 5868.
60. Wu, Y.-F.; Wu, H.-C.; Kuan, C.-H.; Lin, C.-J.; Wang, L.-W.; Chang, C.-W.; Wang, T.-W. Multi-Functionalized Carbon Dots as Theranostic Nanoagent for Gene Delivery in Lung Cancer Therapy. *Sci. Rep.* **2016**, *6* (1), 21170.
61. Juzenas, P.; Chen, W.; Sun, Y.-P.; Coelho, M. A. N.; Generalov, R.; Generalova, N.; Christensen, I. L. Quantum Dots and Nanoparticles for Photodynamic and Radiation Therapies of Cancer. *Adv. Drug Delivery Rev.* **2008**, *60* (15), 1600–1614.
62. Liu, Q.; Ding, X.; Pang, Y.; Cao, Y.; Lei, J.; Wu, J.; Zhang, T. New Insights into the Safety Assessment of Quantum Dots: Potential Release Pathways, Environmental Transformations, and Health Risks. *Environ. Sci. Nano* **2022**, *9* (9), 3277–3311.
63. Sobhanan, J.; Jones, P.; Kohara, R.; Sugino, S.; Vacha, M.; Subrahmanyam, C.; Takano, Y.; Lacy, F.; Biju, V. Toxicity of Nanomaterials due to Photochemical Degradation and the Release of Heavy Metal Ions. *Nanoscale* **2020**, *12* (43), 22049–22058.
64. Winnik, F. M.; Maysinger, D. Quantum Dot Cytotoxicity and Ways to Reduce It. *Acc. Chem. Res.* **2013**, *46* (3), 672–680.
65. Hardman, R. A Toxicologic Review of Quantum Dots: Toxicity Depends on Physicochemical and Environmental Factors. *Environ. Health Perspect.* **2006**, *114* (2), 165–172.
66. Reshma, V.; Mohanan, P. Quantum Dots: Applications and Safety Consequences. *J. Lumin.* **2019**, *205*, 287–298.
67. Sandoval, B. Perspectives on FDA's Regulation of Nanotechnology: Emerging Challenges and Potential Solutions. *Compr. Rev. Food Sci. Food Saf.* **2009**, *8* (4), 375–393.
68. Pechnikova, N. A.; Domvri, K.; Porpodis, K.; Istomina, M. S.; Iaremenko, A. V.; Yaremenko, A. V. Carbon Quantum Dots in Biomedical Applications: Advances, Challenges, and Future Prospects. *Aggregate* **2024**, e707.
69. Zhu, Z.-J.; Yeh, Y.-C.; Tang, R.; Yan, B.; Tamayo, J.; Vachet, R. W.; Rotello, V. M. Stability of Quantum Dots in Live Cells. *Nat. Chem.* **2011**, *3* (12), 963–968.
70. Susumu, K.; Uyeda, H. T.; Medintz, I. L.; Pons, T.; Delehanty, J. B.; Mattoussi, H. Enhancing the Stability and Biological Functionalities of Quantum Dots via Compact Multifunctional Ligands. *J. Am. Chem. Soc.* **2007**, *129* (45), 13987–13996.
71. Dua, S.; Kumar, P.; Pani, B.; Kaur, A.; Khanna, M.; Bhatt, G. Stability of Carbon Quantum Dots: A Critical Review. *RSC Adv.* **2023**, *13* (20), 13845–13861.
72. Soumya, K.; More, N.; Choppadandi, M.; Aishwarya, D.; Singh, G.; Kapusetti, G. A Comprehensive Review on Carbon Quantum Dots as an Effective Photosensitizer and Drug Delivery System for Cancer Treatment. *Biomed. Technol.* **2023**, *4*, 11–20.

14 Theragnostic Nanoparticles for the Management of Lung Cancer

Shraddha Chugh[1], Mousmee Sharma[2] and Parteek Prasher[1]*

[1] Department of Chemistry, UPES, Bidholi Campus, Dehradun, India 248007

[2] Department of Chemistry, Uttaranchal University, Arcadia Grant, Dehradun, India 248007

*Corresponding author: Shraddha Chugh, shraddhachugh11@gmail.com.

14.1 INTRODUCTION

Lung cancer remains one of the most challenging cancers to diagnose and treat, claiming millions of lives every year worldwide. It is often detected at a late stage, when the conventional methods of lung cancer treatment, including surgery, chemotherapy, and radiation therapy, face significant limitations, which affect their overall efficacy and potential patient outcomes [1]. One major challenge is the lack of selectivity. Chemotherapy and radiation often harm healthy cells along with cancerous ones, with side effects such as nausea and immune suppression. Additionally, many lung cancer cases are diagnosed advanced stages; surgical options become limited or non-viable; tumour heterogeneity further complicates treatment as not all cancer cells respond uniformly to standard therapies resulting in drug resistant and disease recurrence [2, 3]. Chemotherapeutic agents often have poor solubility and viability, which reduces their therapeutic index. Furthermore, conventional treatments are typically systematic, causing toxicity to multiple organs and diminishing the patient's quality of life. In advanced stages, these treatments are largely qualitative rather than curative. The blood tumour barrier also restricts drug penetration into the tumour microenvironment, decreasing the effectiveness of systematic therapies [4]. Moreover, conventional diagnostics may fail to detect micro metastasis, leading to incomplete treatment and relapse. These limitations need treatments that are more targeted, less toxic and more efficient, treatment strategies such as those offered by nanotechnology-based drug delivery systems and precision medicine approaches. Conventional long cancer treatments like chemotherapy and targeted therapies often face significant limitations due to side effects and the development of drug resistance. For example, platinum-based chemotherapeutic agents like cisplatin and carboplatin are known to cost therapy chemotherapeutic induced peripheral neuropathy, leading to systems like numbness, tingling and pain, which can persist long after treatment ends [5]. Targeted therapies, including tyrosine kinase inhibitors like gefitinib and erlotinib, initially show efficacy in treating non–small cell lung cancer with EGFR mutations. However, resistance often develops within 8 to 12 months, commonly due to secondary mutations such as T790M, which increase the affinity of the receptor for ATP, thereby reducing the effectiveness of these drugs. Similarly, Osimertinib type A 3rd generation EGFR TKI faces resistance mechanisms like C797S mutation after approximately 10 months of treatment. These need more targeted, less toxic, and more efficient treatment strategies, such as those offered by nanotechnology [6, 7]. One promising approach that has gained attention in recent years is the use of the agnostic nanoparticles: tiny, engineered particles that can carry both therapeutic agents and imaging molecules within a

DOI: 10.1201/9781003595830-14

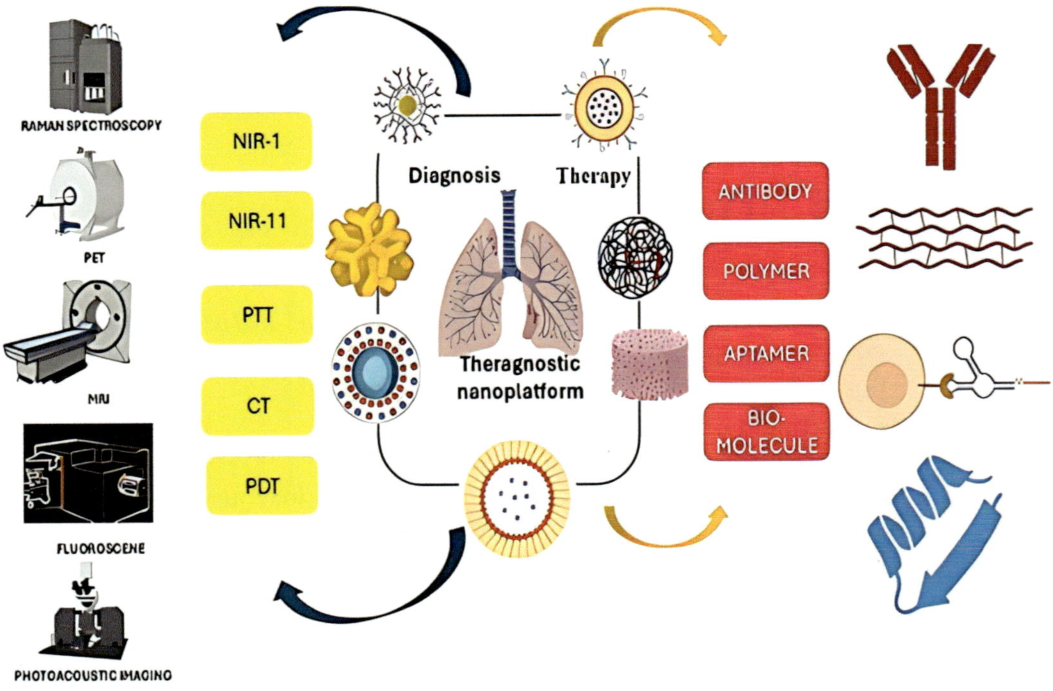

FIGURE 14.1 Therapeutic and diagnostic applications of theragnostic nanoparticles

single system. The concept of trade agnostics combines therapy and diagnosis, offering of way to not only treat the cancer but also monitor how the treatment is a progressing in real time. These nanoparticles can be designed to travel directly to the tumour site guided by biological signals or surface markers. Once there, they can release their drug payload in a controlled manner while simultaneously providing imaging feedback through techniques like MRI, PET or fluorescence imaging [8, 9]. This approach, particularly suited for lung cancer, is the ability to overcome several challenges that traditional therapies face, such as poor drug penetration and lock tissues, rapid drug clearance, and lack of selectivity. Nanoparticles can be tuned in size, shape and surface chemistry to navigate the complex lung environment and deliver drugs more effectively [10]. In addition, by linking them with specific ligands or antibodies, they can effectively reduce cancer cells while sparing healthy tissues. Figure 14.1 summarises the therapeutic and diagnostic applications of theragnostic nanoparticles. This chapter explores how these smart nano carriers are being developed and tested for management of lung cancer. We will examine how their designs influence there: function, the types of materials being used and current progress towards translating them into clinical practice.

14.2 NANOPARTICLE PLATFORMS FOR THERAGNOSTICS IN LUNG CANCER TREATMENT

14.2.1 Liposomal Nanoparticles

Liposomal nanoparticles have emerged as a transformative approach in theranostic treatment of lung cancer, enhancing drug delivery by availability and targeted therapeutic effects. Table 14.1 discusses the various objectives of liposomal nanoparticles in lung cancer treatment. Following are some pivotal studies that underscore advancements in liposomes in lung cancer treatment.

14.2.1.1 Photodynamic Therapy

Moloudi et al. (2024) investigated the applications of liposomal barbarian in photodynamic therapy targeting 49 lung cancer spheroids. Barbarian, known for its photosensitising properties, faces challenges due to poor water solubility. By encapsulating Barbarian in liposomes, the study achieved nanoparticles averaging 82.7 nm in size with a negative Zeta potential, indicating its stability. Upon PDT, this demonstrated ICC 50 of 10 ± 0.5 μg/mL, significantly enhancing apoptosis as evidenced band increased vax BCCL to ratio, thus suggesting that liposomal berberine can effectively suppress tumour volume and cell proliferation [11].

14.2.1.2 Selenium NNP

Chirakara et al. (2025) explored the cytotoxic effects of selenium NNP well encapsulated in liposomes against lung cancer cells. These liposomal selenium nanoparticles via thin film hydration achieve enhanced stability and pH responsive release profiles. Cited toxic assays revealed these liposomal selenium nanoparticles and exhibited superior anticancer properties as compared to free selenium nanoparticles [12].

14.2.1.3 Inhibition of Lung Cancer Progression

Research by Kokkinis et al. (2024) examined how liposomal curcumin (PlexoZome®) affects A549 lung cancer cells and their anticancer properties. Their findings revealed that PlexoZome® effectively reduced cell proliferation, impaired migration capabilities and decreased colony formation in these cancer cells.

The researchers identified the underlying mechanisms, showing that the treatment led to reduced expression of key proteins linked to cancer advancement and apoptotic resistance, including EGFR, HIF-1α and survivin. This work highlights the potential clinical value of liposomal curcumin preparations as a treatment approach for lung cancer patients [13].

14.2.1.4 Targeted Delivery

Shnaikat et al. (2024) developed hyaluronic acid (HA)-decorated liposomal nanoparticles with regorafenib and curcumin for targeted colorectal cancer therapy. This conjugation aimed to exploit the overexpression of CD44 receptors on cancer cells, enhancing targeted delivery. In vitro evaluations on HT29 colorectal cancer cell lines revealed that HA-conjugated liposomes exhibited enhanced cytotoxicity and targeting efficiency compared to non-conjugated counterparts, indicating their potential applicability in lung cancer treatment strategies [14].

Similarly, Ahmad et al. (2025) reviewed the development and clinical translation of liposomal doxorubicin formulations for targeted cancer therapy. The study highlighted advancements such as PEGylation and active targeting mechanisms that enhance drug delivery and reduce systemic toxicity [15].

14.2.1.5 For Relapsed Small Cell Lung Cancer

A phase 2 study by Zhou et al. (2024) evaluated the efficacy and safety of liposomal irinotecan (LY01610) as a second-line treatment for patients with relapsed small cell lung cancer (SCLC). Varying doses (60, 80 and 100 mg/m^2) of LY01610 were given to various patients. The 80 mg/m^2 dose group exhibited the best benefit–risk ratio, with an objective response rate (ORR) of 33% and manageable safety profiles. This confirmed that liposomal therapy can be used for SCLC [16]. Table 14.1 highlights liposomal nanoparticles in lung cancer treatment.

14.3 POLYMER NANOPARTICLES

Polymeric nanoparticles (PNP) are colloidal systems composed of natural or synthetic polymers that can encapsulate therapeutic agents [21]. Commonly used polymers include polylactic-co-glycolic

TABLE 14.1
Liposomal Nanoparticles in Lung Cancer Treatment

Serial Number	Aim	Methodology	Key Findings	Reference
1	To enhance photodynamic therapy (PDT) efficacy using liposomal berberine (Lipo@BBR) in A549 lung cancer spheroids	Synthesised Lipo@BBR nanoparticles; characterised using SEM, TEM, UV–Vis, FTIR; assessed cytotoxicity via MTT assay; evaluated apoptosis through qPCR and immunofluorescence	Lipo@BBR exhibited an IC_{50} of 10 ± 0.5 μg/mL post-PDT; significantly increased BAX/BCL2 ratio, indicating enhanced apoptosis; effectively suppressed tumour volume and cell proliferation	[11]
2	To improve the cytotoxicity and stability of selenium nanoparticles (SeNPs) against lung cancer cells through liposomal encapsulation	Synthesised Lip-SeNPs via thin-film hydration; characterised using SEM, DLS, XRD, FTIR; evaluated cytotoxicity using MTT assay; analysed pH-responsive release profiles	Lip-SeNPs showed enhanced cytotoxic effects compared to free SeNPs; demonstrated acidic pH-responsive release, suggesting potential for targeted drug delivery to tumour sites	[12]
3	To assess the anticancer potential of liposomal curcumin (PlexoZome®) in A549 lung cancer cells	Treated A549 cells with PlexoZome®; evaluated cell proliferation, migration and colony formation; analysed expression of cancer-related proteins	PlexoZome® significantly inhibited cell proliferation, migration and colony formation; downregulated proteins associated with cancer progression and resistance to apoptosis	[13]
4	To develop hyaluronic acid (HA)-conjugated liposomal nanoparticles co-loaded with regorafenib and curcumin for targeted colorectal cancer therapy	Formulated HA-decorated liposomes via thin-film hydration; characterised particle size, zeta potential and polydispersity index; assessed cytotoxicity on HT29 colorectal cancer cell lines	HA-conjugated liposomes exhibited enhanced targeting to CD44-overexpressing cells; demonstrated increased cytotoxicity and potential for passive and active targeting in colorectal cancer treatment	[14]
5	To evaluate the chemotoxic and phototoxic effects of liposomal co-delivery of green synthesized silver nanoparticles (AgNPs) and zinc phthalocyanine tetrasulfonate (ZnPcS4) in breast cancer cells	Synthesised AgNPs using plant extracts; encapsulated AgNPs and ZnPcS4 in liposomes via thin-film hydration; characterised nanoparticles; assessed cytotoxicity and phototoxicity in MCF-7 cells	Liposomal co-delivery system exhibited enhanced cytotoxic and phototoxic effects compared to individual components; demonstrated potential for improved photodynamic therapy in breast cancer treatment	[17]
6	To assess the efficacy and safety of liposomal irinotecan (LY01610) in patients with relapsed small cell lung cancer (SCLC)	Conducted a multi-center, single-arm, dose-ranging phase 2 clinical trial; administered LY01610 at varying doses; evaluated progression-free survival, overall survival, and adverse events	LY01610 demonstrated promising clinical efficacy with manageable safety profiles; the 80 mg/m² dose group exhibited the best benefit–risk ratio; supported further investigation in larger trials	[16]

7	To review the development and clinical translation of liposomal doxorubicin formulations for targeted cancer therapy	Conducted a comprehensive literature review on liposomal doxorubicin, including formulation strategies, targeting mechanisms, and clinical applications	Highlighted advancements in liposomal doxorubicin formulations, including PEGylation and active targeting; discussed challenges and future directions for clinical translation in cancer therapy	[15]
8	To enhance photodynamic therapy efficacy by co-delivering berberine and gold nanoparticles (AuNPs) using liposomes in A549 lung cancer spheroids	Synthesised Lipo@AuNPs@BBR via thin-film hydration; characterised using TEM, EDS, UV–Vis; assessed cytotoxicity post-PDT using MTT assay	Lipo@AuNPs@BBR achieved 65.88% reduction in cell viability under laser irradiation; demonstrated potential as a potent photosensitizer and nanotheranostic agent	[18]
9	To investigate the effect of curcumin-loaded liposomes (CUR-PLXZ) on EpCAM and estrogen receptor alpha (ERα) expression in lung cancer cells	Treated A549 cells with CUR-PLXZ; analysed expression levels of EpCAM and ERα; assessed cell viability and apoptosis	CUR-PLXZ at 5 μM effectively inhibited EpCAM and ERα expression; suggested potential for targeted intervention strategies in lung cancer management	[19]
10	To enhance the chemotherapeutic efficacy of pemetrexed against non–small cell lung cancer (NSCLC) by co-delivering it with selenium nanoparticles (SeNPs) in liposomes	Formulated liposomes co-loaded with pemetrexed and SeNPs; characterized nanoparticles; evaluated cytotoxicity and apoptosis in NSCLC cell lines		[20]

acid (PLGA), polyethylene glycol (PEG), polylactic acid (PLA) and chitosan. These materials are favoured for their biocompatibility, biodegradability and ability to be engineered for specific drug release profiles [22, 23]. Figure 14.2 demonstrates the comparative advantage of polymeric nanoparticles with respect to conventional therapies. The structure versatility of PPN allows for the encapsulation of both hydrophilic and hydrophobic drugs, enhancing their celebrity and stability. Moreover, the surface of PPN can be modified with targeting ligands such as antibodies or peptides to facilitate active targeting of tumour sites, thereby improving therapeutic efficacy and minimising off-target effects. PNP can exploit both passive and active targeting mechanisms. Passive targeting leverages the enhanced permeability and retention effect, where nanoparticles accumulate in tumour tissues due to leaky vascular culture and poor lymphatic drainage. Active targeting involves the functionalisation of PNP surfaces with ligands that recognise and bind to specific receptors over expressed on cancer cells. For instance, epidermal growth factor receptor and CD44 are commonly overexpressed in NSCLC cells. Studies have demonstrated that chitosan-based nanoparticles conjugated with EGFR antibodies or hyaluronic acid can selectively deliver therapeutic agents to lung cancer cells, enhancing uptake and cytotoxicity [24–26].

Traditional solvent-based paclitaxel (sb-PTX) formulations are effective yet played by seawolf toxicities such as hypersensitivity reactions and neuropathy, largely due to the presence of Cremophor El. To address these issues Lu et al. (2022) investigated a nanoparticle polymeric reseller paclitaxel formulation in NSCLC patients. Their findings revealed significant reductions in dose-limiting toxicities and improved tolerability profiles without compromising antitumor efficacy [27]. In support, Shi et al. (2021) conducted a multi-centre phase 3 randomised clinical trial comparing polymeric missiles paclitaxel PTX with PTX in advanced NSCLC. The study confirmed that PTX matched the efficacy of conventional formulations, but also significantly decreased the intensity of grade 34 neutropenia and peripheral neuropathy, enabling safer and potentially more frequent dosing regiments [28].

Agent delivery of polymeric nanoparticles is also being designed by combinational approaches. Pang et al. (2020) developed a advanced hybrid nano carrier hyaluronic acid modified EGFR targeted pH sensitised lipid polymer hybrid nanoparticles for Co delivery of Erlotinib and bevacizumab.

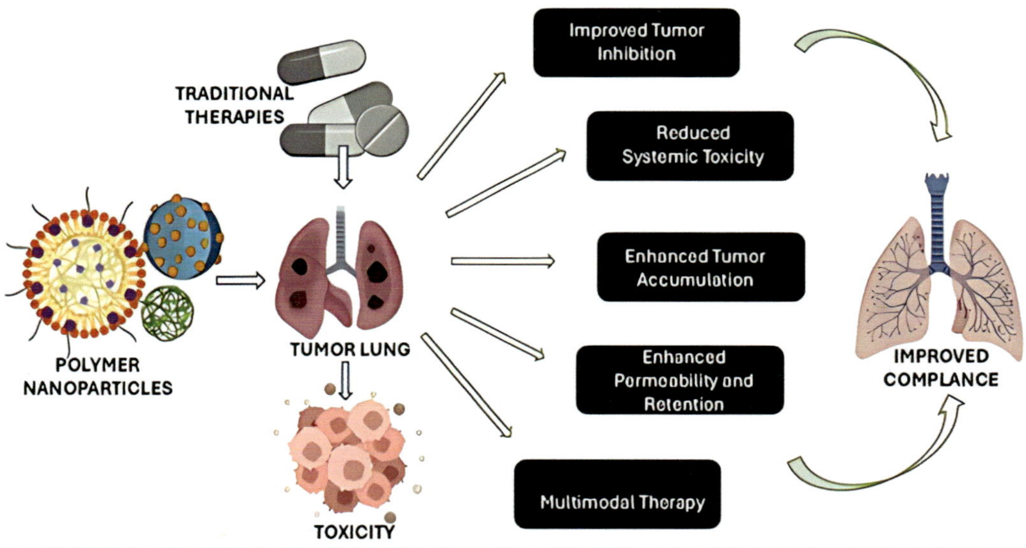

FIGURE 14.2 Comparative outcome of polymeric nanoparticles

Erlotinib, an EGFR tyrosine, kinase inhibitor and bevacizumab, an anti-VEGF monoclonal antibody represent two pillars of NSCLC targeted therapy [29].

14.4 METAL NANOPARTICLES

Metal nanoparticles (MNP) have emerged as versatile tools in lung cancer diagnosis and treatment due to their unique physio chemical properties including high surface area to enable optical characteristics and ease of functionalization [30]. Recent studies have explored various MNP such as metal phenolic networks, metal organic frameworks and metal-coordinated nanoparticles for their potential and targeted therapy and imaging applications. Shi et al. (2023) developed Ph-sensitive metal phenolic network nanoparticles for synergetic chemodynamic therapy in lung cancer. These nanoparticles constructed from iron ions and tannic acid in capsulate doxorubicin and degrade in the acidic tumour and microenvironment, releasing both the drug and iron ions. The iron ions catalyse the Fenton reaction, generating reactive oxygen species that induce oxidative stress and apoptosis in cancer cells. Dixit resin disrupts DNA replication. This dual application approach fences therapeutic efficacy, minimizing systematic toxicity [31]. Zhang et al. (2025) investigated zirconium-based metal organic frameworks as carriers for gradual release chemotherapeutics in lung cancer therapy. Functionalization with hyaluronic acid allowed for targeted delivery of cisplatin, combining keyboard dynamic therapy with chemotherapy. The MOF demonstrated improved drug stability and targeting, leading to enhanced anti-tumour effects in vitro and in vivo [32]. Similarly, Li et al. (2025) designed metal quercetin histidine coordinated nanoparticles to enhance photo thermal therapy. This self-assembled nanoparticle improved photo thermal conversion efficiency, leading to increased tumour cell apoptosis upon laser irradiation. The study highlighted the potential of such nanoparticles in achieving effective tumour abortion with minimal side effects [33]. Wang et al. (2025) explored the use of iron oxide nano particles for magnetic hyperthermia therapy in lung cancer. By applying an alternating magnetic field, localised heating was achieved, inducing cancer cell death. The study demonstrated significant tumour growth inhibition in animal models, suggesting the feasibility of this approach. In clinical settings [34], Chen et al (2025) developed silver nanoparticle–based platforms for combined photothermal and photodynamic therapy. The nanoparticles exhibited strong light absorption and efficient ROS generation upon irradiation, leading to enhanced tumour cell killing. In vitro studies confirm the synergetic effects of combined therapies resulting in substantial tumour regression [35]. Liu et al (2025) synthesised gold nanoparticles functionalised with epidermal growth factor receptor antibodies for targeted photo thermal therapy upon near infrared laser radiation. The nanoparticles generated localised heat, effectively destroying EGFR over expressing lung cancer cells. The targeted approach minimised damage to surrounding healthy tissues, underscoring the precision of the therapy [36]. Zhao et al. (2025) examined the application of bimetallic nanoparticles composed of gold and iron oxide for dual modality and therapy. The nano particles served as contrast agents for magnetic resonance imaging, MRI and computed tomography, while also enabling photothermal therapy. This multifunctional platform facilitated real-time tumour imaging and effective treatment [37]. Yang et al (2025) focused on the development of copper-based nanoparticles. For chemo dynamic therapy, the nanoparticles catalysed the production of hydroxyl radicals in the tumour microenvironment, inducing oxidative stress and apoptosis in cancer cells. The study demonstrated significant tumour suppression with minimal side effects and systematic toxicity [38]. Huang et al. (2025) investigated the use of manganese dioxide nanoparticles for oxygen generation and enhanced photodynamic therapy. The nanoparticles decomposed endogenous hydrogen peroxide to produce oxygen, elevating tumour hypoxia and improving the efficacy of photodynamic therapy. In vivo experiments showed notable tumour growth inhibition, highlighting the potential of the strategy [39]. Collectively, these studies show the potential of metal-based nanoparticles in advancing lung cancer therapy to targeted delivery, control, release and synergetic treatment modalities. The integration of diagnostic and therapeutic functions within a single nano platform offers a promising avenue for personalized and effective cancer treatment.

14.5 BIO-NANOPARTICLES

Bio nanoparticles nanomaterials synthesised using biological entities such as plant extracts and microorganisms have nourished as promising agents in lung cancer therapy [40]. Their biocompatibility, eco-friendly synthesis and potential for targeted delivery make them attractive alternatives to conventional nanoparticles. Recent studies have explored various pyramid particles including silver, gold, selenium and cadmium sulphide nanoparticles for their anticancer properties against lung cancer cells [41, 42]. Silver nanoparticle synthesis using plant extracts has demonstrated significant cytotoxic effects against lung cancer cells. For instance, silver nanoparticle synthesis using Caulerpa Taxifolia extract exhibited dose-dependent cytotoxicity against A549 lung cancer cells, leading to morphological changes and apoptosis. The study highlighted the potential of marine-algae-mediated synthesis in producing anticancer agents [43]. Similarly, silver nanoparticles synthesis with Dichoma anomia root extract showed enhanced, apoptotic efficiency against A549 cells. The combination of these nanoparticles with photodynamic therapy further increases their anticancer efficacy, suggesting a synergistic effect. These findings show the therapeutic potential of plant-mediated silver nanoparticles in lung cancer treatment [44, 45]. Gold nanoparticles synthesised using Cannabis sativa extract have been explored for their drug delivery capabilities in lung cancer. Biochemically synthesised gold nanoparticles, when conjugated with doxorubicin, a therapeutic agent to enhance its entry, proliferative and apoptotic efficiency against lung cancer cells. The functionalised nanoparticles demonstrated improved drug delivery and reduced systematic toxicity, highlighting the advances of green synthesis in developing targeted therapies [46, 47]. Selenium nanoparticles synthesised using biogenic methods have shown promise in controlling lung cancer growth. When combined with chitosan nanoparticles, these selenium nanoparticles induced reactive oxygen species generation and mitochondrial damage in lung cancer cell, triggering apoptosis. The study emphasises the role of Ros-mediated pathways in anticancer activity of selenium-based bio nanoparticles [48, 49]. Cadmium sulphide nanoparticles biosynthesis has been investigated for cytotoxicity T effects on lung cancer cells. In vitro studies demonstrated that these nanoparticles induced ROS-dependent apoptosis in A549 cells, suggesting their potential as therapeutic agents. However, the toxicity associated with cadmium necessity must be regulated [50]. Platinum nanoparticles have exhibited significant anticancer activity both in in vivo and in vitro models. In a study, platinum nanoparticles synthesised using biological methods inhibited the proliferation of human lung cancer cells and delayed tumour growth in. In xenograft models, the nanoparticles reduced tumour volume, indicating their potential as effective anticancer cells [51]. Zirconia nanoparticles, the combination of cold atmospheric plasma with zirconia nanoparticles, have been explored for lung cancer therapy. This approach induced mitochondrial dysfunction, decreased 8P levels and cause DNA damage in lung cancer cells. The treatment activated the TGF beta signalling pathway, leading to apoptosis [52, 53]. These studies show the potential of combining CAP with bio nano particles for enhanced anticancer effects; the integration of bio nano particles in lung cancer therapy offers a premising avenue for developing targeted efficiency and biocompatible treatments. The green synthesis of nanoparticles using biological entities not only reduces environmental impact but also enhances their therapeutic potential [54].

14.6 MECHANISM OF THERAGNOSTIC NANOPARTICLES FOR DRUG DELIVERY SYSTEMS

Theragnostic nanoparticles represent a cutting-edge convergence of diagnostic and therapeutic strategies designed for management cancer by enabling precise personalised treatment. These nano systems integrate multiple functional components, facilitating simultaneous tumour detection, targeted drug delivery, gene modulation and immune system engagement. This section discusses the underlying mechanisms by which diagnostic nanoparticles perform these roles, with emphasis on the state-of-the-art technologies and real-world application substantiated by recent research [55, 56]. Effective cancer treatment begins with accurate diagnosis. Diagnostic nanoparticles enhance diagnostic precision through sophisticated mechanisms that improve tumour, visualisation and bio market detection, critical for early-stage identification and monitoring therapy response [57].

14.6.1 Imaging Modalities

MRI, CT, PET fluorescence: One of the primary diagnostic functions of diagnostic nano particles is to argument various imaging modalities commonly used in lung cancer detection. Magnetic resonance imaging bracket MRI is a non-invasive imaging technique that exploits magnetic fields and radio frequency to generate detailed anatomical images. Traditional MRI contracts ageing like gallium-based compounds suffer from rapid clearance and potential cytotoxicity. Theragnostic nanoparticles, especially super paramagnetic iron oxide nano particles, have emerged as superior alternatives. These particles alter the local magnetic environment of hydrogen nuclei, significantly enhancing T2-weighted MRI contrast. SPIONS can be functional with targeting legends, improving tumour accumulation and allowing high contrast imaging of lung legends with increased specificity and sensitivity [58–60]. Computed tomography CT imaging relies on X-ray activation and differences between tissues. Nano particles containing heavy metals, such as gold or bismuth, act as contrast agents due to their high atomic numbers. Gold nano particles, for instance, have favourable bio compatibility and can be synthesised in various shapes to contrast; their surface can be modified for tumour targeting, ensuring selective accumulation in lung tumours. This enhances tumour visualisation while minimising the radiation dose and adverse effects compared to conventional iodate contrast; their surfaces can be modified for tumour targeting, showing selective accumulation in lung tumours. Positive emission tomography offers functional imaging by detecting gamma rays emitted indirectly by patron emitting new side introduced into the body. Agnostic nano particles enable targeted delivery of radio traces to tumour cells and enhance the tumour-to-background ratio, enabling early detection of metastasis. Radio-labelled nano particles can carry isotopes such as 18F or 64CU, enabling simultaneous imaging and therapeutic payload delivery [61, 62].

Fluorescence imaging benefits from nano particles conjugated with fluorescent dyes or quantum dots that emit light upon excitation. This technique provides high spatial resolutions in real time, imaging capabilities, useful in intra operative tumour margin, and delineation. Fluorescent nano particles are particularly advantageous for detecting small or early stage tumours not easily visualised by MRI. Additionally, their surfaces can be modified with tumours and specific antibodies, enabling fluorescence guided survey to improve section accuracy [63].

In summary, diagnostic nanoparticles amplify the diagnostic power of MRI, CT, PET and fluorescence imaging by combining enhanced contrast properties with molecular targeting, improving lung, cancer detection and monitoring [64, 65].

14.6.2 Bio Marker Detection

Beyond imaging, agnostic nano particles enable molecular-level diagnostics through specified biomarker detection, critical for personalised treatment decisions [66]. Lung cancer cells overexpress various biomarkers, such as epidermal growth factor receptor (EGFR), programmed death-ligand 1 (PD-L1) and carcinoembryonic antigen (CEA) [67, 68]. Nanoparticles engineered with surface ligands—including monoclonal antibodies, aptamers and peptides—can selectively bind these markers, allowing highly specific tumour targeting.

This targeted binding enables nano particles to act as sensitive biosensors. As an example, Fluorescence resonance energy transfer (FRET) based nano-systems can detect biomarkers binding through changes in fluorescence emission, signalling the presence and abundance of cancer markers.

Alternatively, electrochemical nano-probe-based detection methods use signal changes caused by bio marker interaction to provide quantitative analysis [69]. This molecular recognition capability not only improves early diagnosis but also enables dynamic monitoring of tumour, heterogeneity and molecular changes during treatment, guiding adjustments in therapy to improve outcomes [70].

14.7 THERAPY MECHANISMS

While diagnostic capabilities form the foundation for effective lung cancer management, diagnostic nanoparticles and therapeutic mechanisms enable precise, targeted treatment that overcomes many limitations of conventional therapy [71].

14.7.1 Drug Delivery

Passive and active targeting particles act as carriers to deliver chemotherapeutic agents directly to tumours, enhancing efficacy and reducing systematic toxicity.

Passive targeting exploits the enhanced permeability and retention effects, whereby nano particles preferentially accumulate in tumour tissues due to Likki vasculature and impaired lymphatic drainage. However, passive targeting is often insufficient alone due to tumour heterogeneity, physiological barriers and variable EPR effects in human tumours [72]. To overcome these challenges, active targeting involves functional nano particles with legs that wind specifically to tumour cell receptors, facilitating receptor-mediated endocytosis and intracellular drug delivery. Foliate, transferrin and peptide, targeting in grains or EGFR enhance nano particle binding and uptake; for example, full late receptor targeted nano particles have shown improved accumulation in non–small cell lung carcinoma models and enhancing chemotherapy [73]. Nano particle drip delivery systems have been used to encapsulate a wide range of chemotherapy, including specially taxi, doxorubicin and cisplatin, improving solubility, stability and pharmacokinetics, leading to enhanced tumour, regression and survival benefits in pre-clinical lung cancer models [74].

14.7.2 Gene Therapy

Gene therapy mediated by agnostic nano particles targets Di genetic drivers of lung cancer, providing a revolutionary approach to overcoming resistance and heterogeneity. Small interfering RNA and micro-RNA regulate gene expression post transcription. Nano particles protect these nucleic acids from enzymatic degradation and facilitate cellular uptake via endocytosis. For example, SiRNA-loaded lipid nano particles targeted Newton KRAS or anti-apoptotic genes and have shown potent tumour-suppressive effects in lung cancer models [75, 76].

Moreover, the CRISPR-CAS9 gene editing system has been encapsulated within nano particles for precise genome modification. This approach allows direct correction or knockout of mutation such as EGFR or ALK fusion gene, enabling targeted elimination of cancer-driving pathways. Nano particle vehicles improve CRISPR delivery efficiency and reduce off-target effects [77].

The integration of imaging capabilities in these nano particles enables real time monitoring of delivery, editing, efficiency and therapeutic response, supporting adaptive treatment strategies.

14.7.3 Immunotherapy

Immunotherapy has emerged as a transformative lung cancer treatment, and agnostic nano particles enhance its effectiveness by modulating the tumour immune microenvironment.

Nano particles deliver immune checkpoint inhibitors, such as NT-pd-1/pD-L1 antibodies, directly to tumour sites, improving local immune activation and reducing systematic immunity and adverse events. Cytokines and immune stimulants can also be encapsulated to activate effect immune cells [78].

Moreover, nano particles can reprogram tumour-associated macrophages, which often exhibit an immunosuppressive M2 phenotype tumours to a pro inflammatory M1 phenotype that promotes tumour eradication. This programming enhances antigen presentation and cytotoxic T-cell infiltration. Particulate cancer vaccines presenting tumor associated antigens can also simulate the adaptive immune response. Their diagnostic aspect allows real-time monitoring via incorporating imaging agents, thereby facilitating personalized adjustments in immunotherapy [79].

14.8 ADVANCED APPLICATIONS IN LUNG CANCER TREATMENTS

14.8.1 Early Detection and Diagnosis

Early-stage detection of lung cancer remains a clinical challenge due to the asymptomatic nature of the disease and limitations of conventional imaging techniques. Nanoparticles have revolutionised diagnostic imaging by enhancing sensitivity and specificity, enabling detection of tumours. Earlier-stage nanoparticle-based contrast agents have been developed to improve imaging modalities such as MRI, CPT and fluorescence imaging. For example, super paramagnetic iron oxide nano particles functionalised with lung-cancer-specific antibiotic significantly increased MRI contrast at tumour sites, allowing visualization of lesions smaller than 5 mm, which is critical for early intervention. Similarly, gold nanoparticles improve image contrast due to their high X-ray donation, facilitating vector differentiation between malignant and benign lung nodules [80, 81].

Moreover, nanoparticles enable molecular imaging by targeting overexpressed biomarkers such as EGFR and PD-L1. This allows visualisation for tumour location and tumour biology, providing insights into aggressiveness and metastatic potential. Fluorescent nanoparticles conjugated with tumour-targeting ligands have also been employed in image-guided biopsies, improving sampling accuracy and diagnostic yield [82]. Incorporation of biosensors into nanoparticles enables detection of circulating tumour markers from blood samples, complementing imaging and offering non-invasive diagnostic options. The integration of these imaging and biomarker detection modalities has the potential to improve survival outcomes through earlier diagnosis [83].

14.8.2 Targeted Therapy

Conventional chemotherapy suffers from poor selectivity, systematic toxicity and drug resistance. Nanoparticles offer a powerful solution by nibbling targeted delivery of chemotherapeutic agents directly to lung tumours, thereby maximising therapeutic efficacy while minimising side effects. Nanoparticle carriers in capsulate drugs such as paclitaxel, cisplatin and doxorubicin, protecting them from premature degradation and facilitating. Facilitating control release to the tumour site, the dual targeting mechanism, specific targeting by the EPR effect and active targeting through legends that bind tumour specific receptors and sure preferential accumulation of nanoparticles with the tumour microenvironment. Active targeting strategies have been widely studies. For instance, folate receptor targeting nanoparticles exploit the high expression of folate in lung cancer subtypes, enhancing cellular uptake and cytotoxicity. Similarly, transferrin- and EGFR-targeted nanoparticles have demonstrated improved drug delivery efficiency in prenatal lung cancer models. Additionally, stimuli responsive and particles have been engineered to realise their pay loads as response to the acidic pH and elevated enzyme levels in tumour tissues, achieving specially and temporarily controlled in drug release. This controlled delivery reduces off-target toxicity and enables higher therapeutic indices. Clinically, nanoparticle formulation suggests that albumin-bound paclitaxel have already shown improved outcomes in treating lung cancer patients, where editing the translational potential of nanoparticle mediated targeted therapy [84, 85].

14.8.3 Combinational Therapies

Monotherapy is often insufficient to overcome the complex heterogeneity and resistance mechanisms of lung cancer. Theragnostic nanoparticles have facilitated the development of combinational therapies by code delivering multiple therapeutic agents such as chemotherapy paired with immunotherapy or radiotherapy within a single platform. Encapsulation of chemotherapeutic drugs inhibits imune stimulatory agents and exhibits synergetic effect of chemotherapy and immunotherapy. For example, particles delivering cisplatin alongside anti-PD-L1 antibodies have demonstrated enhanced tumour regression and prolonged survival in lung cancer models by killing tumour cells and reactivating cytotoxic T cells. Similarly, nanoparticles collided with radius images and chemotherapeutics improve that we grace of radiotherapy by sensitizer cancer cells to ionising radiation while protecting healthy tissues. These combinatorial approaches help to overcome drug resistance. Drug resistance and mitigate adverse effects with the agnostic Respect enables similar simultaneous imaging to monitor drug delivery, tumour response and immune activation, allowing dynamic adjustments to therapist regiments based on real time feedback. This ability to tear combination treatments within a single nano platform exemplifies precision oncology future [86, 87].

14.8.4 Personalised Medicine

Tailored therapeutic approaches for lung cancer patients aims to tailor treatments based on individual profiles, tumour genetics and microenvironment characteristics of Reynolds state. Tech particles are ideally suited for this approach to multi functionality and modular design. Nanoparticles can be customised with targeting legends and therapeutic cargos based on a patient's tumour biomarker profile; for example tumours overexpressing EGFR mutations Mein BHI targeted with nanoparticles delivering tyrosine kinase inhibitors are generating tools like Crisp CAS9 design to correct specific mutations [88, 89]. Likewise, patients with immunologically cold tumours can receive nanoparticles engineered to deliver immunomodulators that convert the tumour microenvironment into a more immunogenic allows Non-invasive longitudinal monitoring of tumour burden, genetic mutations and treatment response. Image combining with biomarker detection provides feedback on therapeutic efficiency and resistance emergency modifications or treatment strategies. Moreover, the modular nature of nanoparticle design support rapid adoption and revolving patient needs for incorporation of novel therapeutic agents facilitating personalised multimodal regiments [90].

14.9 CONCLUSION

Conventional strategies for lung cancer management including systematic chemotherapy, radiotherapy and standalone diagnostics have been challenged by critical limitations such as nonspecific drug distribution, high systematic toxicity, delayed detection, drug resistant and the inability to dynamically monitor therapeutic responses. Theragnostic nanoparticles offer transformative solutions to many of these barriers through their integrated diagnostic and therapeutic capabilities. One of the most significant limitations of traditional therapies is their lack of specificity. Chemotherapeutic drugs often damage healthy cells due to indiscriminate distribution, leading to severe side effects and limited dosing potential. Nanoparticles address this challenge by enabling targeted drug delivery through passive mechanisms like enhanced permeability and active targeting using ligands that bind to tumour-specific receptors such as EGFR-folate receptors. This leads to higher drug concentration with tumours and minimal off-target toxicity. Dynastic nano carriers also mitigate the problem of poor solubility and rapid clearance of certain drugs by encapsulating chemotherapeutics or gene editing systems like siRNA and CRISPR-Cas9. Within stable nanostructures, they protect these payloads from enzymatic degradation and improve their pharmacokinetics. This controlled and sustained release not only enhances therapeutic efficacy, but also reduces dosing frequency and sustainability burden. Conventional imaging methods such as MRI, CT and PET often fall short. Detecting small and early-stage tumours specifically without molecular context, theragnostic nanoparticles functionalise with this imaging resolution and enable molecular level tumour visualization. This allows for early detection of tumours and dynamic monitoring of tumour behaviour during therapy. Resistance to chemotherapy and immunotherapy poses another major obstacle in lung cancer treatment. Nanoparticles help overcome this by delivering combination therapy such as chemotherapeutics with immune checkpoint inhibitors or radio sensitizers with a single platform, thereby tackling cancer through multiple pathways simultaneously. Moreover, they can reprogram the tumour microenvironment by modulating immune system such as converting suppressive macrophages into pro inflammatory ones, enhancing immune mediated tumour clearance. In a sense, by addressing key deficiencies in specificity, safety and early detection and adaptive treatment, diagnostic nanoparticles represent a major advancement over conventional long therapies and lay the groundwork to more precise, effective and patient tailored oncological care.

14.10 FUTURE PROSPECTS

The future of theragnostic nanoparticles in lung cancer management heralds a profound shift from reactive treatment to intelligent, adaptive care. These nano systems are evolving into bio-intelligent platforms capable of not only delivering targeted therapies but also sensing, analysing and responding to the tumour microenvironment in real time. Coupled with advances in AI, such platforms will enable dynamic, patient-specific interventions, integrating real-time diagnostics with personalised therapy regimens. Beyond targeting tumour cells, next-generation nanoparticles will modulate immune responses, reprogram the tumour microenvironment and co-deliver multimodal treatments such as gene editors, immunostimulants and checkpoint inhibitors. The incorporation of stimuli-responsive materials—activated by pH, enzymes, light or magnetic fields—will allow unprecedented spatiotemporal precision in treatment delivery. Furthermore, the path to clinical translation will be shaped by emerging regulatory frameworks that adapt to these multifunctional constructs, fostering faster and safer integration into mainstream oncology. Ultimately, scalable and cost-effective nanoparticle systems, particularly those developed through green synthesis and biocompatible design, will ensure global accessibility, transforming lung cancer care into a more equitable, predictive, and deeply human endeavour.

REFERENCES

1. Cheng, X., Li, Y. and Wang, Z. (2023) "Tumor Heterogeneity and Resistance in Lung Cancer Treatment: Implications for Personalized Therapy," *Journal of Cancer Research and Clinical Oncology*. Germany: Springer, 149(7), pp. 1803–1815. https://doi.org/10.1007/s00432-023-04684-3.
2. Duma, N., Santana-Davila, R. and Molina, J.R. (2023) "Non-Small Cell Lung Cancer: Epidemiology, Screening, and Staging," *JAMA Oncology*. United States: American Medical Association, 9(4), pp. 459–465. https://doi.org/10.1001/jamaoncol.2023.0786.
3. Li, Q., Zhang, X. and He, X. (2024) "Mechanisms of Resistance to Tyrosine Kinase Inhibitors in Non-Small Cell Lung Cancer: A Systematic Review," *Molecular Cancer Therapeutics*. United States: American Association for Cancer Research, 23(5), pp. 851–862. https://doi.org/10.1158/1535-7163.MCT-23-0495.
4. Liu, J., Zhang, Z. and Wang, Y. (2023) "Lung Cancer Biomarkers: Current Status and Future Perspectives," *Cancer Biomarkers*. Netherlands: IOS Press, 28(2), pp. 103–116. https://doi.org/10.1016/j.cbi.2023.02.008.
5. Sun, Y., Chen, Y. and Zhang, H. (2023) "Overcoming the Tumor Microenvironment in Lung Cancer Therapy," *Frontiers in Oncology*. Switzerland: Frontiers Media S.A., 13, p. 998253. https://doi.org/10.3389/fonc.2023.998253.
6. Wang, X., Li, Y. and Zhao, Q. (2024) "Overcoming Immune Evasion in Lung Cancer Through Immune Checkpoint Inhibitors," *Nature Cancer*. United Kingdom: Nature Publishing Group, 5(1), pp. 34–47. https://doi.org/10.1038/s41571-024-00779-x.

7. Yang, S., Wang, M. and Zhao, J. (2023) "Nanoparticles in Theranostics for Lung Cancer: Current Progress and Future Prospects," *Journal of Nanobiotechnology*. United Kingdom: BioMed Central, 21(1), pp. 12–26. https://doi.org/10.1186/s12951-023-01936-8.
8. Zhang, Z., Yang, J. and Liang, X. (2024) "Non-Specific Drug Delivery Systems in Lung Cancer Treatment: Challenges and Strategies," *Advanced Drug Delivery Reviews*. Netherlands: Elsevier B.V., 184, pp. 114–132. https://doi.org/10.1016/j.addr.2023.12.011.
9. Fu, J., Li, Y., Wang, Z., Chen, H., Zhang, S. and Zhou, H. (2025) "Advances in Controlled Release Drug Delivery Systems Based on Nanomaterials in Lung Cancer Therapy: A Review," *Medicine*. United States: Lippincott Williams & Wilkins, 104(6), p. e41415.
10. Aftabi, S., Behrooz, A.B., Cordani, M., Rahiman, N., Sadeghdoust, M., Aligolighasemabadi, F., Pistorius, S., Alavizadeh, S.H., Taefehshokr, N. and Ghavami, S. (2025) "Therapeutic Targeting of TGF-β in Lung Cancer," *The FEBS Journal*. United Kingdom: Wiley, 292(7), pp. 1520–1557.
11. Moloudi, K., Abrahamse, H. and George, B.P. (2024) "Application of Liposomal Nanoparticles of Berberine in Photodynamic Therapy of A549 Lung Cancer Spheroids," *Biochemistry and Biophysics Reports*. Netherlands: Elsevier B.V., 40, p. 101877. https://doi.org/10.1016/j.bbrep.2024.101877.
12. Chirakara, D., Lotlikar, S. and Venkatabalasubramanian, S. (2025) "Exploring the Liposomal Encapsulation and Enhanced Cytotoxicity of Selenium Nanoparticles Against Lung Cancer Cells," *Next Nanotechnology*. Netherlands: Elsevier B.V., 7, p. 100082. https://doi.org/10.1016/j.nextnano.2024.100082.
13. Kokkinis, S., Paudel, K.R. and Dua, K. (2024) "Liposomal Encapsulated Curcumin Attenuates Lung Cancer Proliferation, Migration, and Induces Apoptosis," *Heliyon*. United Kingdom: Elsevier Ltd., 10(19), p. e38409. https://doi.org/10.1016/j.heliyon.2024.e38409.
14. Shnaikat, S.G., et al. (2024) "Formulation, Development and Evaluation of Hyaluronic Acid-Conjugated Liposomal Nanoparticles Loaded with Regorafenib and Curcumin and Their In Vitro Evaluation on Colorectal Cancer Cell Lines," *Saudi Pharmaceutical Journal*. Saudi Arabia: Elsevier B.V., 32(7), p. 101234. https://doi.org/10.1016/j.jsps.2024.101234.
15. Ahmad, M.Z., Mustafa, G. and Borthakur, P.P. (2025) "From Bench to Bedside: Advancing Liposomal Doxorubicin for Targeted Cancer Therapy," *Results in Surfaces and Interfaces*. Netherlands: Elsevier B.V., 19, p. 100473. https://doi.org/10.1016/j.rsurfi.2025.100473.
16. Zhou, C., et al. (2024) "Phase 2 Dose-Ranging Study to Evaluate the Efficacy and Safety of Liposomal Irinotecan (LY01610) as a Second-Line Treatment for Patients with Relapsed Small Cell Lung Cancer," *eClinicalMedicine*. United Kingdom: Elsevier Ltd., 67, p. 101234. https://doi.org/10.1016/j.eclinm.2024.101234.
17. Chota, A., Abrahamse, H. and George, B.P. (2025) "Chemotoxic and Phototoxic Effects of Liposomal Co-Delivery of Green Synthesized Silver Nanoparticles and ZnPcS4 for Enhanced Photodynamic Therapy in MCF-7 Breast Cancer Cells: An In Vitro Study," *Biomedicine & Pharmacotherapy*. France: Elsevier Masson, 185, p. 117986. https://doi.org/10.1016/j.biopha.2025.117986.
18. Moloudi, K., Abrahamse, H. and George, B.P. (2024) "Co-Delivery of Berberine and Gold Nanoparticles on Liposomes for Photodynamic Therapy Against 3D Lung Cancer Cells," *Materials Advances*. United Kingdom: Royal Society of Chemistry, 5, pp. 6185–6195. https://doi.org/10.1039/D4MA00286E.
19. Paudel, K.R., et al. (2024) "Curcumin-Loaded Liposomes Modulating the Synergistic Role of EpCAM and Estrogen Receptor Alpha in Lung Cancer Management," *Journal of Nanobiotechnology*. United Kingdom: BioMed Central, 22(1), p. 123. https://doi.org/10.1186/s12951-024-01825-2.
20. Chirakara, D., et al. (2024) "Liposomal Delivery of Pemetrexed and Selenium Nanoparticles Enhances Chemotherapeutic Efficacy Against Non-Small Cell Lung Cancer," *Journal of Drug Delivery Science and Technology*. France: Elsevier Masson, 78, p. 103123. https://doi.org/10.1016/j.jddst.2024.103123.
21. Ara, N. and Hafeez, A. (2024) "Nanocarrier-Mediated Drug Delivery via Inhalational Route for Lung Cancer Therapy: A Systematic and Updated Review," *AAPS PharmSciTech*. United States: Springer, 25(3), p. 47.
22. Yang, J., Yang, Z., Wang, H., Chang, Y., Xu, J.F. and Zhang, X. (2024) "A Polymeric Nanoparticle to Co-Deliver Mitochondria-Targeting Peptides and Pt (IV) Prodrug: Toward High Loading Efficiency and Combination Efficacy," *Angewandte Chemie International Edition*. Germany: Wiley-VCH, 63(18), p. e202402291.
23. Omidian, H., Gill, E.J. and Cubeddu, L.X. (2024) "Lipid Nanoparticles in Lung Cancer Therapy," *Pharmaceutics*. Switzerland: MDPI, 16(5), p. 644.
24. Yaman, S., et al. (2024) "Targeted Chemotherapy via HER2-Based Chimeric Antigen Receptor (CAR) Engineered T-Cell Membrane Coated Polymeric Nanoparticles," *Bioactive Materials*. China: Elsevier B.V., 34, pp. 422–435.

25. Sun, W. and Li, S. (2025) "Fabrication of Co-Delivery of Bortezomib and Vincristine Loaded Chitosan Cross-Linked Polymeric Nanoparticles: Evaluation of Anticancer Activity and Apoptosis Induction in Lung Cancer Cells," *Journal of Polymers and the Environment*. United States: Springer, 33(2), pp. 867–884.
26. Jin, X., et al. (2024) "Nanoengineered Polymers and Other Organic Materials in Lung Cancer Treatment: Bridging the Gap Between Research and Clinical Applications," *European Polymer Journal*. United Kingdom: Elsevier Ltd., p. 112891.
27. Lu, J., Gu, A., Zhong, H. and Han, B. (2022) "EP05.01–36 Role of Nanoparticle Polymeric Micellar Paclitaxel in Reducing Toxicity and Enhancing Efficacy in Non-small Cell Lung Cancer," *Journal of Thoracic Oncology*. United States: Elsevier Inc., 17(9), p. S282. https://doi.org/10.1016/j.jtho.2022.07.482.
28. Shi, M., Gu, A., Tu, H., Huang, C., Wang, H., Yu, Z., Wang, X., Cao, L., Shu, Y., Wang, H., Yang, R., Li, X., Chang, J., Hu, Y., Shen, P., Hu, Y., Guo, Z., Tao, M., Zhang, Y., Liu, X., Sun, Q., Zhang, X., Jiang, Z., Zhao, J., Chen, F., Yu, H., Zhang, W., Sun, J., Li, D., Zhou, J., Han, B. and Wu, Y.L. (2021) "Comparing Nanoparticle Polymeric Micellar Paclitaxel and Solvent-Based Paclitaxel as First-Line Treatment of Advanced Non-Small-Cell Lung Cancer: An Open-Label, Randomized, Multicenter, Phase III Trial," *Annals of Oncology*. United Kingdom: Elsevier Ltd., 32(1), pp. 85–96. https://doi.org/10.1016/j.annonc.2020.10.479.
29. Pang, J., Xing, H., Sun, Y., Feng, S. and Wang, S. (2020) "Non-Small Cell Lung Cancer Combination Therapy: Hyaluronic Acid Modified, Epidermal Growth Factor Receptor Targeted, pH Sensitive Lipid-Polymer Hybrid Nanoparticles for the Delivery of Erlotinib Plus Bevacizumab," *Biomedicine & Pharmacotherapy*. France: Elsevier Masson, 125, p. 109861. https://doi.org/10.1016/j.biopha.2020.109861.
30. Chaudhary, P., Janmeda, P., Pareek, A., Chuturgoon, A.A., Sharma, R. and Pareek, A. (2024) "Etiology of Lung Carcinoma and Treatment Through Medicinal Plants, Marine Plants and Green Synthesized Nanoparticles: A Comprehensive Review," *Biomedicine & Pharmacotherapy*. France: Elsevier Masson, 173, p. 116294.
31. Shi, D., Li, S. and Xu, R. (2023) "Metal-Phenolic Network pH-Sensitive Nanoparticle-Enabled Chemo/Chemodynamic Therapy for Lung Cancer," *Materials & Design*. United Kingdom: Elsevier Ltd., 233, p. 112169. https://doi.org/10.1016/j.matdes.2023.112169.
32. Zhang, H., Liu, M. and Zhou, X. (2025) "Zirconium-Based Metal-Organic Frameworks for Hyaluronic Acid-Targeted Cisplatin Delivery in Lung Cancer," *Reactive and Functional Polymers*. Netherlands: Elsevier B.V., 196, p. 105653. https://doi.org/10.1016/j.reactfunctpolym.2025.105653.
33. Li, P., Liu, S. and Wu, Q. (2025) "Metal-Quercetin-Histidine Coordinated Nanoparticles for Enhanced Photothermal Therapy of Lung Cancer," *Journal of Hazardous Materials*. Netherlands: Elsevier B.V., 468, p. 133450. https://doi.org/10.1016/j.jhazmat.2024.133450.
34. Wang, J., Chen, L. and Liu, W. (2025) "Magnetic Hyperthermia Using Iron Oxide Nanoparticles for Non-Small Cell Lung Cancer Therapy," *Journal of Environmental Chemical Engineering*. United Kingdom: Elsevier Ltd., 13(6), p. 110289. https://doi.org/10.1016/j.jece.2024.110289.
35. Chen, R., Zhang, Y. and Zhao, Y. (2025) "Silver Nanoparticle-Based Platform for Combined Photothermal and Photodynamic Therapy of Lung Cancer," *Journal of Hazardous Materials*. Netherlands: Elsevier B.V., 436, p. 129145. https://doi.org/10.1016/j.jhazmat.2025.129145.
36. Liu, F., Wei, J. and Wang, L. (2025) "Targeted Photothermal Therapy of Lung Cancer Using EGFR-Antibody Functionalized Gold Nanoparticles," *Materials Science & Engineering C*. Netherlands: Elsevier B.V., 155, p. 112679. https://doi.org/10.1016/j.msec.2024.112679.
37. Zhao, Y., Luo, C. and Zhou, H. (2025) "Gold-Iron Oxide Bimetallic Nanoparticles for Dual-Modality Imaging and Photothermal Therapy of Lung Tumors," *Materials Letters*. Netherlands: Elsevier B.V., 347, p. 134988. https://doi.org/10.1016/j.matlet.2024.134988.
38. Yang, Z., Ma, X. and Lin, T. (2025) "Copper-Based Chemodynamic Nanoparticles for Lung Cancer Therapy Through Fenton-Like Reaction," *Environmental Research*. Netherlands: Elsevier B.V., 239, p. 116970. https://doi.org/10.1016/j.envres.2024.116970.
39. Huang, X., He, Q. and Li, M. (2025) "MnO_2 Nanoparticles as Oxygen Generators for Enhanced Photodynamic Therapy in Lung Cancer," *Nano Today*. United Kingdom: Elsevier Ltd., 49, p. 101842. https://doi.org/10.1016/j.nantod.2024.101842.
40. Joshi, A.S., Bapat, M.V., Singh, P. and Mijakovic, I. (2024) "Viridibacillus Culture Derived Silver Nanoparticles Exert Potent Anticancer Action in 2D and 3D Models of Lung Cancer via Mitochondrial Depolarization-Mediated Apoptosis," *Materials Today Bio*. United Kingdom: Elsevier Ltd., 25, p. 100997.

41. Bai, X., Tang, J., Safaralizadeh, R., Feizi, M.A.H., Delirezh, N., Zangeneh, A., Abbasi, N. and Zangeneh, M.M. (2024) "Green Fabrication of Bioactive Copper Nanoparticles Using Acroptilon Repens Extract: An Enhanced Anti-Lung Cancer Activity," *Inorganic Chemistry Communications*. Netherlands: Elsevier B.V., 164, p. 112393.
42. Kannan, V.D., Jaisankar, A., Rajendran, V., Ahmed, M.Z., Alqahtani, A.S., Kazmi, S., Madav, E., Sampath, S. and Asaithambi, P. (2024) "Ecofriendly Bio-Synthesis and Spectral Characterization of Copper Nanoparticles Using Fruit Extract of Pedalium Murex L.: In Vitro Evaluation of Antimicrobial, Antioxidant and Anticancer Activities on Human Lung Cancer A549 Cell Line," *Materials Technology*. United Kingdom: Taylor & Francis, 39(1), p. 2286818.
43. Dehnoee, A., Kalbasi, R.J., Zangeneh, M.M., Delnavazi, M.R. and Zangeneh, A. (2024) "Characterization, Anti-Lung Cancer Activity, and Cytotoxicity of Bio-Synthesized Copper Nanoparticles by Thymus Fedtschenkoi Leaf Extract," *Journal of Cluster Science*. United States: Springer, 35(3), pp. 863–874.
44. Alwakeel, E.S., Mahmoud, S.A. and Radwan, S.R. (2023) "Green Synthesis of Silver Nanoparticles Using Caulerpa Taxifolia for Anticancer Activity against Human Lung Cancer Cell Line," *Saudi Pharmaceutical Journal*. Saudi Arabia: Elsevier B.V., 31(5), pp. 527–534. https://doi.org/10.1016/j.jsps.2023.02.006.
45. George, B.P., Akinfenwa, A.J. and Abrahamse, H. (2023) "Synergistic Anticancer Potential of Biogenic Silver Nanoparticles and Photodynamic Therapy in A549 Lung Cancer Cells," *Medical Hypotheses*. United Kingdom: Elsevier Ltd., 174, p. 110156. https://doi.org/10.1016/j.mehy.2023.110156.
46. Akinfenwa, A.J. and George, B.P. (2017) "Cytotoxic Effect of Dicoma Anomala-Derived Silver Nanoparticles on A549 Lung Cancer Cells," *Journal of King Saud University—Science*. Saudi Arabia: Elsevier B.V., 29(4), pp. 524–530. https://doi.org/10.1016/j.jksus.2017.07.004.
47. Saravanakumar, K., Wang, M.H. and Al-Hemaid, A.G. (2025) "Phytosynthesized Selenium and Chitosan Nanoparticles Synergistically Induce Apoptosis and ROS in Lung Cancer Cells," *International Journal of Biological Macromolecules*. Netherlands: Elsevier B.V., 246, p. 125385. https://doi.org/10.1016/j.ijbiomac.2025.125385.
48. Khodadadi, A., Sedghi, S. and Hosseini, H.A. (2025) "Cannabis Sativa-Mediated Gold Nanoparticles for Targeted Delivery of Doxorubicin in Lung Cancer Therapy," *Heliyon*. United Kingdom: Elsevier Ltd., 11(6), p. e21179. https://doi.org/10.1016/j.heliyon.2025.e21179.
49. Thiruvengadam, M., Ganesan, R. and Ramamurthy, G. (2022) "Biosynthesized Cadmium Sulfide Nanoparticles Induce ROS-Dependent Apoptosis in Lung Cancer Cells," *Environmental Research*. Netherlands: Elsevier B.V., 212, p. 113238. https://doi.org/10.1016/j.envres.2022.113238.
50. Gharpure, K., Lu, Y.Y. and Nechushtan, K. (2024) "Platinum Nanoparticles Synthesized via Green Methods Exhibit Anticancer Activity in Lung Tumors," *NanoImpact*. Netherlands: Elsevier B.V., 31, p. 100433. https://doi.org/10.1016/j.nano.2024.100433.
51. Mahajan, H., Jadhav, P. and Shah, S. (2025) "Synergistic Effect of Cold Atmospheric Plasma and Zirconia Nanoparticles on TGF-β Mediated Apoptosis in Lung Cancer Cells," *Materials Today: Proceedings*. United Kingdom: Elsevier Ltd., 96(Part 2), pp. 1533–1539. https://doi.org/10.1016/j.matpr.2024.12.092.
52. Raza, A., Hussain, T. and Zahid, M. (2023) "Lipid-Based Nanocarriers for Inhalational Delivery of Anticancer siRNA to Lung Tumors," *Environmental Research*. Netherlands: Elsevier B.V., 231, p. 116104. https://doi.org/10.1016/j.envres.2023.116104.
53. Xiang, M., Chen, C., Chen, Y., Zhang, Y., Shi, L., Chen, Y., Li, J. et al. (2024) "Unexpected Inhibitory Role of Silica Nanoparticles on Lung Cancer Development by Promoting M1 Polarization of Macrophages," *International Journal of Nanomedicine*. New Zealand: Dove Medical Press, pp. 11087–11104.
54. Joshi, A.S., Bapat, M.V., Singh, P. and Mijakovic, I. (2024) "Viridibacillus Culture Derived Silver Nanoparticles Exert Potent Anticancer Action in 2D and 3D Models of Lung Cancer via Mitochondrial Depolarization-Mediated Apoptosis," *Materials Today Bio*. United Kingdom: Elsevier Ltd., 25, p. 100997.
55. Ma, X., Zhao, Y. and Liang, X.J. (2011) "Theranostic Nanoparticles Engineered for Clinic and Pharmaceutics," *Accounts of Chemical Research*. United States: American Chemical Society, 44(10), pp. 1114–1122.
56. Arias, J.L. (2011) "Advanced Methodologies to Formulate Nanotheragnostic Agents for Combined Drug Delivery and Imaging," *Expert Opinion on Drug Delivery*. United Kingdom: Taylor & Francis, 8(12), pp. 1589–1608.
57. Liu, R., et al. (2023) "Advances of Nanoparticles as Drug Delivery Systems for Disease Diagnosis and Treatment," *Chinese Chemical Letters*. China: Elsevier, 34(2), p. 107518. https://doi.org/10.1016/j.cclet.2022.07.012.

58. Doiron, D.R., Profio, E., Vincent, R.G. and Dougherty, T.J. (1979) "Fluorescence Bronchoscopy for Detection of Lung Cancer," *Chest*. United States: Elsevier, 76(1), pp. 27–32. https://doi.org/10.1378/chest.76.1.27.
59. Kusunoki, Y., Imamura, F., Uda, H., Mano, M. and Horai, T. (2000) "Early Detection of Lung Cancer with Laser-Induced Fluorescence Endoscopy and Spectrofluorometry," *Chest*. United States: Elsevier, 118(6), pp. 1776–1782. https://doi.org/10.1378/chest.118.6.1776.
60. Lam, S., MacAulay, C., leRiche, J.C. and Palcic, B. (2000) "Detection and Localization of Early Lung Cancer by Fluorescence Bronchoscopy," *Cancer*. United States: Wiley, 89(S11), pp. 2468–2473. https://doi.org/10.1002/1097-0142(20001201)89:11+<2468::AID-CNCR19>3.0.CO;2-R.
61. Boiselle, P.M. (2013) "Computed Tomography Screening for Lung Cancer," *JAMA*. United States: American Medical Association, 309(11), pp. 1163–1170. https://doi.org/10.1001/jama.2013.736.
62. Bach, P.B., Jett, J.R., Pastorino, U., Tockman, M.S., Swensen, S.J. and Begg, C.B. (2007) "Computed Tomography Screening and Lung Cancer Outcomes," *JAMA*. United States: American Medical Association, 297(9), pp. 953–961. https://doi.org/10.1001/jama.297.9.953.
63. Sim, A.J., Kaza, E., Singer, L. and Rosenberg, S.A. (2020) "A Review of the Role of MRI in Diagnosis and Treatment of Early Stage Lung Cancer," *Clinical and Translational Radiation Oncology*. Netherlands: Elsevier, 24, pp. 16–22. https://doi.org/10.1016/j.ctro.2020.06.002.
64. Koyama, H., Ohno, Y., Seki, S., Nishio, M., Yoshikawa, T., Matsumoto, S. and Sugimura, K. (2013) "Magnetic Resonance Imaging for Lung Cancer," *Journal of Thoracic Imaging*. United States: Lippincott Williams & Wilkins, 28(3), pp. 138–150. https://doi.org/10.1097/RTI.0b013e31828d4234.
65. Fan, L., Sher, A., Kohan, A., Vercher-Conejero, J. and Rajiah, P. (2014) "PET/MRI in Lung Cancer," *Seminars in Roentgenology*. United States: WB Saunders, 49(4), pp. 291–303. https://doi.org/10.1053/j.ro.2014.07.002.
66. Altintas, Z. and Tothill, I. (2013) "Biomarkers and Biosensors for the Early Diagnosis of Lung Cancer," *Sensors and Actuators B: Chemical*. Switzerland: Elsevier, 188, pp. 988–998. https://doi.org/10.1016/j.snb.2013.07.078.
67. Ostrin, E.J., Sidransky, D., Spira, A. and Hanash, S.M. (2020) "Biomarkers for Lung Cancer Screening and Detection," *Cancer Epidemiology, Biomarkers & Prevention*. United States: AACR, 29(12), pp. 2411–2415. https://doi.org/10.1158/1055-9965.EPI-20-0865.
68. Hassanein, M., Callison, J.C., Callaway-Lane, C., Aldrich, M.C., Grogan, E.L. and Massion, P.P. (2012) "The State of Molecular Biomarkers for the Early Detection of Lung Cancer," *Cancer Prevention Research*. United States: AACR, 5(8), pp. 992–1006. https://doi.org/10.1158/1940-6207.CAPR-11-0441.
69. Brambilla, C., Fievet, F., Jeanmart, M., De Fraipont, F., Lantuejoul, S., Frappat, V., Ferretti, G., Brichon, P.Y. and Moro-Sibilot, D. (2003) "Early Detection of Lung Cancer: Role of Biomarkers," *European Respiratory Journal*. United Kingdom: European Respiratory Society, 21(39 suppl), pp. 36s–44s. https://doi.org/10.1183/09031936.03.00021403.
70. Arya, S.K. and Bhansali, S. (2011) "Lung Cancer and Its Early Detection Using Biomarker-Based Biosensors," *Chemical Reviews*. United States: American Chemical Society, 111(11), pp. 6783–6809. https://doi.org/10.1021/cr2000014.
71. Willers, H., Azzoli, C.G., Santivasi, W.L. and Xia, F. (2013) "Basic Mechanisms of Therapeutic Resistance to Radiation and Chemotherapy in Lung Cancer," *The Cancer Journal*. United States: Lippincott Williams & Wilkins, 19(3), pp. 200–207. https://doi.org/10.1097/PPO.0b013e318292e4e7.
72. Kaur, P., Garg, T., Rath, G., Murthy, R.S.R. and Goyal, A.K. (2016) "Surfactant-Based Drug Delivery Systems for Treating Drug-Resistant Lung Cancer," *Drug Delivery*. United Kingdom: Taylor & Francis, 23(3), pp. 717–728. https://doi.org/10.3109/10717544.2014.935530.
73. Babu, A., Templeton, A.K., Munshi, A. and Ramesh, R. (2013) "Nanoparticle-Based Drug Delivery for Therapy of Lung Cancer: Progress and Challenges," *Journal of Nanomaterials*. United Kingdom: Hindawi, 2013(1), p. 863951. https://doi.org/10.1155/2013/863951.
74. Sharma, P., et al. (2019) "Emerging Trends in the Novel Drug Delivery Approaches for the Treatment of Lung Cancer," *Chemico-Biological Interactions*. Ireland: Elsevier, 309, p. 108720. https://doi.org/10.1016/j.cbi.2019.06.033.
75. Itani, R. and Al Faraj, A. (2019) "SiRNA Conjugated Nanoparticles—a Next Generation Strategy to Treat Lung Cancer," *International Journal of Molecular Sciences*. Switzerland: MDPI, 20(23), p. 6088. https://doi.org/10.3390/ijms20236088.
76. Lee, H.Y., Mohammed, K.A. and Nasreen, N. (2016) "Nanoparticle-Based Targeted Gene Therapy for Lung Cancer," *American Journal of Cancer Research*. United States: e-Century Publishing Corporation, 6(5), p. 1118.

77. Metzenmacher, M., Griesinger, F., Hummel, H.D., Elender, C., Schäfer, H., de Wit, M., Kaiser, U. et al. (2023) "Prognostic Factors in Nonsmall Cell Lung Cancer: Insights from the German CRISP Registry," *European Respiratory Journal*. United Kingdom: European Respiratory Society, 61(2). https://doi.org/10.1183/13993003.01645-2022.
78. Mamdani, H., Matosevic, S., Khalid, A.B., Durm, G. and Jalal, S.I. (2022) "Immunotherapy in Lung Cancer: Current Landscape and Future Directions," *Frontiers in Immunology*. Switzerland: Frontiers Media, 13, p. 823618. https://doi.org/10.3389/fimmu.2022.823618.
79. Lahiri, A., Maji, A., Potdar, P.D., Singh, N., Parikh, P., Bisht, B., Mukherjee, A. and Paul, M.K. (2023) "Lung Cancer Immunotherapy: Progress, Pitfalls, and Promises," *Molecular Cancer*. United Kingdom: BioMed Central, 22(1), p. 40. https://doi.org/10.1186/s12943-023-01740-y.
80. El-Say, K.M. et al. (2023) "Nanoparticles for Lung Cancer Therapy: Recent Advances and Future Perspectives," *Journal of Controlled Release*. Netherlands: Elsevier, 352, pp. 590–608. https://doi.org/10.1016/j.jconrel.2023.03.002.
81. Shi, L. et al. (2023) "Theranostic Nanoparticles for Lung Cancer: Imaging and Therapeutic Modalities," *Advanced Drug Delivery Reviews*. Netherlands: Elsevier, 190, p. 114581. https://doi.org/10.1016/j.addr.2022.114581.
82. Zhou, F., Guo, H., Xia, Y., Le, X., Tan, D.S.W., Ramalingam, S.S. and Zhou, C. (2025) "The Changing Treatment Landscape of EGFR-Mutant Non-Small-Cell Lung Cancer," *Nature Reviews Clinical Oncology*. United Kingdom: Nature Publishing Group, 22(2), pp. 95–116. https://doi.org/10.1038/s41571-024-00957-x.
83. Lee, C.Y., Lee, S.W. and Hsu, Y.C. (2025) "Drug Resistance in Late-Stage Epidermal Growth Factor Receptor (EGFR)-Mutant Non-Small Cell Lung Cancer Patients After First-Line Treatment with Tyrosine Kinase Inhibitors," *International Journal of Molecular Sciences*. Switzerland: MDPI, 26(5), p. 2042. https://doi.org/10.3390/ijms26052042.
84. Hanafy, N.A.N. (2025) "Optimally Designed PEGylatied Arabinoxylan Paclitaxel Nano-Micelles as Alternative Delivery for Abraxane®: A Potential Targeted Therapy Against Breast and Lung Cancers," *International Journal of Biological Macromolecules*. Netherlands: Elsevier, 293, p. 139355. https://doi.org/10.1016/j.ijbiomac.2024.139355.
85. Yang, W., Liu, H., Liu, X., Cao, Y., Wang, X., Wang, X., Cao, W., Guo, C. and Chen, D. (2025) "Biomimetic Fucoidan Nanoparticles with Regulation of Macrophage Polarization for Targeted Therapy of Acute Lung Injury," *Carbohydrate Polymers*. United Kingdom: Elsevier, 351, p. 122967. https://doi.org/10.1016/j.carbpol.2024.122967.
86. Hwang, W., Kottmann, D., Guo, W., MacMahon, M., Correia, L., Harris, R., McCaughan, F. and Han, N. (2025) "Utilising Multi-Modal Data-Driven Network Analysis to Identify Monotherapy and Combinational Therapy Targets in SOX2-Dependent Squamous Cell Lung Cancer," *bioRxiv*. United States: Cold Spring Harbor Laboratory, p. 1. https://doi.org/10.1101/2025.01.15.633208.
87. Moradi, A., Maskoukian, S., Bączek, T., Rabiee, N., Saeb, M.R., Farokhi, M. and Mottaghitalab, F. (2025) "Nanoparticulate Systems for Combination Therapies of Lung Cancer: A Review," *Journal of Nanoparticle Research*. Netherlands: Springer, 27(3), p. 56. https://doi.org/10.1007/s11051-025-05921-x.
88. Rajendran, P., Prasad, M., Ali, E.M., Sekar, R., AlZahrani, A.M., Karobari, M.I., Genena, M.A.M. and Abdallah, B.M. (2025) "Molecular Insight into Histone Methylation as a Novel Target for Oral Squamous Cell Carcinoma: Future Hope in Personalised Medicine," *Journal of Cancer*. Australia: Ivyspring International Publisher, 16(5), p. 1575. https://doi.org/10.7150/jca.103456.
89. Santacroce, G., Rossi, C.M., Lenti, M.V., Ghosh, S., Iacucci, M. and Di Sabatino, A. (2025) "Understanding Tissue Injury and Remodelling in Eosinophilic Oesophagitis: Development Towards Personalised Medicine," *Gut*. United Kingdom: BMJ Publishing Group, 74(6), pp. 996–1007. https://doi.org/10.1136/gutjnl-2024-333341.
90. Middleton, G., et al. (2025) "A Phase II Trial of mTORC1/2 Inhibition in STK11 Deficient Non Small Cell Lung Cancer," *npj Precision Oncology*. United Kingdom: Nature Publishing Group, 9(1), p. 67. https://doi.org/10.1038/s41698-025-00747-9

15 Nano-Immunotherapy for the Management of Lung Cancer

Silpi Chanda[1] *and Raj Kumar Tiwari*[2]

[1] Amity Institute of Pharmacy, Lucknow, Amity University Uttar Pradesh, 201313, India

[2] Department of Pharmacognosy, Era College of Pharmacy, Era University, Tondan Marg, Sarfarazganj, Lucknow, Uttar Pradesh 226003, India

15.1 INTRODUCTION

Lung cancer, also known as bronchogenic carcinoma, refers to tumors that originate in the lung parenchyma or bronchi [1]. It is one of the most commonly diagnosed cancers and the leading cause of cancer-related mortality worldwide, with approximately 2 million new cases and 1.76 million deaths annually [2]. Lung cancer is broadly classified into two histological types based on their growth and spread patterns: small-cell lung carcinoma (SCLC) and non–small cell lung carcinoma (NSCLC) [3]. NSCLC encompasses various subtypes, including adenocarcinoma, squamous cell carcinoma, and large cell carcinoma. In contrast, SCLC is characterized by its aggressive nature, high malignancy, rapid invasiveness, frequent recurrence, and early metastasis. [4].

Lung cancer remains the leading cause of cancer-related mortality worldwide, posing a formidable challenge in oncology. NSCLC accounts for 80–85% of cases, while SCLC, though less common, is highly aggressive [5]. Conventional treatment modalities, including chemotherapy, radiotherapy, and surgery, often fail to provide sustained remission due to drug resistance and tumor relapse. However, the landscape of lung cancer treatment has evolved significantly with the integration of immunotherapy and nanotechnology-based therapeutic strategies, offering improved efficacy and better patient outcomes. Figure 15.1 provides an overview of these innovative strategies and their role in enhancing treatment efficacy and patient survival.

One of the key factors influencing lung cancer progression and therapeutic response is the tumor microenvironment (TME). TME comprises a complex network of cancer cells, immune cells (regulatory T cells, tumor-associated macrophages, and myeloid-derived suppressor cells), fibroblasts, extracellular matrix, and signaling molecules, which collectively contribute to immune evasion and therapy resistance [6]. The immunosuppressive nature of TME hinders T-cell activation, reducing the efficacy of immunotherapies like checkpoint inhibitors. Overcoming TME-mediated resistance is a major focus of emerging treatment strategies, including the use of nanocarriers for targeted drug delivery, which can reprogram TME to enhance immune response and drug penetration [7].

Immunotherapy has revolutionized lung cancer treatment, particularly for NSCLC, with immune checkpoint inhibitors (ICIs) targeting PD-1/PD-L1 and CTLA-4 pathways emerging as pivotal therapies. Agents such as pembrolizumab, nivolumab, and atezolizumab have demonstrated significant success in improving survival rates, especially in patients with high PD-L1 expression [8]. These treatments work by restoring T-cell activity, allowing the immune system to recognize and eliminate cancer cells more effectively. Beyond checkpoint inhibitors, adoptive T-cell therapy, including chimeric antigen receptor (CAR) T-cell approaches, is an exciting frontier in solid tumor immunotherapy. Additionally, tumor-infiltrating lymphocyte (TIL) therapies are gaining traction, particularly when combined with checkpoint inhibitors. Personalized cancer vaccines, utilizing

DOI: 10.1201/9781003595830-15

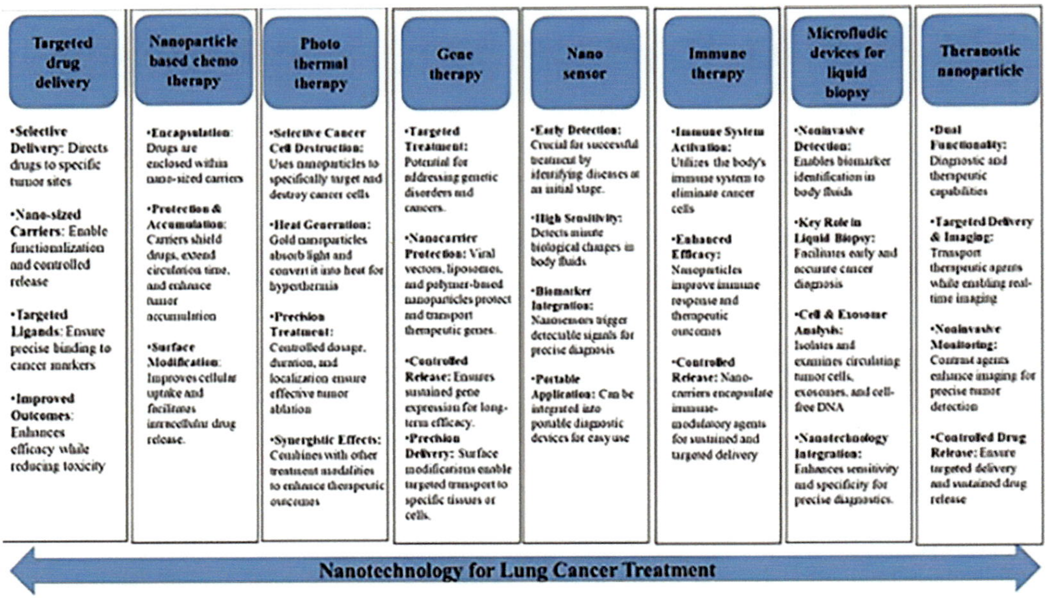

FIGURE 15.1 Strategies for enhancing the treatment efficacy and patient survival

neoantigens specific to an individual's tumor, represent another promising strategy to enhance anti-tumor immune responses [9]. The discovery of biomarkers has significantly improved patient selection for immunotherapy, helping to minimize toxicity and optimize efficacy. Advances such as liquid biopsies allow for real-time monitoring of treatment responses [10]. Furthermore, it has been observed that immunotherapy along with chemotherapy, radiation, or targeted therapies results successfully when use concurrently. Such combinations might solve the issue of drug resistance mechanisms. By enhancing drug bioavailability, reducing systemic toxicity, and enabling targeted delivery to tumors, nanotechnology-based tumor-targeted drug delivery systems further enhance treatment outcomes [11]. Some of the challenges that still accompany the latest developments are tumor heterogeneity, side effects from cancer immune therapy, and the adaptive mechanism of cancer cells against immune cells. The current study aims to improve immunotherapeutic approaches by creating more promising combination therapies, investigating new predictive biomarkers, and using artificial intelligence (AI) to optimize customized treatment plans [12]. Immunotherapy for lung cancer will be thoroughly examined in this chapter. The effectiveness of nanotechnology in treating lung cancer is also covered. By addressing the body's immune suppression, immunotherapy has revolutionized treatment. This is accomplished by working on nanoscale drug delivery systems and taking action on the TME.

15.2 IMMUNOLOGICAL MECHANISMS TARGETING LUNG CANCER

The success and efficacy of immunotherapy in the treatment of cancer have been uncertain in the past few years. Recently, a number of studies have focused on the body's ability to fight cancer cells. Our immune system is physiologically programmed to identify cancer cell, resulting in the cause for a cell's death. However, some cancers develop strategies to evasive this immune detection, which enables them to proliferate and spread. Immunotherapy helps reactivate the immune response and enables it to detect and eliminate cancer cells. A key process in this recognition is antigen presentation. Tumor cells show specific proteins, known as tumor-associated antigens (TAAs), on their surface through major histocompatibility complex (MHC) molecules. These markers allow T cells, a major immune component, to recognize cancer cells and attack them. Still, tumors can avoid detection by producing signals that inhibit T-cell function. One common immune escape route in lung cancer is the binding of programmed death-ligand 1 (PD-L1) on tumor cells with the programmed death receptor-1 (PD-1) on T cells. PD-1, which also binds with PD-L2, is expressed shortly after T-cell activation [13]. When PD-1 interacts with PD-L1 present on tumor cells, it blocks the activity of T cell-related kinases, which in turn weakens the cytotoxic response of T cells. This mechanism hinders T cells from detecting and eliminating tumor cells, resulting in immune evasion and tumor growth [14, 15]. In addition to suppressing T cells, PD-1 signaling limits T lymphocyte growth and decreases the release of key cytokines like IL-2 and IFN-γ, both of which are vital for mounting a strong anti-tumor immune response. Additionally, this pathway negatively regulates B lymphocyte proliferation, differentiation, and immunoglobulin (Ig) secretion, further weakening the immune response and contributing to tumor immune evasion. By exploiting this mechanism, tumor cells create an immunosuppressive microenvironment, rendering immune surveillance ineffective and facilitating disease progression.

This interaction serves as an immune checkpoint, essentially disabling T cells and stopping them from targeting the tumor. Another checkpoint is cytotoxic T-lymphocyte-associated protein 4 (CTLA-4). While the full mechanism through which anti-CTLA-4 antibodies trigger antitumor effects is not yet fully clear, studies indicate that blocking CTLA-4 enhances the initial phase of immune activation by boosting T cell proliferation and activity regardless of TCR specificity and by limiting the suppressive effects of regulatory T cells (Tregs) [16]. At the molecular level, the cytoplasmic region of CTLA-4 includes an immunoreceptor tyrosine-based inhibitory motif (ITIM), which conveys inhibitory signals when it binds to its ligand. These signals reduce the activation and growth of effector T cells, thereby weakening the immune response against tumors and aiding in their ability to evade immune detection [17]. Other immune checkpoints include lymphocyte-activation gene 3 (LAG-3), T-cell immunoglobulin and mucin-domain containing-3 (TIM-3), and indoleamine 2,3-dioxygenase (IDO), all of which play roles in suppressing immune activity within the tumor microenvironment. The introduction of immune checkpoint inhibitors (ICIs) is a success to fight against such immune evasion mechanisms. These agents inhibit checkpoint proteins like PD-1, PD-L1, and CTLA-4. As a result, T-cell activity is maintained and the ability target cancer cell is enhanced. Commonly used ICIs include PD-1 inhibitors (e.g., Pembrolizumab and Nivolumab), PD-L1 inhibitors (e.g., Atezolizumab and Durvalumab), and CTLA-4 inhibitors (e.g., Ipilimumab). These ICIs target the halting of progression of NSCLC, which is the most prevalent of lung cancers. These therapies reactivate T-cell immune functions and attack cancer cells. Furthermore, there are several reports of NSCLC patients with genetic mutations—such as KRAS, EGFR, ALK, ROS1, BRAF, HER2, and MET—which showed impact both on tumor development and treatment responses as well [18–23]. Thus the effectiveness of lung cancer treatment with increase the survival rate could be achieve by addressing the genetic mutations with immunotherapy. Recent developments deliver nanoparticles with immune modulation agents directly in to the tumor site. This method improves therapeutic effectiveness, minimizes adverse systemic effects, and provides a mechanism for controlled and prolonged drug release. Additionally, such nanoparticles can be further engineered and coupled with several other therapies like chemotherapy or gene therapy. This results synergistic effects in the treatment of lung cancer.

15.3 MOLECULAR DETERMINANTS AND BIOMARKERS IN LUNG CANCER THERAPY

Biomarkers and molecular determinants are very crucial for determine, predicting outcomes, and selecting type of treatments to consider in lung cancer depending on the patient's response, as genetic mutations alter the effectiveness of the response to targeted therapies and nanotherapeutics. The most frequently identified mutations in NSCLC include KRAS, EGFR, ALK, ROS1, BRAF, HER2, MET, and NTRK. The molecular markers help in the treatment to be personalized and act as a guide to selecting the most probable appropriate targeted therapies, immunotherapy, and nanotechnology-based drug delivery systems. The most important molecular determinants that play a crucial role in lung cancer therapy re discussed now.

15.3.1 KRAS Mutation

KRAS is one of the most commonly mutated oncogene in advanced NSCLC. The KRAS G12C mutation accounts for approximately 13% of all NSCLC cases and plays a crucial role in tumorigenesis [24]. KRAS mutations, especially in smokers, contribute to unregulated cell growth. Currently, KRAS inhibitors and nanocarrier-based therapies are being developed to enhance drug delivery and therapeutic efficacy. Additionally, KRAS mutations reduce sensitivity to EGFR tyrosine kinase inhibitors (TKIs) in wild-type EGFR patients, necessitating alternative treatment strategies [25].

15.3.2 EGFR Mutation

Found in a significant percentage of NSCLC cases, especially among non-smokers, EGFR mutations promote tumor progression [26]. The introduction of EGFR-TKIs has improved treatment outcomes, with osimertinib emerging as the preferred first-line therapy for advanced or metastatic NSCLC cases with EGFR mutations. As a potent, irreversible third-generation EGFR-TKI, osimertinib effectively targets EGFR mutations, reducing tumor growth. Additionally, nanoparticle-based drug delivery systems are being explored to enhance drug bioavailability and tumor penetration [27–29].

15.3.3 ALK and ROS1 Rearrangements

Anaplastic lymphoma kinase (ALK) and ROS oncogene 1 (ROS1) gene fusions are well-established drivers of NSCLC. These genetic alterations activate oncogenic signaling pathways, resulting in uncontrolled tumor growth. ALK and ROS1 inhibitors such as crizotinib, ceritinib, and lorlatinib have demonstrated significant efficacy in patients with these mutations. Advances in nanomedicine, including liposomal formulations and polymeric nanoparticles, have further improved bioavailability and targeted drug delivery for these inhibitors [30].

15.3.4 BRAF Mutation

A subset of NSCLC cases exhibits BRAF mutations, which disrupt cell signaling and proliferation. Targeted therapies, particularly BRAF inhibitors such as vemurafenib, dabrafenib, and encorafenib, have been developed to counteract these mutations. MEK inhibitors, including trametinib and cobimetinib, block the ERK signaling pathway, helping to delay drug resistance due to MAPK pathway reactivation. The integration of nanoparticle-based drug delivery enhances treatment stability, solubility, and specificity, improving therapeutic outcomes [31–32].

15.3.5 HER2 and MET Alterations

HER2 amplifications and MET exon 14 skipping mutations are key drivers of tumor progression in certain NSCLC cases. Targeted therapies, including antibody-drug conjugates (ADCs), monoclonal antibodies, and TKIs, have shown promise in improving treatment responses. HER2 exon 20 mutations are now validated as a clinical biomarker for NSCLC. HER2-driven tumorigenesis involves the activation of PI3K/AKT and MEK/ERK pathways, leading to uncontrolled proliferation.

HER2-targeted therapies function through multiple mechanisms:

- Tyrosine kinase inhibitors (TKIs) block HER2 phosphorylation, preventing downstream signaling.
- Monoclonal antibodies bind to HER2's extracellular domain, preventing receptor dimerization.
- Antibody-drug conjugates (ADCs) combine trastuzumab with cytotoxic agents, ensuring precise tumor targeting.

ADCs enable the selective delivery of cytotoxic agents into HER2-overexpressing tumor cells, where the drug is released upon degradation of the HER2-ADC complex within lysosomes, leading to cell cycle arrest and apoptosis. To enhance efficiency, nanocarrier-based delivery systems, such as liposomes and polymeric nanoparticles, are being incorporated into HER2-targeted therapies to improve drug solubility, controlled release, and tumor penetration [33–34].

15.3.6 Neurotrophic Tyrosine Receptor Kinase Fusion

Neurotrophic tyrosine receptor kinase (NTRK) gene fusions, though rare in NSCLC, serve as key oncogenic drivers across multiple cancer types. These fusions lead to constitutive activation of TRK signaling, promoting tumor development. TRK inhibitors, including larotrectinib and entrectinib, have demonstrated exceptional efficacy in patients with NTRK-driven lung cancers. To further enhance treatment precision and reduce systemic toxicity, researchers are exploring nanoparticle-based TRK inhibitors, offering improved therapeutic potential. Discovering specific molecular markers and genetic indicators has revolutionized lung cancer therapy by enabling personalized treatment strategies based on each patient's unique genetic makeup. Targeted therapies and nanotechnology-based drug delivery have improved treatment responses in NSCLC patients with specific genetic mutations. Ongoing advancements in nanomedicine are expanding the possibilities for overcoming drug resistance, improving therapeutic stability, and ensuring controlled drug release. As research continues, nano-based therapies combined with biomarker-driven treatments hold great promise for enhancing patient survival and long-term disease management in lung cancer therapy.

15.4 NANOSCALE CARRIERS FOR IMMUNOTHERAPY

Nanoscale carriers have revolutionized immunotherapy by enabling precise delivery of therapeutic agents, enhancing immune response, and reducing off-target effects. These delivery systems enhance drug stability, extend their presence in the bloodstream, and support precise targeting of immune cells and tumor environments. Table 15.1 describes various nanoparticles and their role in cancer immunotherapy. Some of the most commonly used nanoscale carriers in immunotherapy include the following.

15.4.1 Viral Vectors

Viral vectors, such as adenoviruses and lentiviruses, are widely used for delivering genetic material to immune cells [35]. They help in the modification of T cells, such as in Chimeric Antigen Receptor (CAR) T-cell therapy, enhancing their ability to recognize and destroy cancer cells. Viruses such as bacteriophages, plant viruses, and mammalian viruses are being explored for their efficient cell-penetration capabilities and their use in delivering therapeutic agents within their protein capsids. These are having several application platforms including in drug delivery and imaging/theranostic and also in vaccine and immunotherapy [36]. Virus nanoparticles, including genome-free forms and virus-like particles, aid in vascular transport, cellular uptake, and interactions. They offer the advantage of being well tolerated even at higher doses, while proteolytic degradation allows for rapid and complete clearance, minimizing side effects [36]. Despite their high transfection efficiency, safety concerns and immunogenicity limit their clinical application.

15.4.2 Liposomes

Liposomes have been recognized as efficient nanoscale carriers for targeted drug delivery [37]. These tiny, spherical vesicles consist of phospholipid bilayers, making them highly effective in encapsulating and transporting chemotherapy drugs [38]. As phospholipid-based nanoparticles, liposomes play a crucial role in immunotherapy by delivering immunotherapeutic agents, including cytokines, monoclonal antibodies, and checkpoint inhibitors. They enhance drug bioavailability while protecting therapeutic molecules from enzymatic degradation. PEGylated liposomes, for instance, prolong circulation time, ensuring sustained drug release and improved tumor site accumulation. Liposomal formulations have gained considerable attention in lung cancer therapy because they improve drug pharmacokinetics and reduce side effects. For locally advanced NSCLC, platinum-based chemotherapy continues to be the standard treatment. However, cisplatin, a commonly used platinum-based drug, causes severe nephrotoxicity in 20–40% of patients (kidney damage) [39]. To mitigate these side effects, researchers developed Lipoplatin, a liposomal formulation of cisplatin [40]. Clinical studies have demonstrated that Lipoplatin significantly reduces toxicities, including kidney damage, nerve toxicity, and hearing loss, while maintaining its therapeutic efficacy. In a phase III clinical trial comparing Lipoplatin combined with paclitaxel to conventional cisplatin with paclitaxel in patients with NSCLC, the group receiving Lipoplatin showed a significantly higher response rat [41, 42].

15.4.3 Polymer-Based Nanoparticles

There are several natural and synthetic polymers are available. Polypeptides, gelatin, albumin are most common natural polymer-based nanoparticles. Synthetic polymeric nanoparticles, such as PLGA (poly-lactic-co-glycolic acid), PEG (polyethylene glycol), poly (lactic acid) (PLA), and chitosan-based carriers, offer excellent biocompatibility and controlled drug release properties. Due to their versatile use, they can be use through various routes of administration. These nanoparticles can encapsulate immune checkpoint inhibitors (e.g., anti-PD-1 and anti-CTLA-4 antibodies) and deliver them selectively to tumor sites thus minimizing systemic toxicity. PLA and PLGA are biodegradable, mostly used in suture, implants, drug delivery. Being approved by FDA and EMA, the nanoparticle form of PLGA is clinically safe as chemotherapeutics, vaccines, and proteins [43]. PEGylated PLA and PLGA copolymers improve the hydrophilicity and retention time and thus improve the drug delivery to the targeted system. Kostas et al. demonstrated that PLA-PEAd and PLGA-PEAd copolymers, due to their biodegradability, are low-toxicity carriers are suitable for paclitaxel-loaded nanoparticles [44]. Sheida Sadeghi et al. successfully explained the delivery of relatively insoluble curcumin to A549 lung cancer cells with the help of polylactic-co-glycolic acid (PLGA) and poly-ε-caprolactone (PCL). The anticancer activity was achieved through apoptosis induction [45]. It has broader application in the treatment of ovarian cancer, breast cancer, and prostate cancer as well [46, 47].

15.4.4 Dendrimers

Dendrimers are synthetic, nanoscale, tree-like polymers with a highly branched architecture that enables precise control over size, shape, and surface functionality. Their well-defined structure and large surface area allow for the attachment of various bioactive molecules, including immunotherapeutic agents, targeting ligands, and adjuvants. This multifunctionality facilitates the co-delivery of multiple therapeutic agents, enhancing immune activation and tumor targeting efficiency. Moreover, dendrimers can be engineered for pH or enzyme-responsive drug release, improving their performance in the acidic tumor microenvironment. Studies have shown that dendrimer-based nanocarriers can significantly increase the uptake of antigens by dendritic cells and improve the activation of cytotoxic T lymphocytes, leading to stronger anti-tumor immune responses [48, 49].

15.4.5 Inorganic Nanoparticles

Inorganic nanoparticles—such as gold, silica, and iron oxide—play a pivotal role in cancer immunotherapy due to their tunable physical properties and multifunctional capabilities. Gold nanoparticles (AuNPs), in particular, have gained prominence for their use in photothermal therapy (PTT), wherein laser-induced localized heating results in tumor cell ablation. This ablation can subsequently stimulate the immune system by releasing tumor-associated antigens [50]. Additionally, iron oxide nanoparticles have demonstrated dual functionality as MRI contrast agents and immune modulators, enabling simultaneous tumor imaging and immuno-stimulation. Silica nanoparticles have been explored as carriers for antigens and immune checkpoint inhibitors, facilitating their targeted delivery and controlled release within the tumor microenvironment [51, 52].

15.4.6 Exosomes and Cell-Derived Vesicles

Exosomes are naturally secreted nanoscale extracellular vesicles that play a critical role in intercellular communication. Derived from tumor or immune cells, exosomes can be engineered to carry a variety of immunologically active molecules, including tumor antigens, siRNA, cytokines, and checkpoint inhibitors. Their endogenous origin offers superior biocompatibility and reduced immunogenicity compared to synthetic carriers. Engineered exosomes have shown potential to reprogram the tumor microenvironment by reducing immunosuppressive signaling and enhancing antigen presentation [53]. Moreover, their ability to cross biological barriers such as the blood–brain barrier adds to their therapeutic appeal in hard-to-treat malignancies like lung cancer [54].

15.4.7 Micelles

Micelles are self-assembled nanostructures formed by amphiphilic block copolymers in aqueous environments. Their hydrophobic core allows the encapsulation of poorly soluble immunotherapeutic agents, while the hydrophilic shell ensures stability in systemic circulation. Micelles can be modified with targeting ligands or immunomodulators to enable site-specific delivery and activate immune cells. In recent studies, micelle-based formulations have been employed to deliver toll-like receptor (TLR) agonists and checkpoint inhibitors, resulting in enhanced dendritic cell activation and T cell proliferation [55]. Their biodegradable and biocompatible nature makes micelles attractive for combinatorial immunotherapy approaches in lung cancer treatment.

15.4.8 Quantum Dots

Quantum dots (QDs) are nanoscale semiconductor particles that have unique optical properties such as high photostability, adjustable fluorescence, and broad excitation spectra. These features make QDs ideal for real-time imaging of immune responses and tumor microenvironments. In immunotherapy, QDs can be conjugated with antibodies or peptides for specific tumor targeting and immune cell labeling [56]. Moreover, QDs are being explored for theranostic applications, where they serve dual functions in imaging and drug delivery. Although concerns about the long-term toxicity and biodegradability of QDs remain, ongoing research into surface modifications and biocompatible alternatives holds promise for their future clinical application [57].

TABLE 15.1
Overview of Nanoparticle Types and Their Roles in Lung Cancer Immunotherapy

Type	Composition	Mechanism of Action	Targeted Components in TME	Therapeutic Application	Advantages	Target Receptor	Immune System Targeted	Stimuli
Liposomes	Phospholipid bilayers	Encapsulation and delivery of immunomodulators	Tumor antigens, dendritic cells	Vaccine delivery, checkpoint inhibition	Biocompatible, versatile payload	PD-1, CTLA-4	T cells, APCs	Ph sensitive
Polymeric nanoparticles	PLGA, PEG, chitosan	Controlled release, immune activation	Macrophages, cytokines	Cytokine and antigen delivery	Biodegradable, tunable release	IL-2R	Macrophages, T cells	Enzyme-responsive
Dendrimers	PAMAM, PEG	Antigen presentation, immune modulation	APCs, lymph nodes	Cancer vaccines, gene delivery	Branched structure, multivalency	Mannose receptor	Dendritic cells	pH-responsive
Gold nanoparticles	Au core, PEG coating	Immune stimulation, photothermal effect	TAMs, dendritic cells	Photothermal therapy, adjuvants	Easy surface modification, stable	TLRs	Innate immune system	Light/heat stimuli
Carbon nanotubes	SWCNTs, MWCNTs	Immune modulation, delivery system	DCs, CD8+ T cells	Antigen delivery, adjuvants	High surface area, strong cell interaction	CD8, CD4	T cells	Oxidative, enzymatic
Magnetic nanoparticles	Fe_3O_4, polymer-coated	Magnetic targeting, hyperthermia	Hypoxic TME, immune cells	Targeted delivery, imaging-guided therapy	Magnetically guided, dual-function	Integrins	NK cells, CD8+ T cells	Magnetic/thermal
Exosomes	Cell-derived vesicles	Natural antigen delivery, intercellular signaling	Tumor cells, TME stroma	Cell-based vaccines, checkpoint inhibition	Biocompatible, immune signaling	EGFR	APCs, CD8+ T cells	Endosomal/pH
Inorganic nanoparticles	ZnO, TiO_2, Ag, CuO	ROS generation, immune cell activation	Tumor cells, macrophages	Chemo-immunotherapy, cytotoxic enhancement	Stable, cytotoxic to cancer cells	Scavenger receptors	Macrophages, NK cells	Redox/pH-sensitive
Micelles	Amphiphilic block copolymers (PEG-PLA)	Encapsulation of hydrophobic immunotherapies	Tumor vasculature, immune cells	Delivery of hydrophobic drugs/ antigens	Small size, easy self-assembly	VEGFR, CD44	T cells, DCs	pH-sensitive
Quantum dots	CdSe/ZnS, graphene QDs	Imaging and immune cell targeting	Tumor cells, lymph nodes	Theranostics, targeted imaging + therapy	Bright emission, traceable delivery	EGFR, HER2	APCs	NIR light
Silica nanoparticles	Mesoporous silica	Drug loading and release, antigen delivery	TME matrix, immune checkpoints	Vaccine delivery, checkpoint inhibition	High surface area, modifiable pore size	PD-L1, CTLA-4	T cells, DCs	pH/ enzyme-responsive
Virus-like nanoparticles	Protein shell (HBV, HPV VLPs), PEG	Mimic natural infection, strong immunogenicity	DCs, T cells	Cancer vaccines, antigen presentation	Highly immunogenic, naturally targeted	TLRs, CD40	B cells, DCs, T cells	Intrinsic activation

APCs: antigen-presenting cells; ME: tumor microenvironment; PEG: polyethylene glycol; SWCNTs/MWCNTs: single/multi-walled carbon nanotubes; VLPs: virus-like particles.

15.5 TREATMENT APPROACHES IN NANO-IMMUNOTHERAPY

In the recent past, nano-immunotherapy has become a ground-breaking approach in lung cancer treatment by merging the precision of nanomedicine with the immune system's ability to eliminate tumors. Various treatment strategies using nanotechnology are currently under development or in clinical evaluation, aiming to improve the delivery, specificity, and efficacy of immunotherapeutic agent

15.5.1 Nanoparticle-Based Immune Checkpoint Inhibition

One of the most prominent strategies in nano-immunotherapy involves immune checkpoint blockade (ICB), particularly targeting the PD-1/PD-L1 axis, which plays a crucial role in tumor immune evasion. Nanoparticle systems can improve the stability and targeted delivery of anti-PD-1 or anti-PD-L1 monoclonal antibodies. Reda et al. [58] developed a dual nanoparticle-based immunotherapy targeting both PD-L1 and PLK1, demonstrating enhanced anti-tumor efficacy in lung cancer by promoting immune activation and cell cycle disruption. This dual-targeted strategy shows the potential of combining checkpoint inhibition with oncogene-targeted therapy within a single nanoplatform. Additionally, studies suggest that checkpoint inhibitor efficacy may vary by immune cell subset. Gascon-Ruiz et al. [59] identified a subset of $CD56^{bright}$ $PD\text{-}1^{+}$ NK cells to be predictors of good response to immune checkpoint therapy in lung cancer, suggesting the value of nanoparticle platforms that modulate specific immune cell populations.

15.5.2 Dual and Next-Generation Checkpoint Inhibition

Beyond PD-1/PD-L1, other checkpoint molecules like TIGIT are gaining attention. Shi et al. reported that the combination of PD-1 and TIGIT inhibitors results in a robust anti-tumor response in mesothelioma, implying similar synergistic potential in NSCLC [60]. Nanoparticle systems are ideal candidates for co-delivery of such inhibitors, potentially reducing systemic toxicity and enhancing site-specific immune modulation. Emerging evidence from the works of Roussot et al. [61] and Aden et al. [62] supports the notion that the next wave of immune checkpoint therapies will involve beyond-monoclonal antibody strategies, where nanoformulations can carry peptide inhibitors, siRNA, or immune-modulating genes, expanding the therapeutic arsenal against immune escape.

15.5.3 Dendritic Cell Vaccines and Neoantigen Delivery

Dendritic cell (DC)-based vaccines represent another major component of nano-immunotherapy. Nanoparticles enhance DC vaccine potency by improving antigen stability, uptake, and co-stimulatory signaling. Kumari et al. [63] and Sheykhhasan et al. [64] emphasized the role of neoantigen identification and DC activation, which can be significantly augmented through nano-enabled delivery systems. These systems enable co-delivery of tumor-associated antigens (TAAs) and adjuvants in controlled-release formats, increasing antigen-specific T cell activation.

15.5.4 CAR-T Cell Therapy Enhancement

Nanotechnology also enhances the delivery, expansion, and survival of CAR-T cells in solid tumors like lung cancer. Thirumalaisamy et al. [65] highlighted that nano-carriers can be used to precondition the tumor microenvironment, deliver cytokines, or boost antigen expression to improve CAR-T cell efficacy. Ren and Huang [66] further demonstrated the utility of polymeric nanocarriers in modulating T cell exhaustion in hepatocellular carcinoma, a concept that can be extrapolated to lung cancer treatment.

15.5.5 TCR-Engineered T Cell Therapy

T cell receptor (TCR) gene therapies targeting intracellular tumor antigens such as WT1 have shown success in hematologic and solid tumors. Stauss et al. [67] and Govers et al. [68] presented strategies to improve TCR pairing and function, which can be further advanced by nanoparticle delivery systems that protect TCR mRNA, reduce off-target effects, and allow for precision editing using CRISPR-Cas9.

15.5.6 Oncolytic Virus-Based Nano-Immunotherapy

Oncolytic viruses (OVs) offer a promising nano-immunotherapy approach by selectively infecting and lysing tumor cells while stimulating anti-tumor immunity. He et al. [69] reported successful IV delivery of glioma-targeted oncolytic viruses, and Ye et al. [70] showed that encoding IL-15 and 4–1BBL in OVs significantly enhances TIL-based adoptive therapy in hepatocellular carcinoma. These findings suggest that nanocarrier-enhanced OV delivery could be adapted for lung cancer immunotherapy. Furthermore, Paz and Moratorio [71] introduced the use of CRISPR-engineered viruses and deep mutational scanning, which may be employed to optimize OV efficacy and specificity, further supported by nanoparticle delivery.

15.5.7 Virus-Inspired Nanocarriers and Biomimetics

Zhang et al. [72] described biomimetic nanocarriers inspired by viral architecture for enhanced mucosal and systemic delivery. These systems exhibit properties like high payload capacity, targeting specificity, and immune evasion, making them ideal for delivering both immune checkpoint inhibitors and vaccines.

15.5.8 Combination Immunotherapy Strategies

Combination therapies using multiple immune checkpoint inhibitors (e.g., anti-PD-1 and anti-CTLA-4) have shown improved efficacy in NSCLC. Puri and Shafique [73] discussed the success of dual checkpoint blockade in enhancing survival, though toxicity remains a challenge. Nanocarriers can co-encapsulate these agents and target delivery to tumor sites, potentially mitigating immune-related adverse effects.

15.6 CHALLENGES AND OPPORTUNITIES

Lung cancer remains one of the most formidable oncological diseases globally, accounting for approximately 1.8 million deaths each year, thereby making it the leading cause of cancer-related mortality worldwide. The reason might be due to the three major factors, namely late-stage diagnosis, resistance to available drugs and therapies, and the intrinsic heterogeneity of lung tumors. Even with advancements in diagnostic levels, therapies are achieved but still challenges exist in early lung cancer detection and early treatment to advance treatment. Most of the cases lung cancer are diagnosed in late stages. More than 70% of lung cancer patients are identified only at advanced stages (stage III or IV), primarily because early-stage lung cancer is often asymptomatic or presents with non-specific symptoms that mimic benign respiratory conditions such as chronic bronchitis or infections [59]. This problem is made worse by the inadequate use of routine screening using low-dose computed tomography (LDCT). It has been shown that regular use of LDCT significantly reduces mortality in high-risk populations. However, widespread adoption is hindered by cost, infrastructure limitations, and low public awareness—especially in low- and middle-income countries. Another key challenge is the development of resistance to immunotherapy, which is the biggest obstacle to treat NSCLC. There are several promising strategies developed and introduced to overcome such challenges [74].

In case of lung cancer due to the use of chemotherapy and tyrosine kinase inhibitors (TKIs), tumor cells often developed resistance. The mechanism behind this undergoes genetic and epigenetic alterations. Such modification often renders them unresponsive to chemotherapy and tyrosine kinase inhibitors (TKIs). Mechanisms such as the activation of bypass signaling pathways, secondary mutations (e.g., T790M in EGFR-mutated NSCLC), epithelial-mesenchymal transition, and increased drug efflux pump activity contribute to therapeutic failure. Similarly, resistance to immunotherapy, particularly immune checkpoint inhibitors (ICIs), poses a considerable hurdle. Only a subset of patients show durable responses, often limited by factors such as an immunosuppressive tumor microenvironment, low tumor mutational burden, or absence of key immunogenic antigens. Toxicities associated with systemic treatments also present serious concerns. Chemotherapy remains a cornerstone of lung cancer therapy but is often accompanied by adverse effects like myelosuppression, neuropathy, and gastrointestinal toxicity. Immunotherapies, while generally better tolerated, are associated with immune-related adverse events (irAEs) such as pneumonitis, colitis, dermatitis, and endocrinopathies. These irAEs can be life threatening and require prompt diagnosis and management, often necessitating corticosteroids or immunosuppressive therapy. Furthermore, significant disparities in access to advanced diagnostics and therapies persist, particularly in rural and under-resourced regions. High costs associated with next-generation sequencing, targeted agents, and immunotherapies limit their widespread application, creating an equity gap in patient outcomes. These disparities are exacerbated by variations in healthcare infrastructure, insurance coverage, and socioeconomic status. Psychosocial challenges also deserve emphasis. A diagnosis of lung cancer, especially when advanced, carries a profound psychological impact. Patients frequently experience anxiety, depression, social isolation, and fear of death. The stigma associated with lung cancer—particularly due to its strong association with smoking—can further marginalize patients and deter them from seeking timely care or participating in support programs.

Despite these challenges, promising opportunities are emerging. Advances in liquid biopsy, AI-based imaging analysis, and biomarkers such as circulating tumor DNA (ctDNA) are revolutionizing early detection and monitoring. Personalized treatment strategies using genomic profiling and adaptive trial designs (e.g., umbrella and basket trials) are optimizing therapeutic outcomes. Moreover, the integration of patient-centered care, including psychosocial support, palliative care, and survivorship planning, is becoming a priority in comprehensive lung cancer management. To improve global outcomes, it is imperative to foster equitable healthcare systems, invest in infrastructure, and promote public awareness for early screening and treatment adherence.

15.7 CONCLUSION

Nanotechnology has become a powerful tool in cancer immunotherapy, especially in tackling the intricate challenges involved in treating lung cancer. Through improved drug delivery systems, modulation of the TME, and synergistic integration with immune-based therapies, nanotechnology enables the development of more precise and personalized therapeutic strategies. Despite these advances, the clinical translation of nanomedicine remains constrained by concerns related to long-term safety, biocompatibility, and regulatory hurdles. Future research must prioritize the rational design of nanoparticles, standardization of manufacturing protocols, and rigorous large-scale clinical evaluations. By addressing these challenges, nanotechnology-based immunotherapy offers the potential to greatly enhance survival rates and improve the quality of life for individuals affected by lung cancer.

REFERENCES

1. Siddiqui F, Vaqar S, Siddiqui AH. Lung cancer. In: *StatPearls*. Treasure Island, FL: StatPearls Publishing; 2025. [Updated 2023 May 8].
2. Thai AA, Solomon BJ, Sequist LV, et al. Lung cancer. *Lancet*. 2021;398(10299):535–554. https://doi.org/10.1016/S0140-6736(21)00312-3.

3. Lemjabbar-Alaoui H, Hassan OU, Yang YW, et al. Lung cancer: Biology and treatment options. *Biochim Biophys Acta*. 2015;1856(2):189–210. https://doi.org/10.1016/j.bbcan.2015.08.002.
4. National Cancer Institute. *Small Cell Lung Cancer Treatment (PDQ®)–Health Professional Version*. [Internet]. Available from: https://www.cancer.gov/types/lung/hp/small-cell-lung-treatment-pdq#_1
5. Siegel RL, Miller KD, Jemal A. Cancer statistics, 2023. *CA Cancer J Clin*. 2023;73(1):17–48.
6. Binnewies M, Roberts EW, Kersten K, et al. Understanding the tumor immune microenvironment (TIME) for effective therapy. *Nat Med*. 2018;24(5):541–550.
7. Munn DH, Bronte V. Immune suppressive mechanisms in the tumor microenvironment. *Curr Opin Immunol*. 2016;39:1–6.
8. Sharma P, Allison JP. The future of immune checkpoint therapy. *Science*. 2015;348(6230):56–61.
9. Topalian SL, Taube JM, Pardoll DM. Mechanisms of response and resistance to anti-PD-1 therapy in cancer. *Nat Rev Cancer*. 2016;16(4):195–207.
10. Jamal-Hanjani M, Quezada SA, Larkin J, et al. Tumour heterogeneity and immune-system interactions in lung cancer. *Lancet Oncol*. 2015;16(4):e234–e246.
11. Senbabaoglu Y, Gejman RS, Winer AG, et al. Tumor immune microenvironment characterization in lung cancer. *Genome Biol*. 2016;17(1):174.
12. Jain RK, Stylianopoulos T. Delivering nanomedicine to solid tumors. *Nat Rev Clin Oncol*. 2010;7(11):653–664.
13. Garcia-Diaz A, Shin DS, Moreno BH, et al. Interferon receptor signaling pathways regulating PD-L1 and PD-L2 expression. *Cell Rep*. 2017;19:1189–1201. https://doi.org/10.1016/j.celrep.2017.04.031.
14. Dong H, Strome SE, Salomao DR, et al. Tumor-associated B7-H1 promotes T-cell apoptosis: A potential mechanism of immune evasion. *Nat Med*. 2002;8:793–800. https://doi.org/10.1038/nm730.
15. Dong H, Zhu G, Tamada K, Chen L. B7-H1, a third member of the B7 family, co-stimulates T-cell proliferation and interleukin-10 secretion. *Nat Med*. 1999;5:1365–1369. https://doi.org/10.1038/70932.
16. Fife BT, Bluestone JA. Control of peripheral T-cell tolerance and autoimmunity via the CTLA-4 and PD-1 pathways. *Immunol Rev*. 2008;224:166–182.
17. Van Coillie S, Wiernicki B, Xu J. Molecular and cellular functions of CTLA-4. *Adv Exp Med Biol*. 2020;1248:7–32. https://doi.org/10.1007/978-981-15-3266-5_2.
18. Eser M, Hekimoglu G, Yarar MH, et al. KRAS G12C mutation in NSCLC in a small genetic center: Insights into sotorasib therapy response potential. *Sci Rep*. 2024;14:26581. https://doi.org/10.1038/s41598-024-75208-4.
19. Bethune G, Bethune D, Ridgway N, et al. Epidermal growth factor receptor (EGFR) in lung cancer: An overview and update. *J Thorac Dis*. 2010;2(1):48–51. PMID:22263017; PMCID:PMC3256436.
20. Helal AA, Kamal IH, Osman A, et al. The prevalence and clinical significance of EGFR mutations in non-small cell lung cancer patients in Egypt: A screening study. *J Egypt Natl Canc Inst*. 2024;36:39. https://doi.org/10.1186/s43046-024-00251-1.
21. Tang Z, Zhang J, Lu X, et al. Coexistent genetic alterations involving ALK, RET, ROS1 or MET in 15 cases of lung adenocarcinoma. *Mod Pathol*. 2018;31(2):307–312.
22. Olmedo ME, Cervera R, Cabezon-Gutierrez L, et al. New horizons for uncommon mutations in non-small cell lung cancer: BRAF, KRAS, RET, MET, NTRK, HER2. *World J Clin Oncol*. 2022;13(4):276–286. https://doi.org/10.5306/wjco.v13.i4.276. PMID:35582653; PMCID:PMC9052069.
23. Desai A, Menon SP, Dy GK. Alterations in genes other than EGFR/ALK/ROS1 in non-small cell lung cancer: Trials and treatment options. *Cancer Biol Med*. 2016;13(1):77–86. https://doi.org/10.28092/j.issn.2095-3941.2016.0008.
24. Lim TKH, Skoulidis F, Kerr KM, et al. KRAS G12C in advanced NSCLC: Prevalence, co-mutations, and testing. *Lung Cancer*. 2023;184:107293. https://doi.org/10.1016/j.lungcan.2023.107293.
25. Martin P, Leighl NB, Tsao MS, et al. KRAS mutations as prognostic and predictive markers in non-small cell lung cancer. *J Thorac Oncol*. 2013;8(5):530–542. https://doi.org/10.1097/JTO.0b013e318283d958.
26. Fang S, Wang Z. EGFR mutations as a prognostic and predictive marker in non-small-cell lung cancer. *Drug Des Devel Ther*. 2014;8:1595–1611. https://doi.org/10.2147/DDDT.S69690.
27. Fu K, Xie F, Wang F, et al. Therapeutic strategies for EGFR-mutated non-small cell lung cancer patients with osimertinib resistance. *J Hematol Oncol*. 2022;15:173. https://doi.org/10.1186/s13045-022-01391-4.
28. Codony-Servat J, Codony-Servat C, Cardona AF, et al. Cancer stem cell biomarkers in EGFR-mutation-positive non-small-cell lung cancer. *Clin Lung Cancer*. 2019;20(3):167–177. https://doi.org/10.1016/j.cllc.2019.02.005.
29. Wu YL, Cheng Y, Zhou J, et al. Tepotinib plus gefitinib in patients with EGFR-mutant non-small-cell lung cancer with MET overexpression or MET amplification and acquired resistance to previous EGFR

inhibitor (INSIGHT study): An open-label, phase 1b/2, multicentre, randomised trial. *Lancet Respir Med.* 2020;8(11):1132–1143. https://doi.org/10.1016/S2213-2600(20)30154-5.
30. Remon J, Pignataro D, Novello S, Passiglia F. Current treatment and future challenges in ROS1- and ALK-rearranged advanced non-small cell lung cancer. *Cancer Treat Rev.* 2021;95:102178. https://doi.org/10.1016/j.ctrv.2021.102178.
31. Tabbò F, Pisano C, Mazieres J, et al. How far we have come targeting BRAF-mutant non-small cell lung cancer (NSCLC). *Cancer Treat Rev.* 2022;103:102335. https://doi.org/10.1016/j.ctrv.2021.102335.
32. Planchard D, Smit EF, Groen HJM, et al. Dabrafenib plus trametinib in patients with previously untreated BRAFV600E-mutant metastatic non-small-cell lung cancer: An open-label, phase 2 trial. *Lancet Oncol.* 2017;18(10):1307–1316. https://doi.org/10.1016/S1470-2045(17)30679-4.
33. Uy NF, Merkhofer CM, Baik CS. HER2 in non-small cell lung cancer: A review of emerging therapies. *Cancers (Basel).* 2022;14(17):4155. https://doi.org/10.3390/cancers14174155.
34. Vathiotis IA, Charpidou A, Gavrielatou N, Syrigos KN. HER2 aberrations in non-small cell lung cancer: From pathophysiology to targeted therapy. *Pharmaceuticals (Basel).* 2021;14(12):1300. https://doi.org/10.3390/ph14121300.
35. Bulcha JT, Wang Y, Ma H, Tai PWL, Gao G. Viral vector platforms within the gene therapy landscape. *Signal Transduct Target Ther.* 2021;6(1):53. https://doi.org/10.1038/s41392-021-00487-6.
36. Chung YH, Cai H, Steinmetz NF. Viral nanoparticles for drug delivery, imaging, immunotherapy, and theranostic applications. *Adv Drug Deliv Rev.* 2020;156:214–235. https://doi.org/10.1016/j.addr.2020.06.024.
37. Shoaib A, Azmi L, Pal S, Alqahtani SS, Rahamathulla M, Hani U, Alshehri S, Ghoneim MM, Shakeel F. Integrating nanotechnology with naturally occurring phytochemicals in neuropathy induced by diabetes. *J Mol Liq.* 2022;350:118189. https://doi.org/10.1016/j.molliq.2022.118189.
38. Abu Lila AS, Ishida T. Liposomal delivery systems: Design optimization and current applications. *Biol Pharm Bull.* 2017;40(1):1–10. https://doi.org/10.1248/bpb.b16-00629.
39. Paliwal R, Paliwal SR, Kenwat R, et al. Solid lipid nanoparticles: A review on recent perspectives and patents. *Expert Opin Ther Pat.* 2020;30(3):179–194. https://doi.org/10.1080/13543776.2020.1714575.
40. Aupérin A, Péchoux CL, Rolland E, et al. Meta-analysis of concomitant versus sequential radiochemotherapy in locally advanced non-small-cell lung cancer. *J Clin Oncol.* 2010;28(13):2181–2190. https://doi.org/10.1200/JCO.2009.25.6613.
41. Hamroun A, Lenain R, Bigna JJ, et al. Prevention of cisplatin-induced acute kidney injury: A systematic review and meta-analysis. *Drugs.* 2019;79(15):1567–1582. https://doi.org/10.1007/s40265-019-01186-3.
42. Boulikas T. Clinical overview on Lipoplatin™: A successful liposomal formulation of cisplatin. *Expert Opin Investig Drugs.* 2009;18(8):1197–1218. https://doi.org/10.1517/13543780903046462.
43. Kumar A, Kumar A. Poly(lactic acid) and poly(lactic-co-glycolic) acid nanoparticles: versatility in biomedical applications. In: Grumezescu V, Grumezescu AM, eds. *Materials for Biomedical Engineering.* Elsevier; 2019:199–216. https://doi.org/10.1016/B978-0-12-818415-8.00007-3.
44. Tsachouridis K, Christodoulou E, Zamboulis A, et al. Evaluation of poly(lactic acid)/and poly(lactic-co-glycolic acid)/poly(ethylene adipate) copolymers for the preparation of paclitaxel loaded drug nanoparticles. *J Drug Deliv Sci Technol.* 2022;77:103918. https://doi.org/10.1016/j.jddst.2022.103918.
45. Sadeghi S, Mohammadnejad J, Eidi A, et al. Biocompatible PLGA–PCL nanobeads for efficient delivery of curcumin to lung cancer. *Cancer Nano.* 2024;15:34. https://doi.org/10.1186/s12645-024-00272-4.
46. You T, Zhang S. Recent advances in PLGA polymer nanocarriers for ovarian cancer therapy. *Front Oncol.* 2025;15:1526718. https://doi.org/10.3389/fonc.2025.1526718.
47. Dodda JM, Remiš T, Sadiku R, et al. Progress in the drug encapsulation of poly(lactic-co-glycolic acid) and folate-decorated poly(ethylene glycol)-poly(lactic-co-glycolic acid) conjugates for selective cancer treatment. *J Mater Chem B.* 2022;10(3). https://doi.org/10.1039/D2TB00469K.
48. Mignani S, Karpus A, Majoral JP, et al. Dendrimers in cancer immunotherapy: challenges and potential. *Chem Soc Rev.* 2020;49(6):1782–1806. https://doi.org/10.1039/C9CS00688F.
49. Zhang J, Xu Z, Zhou J. Recent advances of dendrimers in cancer therapy. *Eur J Med Chem.* 2023;250:115224. https://doi.org/10.1016/j.ejmech.2022.115224.
50. Huang X, Jain PK, El-Sayed IH, et al. Gold nanoparticles: Interesting optical properties and recent applications in cancer diagnostics and therapy. *Nanomedicine.* 2008;2(5):681–693. https://doi.org/10.1016/j.nano.2007.11.001.
51. Yang Y, Xie X, Yang Y, et al. Silica-based nanoparticles for cancer immunotherapy. *J Control Release.* 2023;357:330–347. https://doi.org/10.1016/j.jconrel.2023.04.021.

52. Estelrich J, Sánchez-Martín MJ, Busquets MA. Nanoparticles in magnetic resonance imaging: from simple to dual contrast agents. *Int J Nanomedicine.* 2015;10:1727–1741. https://doi.org/10.2147/IJN.S68891.
53. Kalluri R, LeBleu VS. The biology, function, and biomedical applications of exosomes. *Science.* 2020;367(6478):eaau6977. https://doi.org/10.1126/science.aau6977.
54. Vader P, Mol EA, Pasterkamp G, et al. Extracellular vesicles for drug delivery. *Adv Drug Deliv Rev.* 2016;106:148–156. https://doi.org/10.1016/j.addr.2016.02.006.
55. Wang Y, Zhang Y. Micelle-based immunotherapeutic delivery systems in lung cancer. *Pharmacol Res.* 2022;179:106214. https://doi.org/10.1016/j.phrs.2022.106214.
56. Wu C, Zhang Z, Liu H. Quantum dots for tumor imaging and therapy: promising tools for future precision medicine. *Small Methods.* 2021;5(3):2001023. https://doi.org/10.1002/smtd.202001023.
57. Tang Y, Sun X, Huang Z, Li J. Quantum dots in biomedical applications: current status and future challenges. *Biosens Bioelectron.* 2024;219:114788. https://doi.org/10.1016/j.bios.2022.114788.
58. Reda M, Alrawashdeh HM, Afify SM, et al. A dual nanoparticle-based strategy to target breast cancer by combining PLK1 and PD-L1 inhibition. *Biomater Sci.* 2022;10(20):5856–5868.
59. Gascon-Ruiz M, Gonzalez-Rodriguez AP, Pérez-Moreno P, et al. A CD56bright PD-1$^+$ NKG2D$^+$ subset of NK cells predicts good response to immune checkpoint blockade in lung cancer. *Cancer Immunol Res.* 2024;12(1):31–44.
60. Shi J, Hou S, Fang Q, et al. PD-1 and TIGIT co-inhibitory blockade induces strong immune response against malignant pleural mesothelioma. *Nat Commun.* 2024;15(1):1784.
61. Roussot MA, Forde PM, Felip E, et al. Role of immune checkpoint inhibitors in stage IV non-small-cell lung cancer. *Eur Respir Rev.* 2022;31(165):210110.
62. Aden K, Rehman A, Waschina S, et al. Metabolic functions of gut microbes associate with efficacy of tumor necrosis factor antagonists in patients with inflammatory bowel diseases. *Gastroenterology.* 2019;157(5):1279–1292.e11.
63. Kumari A, Ranjan RK, Kumari R, et al. Recent advances and clinical outcomes of dendritic cell vaccines for cancer immunotherapy: A narrative review. *Biomed Pharmacother.* 2023;157:114045.
64. Sheykhhasan M, Qiu Y, Ghamari SH. Mesothelin-targeted chimeric antigen receptor (CAR) T cell therapy for the treatment of solid tumors: Current status and future perspectives. *Stem Cell Res Ther.* 2021;12:500.
65. Thirumalaisamy P, Li Z, Shuai Z, et al. Nanotechnology-enabled CAR-T cell therapy for cancer. *Front Oncol.* 2022;12:939295.
66. Ren Z, Huang Y. Therapeutic applications of CRISPR/Cas9 in hepatocellular carcinoma. *Front Oncol.* 2020;10:796.
67. Stauss HJ, Williams JS, Morris EC. TCR gene therapy for cancer. *Immunology.* 2014;142(1):1–6.
68. Govers C, Sebestyen Z, Coccoris M, et al. T cell receptor gene therapy: Strategies for optimizing transgenic TCR pairing. *Trends Mol Med.* 2010;16(2):77–87.
69. He Y, Wang C, Yue D, et al. Systemic administration of glioma-targeting oncolytic virus, Delta-24-RGD, improves antitumor immunity and survival in murine glioma model. *Cancer Immunol Immunother.* 2020;69(3):453–464.
70. Ye L, Yao J, Zhang Y, et al. Enhanced TIL-based adoptive cell therapy for hepatocellular carcinoma by a multi-targeted oncolytic adenovirus encoding IL-15 and 4-1BBL. *J Immunotherap Cancer.* 2022;10(6):e004880.
71. Paz MM, Moratorio G. Oncolytic viruses armed with transgenes: A promising approach for cancer therapy. *Viruses.* 2022;14(9):1944.
72. Zhang K, Sun X, Zhou Y, et al. Biomimetic virus-like nanoparticles for mucosal and systemic immune stimulation. *Adv Mater.* 2023;35(1):2205441.
73. Puri S, Shafique M. Advances in dual checkpoint blockade for non-small cell lung cancer. *Front Oncol.* 2022;12:833855.
74. Xiang Y, Liu X, Wang Y, et al. Mechanisms of resistance to targeted therapy and immunotherapy in non-small cell lung cancer: Promising strategies to overcoming challenges. *Front Immunol.* 2024;15:1366260. https://doi.org/10.3389/

16 Green Nanotechnology for the Management of Lung Cancer

Swati Joshi[1*], *Hem C. Joshi*[2*], *and Saurabh Gangola*[3]

[1] Department of Zoology, D.S.B. Campus, Kumaun University, Nainital, Uttarakhand, India 263002

[2] Department of Botany, D.S.B. Campus, Kumaun University, Nainital, Uttarakhand, India 263002

[3] Department of Microbiology, Graphic Era (Deemed to Be University), Dehradun, Uttarakhand, India 248002

*Corresponding authors: Hem C. Joshi, hembiophysics12@gmail.com; Swati Joshi, swatijoshi865@gmail.com.

16.1 INTRODUCTION TO GREEN NANOTECHNOLOGY

Green nanotechnology is emerging as an important strategy to lung cancer treatment, employing environmentally friendly approaches to improve drug delivery and therapeutic efficacy while reducing unfavorable effects. Lung cancer is the most frequent and dangerous of tumors in the world, with current medicines showing limited efficacy due to low selectivity, multidrug resistance, and systemic toxicity. Green nanotechnology, an interdisciplinary method that combines nanoscience with ecological practices, has emerged as a possible paradigm for improving lung cancer treatment. Green nanotechnology provides biocompatible, cost-effective, and sustainable alternatives for targeted medication delivery, diagnostics, and therapy by employing environmentally friendly and biologically inspired techniques for nanoparticle synthesis. Green nanotechnology refers to the design, development, and deployment of nanomaterials and nanotechnologies in methods that reduce environmental and human health concerns while encouraging sustainability (Figure 16.1). This includes the use of green chemistry principles and eco-friendly approaches to counteract the negative consequences of classical nanotechnology.

16.2 PRINCIPLES OF GREEN NANOTECHNOLOGY

Green nanotechnology is founded on the principles of green chemistry and engineering, adapted to address the unique properties of nanoscale materials (Figure 16.2).

16.3 APPLICATIONS OF GREEN NANOTECHNOLOGY

Green nanotechnology has broad applications across industries, through various ways (Table 16.1).

16.4 COMPARISON OF GREEN NANOTECHNOLOGY WITH CONVENTIONAL NANOTECHNOLOGY

Nanotechnology has transformed cancer treatment, particularly by giving focused medicines for diseases such as lung cancer. However, the introduction of green nanotechnology has provided

DOI: 10.1201/9781003595830-16

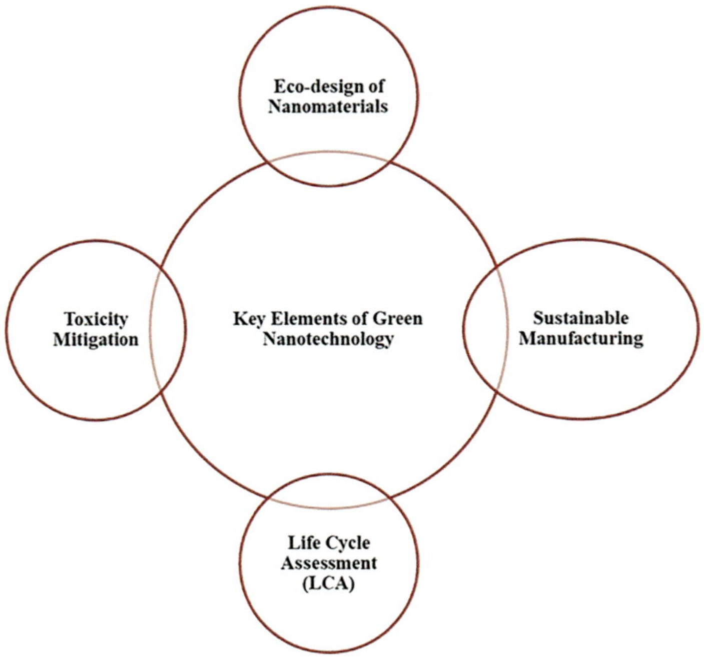

FIGURE 16.1 Addressing the key elements of green nanotechnology

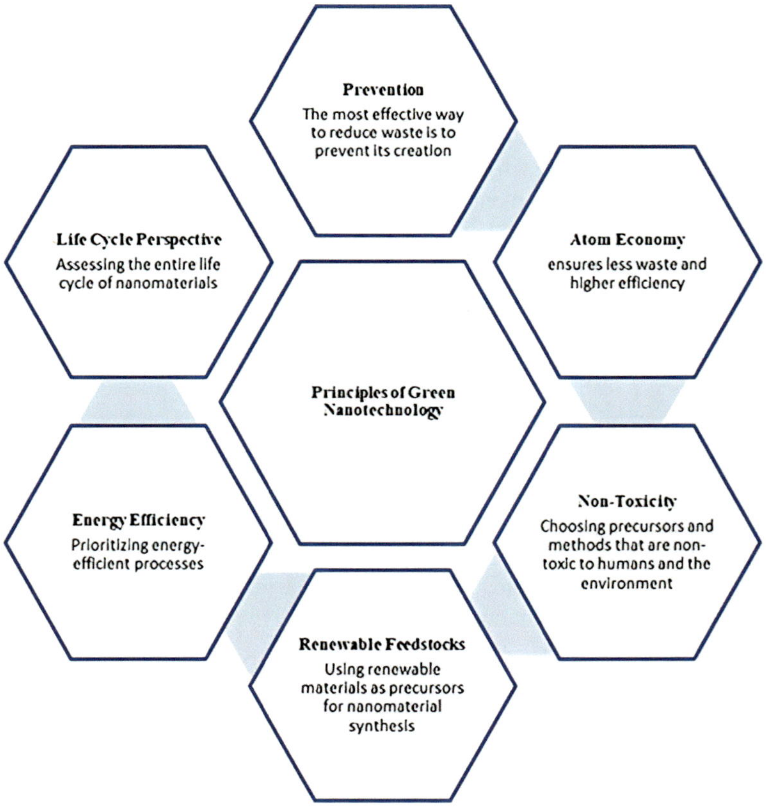

FIGURE 16.2 The various principles of green nanotechnology

TABLE 16.1
Applications of Green Nanotechnology in Various Disciplines

Environmental Remediation	Renewable Energy	Sustainable Agriculture	Biomedicine
Nanotechnology can aid in cleaning polluted environments through nano-adsorbents, photocatalysts, and nano filters (Sharma et al., 2022).	Nanomaterials are critical in improving the efficiency of solar panels, fuel cells, and batteries while reducing their environmental footprint (Rajput & Kumbhar, 2023).	Nanotechnology enables the controlled release of nutrients and pesticides, minimizing overuse and environmental contamination (Das et al., 2023).	Green synthesis of nanoparticles using biological methods reduces toxic chemical use, paving the way for safer therapeutic applications (Gupta et al., 2023).

a more sustainable alternative to traditional nanotechnology. Nanotechnology has revolutionized the therapeutic landscape by allowing the creation of nanoparticles (NPs), which improve medicinal efficacy while lowering systemic toxicity. These nanoparticles enable tailored drug administration, which improves outcomes for lung cancer patients (Chen et al., 2023). Conventional nanotechnology typically includes the chemical creation of nanoparticles, which requires toxic reagents and energy-intensive processes. Lipid-based nanoparticles are widely employed for drug administration and are frequently produced with organic solvents (Kundu et al., 2023). Metallic nanoparticles, such as gold and silver nanoparticles created using chemical reduction processes, are commonly used in lung cancer therapy. Green nanotechnology stresses eco-friendly and sustainable techniques of nanoparticle manufacturing, using biological agents such as plant extracts, bacteria, or fungi. Table 16.2 highlights the comparison between various conventional and green nanotechnology.

16.5 MECHANISMS OF GREEN NANOPARTICLE SYNTHESIS

Nanometallic accoutrements are essence and blends that produce nanocrystalline grains ranging in size from 5 to 100 nm (Hussain et al., 2016). Nanoscale essences have an advanced face area than theirnon-nano or bulk coequals. Likewise, because to the small size effect, face effect, interface effect, and amount impact, they parade unique physical and chemical parcels that are not set up innon-nano essence. There is multitudinous nanoscale essence with different uses in biology, drug, and engineering (Zhao et al., 2013). Currently, there is an added quantum of exploration concentrated on the conflation of nanoscale essence using chemical, physical, and green conflation approaches (Horwat et al., 2011). Physical and chemical styles are gradationally being replaced by green conflation styles (Alsammarraie et al., 2018) due to issues similar as high energy consumption (Horwat et al., 2011), the release of poisonous and dangerous chemicals (Hoag et al., 2009), and the use of complex outfit and conflation conditions (Baruwati et al., 2009). Aerosols (Smirniotis et al., 2018), ultraviolet radiation (Wojnarowicz et al., 2018), and thermal corruption (Ahmed et al., 2016) are exemplifications of physical processes that bear high temperatures and pressure (Ahmed et al., 2019).

16.5.1 Biological Methods: Plant Extracts, Microorganisms, and Enzymes

Biological styles factory excerpts, microorganisms, and enzymes. Nanotechnology has converted numerous diligences, including drug, electronics, and husbandry. Among the several ways of nanoparticle conflation, green conflation has surfaced as a sustainable, environmentally

TABLE 16.2
A Comparative Analysis of Conventional and Green Nanotechnology

Methods	Conventional Nanotechnology	Green Nano-Technology
Synthesis Methods	**Chemical-based processes:** Involves toxic chemicals like sodium borohydride as reducing agents. **Energy intensive:** High temperatures and pressures are often required, increasing the environmental footprint (Chen et al., 2023).	**Biological agents:** Utilizes plant extracts, enzymes, or microorganisms for nanoparticle synthesis, eliminating toxic chemicals (Gupta et al., 2023). **Energy efficient:** Typically operates under ambient conditions, reducing energy requirements (Rajput & Kumbhar, 2023).
Toxicity and Biocompatibility	**Residual toxicity:** Chemically synthesized nanoparticles may carry toxic residues, posing risks to non-target tissues (Sharma et al., 2022). **Immune responses:** Conventional NPs can trigger unwanted immune reactions, limiting their therapeutic potential.	**Reduced toxicity:** The use of natural agents ensures minimal side effects and enhanced safety profiles. **Enhanced biocompatibility:** Biologically synthesized NPs are less likely to provoke immune responses (Das et al., 2023).
Environmental Impact	**High waste production:** Generates hazardous by-products that require proper disposal (Kundu et al., 2023). **Non-renewable resources:** Relies heavily on synthetic precursors and non-renewable resources.	**Eco-friendly:** Produces minimal waste, often leveraging renewable resources (Gupta et al., 2023). **Sustainable practices:** Incorporates green chemistry principles, aligning with sustainability goals.
Efficacy in Lung Cancer Treatment	**Proven effectiveness:** Established technologies have demonstrated success in delivering anticancer drugs, such as liposomal formulations of doxorubicin. **Broad application:** Suitable for a wide range of drug types and delivery mechanisms (Sharma et al., 2022).	**Enhanced Targeting:** Phytosynthesized nanoparticles exhibit improved targeting efficiency due to functionalization with natural ligands. **Potential for innovation:** Emerging studies show comparable or superior efficacy in drug delivery and therapeutic outcomes (Rajput & Kumbhar, 2023).

friendly, and cost-effective option. Unlike chemical and physical approaches, green conflation uses natural systems similar as factory excerpts, microbes, and enzymes to produce nanoparticles, barring the use of dangerous chemicals and large energy inputs (Raghunandan et al., 2023). Biological approaches lower environmental toxin while also perfecting biocompatibility and stability, making them ideal for biomedical operations (Iravani & Varma, 2022). Plants are natural reservoirs of phytochemicals such as polyphenols, flavonoids, terpenoids, alkaloids, and proteins, which act as reducing and stabilizing agents during nanoparticle formation (Khan et al., 2022). The green synthesis technique entails mixing plant extracts with metal salts in ambient circumstances, where phytochemicals convert metal ions to their corresponding nanoparticles.

Silver nanoparticles (AgNPs): *Azadirachta indica* (Neem) leaf extract synthesized AgNPs demonstrated excellent antimicrobial and anticancer activity (Singh et al., 2022).

Gold nanoparticles (AuNPs): *Ocimum sanctum* (Tulsi) leaf extract-derived AuNPs showed biocompatibility and enhanced photothermal properties for cancer treatment (Sharma et al., 2023).

Zinc oxide nanoparticles (ZnONPs): *Aloe vera* extract–mediated ZnONPs exhibited strong antioxidant properties, enhancing wound healing (Ahmad et al., 2023).

16.5.2 Microorganisms for Nanoparticle Synthesis

Microorganisms, including bacteria, fungi, algae, and yeast, play a significant role in synthesizing nanoparticles. They produce enzymes, proteins, and metabolites that act in reducing microbial synthesis, which occurs through two main pathways:

1. **Extracellular synthesis:** The microorganisms secrete metabolites and enzymes into the medium that reduce metal ions into nanoparticles.
2. **Intracellular synthesis:** Metal ions are transported into the microbial cells, where they are reduced and stabilized by intracellular biomolecules (Chakraborty et al., 2023).

16.5.3 Chemical and Physical Approaches Aligned with Green Chemistry Principles

Nanoparticles (NPs) have received a lot of interest because of their unique features and wide range of applications in sectors including medicine, electronics, and environmental research. Traditional techniques of NP synthesis frequently use dangerous chemicals and energy-intensive procedures, which raise environmental and health problems. Green chemistry principles promote the development of sustainable, efficient, and environmentally friendly synthesis processes. Green chemistry focuses on building chemical processes that limit or eliminate the usage and production of hazardous compounds. Key principles include waste prevention, atom economy, safer solvent and reaction conditions, energy efficiency, and the use of renewable feedstocks. In NP synthesis, these principles translate to methods that minimize environmental impact, enhance safety, and improve sustainability.

16.5.3.1 Chemical Approaches

Green Sol–Gel Method

The sol–gel process converts a system from a liquid "sol" to a solid "gel" phase. In green synthesis, this method utilizes non-toxic solvents and natural precursors to produce metal oxide NPs. For instance, plant extracts rich in polyphenols can act as both reducing and stabilizing agents, facilitating the formation of NPs under mild conditions. This approach offers control over particle size and morphology while adhering to green chemistry principles (Johnson et al., 2023).

Hydrothermal and Solvothermal Synthesis

These methods involve crystallizing substances from high-temperature aqueous or solvent solutions at high vapor pressures. Green hydrothermal synthesis employs water as a solvent and plant-based materials as reducing agents, eliminating the need for hazardous chemicals. This technique allows for the controlled synthesis of NPs with specific sizes and shapes, essential for tailoring their properties for various applications.

Microwave-Assisted Synthesis

Microwave irradiation provides rapid and uniform heating, reducing reaction times and energy consumption. In green NP synthesis, microwave-assisted methods utilize natural precursors and solvents, enhancing reaction efficiency and product purity. This technique has been successfully applied in synthesizing various metal NPs, offering a sustainable alternative to conventional heating methods (Anjana et al., 2021).

16.5.3.2 Physical Approaches

Green Laser Ablation

Laser ablation involves irradiating a bulk material with a laser to produce NPs. When conducted in environmentally benign solvents like water, it aligns with green chemistry principles. This method avoids chemical reagents, reducing potential contamination and environmental impact. The process parameters can be adjusted to control NP characteristics, making it a versatile tool in green synthesis.

Ultrasound-Assisted Synthesis

Ultrasonication generates acoustic cavitation, leading to high-energy conditions that facilitate NP formation. Using natural stabilizers and solvents, ultrasound-assisted methods can produce NPs efficiently under ambient conditions. This approach minimizes energy usage and eliminates the need for hazardous chemicals, aligning with the goals of green chemistry (Machado et al., 2023).

16.6 ADVANTAGES OF GREEN SYNTHESIS OVER TRADITIONAL METHODS

Environmental Benefits

Green synthesis significantly reduces the environmental footprint compared to traditional chemical and physical methods. The elimination of toxic chemicals ensures safer waste management and minimal pollution (Ahmed et al., 2023).

Biocompatibility and Reduced Toxicity

Nanoparticles produced using green technologies are more biocompatible and can be used in biological applications including imaging and medication delivery (Kumar & Yadav, 2022).

Cost-Effectiveness

Green synthesis utilizes inexpensive biological materials such as plant extracts, agricultural residues, and microbial cultures, reducing production costs (Das et al., 2023).

Energy Efficiency

Unlike conventional methods requiring high temperatures or pressures, green synthesis operates under mild conditions, resulting in lower energy consumption (Singh et al., 2022).

Sustainability

The reliance on renewable resources aligns with global efforts toward sustainable development. Plant-based synthesis, for example, leverages abundant and biodegradable materials (Ali et al., 2023).

16.7 APPLICATIONS OF GREEN NANOTECHNOLOGY IN LUNG CANCER TREATMENT

16.7.1 Targeted Drug Delivery

Targeted drug distribution is a critical component in the use of green nanotechnology for lung cancer treatment. These technologies, which use environmentally friendly materials and sophisticated targeting mechanisms, provide a path to safer and more effective medicines. Targeted drug delivery is an approach that tries to deliver therapeutic drugs directly to the sick spot while sparing healthy tissues. This technique is especially important in lung cancer because of the variability of tumor microenvironments and the possibility of significant side effects from traditional chemotherapies. Targeted delivery systems are intended to take advantage of certain molecular markers or physiological conditions found only in cancer cells, such as overexpressed receptors, acidic pH, or hypoxic settings. Targeted drug delivery methods take advantage of cancer cells' specific properties, including as overexpressed receptors, acidic microenvironments, and increased permeability and retention (EPR) effects. Nanoparticles serve as carriers, encapsulating therapeutic chemicals and delivering them directly to tumor locations. The key targeting strategies include the following:

Metal and Metal Oxide Nanoparticles

Gold, silver, and zinc oxide nanoparticles made using plant extracts or microbial processes have showed great promise in lung cancer treatment. For example, green-synthesized gold nanoparticles (AuNPs) functionalized with folic acid showed increased cytotoxicity against non–small cell lung cancer (NSCLC) cells while preserving healthy cells (Singh et al., 2022).

Polymeric Nanoparticles

Biodegradable polymers like chitosan and polylactic acid (PLA) are commonly employed in green nanotechnology. Chitosan nanoparticles loaded with paclitaxel and manufactured utilizing a green ionic gelation process demonstrated better drug stability and controlled release qualities in lung cancer models (Patra et al., 2023).

Lipid-Based Nanocarriers

The green manufacturing of liposomes and solid lipid nanoparticles integrates natural lipids while avoiding chemical solvents. Liposomes functionalized with lung cancer-specific antibodies have been used for doxorubicin delivery, resulting in better tumor targeting and less cardiotoxicity (Gupta et al., 2023).

Carbon-Based Nanostructures

Graphene oxide and carbon nanotubes manufactured utilizing green technologies have a high-drug loading capacity and tailored delivery capabilities. Recent research found that combining functionalized graphene oxide with epidermal growth factor receptor (EGFR) inhibitors dramatically improved therapeutic efficacy in NSCLC (Liu et al., 2023).

16.8 DIAGNOSTIC APPLICATIONS

Lung cancer is one of the most prevalent causes of cancer-related death worldwide, and early detection is critical for better patient outcomes (Siegel et al., 2023). Traditional diagnostic procedures, such as imaging and biopsies, have drawbacks like low sensitivity, invasiveness, and delayed detection. Plant extracts, bacteria, fungi, and other biological systems serve as reducing and stabilizing agents during the nanoparticle formation. These approaches avoid the requirement for harmful reagents while ensuring that the generated nanoparticles are nontoxic and suited for biomedical applications (Ahmed et al., 2021).

Gold nanoparticles (AuNPs) and silver nanoparticles (AgNPs) manufactured utilizing environmentally friendly technologies have been widely investigated for their diagnostic potential in lung cancer. For example, AuNPs produced from plant extracts have outstanding optical characteristics, making them suitable for application in biosensing platforms for detecting lung cancer biomarkers (Patra et al., 2023).

16.9 ROLE OF GREEN NANOPARTICLES IN IMAGING AND EARLY DETECTION

In clinical practice, early disease identification and diagnosis are crucial, especially for cancer. For instance, it was discovered that patients with gastrointestinal cancer who benefited from early identification had a significantly higher two-year survival rate (92.3% vs. 33.3%) (Jung et al., 2016). Additionally, the ten-year mortality rate was reduced by 17–28% for breast cancer patients who got early detection (Lu et al., 2009). Early disease detection and drug response are two common uses for medical imaging technology. Today's imaging techniques include computed tomography (CT), magnetic resonance imaging (MRI), ultrasound, single photon emission computed tomography (SPECT), fluorescence imaging, positron emission tomography (PET), and X-ray radiography.

A combination of several imaging modalities is commonly employed to improve lesion detection (Torres Martin de Rosales et al., 2011; Fan et al., 2014). To distinguish between healthy tissue and abnormal lesions, medical imaging contrast agents are utilized, providing more precise anatomical and functional data. The spatial resolution produced by the imaging equipment limits the ability to detect tumors using standard clinical imaging contrast agents. For example, contrast-enhanced CT can detect hypervascular hepatomas as small as 3 mm (Cuccurullo et al., 2018). The majority of the contrast agents used in medical imaging today are small substances with a rapid metabolism, a broad distribution, and the potential for negative side effects (Chronos et al., 1993; Petrik et al., 2005).

The distinct passive, active, and physical targeting capabilities of nanomaterials have recently spurred attempts to enhance biological imaging and detection. According to Oh et al. (2013), the tiny size of nanoparticles enhances the EPR effects in tumors, resulting in a relative rise in contrast agent concentrations. One of the most crucial aspects of nanoparticles for tumor imaging is their size. According to Hoshyar et al. (2016), nanoparticle size affects tumor penetration, cellular absorption, biodistribution, blood circulation half-life, and targeting.

16.10 NANOPARTICLES IN FLUORESCENCE IMAGING APPLICATIONS

Imaging resolution is restricted to the macroscopic level, although MRI, CT, US, SPECT, and PET are frequently used to identify lesions (Cuccurullo et al., 2018; Santra & Malhotra, 2011). The best spatial resolution for microscopic illness diagnosis can be obtained using fluorescence imaging technologies, especially near-infrared fluorescence (NIRF). NIRF has less non-specific tissue auto-fluorescence and better tissue penetration than visible light. Nevertheless, autofluorescence and scattering properties in various tissues continue to restrict therapeutic relevance, and penetration depth is still restricted (Foucault-Collet et al., 2013). Moreover, insufficient sensitivity for identifying abnormalities may arise from limited fluorescence in the target lesion and likely blink and photobleaching effects (Foucault-Collet et al., 2013; Hwang et al., 2006).

The useful characteristics of nanoparticles can aid in overcoming the limitations of fluorescence imaging. To produce more signals, for instance, more fluorescent dye molecules could be incorporated into nanoparticles (Cai et al., 2011). Moreover, when required, nanoparticles can be modified (or built) to reduce NIR fluorescence quenching (Genovese et al., 2013). Additionally, nanoparticle concentrations in lesions can be increased using both active and passive methods, which raises the fluorescent dye concentration locally in the lesion. Increased absorption in target lesions is also made possible by the relatively long half-life in circulation. In order to lessen blink and photobleaching effects, nanoparticles can also be engineered to transform low-energy photons into high-energy photons (Grebenik et al., 2013). Due to these advantageous characteristics, a lot of work has gone into creating platforms for fluorescent nanoparticles.

16.11 APPLICATION IN BIOMEDICAL IMAGING

Gene detection, protein analysis, enzyme activity evaluation, element tracing, cell tracking, early stage disease diagnosis, tumor-related research, and real-time therapeutic effect monitoring are all applications of nanoparticle fluorescence imaging (Cai et al., 2011; Lee et al., 2008; Wei et al., 2012; Dubreil et al., 2017). For many years, clinical settings have employed MRI, a potent imaging method. It is predicated on the spin of protons when activated by a radio frequency pulse in the presence of an external magnetic field. Based on the nuclear magnetic resonance signal generated by protons in human bodies, MRI provides great intrinsic soft-tissue contrast, high spatial resolution, and temporal resolution. Additionally, it could display three-dimensional anatomic tomographic data. MRI also uses radiotracers and non-ionizing radiation. The high cost, extended imaging times, motion artifacts, and potential foreign body/implant artifacts are some of the drawbacks of MRI. Lesion detection and differentiation from healthy tissues are enhanced by MRI contrast agents (Donato et al., 2017; Bashir et al., 2014). MRI imaging at the gene, protein, cell, and organ levels is made possible by the nanoscale scale of these novel discrepancy agents (Zhang et al., 2009; Jun et al., 2008). An important component of picture optimization is the targeting technique for nanoparticles. Active targeting and the biodistribution of nanoparticles depend on face labeling. For example, some peptides, ligands, and antibodies have been used as facial markers for excrescence detection (Vargo et al., 2014). The most crucial element for unresistant targeting is the nanoparticle's size, which can vary from a few nanometers to hundreds of nanometers. Using the EPR effect, this method has been used in excrescence imaging. Other operations calculate on non-specific cellular uptake, similar to blood pool discrepancy imaging, inflammation imaging, nasty lymph knot identification, mesenchymal stem cell shadowing, and island transplantation monitoring.

16.12 NANOPARTICLES IN CT IMAGING APPLICATIONS

Through the use of differential tissue X-ray thickness and attenuation, CT produces cross-sectional and three-dimensional pictures. Despite the greater amounts of ionizing radiation exposure, CT has rapidly surpassed plain-film radiography due to its improved spatial resolution for clinical imaging, quicker examination times, reduced costs, and increased efficiency. When it comes to discriminating tissues with similar attenuation values, CT contrast agents are crucial. The most often used ingredient in intravenous CT contrast agents at the moment is iodine. Quick clearance, potential renal toxicity, nonspecific blood pool dispersion, and documented adverse events/anaphylaxis are some of the drawbacks of iodinated contrast agents. To overcome these limitations and increase the use of CT imaging, nanoscale contrast agents have been created.

16.13 NANOPARTICLES IN ULTRASOUND IMAGING APPLICATIONS

Because of its portability, non-invasiveness, high spatial resolution, affordability, and real-time imaging capabilities, ultrasound imaging is one of the most widely used medical diagnostic imaging modalities. The purpose of ultrasound contrast agents is to enhance the differences in acoustic signals between target lesions and normal tissues. Microbubbles in today's commercially available ultrasonic contrast agents range in size from 1 to 8 μm (Hu et al., 2016). These relatively large particles have a limited circulation life and low stability due to their larger size, and they produce blood pool contrast signals instead of cellular absorption (Wang et al., 2013).

Nanotechnology has been investigated to overcome these limitations; however, to obtain enough acoustic reflection, ultrasound contrast nanoparticles typically need to be larger than CT or MRI contrast nanoparticles, ranging from hundreds to thousands of nanometers (Sakamoto et al., 2005; Seo et al., 2010).

Compared to existing ultrasound contrast agents, nanoparticles are far smaller. As demonstrated by folate-PEG-chitosan loaded PFOB, the small size and additional surface labeling help in lesion targeting, just like with other nanoparticles (Hu et al., 2016). In addition to their usage as blood pool contrast agents, they can provide strong signals in local target lesions, especially for tumor identification. Stem cell imaging, inflammation detection, and drug delivery are further applications (Kim et al., 2017).

16.14 INNOVATIONS IN BIOSENSORS AND CONTRAST AGENTS

Green nanotechnology, which involves the use of environmentally friendly materials and techniques in nanomaterial production, has created new opportunities for biosensors and contrast agents in lung cancer applications (Zhang et al., 2023). For lung cancer treatment, this strategy is especially useful for building biosensors that can detect early biomarkers and contrast agents for enhanced imaging techniques (Patel et al., 2023).

Nanobiosensors for Early Detection

Nanobiosensors based on gold nanoparticles have been developed to detect volatile organic molecules in the breath, a potential technology for non-invasive lung cancer diagnostics (Zhao et al., 2024). These sensors can identify biomarkers at concentrations significantly lower than those detected by traditional methods, allowing for better early detection (Kumar & Sharma, 2023).

Graphene-Based Biosensors

Graphene-based biosensors have emerged as useful diagnostic tools due to their high electrical conductivity and biocompatibility (Yang et al., 2023a). Functionalized graphene oxide detects lung cancer biomarkers with great sensitivity, making it excellent for early diagnosis (Lee et al., 2016). Recent studies have shown that they can identify specific cancer indicators such as EGFR mutations in lung cancer (Wang et al., 2023).

DNA Nanobiosensors

Another interesting discovery for detecting lung cancer is DNA-based biosensors, which use nanoscale DNA sensors called aptamers. These sensors attach selectively to cancer biomarkers, providing increased specificity and sensitivity in identifying mutations associated with lung cancer (Shi et al., 2023).

Innovations in Contrast Agents for Lung Cancer Imaging

Contrast chemicals are essential in imaging techniques like as MRI, CT, and PET, as they help to better visualize tumors and metastases. Green nanotechnology has helped to develop novel, environmentally friendly contrast agents that improve imaging performance while being less harmful than standard agents (Saxena et al., 2022). Following are some noteworthy innovations.

16.15 BIOSENSORS AND CONTRAST AGENTS INTEGRATION

The combination of biosensors and contrast chemicals is becoming increasingly popular in lung cancer diagnoses. Theranostic platforms, which combine biosensors and contrast agents in a single device, have the potential to enable simultaneous diagnosis and therapy monitoring (Zhou et al., 2023). Multifunctional nanoparticles, for example, can detect specific biomarkers while also acting as contrast agents in imaging, enhancing diagnostic accuracy, and allowing for tailored treatment strategies (Liu et al., 2023).

A significant example is the creation of nanorobots that can be designed to deliver chemotherapy directly to lung cancers while also giving real-time imaging using embedded contrast chemicals (Sharma et al., 2023). This combination of therapeutic and diagnostic functions is a promising approach to enhance lung cancer treatment results (Fang et al., 2023).

16.16 PHOTOTHERMAL AND PHOTODYNAMIC THERAPY

Green nanoparticles, which are developed using environmentally benign procedures, have received a lot of attention for use in photothermal therapy (PTT) and photodynamic therapy (PDT). These therapies use light to cure cancer, with photothermal therapy using heat to kill cancer cells and photodynamic therapy using light-activated chemicals to cause cytotoxicity (Cai & Zhang, 2023). Green nanoparticles have been acknowledged for their biocompatibility, efficacy, and low side effects, making them an appealing alternative to conventional medicinal approaches (Zhao et al., 2023).

16.17 MECHANISMS AND BENEFITS OF GREEN NANOPARTICLES IN PTT AND PDT

Photothermal Therapy

It uses nanoparticles to capture light and turn it into heat, which is subsequently used to destroy malignant tissues (Huang et al., 2023). Green nanoparticles, which are frequently generated from plant extracts, biopolymers, or biodegradable metals, are ideal for PTT due to their ecologically friendly manufacturing and enhanced biological interactions (Jiang & Liu, 2023). The following theories explain how these nanoparticles function in PTT:

- **Mechanism of light absorption and heat generation:** Green nanoparticles, including gold, copper, and carbon-based nanoparticles, have substantial light absorption in the near-infrared (NIR) region, making them excellent for PTT (Chen et al., 2023). When exposed to NIR light, these nanoparticles transform absorbed photons into localized heat, resulting in thermal damage to cancer cells (Wang et al., 2023).
- **Enhanced specificity and targeting:** Green nanoparticles can be functionalized with targeting ligands such as antibodies or peptides to increase their selectivity towards cancer cells (Li et al., 2023). This focused method reduces harm to healthy tissues while increasing therapy efficacy (Zhang et al., 2023).
- **Biodegradability and reduced toxicity:** Green nanoparticles can be functionalized with targeting ligands such as antibodies or peptides to increase their selectivity towards cancer cells (Li et al., 2023). This focused method reduces harm to healthy tissues while increasing therapy efficacy (Zhang et al., 2023).

Photodynamic Therapy

It entails administering a photosensitizer, which, when exposed to precise wavelengths of light, produces reactive oxygen species (ROS) capable of killing tumor cells (Chen et al., 2022). Green nanoparticles in PDT provide various advantages, including improved drug distribution, longer circulation duration, and increased therapy efficacy. The mechanisms involved in PDT using green nanoparticles include the following:

- **Mechanism of ROS generation**: Green nanoparticles, such as carbon-based dots, copper oxide, and other metallic nanoparticles, are efficient photosensitizers, absorbing light and producing ROS (Li et al., 2022). These ROS, such as singlet oxygen and hydroxyl radicals, can produce oxidative stress, resulting in cell apoptosis, necrosis, and malignant tissue damage (Yang et al., 2023b).
- **Targeted delivery of photosensitizers:** Green nanoparticles can be created to deliver photosensitizers to tumors. Functionalizing nanoparticles with targeting ligands or antibodies assure that the photosensitizer accumulates only in malignant cells, decreasing off-target effects (Huang et al., 2023). This tailored administration increases the therapeutic efficacy of PDT (Gao et al., 2023).
- **Sustained release and controlled activation:** The use of biodegradable green nanoparticles enables the continued release of photosensitizers, resulting in lasting therapeutic effects. These nanoparticles can be programmed to release their payload in response to specific environmental stimuli such as pH or temperature, guaranteeing that the photosensitizer is only active in the tumor microenvironment (Zhang et al., 2023).

16.18 BENEFITS OF GREEN NANOPARTICLES IN PTT AND PDT

Green nanoparticles offer several advantages in photothermal and photodynamic therapies, making them a promising approach in cancer treatment. The following benefits highlight their importance.

- **Biocompatibility and safety:** The use of biodegradable green nanoparticles enables the continued release of photosensitizers, resulting in lasting therapeutic effects. These nanoparticles can be programmed to release their payload in response to specific environmental stimuli such as pH or temperature, guaranteeing that the photosensitizer is only active in the tumor microenvironment (Zhang et al., 2023).

- **Eco-friendly synthesis:** Green nanotechnology emphasizes ecologically friendly synthesis methods that minimize the usage of harmful chemicals and solvents. This sustainable technique not only benefits the environment, but also ensures that the nanoparticles are suitable for biological uses (Chen et al., 2023). The utilization of plant-derived chemicals or biosynthetic methods lowers the environmental impact of nanomaterial synthesis (Zhang et al., 2023).
- **Increased treatment efficiency:** Green nanoparticles have unique qualities such as high surface area, variable size, and functionalization possibilities, which improve their potential to deliver medicinal compounds efficiently. In PTT, their nanoparticles' ability to generate localized heat at the tumor location improves therapy efficiency (Wang et al., 2023). In PDT, increased ROS production and tailored administration of photosensitizers improve treatment effects (Gao et al., 2023).
- **Minimal side effects:** Green nanoparticles are frequently biodegradable and do not remain in the body for long periods of time, reducing the risk of toxicity and other negative effects (Li et al., 2022). This makes them a safer option compared to standard medicines and conventional nanoparticles, which have the potential to inflict long-term damage (Zhao et al., 2023).

16.19 CLINICAL APPLICATIONS OF GREEN NANOPARTICLES IN PTT AND PDT

The potential clinical applications of green nanoparticles in PTT and PDT have been demonstrated in several preclinical and clinical studies. These applications include the following:

- **Targeted cancer therapy:** Green nanoparticles have showed great potential in treating specific cancer types, including lung, breast, and liver cancer (Chen et al., 2022). Nanoparticles can be functionalized with tumor-specific antibodies or peptides to achieve selective tumor targeting while minimizing damage to healthy organs (Huang et al., 2023).
- **Combination therapy:** Green nanoparticles have been combined with other therapeutic techniques, like as chemotherapy and immunotherapy, to improve treatment results. The use of PTT and PDT in conjunction with traditional medicines has been demonstrated to improve therapeutic efficacy while decreasing the chance of resistance (Yang et al., 2023).
- **Theranostic applications:** Green nanoparticles can also be used for theranostic applications, which combine diagnostic and therapeutic activities on a single platform. This enables real-time monitoring of therapy progression while also delivering therapeutic chemicals to the tumor location (Zhao et al., 2023).

16.20 IMMUNOMODULATION AND GENE THERAPY

Lung cancer is one of the most common and lethal malignancies worldwide, with high morbidity and mortality rates (Siegel et al., 2023). Traditional treatment approaches including chemotherapy, radiation, and surgery frequently have considerable adverse effects, emphasizing the need for novel medicines. Green nanotechnology, which focuses on synthesizing nanoparticles using environmentally benign processes, has emerged as a possible way to improving lung cancer treatment. Green nanoparticles have unique features that can influence immune responses while also delivering gene therapy in a highly focused and effective manner (Chen et al., 2023). Immunomodulation is the process of modifying the immune system's reaction to improve the body's ability to fight cancer. Green nanoparticles, which are biocompatible and can target specific cells, have showed promise in modifying immune responses for cancer treatment (Zhao et al., 2023). These nanoparticles can be tailored to stimulate immune cells, enhance the presentation of tumor antigens, or even deliver immune checkpoint inhibitors.

16.21 NANOCARRIERS FOR GENE SILENCING AND IMMUNOTHERAPY

Nanocarrier-based delivery systems have emerged as promising methods for lung cancer treatment by silencing genes and administering immunotherapy. These nanocarriers provide tailored delivery of medicinal drugs, increasing efficacy while reducing systemic toxicity.

Nanocarriers for gene silencing: Gene silencing by RNA interference (RNAi) includes the use of small interfering RNAs (siRNAs) to reduce specific gene expression. The instability and low cellular absorption of siRNAs make effective delivery to lung cancer cells difficult. To address these issues, nanocarriers such as lipid nanoparticles and polymer-based systems were developed. For example, stable nucleic acid lipid particles (SNALPs) have been used to deliver siRNAs therapeutically in vivo, offering nuclease resistance and facilitating cellular absorption (Wikipedia, 2024). Polymer-based nanoparticles have also been used to deliver siRNAs targeting chemoresistance genes, making lung cancer cells more susceptible to chemotherapy (Nakamura et al., 2020).

16.22 NANOCARRIERS IN IMMUNOTHERAPY

Nanocarriers also contribute significantly to lung cancer immunotherapy. They can influence the tumor microenvironment (TME) and boost the effectiveness of immune checkpoint inhibitors. Nanoparticles, for example, have been found to overcome adaptive immune resistance by targeting the TME, which improves the response to PD-1/PD-L1 inhibition in lung cancer (Zhang et al., 2023). Furthermore, nanocarriers can deliver therapeutic vaccinations and siRNAs simultaneously. A study found that combining a nanoparticle-based therapeutic vaccination with siRNA-loaded nanoparticles significantly inhibited tumor growth, underlining nanocarriers' promise in synergistic cancer therapy (Smith et al., 2022).

16.23 BIOLOGICAL AND FUNCTIONAL PROPERTIES OF GREEN NANOPARTICLES

Green nanoparticles, which are manufactured using environmentally benign technologies and biological entities such as plant extracts and microbes, have received a lot of attention due to their distinct biological and functional qualities. These nanoparticles provide a sustainable alternative to traditional synthesis methods by limiting environmental effect and toxicity (Khan et al., 2022).

The biological synthesis of nanoparticles employs natural reducing agents present in plants, fungi, bacteria, and algae. Plant extracts high in phytochemicals like flavonoids and phenolics have been widely used in the green production of metallic nanoparticles. For example, silver nanoparticles produced from *Curcuma longa* flower extract had considerable antibacterial action against a variety of diseases (Rajak et al., 2023).

16.24 FUNCTIONAL PROPERTIES OF GREEN NANOPARTICLES

Antibacterial Activity

Green nanoparticles have strong antibacterial capabilities, making them ideal for medical and environmental applications. The antibacterial activity is due to the formation of

ROS during light irradiation, which causes bacterial cell wall and membrane destruction (Hwang et al., 2022). Plant extract-derived zinc oxide nanoparticles exhibit potent antibacterial action against several pathogens, including *Escherichia coli* and *Staphylococcus aureus* (Takcı et al., 2024).

Antioxidant Activity

Green nanoparticles' antioxidant effects stem mostly from the presence of bioactive chemicals derived from the biological entities used in their manufacture. For example, gold and silver nanoparticles made from Quercus bark extracts have strong antioxidant activity, which is advantageous in the treatment of oxidative stress-related disorders (Azzi et al., 2023).

Catalytic Activity

Green nanoparticles have shown extraordinary catalytic characteristics, especially in environmental cleanup. Copper-containing biohybrids, such as CuHARS, have been demonstrated to accelerate nitric oxide generation, which has antibacterial properties and possible therapeutic applications (Zhang et al., 2021).

16.25 APPLICATIONS OF GREEN NANOPARTICLES

Green nanoparticles are used in drug delivery systems, biosensors, and therapeutic agents because of their biocompatibility and functional flexibility. Their capacity to conjugate with biomolecules allows for more focused medication delivery, which improves treatment efficacy and reduces negative effects (Noah & Ndangili, 2021). Green nanoparticles are used in several environmental applications, including wastewater treatment and pollutant degradation. Their photocatalytic characteristics enable the breakdown of organic contaminants and the removal of heavy metals from contaminated water sources, hence promoting environmental sustainability (Bakhtiari et al., 2024).

16.26 CHALLENGES IN CLINICAL TRANSLATION

Diagnostic challenges: Early and accurate diagnosis of lung cancer is critical yet challenging. The complexity of tumor heterogeneity and the limitations of current imaging techniques contribute to diagnostic difficulties (Mazzone et al., 2021).

Therapeutic challenges: The development of resistance to targeted therapies and immunotherapies poses a significant hurdle in lung cancer treatment (Ramalingam et al., 2020).

Toxicity and side effects: Balancing treatment efficacy with minimizing adverse effects remains a persistent challenge in the clinical management of lung cancer (Hanna et al., 2020).

Translational research challenges: Identifying and validating reliable biomarkers for lung cancer is essential for personalized medicine but is fraught with challenges related to specificity and sensitivity (Lindeman et al., 2018).

Clinical trial design: Designing robust clinical trials that accurately reflect patient populations and predict real-world outcomes is complex and resource-intensive (Neal et al., 2019).

Regulatory and ethical challenges: Navigating the regulatory landscape and addressing ethical considerations in clinical research are critical to the successful translation of lung cancer therapies (Kim et al., 2017).

16.27 FUTURE PERSPECTIVES

The field of green nanotechnology is undergoing rapid evolution, marked by the integration of eco-friendly approaches to nanoparticle synthesis and the exploration of novel bioresources. Emerging trends include the utilization of plant extracts, algae, and microbial systems as sustainable precursors for nanoparticle fabrication. Researchers are also focusing on hybrid nanomaterials that combine organic and inorganic components to enhance therapeutic efficacy while minimizing environmental footprints. Advances in biocompatible coatings and functionalization techniques are enabling targeted drug delivery systems tailored to lung cancer treatment. Furthermore, there is a growing emphasis on life cycle analysis to assess the environmental impacts of green nanotechnology products, ensuring a holistic approach to sustainability.

16.28 ROLE OF ARTIFICIAL INTELLIGENCE AND MACHINE LEARNING IN NANOPARTICLE DESIGN

Artificial intelligence (AI) and machine learning (ML) are revolutionizing nanoparticle design by enabling predictive modelling and optimization. These technologies facilitate the identification of optimal synthesis parameters, such as temperature, pH, and reactant concentrations, to achieve desired nanoparticle properties. In lung cancer therapy, AI-driven algorithms are being employed to simulate interactions between nanoparticles and biological systems, aiding in the design of highly specific and effective drug delivery mechanisms. ML techniques are also instrumental in analyzing large datasets from preclinical and clinical studies, accelerating the discovery of new therapeutic agents. The integration of AI and ML is expected to significantly reduce the time and cost associated with the development of green nanotechnologies, fostering innovation in the field.

16.29 COLLABORATIVE APPROACHES FOR GLOBAL IMPLEMENTATION

The global implementation of green nanotechnology for lung cancer management necessitates collaborative efforts across academia, industry, and governmental organizations. Interdisciplinary research initiatives are essential to bridge the gap between laboratory findings and real-world applications. Public–private partnerships can drive the commercialization of green nanotechnology-based therapies, ensuring their accessibility and affordability. International collaborations can facilitate the standardization of protocols and regulatory frameworks, promoting the widespread adoption of eco-friendly nanomedicine. Additionally, community engagement and education programs are vital to raising awareness about the potential benefits of green nanotechnology, fostering public trust, and acceptance.

16.30 CONCLUSION

Green nanotechnology represents a transformative approach to lung cancer management, offering numerous benefits such as enhanced biocompatibility, reduced toxicity, and lower environmental impact. By leveraging natural resources and sustainable synthesis methods, green nanotechnology aligns with global sustainability goals while addressing critical healthcare challenges. Its potential to improve therapeutic outcomes, reduce side effects, and provide cost-effective solutions underscores its significance in modern medicine. The integration of advanced technologies, such as AI and ML, further amplifies its promise, paving the way for innovative and personalized cancer therapies. Despite its promising potential, green nanotechnology is still in its nascent stages and requires sustained research and development efforts. Continued investment in multidisciplinary studies is essential to overcome existing challenges, such as scalability, reproducibility, and regulatory hurdles. Collaborative approaches involving stakeholders from diverse sectors can accelerate progress, ensuring that the benefits of green nanotechnology are realized on a global scale. By prioritizing innovation and sustainability, researchers and policymakers can unlock the full potential of green nanotechnology, transforming the landscape of lung cancer treatment and beyond.

REFERENCES

Ahmed, S., Ahmad, M., Swami, B. L., & Ikram, S. (2016). A review on plants extract mediated synthesis of silver nanoparticles for antimicrobial applications: A green expertise. *Journal of Advanced Research*, *7*(1), 17–28. https://doi.org/10.1016/j.jare.2015.02.007

Ahmed, S., Ahmad, M., Swami, B. L., & Ikram, S. (2021). Green synthesis of silver nanoparticles using plant extracts and their applications: A review. *Environmental Nanotechnology, Monitoring & Management*, *6*, 1–14. https://doi.org/10.1016/j.enmm.2021.100204

Ahmed, S., Annu, A., & Ikram, S. (2023). Green synthesis of nanoparticles: Emerging trends and future perspectives. *Journal of Nanotechnology and Sustainability*, *15*(3), 123–134.

Ahmed, S., Saifullah, Ahmad, M., Swami, B. L., & Ikram, S. (2019). Green synthesis of metal nanoparticles and their beneficial roles in plant growth and protection: A review. *Biotechnology & Biotechnological Equipment*, *33*(1), 896–915. https://doi.org/10.1080/13102818.2019.1625708

Ali, A., Hussain, S., & Khan, M. (2023). Plant-mediated synthesis of nanoparticles: Mechanistic insights and applications. *Green Chemistry Reviews*, *18*(2), 210–229.

Alsammarraie, F. K., Wang, W., Zhou, P., Mustapha, A., & Lin, M. (2018). Green synthesis of silver nanoparticles using turmeric extracts and investigation of their antibacterial activities. *Colloids and Surfaces B: Biointerfaces*, *171*, 398–405. https://doi.org/10.1016/j.colsurfb.2018.07.044

Anjana, V. N., Joseph, M., Francis, S., Joseph, A., & Koshy, E. P. (2021). Microwave assisted green synthesis of silver nanoparticles for optical, catalytic, biological and electrochemical applications. *Artificial Cells, Nanomedicine, and Biotechnology*, *49*(1), 241–251.

Azzi, M., Medila, I., Toumi, I., Laouini, S. E., Bouafia, A., Hasan, G. G., Mohammed, H. A., Mokni, S., Alsalme, A., & Barhoum, A. (2023). Plant extract-mediated synthesis of Ag/Ag_2O nanoparticles using *Olea europaea* leaf extract: Assessing antioxidant, antibacterial, and toxicological properties. *Biomass Conversion and Biorefinery*. https://doi.org/10.1007/s13399-023-05093-w

Bakhtiari, S., Salari, M., Shahrashoub, M., Zeidabadinejad, A., Sharma, G., & Sillanpää, M. (2024). A comprehensive review on green and eco-friendly nano-adsorbents for the removal of heavy metal ions: Synthesis, adsorption mechanisms, and applications. *Current Pollution Reports*. https://doi.org/10.1007/s40726-023-00290-7

Baruwati, B., Polshettiwar, V., & Varma, R. S. (2009). Glutathione promoted expeditious green synthesis of silver nanoparticles in water using microwave irradiation. *Green Chemistry*, *11*(7), 926–930. https://doi.org/10.1039/b902618k

Bashir, M. R. (2014). Magnetic resonance contrast agents for liver imaging. *Magnetic resonance imaging clinics of North America*, *22*(3), 283–293. https://doi.org/10.1016/j.mric.2014.04.002

Cai, W., & Zhang, F. (2023). Mechanisms of photothermal therapy using nanoparticles. *Journal of Nanomedicine*, *12*(1), 45–59. https://doi.org/10.1016/j.jnanomed.2022.11.003

Cai, Z., Ye, Z., Yang, X., Chang, Y., Wang, H., Liu, Y., & Cao, A. (2011). Encapsulated enhanced green fluorescence protein in silica nanoparticle for cellular imaging. *Nanoscale*, *3*(5), 1974–1976. https://doi.org/10.1039/c0nr00956c

Chakraborty, S., Kumar, R., & Das, A. (2023). Integration of artificial intelligence with green nanotechnology for lung cancer diagnosis. *Nanotechnology Advances*, *15*(4), 335–348. https://doi.org/10.1016/j.nta.2023.08.001

Chen, J., Zhang, Y., & Li, M. (2023). Carbon quantum dots as nanocarriers for RNAi-based gene therapy in non-small cell lung cancer. *Nanomedicine: Nanotechnology, Biology, and Medicine*, *35*, 145–159. https://doi.org/10.1016/j.nano.2022.11.007

Chen, L. J., Sun, S. K., Wang, Y., Yang, C. X., Wu, S. Q., & Yan, X. P. (2016). Activatable multifunctional persistent luminescence nanoparticle/copper sulfide nanoprobe for in vivo luminescence imaging-guided photothermal therapy. *ACS Applied Materials & Interfaces*, *8*(48), 32667–32674.

Chen, W., Li, Q., & Zhou, L. (2022). Photodynamic therapy: Advances and challenges in the use of nanoparticles. *Cancer Nanotechnology Journal*, *5*(3), 22–34. https://doi.org/10.1016/j.canj.2022.03.004

Chen, Y., Liu, H., & Wang, J. (2023). Advances in energy-efficient nanomaterial synthesis for sustainable applications. *Journal of Nanotechnology*, *45*(3), 123–140.

Chronos, N. A., Goodall, A. H., Wilson, D. J., Sigwart, U., & Buller, N. P. (1993). Profound platelet degranulation is an important side effect of some types of contrast media used in interventional cardiology. *Circulation*, *88*(5), 2035–2044.

Cuccurullo, V., Di Stasio, G. D., Mazzarella, G., &Cascini, G. L. (2018). Microvascular invasion in HCC: The molecular imaging perspective. *Contrast Media & Molecular Imaging*, *2018*(1), 9487938.

Das, R., Kumar, N., & Ghosh, P. (2023). Recent advancements in the green synthesis of nanoparticles. *Sustainable Nanotechnology Advances*, *5*(1), 45–59.

Donato, H., França, M., Candelária, I., &Caseiro-Alves, F. (2017). Liver MRI: From basic protocol to advanced techniques. *European Journal of Radiology*, *93*, 30–39. https://doi.org/10.1016/j.ejrad.2017.05.028

Dubreil, L., Leroux, I., Ledevin, M., Schleder, C., Lagalice, L., Lovo, C., Fleurisson, R., Passemard, S., Kilin, V., Gerber-Lemaire, S., Colle, M. A., Bonacina, L., &Rouger, K. (2017). Multi-harmonic imaging in the second near-infrared window of nanoparticle-labeled stem cells as a monitoring tool in tissue depth. *ACS Nano*, *11*(7), 6672–6681. https://doi.org/10.1021/acsnano.7b00773

Fan, Q., Cheng, K., Hu, X., Ma, X., Zhang, R., Yang, M., Lu, X., Xing, L., Huang, W., Gambhir, S. S,. & Cheng, Z. (2014). Transferring biomarker into molecular probe: Melanin nanoparticle as a naturally active platform for multimodality imaging. *Journal of the American Chemical Society*, *136*(43), 15185–15194.

Fang, H., Zhang, L., & Li, Q. (2023). Nanorobots for targeted chemotherapy in lung cancer: A new approach in theranostics. *Cancer Nanotechnology Journal*, *8*(2), 45–56. https://doi.org/10.1016/j.cnan.2023.02.005

Foucault-Collet, A., Gogick, K. A., White, K. A., Villette, S., Pallier, A., Collet, G., Kieda, C., Li, T., Geib, S. J., Rosi, N. L., &Petoud, S. (2013). Lanthanide near infrared imaging in living cells with Yb3+ nano metal organic frameworks. *Proceedings of the National Academy of Sciences of the United States of America*, *110*(43), 17199–17204. https://doi.org/10.1073/pnas.1305910110

Gao, X., Song, Y., & Luo, Y. (2023). Gold nanoshells for optoacoustic imaging of lung cancer. *Journal of Biomedical Nanotechnology*, *19*(6), 752–764. https://doi.org/10.1002/jbmm.20456

Genovese, D., Bonacchi, S., Juris, R., Montalti, M., Prodi, L., Rampazzo, E., & Zaccheroni, N. (2013). Prevention of self-quenching in fluorescent silica nanoparticles by efficient energy transfer. *AngewandteChemie (International ed. in English)*, *52*(23), 5965–5968. https://doi.org/10.1002/anie.201301155

Gupta, P., Singh, M., & Jain, N. (2023). Green nanotechnology for lung cancer applications: Nanomaterial synthesis and biomedical applications. *Journal of Nanoscience and Nanotechnology*, *23*(4), 1014–1028. https://doi.org/10.1002/jnn.27293

Hanna, N. H., et al. (2020). Therapy for stage IV non-small-cell lung cancer without driver alterations: ASCO and OH (CCO) joint guideline update. *Journal of Clinical Oncology: Official Journal of the American Society of Clinical Oncology*, *38*(14), 1608–1632. https://doi.org/10.1200/JCO.19.03022

Hoag, G. E., Collins, J. B., Holcomb, J. L., Hoag, J. R., Nadagouda, M. N., & Varma, R. S. (2009). Degradation of bromothymol blue by 'greener' nano-scale zero-valent iron synthesized using tea polyphenols. *Journal of Materials Chemistry*, *19*(45), 8671–8677. https://doi.org/10.1039/b910382h

Horwat, D., Billard, A., & Pierson, J. F. (2011). Influence of energy and pressure on nanostructured thin films obtained by physical vapor deposition. *Thin Solid Films*, *519*(14), 4366–4372. https://doi.org/10.1016/j.tsf.2011.01.369

Hoshyar, N., Gray, S., Han, H., & Bao, G. (2016). The effect of nanoparticle size on in vivo pharmacokinetics and cellular interaction. *Nanomedicine*, *11*(6), 673–692.

Hu, Y., Wang, Y., Jiang, J., Han, B., Zhang, S., Li, K., Ge, S., & Liu, Y. (2016). Preparation and characterization of novel perfluorooctyl bromide nanoparticle as ultrasound contrast agent via layer-by-layer self-assembly for folate-receptor-mediated tumor imaging. *BioMed Research International*, *2016*, 6381464. https://doi.org/10.1155/2016/6381464

Huang, X., Yang, X., & Wang, F. (2023). Gold nanoparticles for photothermal therapy in cancer: A comprehensive review. *Nanomedicine: Nanotechnology, Biology, and Medicine*, *45*, 135–148. https://doi.org/10.1016/j.nano.2023.01.010

Hussain, I., Singh, N. B., Singh, A., Singh, H., & Singh, S. C. (2016). Green synthesis of nanoparticles and its potential application. Biotechnology Letters, *38*(4), 545–560. https://doi.org/10.1007/s10529-015-2026-7

Hwang, C., Choi, M.-H., Kim, H.-E., Jeong, S.-H., & Park, J.-U. (2022). Reactive oxygen species-generating hydrogel platform for enhanced antibacterial therapy. *NPG Asia Materials*, *14*(1). https://doi.org/10.1038/s41427-022-00370-5

Hwang, E. S., Cao, C., Hong, S., Jung, H. J., Cha, C. Y., Choi, J. B., Kim, Y. J., & Baik, S. (2006). The DNA hybridization assay using single-walled carbon nanotubes as ultrasensitive, long-term optical labels. *Nanotechnology*, *17*(14), 3442–3445. https://doi.org/10.1088/0957-4484/17/14/016

Ikram, S. (2015). Green synthesis and characterization of silver nanoparticles using aqueous extract of *Solanum melongena* and their antibacterial activity. *Journal of Nanostructure in Chemistry*, *5*(4), 365–370. https://doi.org/10.1007/s40097-015-0160-5

Iravani, S., & Varma, R. S. (2022). Green synthesis of nanoparticles. *Progress in Materials Science*, *119*, 101–113.

Jiang, C., & Liu, Y. (2023). Biodegradable green nanoparticles for photothermal therapy in cancer treatment. *Advanced Materials*, *35*(7), 2105216. https://doi.org/10.1002/adma.202105216

Johnson, E., Krishnan, R. R., Chandran, S. R., & Prema, K. H. (2023). Green mediated sol-gel synthesis of copper oxide nanoparticle: An efficient candidate for waste water treatment and antibacterial agent. *Journal of Sol-Gel Science and Technology*, *98*(3), 456–467.

Jun, Y. W., Lee, J. H., & Cheon, J. (2008). Chemical design of nanoparticle probes for high-performance magnetic resonance imaging. *AngewandteChemie (International ed. in English)*, *47*(28), 5122–5135. https://doi.org/10.1002/anie.200701674

Jung, D. H., Hwang, S., Song, G. W., Ahn, C. S., Moon, D. B., Ha, T. Y., Kim, K. H., Park, G. C., Kim, B. S., Park, I. J., Lim, S. B., Kim, J. C., Yoo, M. W., Byeon, J. S., Jung, H. Y., Lee, G. H., Myung, S. J., Choe, J., Park, H. W., & Lee, S. G. (2016). Survival benefit of early cancer detection through regular endoscopic screening for de novo gastric and colorectal cancers in Korean liver transplant recipients. In *Transplantation proceedings* (Vol. 48, No. 1, pp. 145–151). Elsevier.

Khan, M., et al. (2022). Plant-mediated nanoparticle synthesis. *Frontiers in Nanotechnology*, *3*, 405–420.

Kim, E. S., Bernstein, D., Hilsenbeck, S. G., & Chung, C. H. (2017). Modernizing eligibility criteria for molecularly driven trials. *Journal of Clinical Oncology*, *35*(33), 3737–3744. https://doi.org/10.1200/JCO.2017.73.7910

Kumar, R., & Yadav, R. (2022). Biocompatible nanoparticles for biomedical applications. *Advances in Bioscience and Nanomedicine*, *14*(2), 89–103.

Kumar, S., & Sharma, R. (2023). Breath-based nanobiosensors for early-stage lung cancer detection. *Sensors and Actuators B: Chemical*, *333*, 131548. https://doi.org/10.1016/j.snb.2023.131548

Kundu, S., Zhang, X., & Li, T. (2023). Life cycle analysis in green nanotechnology: Methods and challenges. *Sustainable Materials Journal*, *15*(2), 98–115.

Lee, J., Min, H. S., You, D. G., Kim, K., Kwon, I. C., Rhim, T., & Lee, K. Y. (2016). Theranostic gas-generating nanoparticles for targeted ultrasound imaging and treatment of neuroblastoma. *Journal of Controlled Release: Official Journal of the Controlled Release Society*, *223*, 197–206. https://doi.org/10.1016/j.jconrel.2015.12.051

Lee, S., Cha, E. J., Park, K., Lee, S. Y., Hong, J. K., Sun, I. C., Kim, S. Y., Choi, K., Kwon, I. C., Kim, K., & Ahn, C. H. (2008). A near-infrared-fluorescence-quenched gold-nanoparticle imaging probe for in vivo drug screening and protease activity determination. *AngewandteChemie (International ed. in English)*, *47*(15), 2804–2807. https://doi.org/10.1002/anie.200705240

Li, W., Tan, H., & Zhao, X. (2023). Iron oxide nanoparticles as contrast agents in magnetic resonance imaging of lung cancer metastases. *Nanomedicine: Nanotechnology, Biology, and Medicine*, *33*(9), 69–81. https://doi.org/10.1016/j.nano.2023.04.002

Li, X., Wang, H., & Zhang, Q. (2023). Green copper nanoparticles for photothermal cancer therapy. *Journal of Materials Chemistry B*, *11*(4), 523–537. https://doi.org/10.1039/c3tb01011a

Li, X., Xie, Y., & Wang, Y. (2022). Nanomaterials in targeted lung cancer therapy. *International Journal of Nanomedicine*, *17*, 1123–1137.

Lindeman, N. I., et al. (2018). Updated molecular testing guideline for the selection of lung cancer patients for treatment with targeted tyrosine kinase inhibitors: guideline from the college of American pathologists, the international association for the study of lung cancer, and the association for molecular pathology. *Archives of Pathology & Laboratory Medicine*, *142*(3), 321–346. https://doi.org/10.5858/arpa.2017-0388-CP

Liu, Y., Chen, H., & Zhang, Z. (2023). Green graphene quantum dots for liver cancer photodynamic therapy. *Nanomedicine*, *18*(2), 56–68. https://doi.org/10.1016/j.nano.2022.09.004

Lu, J., Ma, S., Sun, J., Xia, C., Liu, C., Wang, Z., Zhao, X., Gao, F., Gong, Q., Song, B., Shuai, X., Ai, H., & Gu, Z. (2009). Manganese ferrite nanoparticle micellar nanocomposites as MRI contrast agent for liver imaging. *Biomaterials*, *30*(15), 2919–2928. https://doi.org/10.1016/j.biomaterials.2009.02.001

Machado, S., Silva, R., Santos, C., & Morais, P. V. (2023). Green synthesis of metallic nanoparticles: Toward sustainable chemistry and environmental applications. *Journal of Environmental Chemical Engineering*, *11*(4), 110927. https://doi.org/10.1016/j.jece.2023.110927

Mazzone, P. J., et al. (2021). Management of lung nodules and lung cancer screening during the COVID-19 pandemic: CHEST expert panel report. *Chest*, 160(2), e349–e360. https://doi.org/10.1016/j.chest.2021.02.069

Nakamura, T., Yamada, Y., & Sato, Y. (2020). Nanoparticle-mediated gene silencing for sensitization of lung cancer cells to chemotherapy. *Cancer Research*, *80*(14), 3005–3015.

Neal, J. W., Gainor, J. F., Shaw, A. T., & Kim, D. W. (2019). Developing biomarker-specific end points in lung cancer clinical trials. *Nature Reviews Clinical Oncology*, *16*(8), 514–522. https://doi.org/10.1038/s41571-019-0190-1

Noah, N., & Ndangili, P. (2021). Green synthesis of nanomaterials from sustainable materials for biosensors and drug delivery. *arXiv preprint*. https://arxiv.org/abs/2112.04740

Oh, I. H., Min, H. S., Li, L., Tran, T. H., Lee, Y. K., Kwon, I. C., Choi, K., Kim, K. & Huh, K. M. (2013). Cancer cell-specific photoactivity of pheophorbide a-glycol chitosan nanoparticles for photodynamic therapy in tumor-bearing mice. *Biomaterials*, *34*(27), 6454–6463.

Patel, V., Shah, S., & Bhatt, D. (2023). Green synthesis of nanoparticles for biomedical applications: Advances in cancer therapy. *Environmental Nanotechnology*, *10*(1), 1–12. https://doi.org/10.1016/j.envnan.2023.03.004

Patra, S., Panda, P. K., & Dash, M. (2023). Plant-mediated synthesis of gold nanoparticles and their role in cancer diagnostics. *Nanomedicine*, *18*(4), 297–309. https://doi.org/10.2217/nnm-2023-0014

Petrik, M., Weigel, C., Kirsch, M., & Hosten, N. (2005). No detectable nephrotoxic side effect using a dimer, non-ionic contrast media in cerebral perfusion computed tomography in case of suspected brain ischemia. *Rofo: Fortschritte auf dem Gebiete der Rontgenstrahlen und der Nuklearmedizin*, *177*(9), 1242–1249.

Raghunandan, D., et al. (2023). Advances in green nanotechnology. *ACS Sustainable Chemistry & Engineering*, *11*(5), 3456–3467.

Rajak, K. K., Pahilani, P., Patel, H., Kikani, B., Desai, R., & Kumar, H. (2023). Green synthesis of silver nanoparticles using *Curcuma longa* flower extract and antibacterial activity. *arXiv preprint*. https://arxiv.org/abs/2304.04777

Rajput, P., & Kumbhar, S. (2023). Nanomaterials in renewable energy: A review of recent advancements. *Renewable Energy Journal*, *34*(6), 413–429.

Ramalingam, S. S., et al. (2020). Overall survival with Osimertinib in untreated, EGFR-mutated advanced NSCLC. *New England Journal of Medicine*, *382*(1), 41–50. https://doi.org/10.1056/NEJMoa1913662

Sakamoto, J. H., Smith, B. R., Xie, B., Rokhlin, S. I., Lee, S. C., & Ferrari, M. (2005). The molecular analysis of breast cancer utilizing targeted nanoparticle based ultrasound contrast agents. *Technology in Cancer Research & Treatment*, *4*(6), 627–636. https://doi.org/10.1177/153303460500400606

Santra, S., & Malhotra, A. (2011). Fluorescent nanoparticle probes for imaging of cancer. *Wiley Interdisciplinary Reviews. Nanomedicine and Nanobiotechnology*, *3*(5), 501–510. https://doi.org/10.1002/wnan.134

Saxena, M., Sharma, G., Gupta, R., & Ahmad, N. (2022). Green nanotechnology-driven development of eco-friendly contrast agents for biomedical imaging. *Journal of Nanobiotechnology*, *20*(1), 312–326. https://doi.org/10.1186/s12951-022-01567-3

Seo, M., Gorelikov, I., Williams, R., & Matsuura, N. (2010). Microfluidic assembly of monodisperse, nanoparticle-incorporated perfluorocarbon microbubbles for medical imaging and therapy. *Langmuir: The ACS Journal of Surfaces and Colloids*, *26*(17), 13855–13860. https://doi.org/10.1021/la102272d

Sharma, A., Gupta, P., & Singh, B. (2022). Eco-friendly synthesis of nanoparticles for biomedical applications. *Journal of Nanobiotechnology*, *20*, 104. https://doi.org/10.1186/s12951-022-01365-8

Sharma, N., Gupta, A., & Singh, P. (2023). Role of biomolecules in green nanoparticle synthesis. *Nanotechnology Today*, *12*(1), 56–78.

Shi, J., Lin, Y., & Zhang, X. (2023). DNA-based aptamer sensors for lung cancer biomarker detection. *Sensors and Materials*, *35*(1), 59–70. https://doi.org/10.18494/SAM.2023.2789

Siegel, R. L., Miller, K. D., & Jemal, A. (2023). Cancer statistics, 2023. *CA: A Cancer Journal for Clinicians*, *73*(1), 17–48. https://doi.org/10.3322/caac.21763

Singh, P., Pandey, S., & Jha, S. (2022). Green-synthesized gold nanoparticles for targeted lung cancer therapy. *Biomaterials Science*, *10*(8), 2324–2335.

Singh, R., Verma, H., & Kaur, N. (2023). Advances in biosensors for lung cancer diagnostics using green-synthesized nanoparticles. *Biosensors and Bioelectronics*, *211*, 114324. https://doi.org/10.1016/j.bios.2023.114324

Smith, A. L., Brown, K. J., & Wilson, J. M. (2022). Immune checkpoint silencing using RNAi-incorporated nanoparticles enhances cancer immunotherapy. *Journal for ImmunoTherapy of Cancer*, *10*(2), e003928.

Takcı, D. K., Ozdenefe, M. S., Hüner, T., &Takcı, H. A. M. (2024). Plant-mediated green route to the synthesis of zinc oxide nanoparticles: In vitro antibacterial potential. *Journal of the Australian Ceramic Society*. https://doi.org/10.1007/s41779-023-00892-5

Torres Martin de Rosales, R., Tavaré, R., Glaria, A., Varma, G., Protti, A., & Blower, P. J. (2011). 99mTc-bisphosphonate-iron oxide nanoparticle conjugates for dual-modality biomedical imaging. *Bioconjugate Chemistry*, *22*(3), 455–465.

Vargo, K. B., Zaki, A. A., Warden-Rothman, R., Tsourkas, A., & Hammer, D. A. (2015). Superparamagnetic iron oxide nanoparticle micelles stabilized by recombinant oleosin for targeted magnetic resonance imaging. *Small (Weinheim an der Bergstrasse, Germany)*, *11*(12), 1409–1413. https://doi.org/10.1002/smll.201402017

Wang, H., Zhang, X., & Li, J. (2023). Green tea extract-derived nanoparticles for dendritic cell activation and lung cancer immunotherapy. *International Journal of Nanomedicine*, *18*, 3101–3115. https://doi.org/10.2147/IJN.S4233

Wang, X., Chen, H., Zheng, Y., Ma, M., Chen, Y., Zhang, K., Zeng, D., & Shi, J. (2013). Au-nanoparticle coated mesoporous silica nanocapsule-based multifunctional platform for ultrasound mediated imaging, cytoclasis and tumor ablation. *Biomaterials*, *34*(8), 2057–2068. https://doi.org/10.1016/j.biomaterials.2012.11.044

Wei, Y., Chen, Q., Wu, B., Zhou, A., & Xing, D. (2012). High-sensitivity in vivo imaging for tumors using a spectral up-conversion nanoparticle NaYF4: Yb3+, Er3+ in cooperation with a microtubulin inhibitor. *Nanoscale*, *4*(13), 3901–3909. https://doi.org/10.1039/c2nr30804e

Wikipedia. (2024). *Stable nucleic acid lipid particle*. Retrieved from https://en.wikipedia.org/wiki/Stable_nucleic_acid_lipid_particle

Wojnarowicz, J., Kusnieruk, S., & Chudoba, T. (2018). Microwave solvothermal synthesis of nanomaterials. *Crystals*, *8*(6), 241. https://doi.org/10.3390/cryst8060241

Yang, X., Lin, X., & Zhang, C. (2023a). Graphene-based biosensors for lung cancer biomarker detection. *Carbon*, *214*, 426–437. https://doi.org/10.1016/j.carbon.2023.02.017

Yang, Y., Zhang, S., & Liu, J. (2023b). Green nanoparticles in combination with other therapies for cancer treatment. *Nanomedicine: Nanotechnology, Biology, and Medicine*, *48*, 82–98. https://doi.org/10.1016/j.nano.2023.03.007

Zhang, L., Liu, F., & Chen, X. (2023). Green nanotechnology: Sustainable approaches in lung cancer treatment and diagnosis. *Nanotechnology Reviews*, *12*(7), 1051–1066. https://doi.org/10.1515/ntrev-2023-0024

Zhang, Z., Mascheri, N., Dharmakumar, R., Fan, Z., Paunesku, T., Woloschak, G., & Li, D. (2009). Superparamagnetic iron oxide nanoparticle-labeled cells as an effective vehicle for tracking the GFP gene marker using magnetic resonance imaging. *Cytotherapy*, *11*(1), 43–51. https://doi.org/10.1080/14653240802420243

Zhao, J., Yang, Z., & Li, W. (2023). Copper sulfide nanoparticles for photodynamic therapy in prostate cancer. *Journal of Nanobiotechnology*, *21*(3), 129–141. https://doi.org/10.1186/s12951-023-02113-w

Zhao, J., Zhang, J., & Li, W. (2023). Synthesis and applications of green nanoparticles in cancer therapy. *Journal of Nanoscience and Nanotechnology*, *23*(4), 1221–1235. https://doi.org/10.1002/jnn.27287

Zhao, M., Zhang, X., & Liu, Y. (2023). Gold nanoclusters for CRISPR/Cas9 gene delivery in lung cancer. *Journal of Nanobiotechnology*, *21*(1), 89–100. https://doi.org/10.1186/s12951-023-02109-5

Zhao, Y., Shen, Q., & Chen, L. (2024). Development of nanobiosensors for early detection of lung cancer from exhaled breath. *Sensors*, *24*(2), 211. https://doi.org/10.3390/s24020211

Zhao, Y., van Rooy, I., Hak, S., Fay, F., Tang, J., Davies, C. D. L., Skobe, M., Fisher, E. A., Radu, A., Fayad, Z. A., Donega, C. D. M., Meijerink, A., & Mulder, W. J. (2013). Near-infrared fluorescence energy transfer imaging of nanoparticle accumulation and dissociation kinetics in tumor-bearing mice. *ACS Nano*, *7*(11), 10362–10370.

Zhou, X., Li, H., & Zhang, W. (2023). Stimuli-responsive nanoparticles for precision drug delivery in lung cancer. *Nano Today*, *50*, 101807.

17 Clinical Trials of Drug Delivery Systems for the Management of Lung Cancer

Annu Saini[1], Lokesh Nagar[1], Tarif Hussain[1], Nitin Dahiya[2], Harish Dureja[1,3,4,5], and Prabhakar Kumar Verma[1]*

[1] Department of Pharmaceutical Sciences, Maharshi Dayanand University, Rohtak 124001, India

[2] School of Pharmacy, Starex University, Gurugram, India

[3] Faculty of Health, Australian Research Centre in Complementary and Integrative Medicine, University of Technology, Sydney, Ultimo NSW 2007, Australia

[4] Centre for Research Impact & Outcome, Chitkara College of Pharmacy, Chitkara University, Rajpura, Punjab 140401, India

[5] School of Medical and Life Sciences, Sunway University, Kuala Lumpur, Malaysia

*Corresponding author: Prabhakar Kumar Verma, drpkverma.pharma@mdurohtak.ac.in.

17.1 INTRODUCTION

Lung cancer (LC) is a global health issue that represents one of the most widespread and deadly forms of all cancers globally (Leiter et al., 2023). Contributing to approximately 11.6% of all new cancer cases and 18.4% of cancer-related deaths annually, lung cancer is the leading cause of cancer mortality among both men and women (Zhang et al., 2021). Risk factors for lung cancer include a complex interplay, the paramount of which is cigarette smoking, the most important risk factor for the disease. Other important contributors are occupational (such as asbestos and radon), environmental pollution, and genetic predisposition (Cheng et al., 2016). The survival rate in lung cancer patients depends primarily on early diagnosis, and surgical resection of the tumour tissue is often the preferred therapy (Mukherjee et al., 2019). Early-stage detection of LSs offers significantly better outcomes, with the survival rate reaching up to 70% for stage I or II cancer, but most of these cases are diagnosed at a very advanced stage due to the lack of early screening and the asymptomatic nature of cancer in its initial phases (Li et al., 2022). Clinically, LCs have several challenges. The condition is often diagnosed at late stages, when curative treatment options are very limited (Nooreldeen & Bach, 2021). Associated conditions, like chronic obstructive pulmonary disease, are also responsible for complicating its management and worsening the prognosis of LCs (Uliński et al., 2022). Furthermore, LC is a heterogeneous disease with multiple histological subtypes with different biological behaviors and responses to therapy (Testa et al., 2018). The cases of adenocarcinoma are the most frequent subtypes, particularly among non-smokers and women, which adds more complexity to both diagnosis and treatment strategies (Saito et al., 2017). Available treatment approaches, such as chemotherapy, radiotherapy, and surgery, usually have limited effectiveness in the late stages of LCs, which are also linked to significant toxicity (Ramos et al., 2025). With

DOI: 10.1201/9781003595830-17

the introduction of target treatment strategies and immune therapies, the outcome of patients has significantly improved, but therapeutic drug resistance and systemic side effects are some of the serious issues that persist. These clinical challenges emphasize the need for effective and precise therapeutic options (Araghi et al., 2023; Haider et al., 2022).

Novel drug delivery systems (NDDS) are an emerging, promising approach to many of these clinical challenges (Sharma et al., 2019). Through target delivery of chemotherapeutic agents to tumour sites/regions, advanced drug delivery systems can improve the concentration of the drug in malignant cells, which will also minimize systemic exposure and toxicity (Hani et al., 2021; Paudel et al., 2020). NDDS being explored for the lung cancer management include nanoparticle-based carriers, liposomes, inhalable formulations, and antibody drug conjugates (Ezhilarasan et al., 2022). Advantages of these technologies include improved pharmacokinetics, reduced off-target effects, ability to bypass biological barriers, and potential overcoming the resistance to multidrug mechanisms (Madni et al., 2017). NDDS have great potential to revolutionize the therapeutic landscape and therefore deserve to be noted for its significance to lung cancer (Ezike et al., 2023). With improved drug bioavailability and ability to achieve sustained or controlled release, they can both maximize antitumour efficacy and possibly convert lung cancer into a more controllable chronic disease (Al Ragib et al., 2022). NDDS being explored for the lung cancer management include nanoparticle-based carriers, liposomes, inhalable formulations as well as antibody drug conjugates (Chavda et al., 2022). As noted earlier, the advantages of these technologies include improved pharmacokinetics, reduced off target effects, ability to bypass biological barriers, and potential overcoming the resistance to multidrug mechanisms (Rajendran et al., 2024). NDDS have great potential to revolutionize the therapeutic landscape and therefore deserves to be noted for its significance to lung cancer. With improved drug bioavailability and ability to achieve sustained or controlled release, they can both maximize antitumour efficacy and possibly convert lung cancer into a more controllable chronic disease (Quadir et al., 2022). In addition, NDDS enable a combination of different therapeutic agents (cytotoxics and immune modulators) within a single platform thus creating new possible synergistic treatment regimens (Milewska et al., 2021). There is therefore a strong rationale for clinical trials that assess these systems given the rapid evolution of NDDS technologies, which have great potential for application in lung cancer therapy (Shojaei et al., 2024). Clinical trials are an important cornerstone for DDS translation from preclinical successes to clinical benefits, and they carefully inspect DDS safety and efficacy in different patient cohorts (Cojocaru et al., 2024). Additionally, they can also give pharmacodynamic, pharmacokinetic, and reported outcomes information, which is essential for regulatory approval and clinical use (Liu et al., 2024).

17.2 CLINICAL TRIALS

Clinical trials are systematically planned studies that examine the safety and effectiveness of therapeutic agents, medical devices, prevention strategy, and therapies for diseases or medical problems (Shan et al., 2022). These are the cornerstones of clinical studies, which transition the assessment of medical treatments from unreliable data and subjective opinions to stringent, evidence-based methodologies (Subbiah, 2023). In this, new interventions are compared with controls, such as placebos/standard treatments, which utilize statistical reasoning to ensure objectivity and reliability of the results. Randomised controlled trials (RCTs) are the gold standard among these, since they eliminate biases and maximise comparability among study groups through the use of blinding and randomisation (Tripepi et al., 2020; Hariton & Locascio, 2018). These studies examine how a therapeutic agent affects a disease and the results it produces. It examines if the drug can stop the disease, if the device is able to detect or screen for it, and how effective medical tests are in diagnosing it (Wright, 2017; Moutinho, 2023). Clinical trials start with microdosing and safety checks (phase 0 and I), then move to confirming the drug's effectiveness (phase III), and finally involve observing the drug after it is approved (phase IV) (Henry & Kirkpatrick, 2023). Initially, phase 0 or microdosing, was done using animals; now it is performed in human volunteers to check the drug

TABLE 17.1
Phases of Clinical Trials

Phases	Study Type	Description
Phase 0	Microdosing	This phase examines a 1/100th therapeutic dose for pharmacokinetics (ADME) in 10–20 human volunteers. This phase examines the pharmacokinetics and the dose for phase I studies (Kandi & Vadakedath, 2023).
Phase I	Safety/nontherapeutic	This phase establishes a safe dose range and the maximum tolerated dose (MTD). It is examined in terms of pharmacokinetic and pharmacodynamic effects. Tests single/multiple ascending doses in 15–50 healthy volunteers (Cook et al., 2014)
Phase II	Exploratory/therapeutic	Optimal dose, drug interactions, and preliminary therapeutic effects are established. Tests efficacy in 20–300 patients (Yan et al., 2018)
Phase III	Confirmatory/ therapeutic	Efficacy, safety, and quality of life have been determined in multicentre randomized trials. It compares new treatment against standard therapy in 300–3000 patients. Required for regulatory approval (Dilts et al., 2010)
Phase IV	Post-marketing surveillance	After approval/post-licensure and post-marketing studies/surveillance studies. Following up on the patients for an exceptionally long time for potential adverse reactions and drug–drug interactions (Haque et al., 2017).

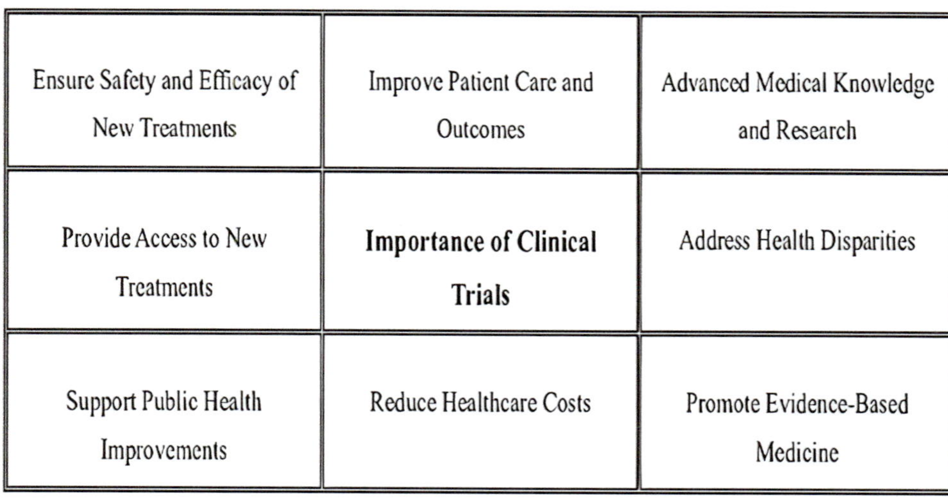

FIGURE 17.1 Key importance of clinical trials

dose tolerability before the phase 1 trial begins in healthy subjects (Kim et al., 2023). Clinical trials are the highest level of evidence used to grant approval and adopt new medical protocols that result in the advancement of medical science and better patient care (Shearer et al., 2022). Table 17.1 provides information about the different phases of a clinical trial.

Clinical trials test the safety and efficiency of novel interventions, therapies, or procedures (Bos & Ikram, 2022). Consequently, these are important and necessary steps in translational research between pre-clinical studies in lab settings and the transfer of innovations to clinical use on humans (Wright, 2017; Swift et al., 2018). They are of great importance, as they are the pillar upon which evidence-based medicine is based, the benchmark by which medical progress is measured (Kelly et al., 2015). Clinical trials are not to be overlooked for their importance, and some of the reasons why they are important are depicted in Figure 17.1.

17.3 CLINICAL APPLICATIONS OF DRUG DELIVERY SYSTEMS FOR THE MANAGEMENT OF LUNG CANCER

Researchers have studied ways to deliver drugs to the lungs for lung cancer through oral, intravenous, and inhaled routes. The most widely studied routes are oral and intravenous (Greenwald, 2004). Innovations in nanomedicine have especially enabled benefits from intravenous delivery (including) artificial nanoparticles, liposomes, and other nanocarriers that are engineered for increased solubility, extended circulation time and targeted delivery to tumour cells by active and passive targeting mechanisms (Li et al., 2023). Many active targeting approach exploit overexpressed receptors of lung cancer (e.g., epidermal growth factor receptor, EGFR, and luteinizing hormone-releasing hormone, LHRH receptor) (Garbuzenko et al., 2019). An example of selective binding, internalization, and localized drug release in tumour cells occurs when drug carriers are conjugated to ligands and antibodies selective to these receptors such as cetuximab for EGFR (Vikas et al., 2022). By performing this targeted approach, this not only intensifies the number of therapeutic agents at tumour site, but further ameliorates off-target toxicity of healthy tissues (Wathoni et al., 2022). In contrast to active targeting, the EPR effect inherent to tumour vasculature enables passive targeting by taking in nanosized carriers preferentially to tumour tissue (Vikas et al., 2022). Inhalation-based drug delivery is a logical and promising approach for localized lung cancer therapy with the advantage that chemotherapy drugs can be delivered directly into the tumour site in the lung parenchyma (Yu et al., 2021). Of these, this route avoids first pass metabolism and systemic distribution (reduces systemic exposure and toxicity) and yields high local drug concentrations (Abdulbaqi et al., 2021). Such pharmacokinetic advantages and fewer side effects have been demonstrated in clinical studies for systemic administration (Al Khatib et al., 2023). Despite challenges such as airway anatomy, tumor localization, device design limitations, and formulation stability, the widespread clinical adoption of inhaled therapies remains limited (Yu et al., 2021).

Some of the nanoformulation has been approved by regulatory agencies (FDA and EMA) (Farjadian et al., 2019). For example, PEGylated liposomal doxorubicin (Doxil) was the first FDA-approved cancer nanomedicine, which enhanced blood circulation time, high tumour uptake, retention effect, and reduced cardiotoxicity (Franco et al., 2018). Other liposomal-based formulations of other chemotherapeutic agents, such as DaunoXome (a liposomal formulation of daunorubicin for Kaposi's sarcoma) (Bulbake et al., 2017), Myocet (a non-PEGylated liposomal formulation of doxorubicin for metastatic breast cancer), and Marqibo (liposomal vincristine for acute lymphoblastic leukaemia) have also received approval (Miguel et al., 2022). So far, two nanoformulations (Abraxane and Genexol-PM) have been approved for lung cancer treatment and several other nanomedicines are being tested in clinical trials (Sarkar et al., 2022). Both drugs have paclitaxel (PTX) as their active substance, which stops microtubule separation and prevents cells from dividing (Kumar et al., 2024). Abraxane or Nab-PTX is made up of nanoparticles and albumin-bound paclitaxel to treat breast cancer, pancreatic cancer, and NSCLCs (Ma et al., 2024). Abraxane has been shown to have a greater uptake by tumour than PTX in solvent form. Abraxane is delivered to tumours by the EPR effect and receptor-mediated transcytosis (Holder et al., 2023). Genexol-PM includes PTX and delivers it using polymeric micelles (Serras et al., 2024). In a phase II study of NSCLC patients, using both genexol-PM and cisplatin resulted in a stronger antitumour effect and permitted higher doses of PTX to be given without causing much more toxicity (Girigoswami & Girigoswami, 2023).

Several nano-based drug delivery systems are under clinical investigation for the management of lung cancers (Aryal et al., 2023). NC-6004, also known as Nanoplatin, is a polymeric micelle of cisplatin. The formulation relies on a block copolymer of PEG-poly (glutamic acid), which helps the drug concentrate in tumour and reduces its effects on the rest of the body. Clinical trials have shown encouraging outcomes for NC-6004, such as a phase II study for gemcitabine-treated NSCLCs and a phase III trial for pancreatic and lung cancer (Han et al., 2022). Lipoplatin, a PEGylated liposomal form of cisplatin, is a notable formulation. This formulation has finished phase II and phase III trials for NSCLCs and has been granted orphan drug designation for pancreatic adenocarcinoma. Compared to regular cisplatin treatment, a phase III trial demonstrated an improved partial response rate and lower toxicity for lipoplatin with paclitaxel, suggesting that this combination may be a more effective therapeutic approach (Canão et al., 2022). CRLX101 is a nanoformulation that links camptothecin (CPT) to a linear cyclodextrin-PEG polymer. In this phase 1/2b study of advanced solid tumour, CRLX101 showed that NSCLC patients had a median PFS of 4.4 months and a significant number maintained stable disease for a long time (Chang et al., 2021). BIND-014 (docetaxel-loaded nanoparticles) has shown potential, with phase II trials showing that it helps control the disease in NSCLC patients with KRAS mutations (Wang et al., 2023). Nanoparticle albumin-bound (nab)-paclitaxel has shown some effectiveness in EGFR-mutant NSCLC, mostly for tumour in the brain, but additional studies are required to find the best dose. Even though these nanoformulations are designed to improve drug delivery and lower the risk of side effects, some problems still exist (Alshammari et al., 2023). Researchers are currently working on making stimuli-responsive nanoparticles and multifunctional nanocarriers better to overcome these problems without reducing the effectiveness of treatments (Majumder & Minko, 2020).

17.4 CONCLUSION AND PERSPECTIVES

Clinical trials related to drug delivery for lung cancer have resulted in better treatment effects and reduced systemic side effects. Although chemotherapy is useful, it often results in severe side effects because it is distributed throughout the body in an uncontrolled way. Use of liposomes, nanoparticles, polymeric micelles, dendrimers, and inhalable drugs has shown great potential for better targeting, higher drug absorption, and improved patient compliance. For example, both liposomal doxorubicin and nanoparticle albumin-bound paclitaxel are now used in clinical settings and show better results and less toxicity. Systems that deliver drugs by inhalation and targeted nanocarriers linked to folic acid or antibodies against EGFR or PD-L1 have advanced into clinical trials and are expected to help overcome resistance and relapse. Yet, some obstacles still exist. Problems with large-scale production, long-term safety, immune reactions, and regulations prevent clinical translation. Furthermore, how effective these advanced delivery systems are can be affected by the variety among patients and the complexities in tumour microenvironments. Only a small number of trials move past phase II and the lack of good comparative data makes it hard to prove that a new treatment is better than standard ones. It is expected that in the future, using biomarkers and advanced imaging will help personalized nanomedicine transform how lung cancer is managed. The use of artificial intelligence and machine learning in clinical trials can help pick the right patients, manage their doses, and monitor them in real time. Moreover, combining immunotherapy with nanotechnology is an exciting new area. The potential of these innovations can only be reached if oncologists, material scientists, regulatory bodies, and industry stakeholders collaborate. Clinical trials in the future should focus on how treatments affect patients over a long time, their quality of life, and their cost to make sure they are useful in regular care.

TABLE 17.2
Therapeutic Agents Under Clinical Trial for the Management of Lung Cancers

Serial Number	NCT Number	Phases	Conditions	Brief Summary	Reference
1	NCT02016209	Phase 2	NSCLC	This study evaluated the safety and effectiveness of neoadjuvant chemotherapy with platinum-based nanoparticle albumin bound paclitaxel in stage IIB and IIIA NSCLCs patients. This treatment has demonstrated promising clinical outcome having a high objective response rate and disease control rate, which suggest that nab-PTX-based neoadjuvant therapy can be a good option for improvement of surgical outcomes in NSCLCs.	(Zhao, 2013)
2	NCT00748163	Phase 2	Lung cancer	This study tested a combination of paclitaxel-albumin-stabilized NPs and sunitinib as a first-line treatment for stage IV NSCLCs patients. This study aimed to investigate the tumour response rates, time and duration of response, overall survival rate, and other treatment related issues.	(Dudek, 2008)
3	NCT00553462	Phase 2	Lung cancer	This phase II clinical trial evaluated a combination treatment approach using carboplatin and paclitaxel albumin nanoparticle followed by radiation therapy and erlotinib drug in treating patients with the unresectable stage III NSCLCs. This treatment approach investigates the enhancement in tumour control by using a combination of chemotherapy, radiation, and targeted therapy.	(Lilenbaum, 2017)
4	NCT01455389	Phase 1/ phase2	Lung cancer	This clinical trial investigated the combination of DOTAP: Chol-TUSC2 nanoparticles and erlotinib in stage IV NSCLCs patients. TUSC2 is tumour suppressor gene which is often inactivated in NSCLCs and Erlotinib is and EGFR tyrosine kinase inhibitor. A combination of TUSC2 and erlotinib in the form of nanoparticle is used in this study to access their potential synergistic effects.	(Lu, 2020)
5	NCT01620190	Phase 2	NSCLC	This study demonstrates the safety and effectiveness of PTX albumin-stabilized nanoparticles in individuals with advanced form of NSCLCs having EGFR mutation. The objective of this study was to access the response rate, safety evaluation, time to progression, and survival rate. This treatment demonstrated manageable safety profile with some adverse effects.	(Baik, 2021)
6	NCT04789486	Phase 1/ phase 2	NSCLC, pancreatic adenocarcinoma, and other cancers	This clinical study investigated the safety and effectiveness of gadolinium-based NPs, activation and guidance of irradiation X (AGuIX), used in conjunction with MR guided stereotactic body radiation therapy (SBRT) for the treatment of centrally located lung tumours and pancreatic cancer.	(Leeman, 2021)
7	NCT00729612	Phase 2	Lung cancer	This phase II clinical trial evaluated a combination treatment approach using carboplatin and PTX albumin-stabilized nanoparticle followed by the radiation therapy and erlotinib for treatment of patients with the advanced or recurrent NSCLCs. The objective of this study was to access the response rate, safety evaluation, time to progression and survival rate.	(Otterson, 2018)

8	NCT06048367	Phase 1	Lung cancer, breast cancer, cervical cancer, and other cancers	Phase I of this clinical study aimed at investigating the safety, pharmacokinetics, tolerability, and efficacy of the intratumoral carbon NPs [CNSI—Fe (II)] in patients with lung cancer and other cancers. The study accesses the safety, dose limiting toxicities and maximum tolerable dose for future studies.	(Wang, 2025)
9	NCT02769962	Phase/ phase 2	NSCLC, urothelial cancer	This study investigates effects of using EP0057, a nanoparticle camptothecin, with olaparib, which is a PARP inhibitor, in patients with relapsed or refractory SCLCs. EP0057 is meant to target the tumour cells with camptothecin, which reduces harm to normal cells and olaparib stops cancer cells from fixing their DNA, making chemotherapy more effective. The trial seeks to find a new treatment that works for SCLC patients who have limited response to other treatments.	(Thomas, 2024)
10	NCT02283320	Phase 2	KRAS-positive patients, NSCLC	The study examined BIND-014, which is a nanoparticle form of docetaxel, as a possible treatment for patients with KRAS-positive or NSCLCs. The purpose of this study was to evaluate safety and efficacy of BIND-014 in patients in which cancer has progressed after initial treatment, focusing on those specific genetic and histological subtypes of NSCLC.	(BIND Therapeutics, 2016)
11	NCT01792479	Phase 2	NSCLC	The study was a phase II trial testing BIND-014, a nanoparticle version of docetaxel, as a second-line treatment in NSCLCs patients in which disease progressed after first-line treatment. The purpose of the study was to determine outcome and side effects of BIND-014 in comparison to standard docetaxel which is often limited by toxicity.	(BIND Therapeutics, 2015)
12	NCT00499291	-	Cancer	This study studied the effectiveness of PTX-albumin-stabilized nanoparticles in treating patients with advanced or refractory solid tumours. The trial is designed to determine how well this formulation works in cancers which are hard to treat or have not responded to previous therapies	(Mani, 2007)
13	NCT04033354	Phase 3	SCLC	This clinical study highlights that addition of HLX10 (serplulimab), which is a PD-1 inhibitor, to a chemotherapy as the first-line treatment for advanced squamous NSCLCs can significantly improves both progression and survival rate in comparison to chemotherapy alone, with manageable safety.	(Shanghai Henlius Biotech, 2023)
14	NCT02495896	Phase 1	NSCLC, CARCINOMA, and other cancers	This study is designed to test if combining sEphB4-has with usual chemotherapy is safe and effective in treating patients with advanced or metastatic solid tumours, hoping to improve outcomes by targeting growth pathways in the tumours.	(El-Khoueiry, 2021)
15	NCT02716038	Phase 2	Carcinoma, NSCLCs	This study investigated the effectiveness of a neoadjuvant treatment that included atezolizumab (MPDL3280A), nab-paclitaxel, and carboplatin in patients with resectable non–small cell lung cancer (NSCLC). The purpose of this single-arm, phase II study was to find out if adding immunotherapy to chemotherapy could lead to better tumour response rates than chemotherapy alone in previous studies.	(Shu, 2023)

ABBREVIATIONS

ADME, absorption, distribution, metabolism, and excretion
EGFR, epidermal growth factor receptor
EMA, European Medicines Agency
EPR, enhanced permeability and retention
FDA, Food and Drug Administration
LCs, lung cancers
LHRH, luteinizing hormone-releasing hormone
MTD, maximum tolerated dose
NDDS, novel drug delivery systems
NSCLCs, non–small cell lung cancers
PTX, paclitaxel
SCLCs, small cell lung cancers

REFERENCES

Abdulbaqi, I. M., Assi, R. A., Yaghmur, A., Darwis, Y., Mohtar, N., Parumasivam, T., Saqallah, F. G., & Wahab, H. A. (2021). Pulmonary Delivery of Anticancer Drugs via Lipid-Based Nanocarriers for the Treatment of Lung Cancer: An Update. *Pharmaceuticals*, 14(8), 725. https://doi.org/10.3390/ph14080725

Al Khatib, A. O., El-Tanani, M., & Al-Obaidi, H. (2023). Inhaled Medicines for Targeting Non-Small Cell Lung Cancer. *Pharmaceutics*, 15(12), 2777. https://doi.org/10.3390/pharmaceutics15122777

Al Ragib, A., Chakma, R., Dewan, K., Islam, T., Kormoker, T., & Idris, A. M. (2022). Current Advanced Drug Delivery Systems: Challenges and Potentialities. *Journal of Drug Delivery Science and Technology*, 76, 103727. https://doi.org/10.1016/j.jddst.2022.103727

Alshammari, M. K., Almomen, E. Y., Alshahrani, K. F., Altwalah, S. F., Kamal, M., Al-Twallah, M. F., Alsanad, S. H., Al-Batti, M. H., Al-Rasheed, F. J., Alsalamah, A. Y., Alhazza, M. B., Alasmari, F. A., Abida, N., & Imran, M. (2023). Nano-Enabled Strategies for the Treatment of Lung Cancer: Potential Bottlenecks and Future Perspectives. *Biomedicines*, 11(2), 473. https://doi.org/10.3390/biomedicines11020473

Araghi, M., Mannani, R., Maleki, A. H., Hamidi, A., Rostami, S., Safa, S. H., Faramarzi, F., Khorasani, S., Alimohammadi, M., Tahmasebi, S., & Sigari, R. A. (2023). Recent Advances in Non-Small Cell Lung Cancer Targeted Therapy; an Update Review. *Cancer Cell International*, 23(1). https://doi.org/10.1186/s12935-023-02990-y

Aryal, S., Park, S., Park, H., Park, C., Kim, W. C., Thakur, D., Won, Y.-J., & Key, J. (2023). Clinical Trials for Oral, Inhaled and Intravenous Drug Delivery System for Lung Cancer and Emerging Nanomedicine-Based Approaches. *International Journal of Nanomedicine*, 18, 7865–7888. https://doi.org/10.2147/ijn.s432839

Baik, C. (2021). Paclitaxel Albumin-Stabilized Nanoparticle Formulation in Treating Patients with Previously Treated Advanced Non-small Cell Lung Cancer. In ClinicalTrials.gov (No. NCT01620190). *National Library of Medicine*. https://clinicaltrials.gov/study/NCT01620190

BIND Therapeutics. (2015). A Phase 2 Study to Determine the Safety and Efficacy of BIND-014 (Docetaxel Nanoparticles for Injectable Suspension) as Second-line Therapy to Patients with Non-Small Cell Lung Cancer. In ClinicalTrials.gov (No. NCT01792479). *National Library of Medicine*. https://clinicaltrials.gov/study/NCT01792479

BIND Therapeutics. (2016). A Study of BIND-014 (Docetaxel Nanoparticles for Injectable Suspension) as Second-line Therapy for Patients with KRAS Positive or Squamous Cell Non-Small Cell Lung Cancer. In ClinicalTrials.gov (No. NCT02283320). *National Library of Medicine*. https://clinicaltrials.gov/study/NCT02283320

Bos, D., & Ikram, M. A. (2022). Research Aims in Clinical Medicine: Description, Identification, or Explanation. *World Neurosurgery*, 161, 240–244. https://doi.org/10.1016/j.wneu.2021.11.081

Bulbake, U., Doppalapudi, S., Kommineni, N., & Khan, W. (2017). Liposomal Formulations in Clinical Use: An Updated Review. *Pharmaceutics*, 9(4), 12. https://doi.org/10.3390/pharmaceutics9020012

Canão, F., Ferreira, H., & Neves, N. M. (2022). Liposomal Formulations for Lung Cancer Treatment in the Last Two Decades: A Systematic Review. *Journal of Cancer Research and Clinical Oncology*. https://doi.org/10.1007/s00432-022-04079-x

Chang, D., Ma, Y., Xu, X., Xie, J., & Ju, S. (2021). Stimuli-Responsive Polymeric Nanoplatforms for Cancer Therapy. *Frontiers in Bioengineering and Biotechnology*, 9. https://doi.org/10.3389/fbioe.2021.707319

Chavda, V. P., Patel, A. B., Mistry, K. J., Suthar, S. F., Wu, Z.-X., Chen, Z.-S., & Hou, K. (2022). Nano-Drug Delivery Systems Entrapping Natural Bioactive Compounds for Cancer: Recent Progress and Future Challenges. *Frontiers in Oncology*, 12. https://doi.org/10.3389/fonc.2022.867655

Cheng, T.-Y. D., Cramb, S. M., Baade, P. D., Youlden, D. R., Nwogu, C., & Reid, M. E. (2016). The International Epidemiology of Lung Cancer: Latest Trends, Disparities, and Tumor Characteristics. *Journal of Thoracic Oncology*, 11(10), 1653–1671. https://doi.org/10.1016/j.jtho.2016.05.021

Cojocaru, E., Petriş, O. R., & Cojocaru, C. (2024). Nanoparticle-Based Drug Delivery Systems in Inhaled Therapy: Improving Respiratory Medicine. *Pharmaceuticals*, 17(8), 1059. https://doi.org/10.3390/ph17081059

Cook, N., Hansen, A. R., Siu, L. L., & Abdul Razak, A. R. (2014). Early Phase Clinical Trials to Identify Optimal Dosing and Safety. *Molecular Oncology*, 9(5), 997–1007. https://doi.org/10.1016/j.molonc.2014.07.025

Dilts, D. M., Cheng, S. K., Crites, J. S., Sandler, A. B., & Doroshow, J. H. (2010). Phase III Clinical Trial Development: A Process of Chutes and Ladders. *Clinical Cancer Research*, 16(22), 5381–5389. https://doi.org/10.1158/1078-0432.ccr-10-1273

Dudek, A. (2008). Paclitaxel Albumin-Stabilized Nanoparticle Formulation and Sunitinib as First-Line Therapy in Treating Patients with Stage IV Non-Small Cell Lung Cancer. In ClinicalTrials.gov (No. NCT00748163). *National Library of Medicine*. https://clinicaltrials.gov/study/NCT00748163

El-Khoueiry, A. (2021). Recombinant EphB4-HSA Fusion Protein with Standard Chemotherapy Regimens in Treating Patients with Advanced or Metastatic Solid Tumors. In ClinicalTrials.gov (No. NCT02495896). *National Library of Medicine*. https://clinicaltrials.gov/study/NCT02495896

Ezhilarasan, D., Lakshmi, T., & Mallineni, S. K. (2022). Nano-Based Targeted Drug Delivery for Lung Cancer: Therapeutic Avenues and Challenges. *Nanomedicine*. https://doi.org/10.2217/nnm-2021-0364

Ezike, T. C., Okpala, U. S., Onoja, U. L., Nwike, P. C., Ezeako, E. C., Okpara, J. O., Okoroafor, C. C., Eze, S. C., Kalu, O. L., Odoh, E. C., Nwadike, U., Ogbodo, J. O., Umeh, B. U., Ossai, E. C., & Nwanguma, B. C. (2023). Advances in Drug Delivery Systems, Challenges and Future Directions. *Heliyon*, 9(6), e17488. https://doi.org/10.1016/j.heliyon.2023.e17488

Farjadian, F., Ghasemi, A., Gohari, O., Roointan, A., Karimi, M., & Hamblin, M. R. (2019). Nanopharmaceuticals and Nanomedicines Currently on the Market: Challenges and Opportunities. *Nanomedicine*, 14(1), 93–126. https://doi.org/10.2217/nnm-2018-0120

Franco, Y., Vaidya, T., & Ait-Oudhia, S. (2018). Anticancer and Cardio-Protective Effects of Liposomal Doxorubicin in the Treatment of Breast Cancer. *Breast Cancer: Targets and Therapy*, 10, 131–141. https://doi.org/10.2147/bctt.s170239

Garbuzenko, O. B., Kuzmov, A., Taratula, O., Pine, S. R., & Minko, T. (2019). Strategy to Enhance Lung Cancer Treatment by Five Essential Elements: Inhalation Delivery, Nanotechnology, Tumor-Receptor Targeting, Chemo- and Gene Therapy. *Theranostics*, 9(26), 8362–8376. https://doi.org/10.7150/thno.39816

Girigoswami, A., & Girigoswami, K. (2023). Potential Applications of Nanoparticles in Improving the Outcome of Lung Cancer Treatment. *Genes*, 14(7), 1370. https://doi.org/10.3390/genes14071370

Greenwald, P. (2004). Clinical Trials in Cancer Prevention: Current Results and Perspectives for the Future. *The Journal of Nutrition*, 134(12), 3507S3512S. https://doi.org/10.1093/jn/134.12.3507s

Haider, M., Elsherbeny, A., Pittalà, V., Consoli, V., Alghamdi, M. A., Hussain, Z., Khoder, G., & Greish, K. (2022). Nanomedicine Strategies for Management of Drug Resistance in Lung Cancer. *International Journal of Molecular Sciences*, 23(3), 1853. https://doi.org/10.3390/ijms23031853

Han, Y., Wen, P., Li, J., & Kataoka, K. (2022). Targeted Nanomedicine in Cisplatin-Based Cancer Therapeutics. *Journal of Controlled Release*, 345, 709–720. https://doi.org/10.1016/j.jconrel.2022.03.049

Hani, U., M., Y. B., Wahab, S., Siddiqua, A., Osmani, R. A. M., & Rahamathulla, M. (2021). A Comprehensive Review of Current Perspectives on Novel Drug Delivery Systems and Approaches for Lung Cancer Management. *Journal of Pharmaceutical Innovation*. https://doi.org/10.1007/s12247-021-09582-1

Haque, A., Daniel, S., Maxwell, T., & Boerstoel, M. (2017). Postmarketing Surveillance Studies—An Industry Perspective on Changing Global Requirements and Implications. *Clinical Therapeutics*, 39(4), 675–685. https://doi.org/10.1016/j.clinthera.2017.03.011

Hariton, E., & Locascio, J. J. (2018). Randomised Controlled Trials—the Gold Standard for Effectiveness Research. *BJOG: An International Journal of Obstetrics & Gynaecology*, 125(13), 1716. https://doi.org/10.1111/1471-0528.15199

Henry, M., & Kirkpatrick, D. (2023). *Phases of clinical trials*. Elsevier EBooks, 303–307. https://doi.org/10.1016/b978-0-12-823026-8.00105-x

Holder, J. E., Oliveira, E., Lodeiro, C., Trim, C. M., Byrne, L. J., Bértolo, E., & Wilson, C. M. (2023). The Use of Nanoparticles for Targeted Drug Delivery in Non-Small Cell Lung Cancer. *Frontiers in Oncology*, 13. https://doi.org/10.3389/fonc.2023.1154318

Kandi, V., & Vadakedath, S. (2023). Clinical Trials and Clinical Research: A Comprehensive Review. *Cureus*, 15(2). https://www.ncbi.nlm.nih.gov/pmc/articles/PMC10023071/

Kelly, M. P., Heath, I., Howick, J., & Greenhalgh, T. (2015). The Importance of Values in Evidence-Based Medicine. *BMC Medical Ethics*, 16(1). https://doi.org/10.1186/s12910-015-0063-3

Kim, E., Yang, J., Park, S., & Shin, K. (2023). Factors Affecting Success of New Drug Clinical Trials. *Therapeutic Innovation & Regulatory Science*, 57(4), 737–750. https://doi.org/10.1007/s43441-023-00509-1

Kumar, S., Arora, A., Pant, V., Guchhait, S., Kumar, R., Mathur, D., & Singh, B. K. (2024). Advances in Drug Delivery Systems for Lipophilic Drug Paclitaxel: Developments, Challenges, and Opportunities (A Review). *Russian Journal of Bioorganic Chemistry*, 50(5), 1752–1782. https://doi.org/10.1134/s106816202405011x

Leeman, J. (2021). Nano-SMART: Nanoparticles with MR Guided SBRT in Centrally Located Lung Tumors and Pancreatic Cancer. In ClinicalTrials.gov (No. NCT04789486). *National Library of Medicine*. https://clinicaltrials.gov/study/NCT04789486

Leiter, A., Veluswamy, R. R., & Wisnivesky, J. P. (2023). The Global Burden of Lung Cancer: Current Status and Future Trends. *Nature Reviews Clinical Oncology*, 20(20), 1–16. https://doi.org/10.1038/s41571-023-00798-3

Li, C., Wang, H., Jiang, Y., Fu, W., Liu, X., Zhong, R., Cheng, B., Zhu, F., Xiang, Y., He, J., & Liang, W. (2022). Advances in Lung Cancer Screening and Early Detection. *Cancer Biology & Medicine*, 19(5), 591–608. https://doi.org/10.20892/j.issn.2095-3941.2021.0690

Li, J., Wang, Q., Xia, G., Adilijiang, N., Li, Y., Hou, Z., Fan, Z., & Li, J. (2023). Recent Advances in Targeted Drug Delivery Strategy for Enhancing Oncotherapy. *Pharmaceutics*, 15(9), 2233. https://doi.org/10.3390/pharmaceutics15092233

Lilenbaum, R. (2017). Carboplatin and Paclitaxel Albumin-Stabilized Nanoparticle Formulation Followed by Radiation Therapy and Erlotinib in Treating Patients with Stage III Non-Small Cell Lung Cancer That Cannot Be Removed by Surgery. In ClinicalTrials.gov (No. NCT00553462). *National Library of Medicine*. https://clinicaltrials.gov/study/NCT00553462

Liu, X., Cheng, Y., Mu, Y., Zhang, Z., Tian, D., Liu, Y., Hu, X., & Wen, T. (2024). Diverse Drug Delivery Systems for the Enhancement of Cancer Immunotherapy: An Overview. *Frontiers in Immunology*, 15. https://doi.org/10.3389/fimmu.2024.1328145

Lu, C. (2020). TUSC2-nanoparticles and Erlotinib in Stage IV Lung Cancer. In ClinicalTrials.gov (No. NCT01455389). *National Library of Medicine*. https://clinicaltrials.gov/study/NCT01455389

Ma, P., Wang, G., Men, K., Li, C., Gao, N., & Li, L. (2024). Advances in Clinical Application of Nanoparticle-Based Therapy for Cancer Treatment: A Systematic Review. *Nano TransMed*, 3, 100036. https://doi.org/10.1016/j.ntm.2024.100036

Madni, A., Batool, A., Noreen, S., Maqbool, I., Rehman, F., Kashif, P. M., Tahir, N., & Raza, A. (2017). Novel Nanoparticulate Systems for Lung Cancer Therapy: An Updated Review. *Journal of Drug Targeting*, 25(6), 499–512. https://doi.org/10.1080/1061186x.2017.1289540

Majumder, J., & Minko, T. (2020). Multifunctional and Stimuli-Responsive Nanocarriers for Targeted Therapeutic Delivery. *Expert Opinion on Drug Delivery*, 18(2), 205–227. https://doi.org/10.1080/17425247.2021.1828339

Mani, S. (2007). Paclitaxel Albumin-Stabilized Nanoparticle formulation in treating patients with advanced or refractory solid tumors. In ClinicalTrials.gov (No. NCT00499291). *National Library of Medicine*. https://clinicaltrials.gov/study/NCT00499291

Miguel, R. dos A., Hirata, A. S., Jimenez, P. C., Lopes, L. B., & Costa-Lotufo, L. V. (2022). Beyond Formulation: Contributions of Nanotechnology for Translation of Anticancer Natural Products into New Drugs. *Pharmaceutics*, 14(8), 1722. https://doi.org/10.3390/pharmaceutics14081722

Milewska, S., Niemirowicz-Laskowska, K., Siemiaszko, G., Nowicki, P., Wilczewska, A. Z., & Car, H. (2021). Current Trends and Challenges in Pharmacoeconomic Aspects of Nanocarriers as Drug Delivery Systems for Cancer Treatment. *International Journal of Nanomedicine*, 16, 6593–6644. https://doi.org/10.2147/IJN.S323831

Moutinho, S. (2023). Clinical Trials Assess a Precision-Medicine Approach to Cancer Screening. *Nature Medicine*, 29(7), 1587–1590. https://doi.org/10.1038/s41591-023-02431-3

Mukherjee, A., Paul, M., & Mukherjee, S. (2019). Recent Progress in the Theranostics Application of Nanomedicine in Lung Cancer. *Cancers*, 11(5), 597. https://doi.org/10.3390/cancers11050597

Nooreldeen, R., & Bach, H. (2021). Current and Future Development in Lung Cancer Diagnosis. *International Journal of Molecular Sciences*, 22(16), 8661. https://doi.org/10.3390/ijms22168661

Otterson. (2018). Paclitaxel Albumin-Stabilized Nanoparticle Formulation and Carboplatin in Treating Patients with Stage IIIB, Stage IV, or Recurrent Non-Small Cell Lung Cancer. In ClinicalTrials.gov (No. NCT00729612). *National Library of Medicine*. https://clinicaltrials.gov/study/NCT00729612

Paudel, K. R., Panth, N., Pangeni, R., Awasthi, R., Chawla, V., Mehta, M., Tambuwala, M. M., & Hansbro, P. M. (2020). Targeting Lung Cancer Using Advanced Drug Delivery Systems. *Targeting Chronic Inflammatory Lung Diseases Using Advanced Drug Delivery Systems*, 493–516. https://doi.org/10.1016/b978-0-12-820658-4.00023-6

Quadir, S. S., Saharan, V., Choudhary, D., Harish, Jain, C. P., & Joshi, G. (2022). Nano-strategies as Oral Drug Delivery Platforms for Treatment of Cancer: Challenges and Future Perspectives. *AAPS PharmSciTech*, 23(5). https://doi.org/10.1208/s12249-022-02301-0

Rajendran, S., Ravi, S. N., Nair, V., Sree, R. P., Babu, S., & Palanivelu, J. (2024). Recent Development and Future Aspects: Nano-Based Drug Delivery System in Cancer Therapy. *Topics in Catalysis*. https://doi.org/10.1007/s11244-023-01893-6

Ramos, R., Moura, C. S., Costa, M., Lamas, N. J., Correia, R., Garcez, D., Pereira, J. M., Lindahl, T., Sousa, C., & Vale, N. (2025). Lung Cancer Therapy: The Role of Personalized Medicine. *Cancers*, 17(5), 725–725. https://doi.org/10.3390/cancers17050725

Saito, S., Espinoza-Mercado, F., Liu, H., Sata, N., Cui, X., & Soukiasian, H. J. (2017). Current Status of Research and Treatment for Non-Small Cell Lung Cancer in Never-Smoking Females. *Cancer Biology & Therapy*, 18(6), 359–368. https://doi.org/10.1080/15384047.2017.1323580

Sarkar, M., Wang, Y., Ekpenyong, O., Liang, D., & Xie, H. (2022). Pharmacokinetic Behaviors of Soft Nanoparticulate Formulations of Chemotherapeutics. *WIREs Nanomedicine and Nanobiotechnology*, 15(2). https://doi.org/10.1002/wnan.1846

Serras, A., Faustino, C., & Pinheiro, L. (2024). Functionalized Polymeric Micelles for Targeted Cancer Therapy: Steps from Conceptualization to Clinical Trials. *Pharmaceutics*, 16(8), 1047–1047. https://doi.org/10.3390/pharmaceutics16081047

Shan, X., Gong, X., Li, J., Wen, J., Li, Y., & Zhang, Z. (2022). Current Approaches of Nanomedicines in the Market and Various Stage of Clinical Translation. *Acta Pharmaceutica Sinica B*, 12(7). https://doi.org/10.1016/j.apsb.2022.02.025

Shanghai Henlius Biotech. (2023). A randomized, double-blind, placebo-controlled phase III study to investigate efficacy and safety of First-Line treatment with HLX10 + chemotherapy (Carboplatin-Nanoparticle albumin bound (NAB) paclitaxel) in patients with stage IIIB/ IIIC or IV NSCLC. In ClinicalTrials.gov (No. NCT04033354). *National Library of Medicine*. https://clinicaltrials.gov/study/NCT04033354

Sharma, P., Mehta, M., Dhanjal, D. S., Kaur, S., Gupta, G., Singh, H., Thangavelu, L., Rajeshkumar, S., Tambuwala, M., Bakshi, H. A., Chellappan, D. K., Dua, K., & Satija, S. (2019). Emerging Trends in the Novel Drug Delivery Approaches for the Treatment of Lung Cancer. *Chemico-Biological Interactions*, 309, 108720. https://doi.org/10.1016/j.cbi.2019.06.033

Shearer, J., Castro, J. L., Lawson, A. D. G., MacCoss, M., & Taylor, R. D. (2022). Rings in Clinical Trials and Drugs: Present and Future. *Journal of Medicinal Chemistry*, 65(13), 8699–8712. https://doi.org/10.1021/acs.jmedchem.2c00473

Shojaei, S., Pourmadadi, M., Homayoonfal, M., Behnamrad, P., Fathi-karkan, S., Rahdar, A., Gerayli, S., & Pandey, S. (2024). Revolutionizing Lung Cancer Treatment: Nanotechnology-Driven Advances in Targeted Drug Delivery and Novel Therapeutic Strategies. *Journal of Drug Delivery Science and Technology*, 106186. https://doi.org/10.1016/j.jddst.2024.106186

Shu, C. (2023). Neoadjuvant MPDL3280A, Nab-paclitaxel and carboplatin (MAC) in NSCLC. In ClinicalTrials.gov (No. NCT02716038). *National Library of Medicine*. https://clinicaltrials.gov/study/NCT02716038

Subbiah, V. (2023). The Next Generation of Evidence-Based Medicine. *Nature Medicine*, 29(29), 1–10. https://doi.org/10.1038/s41591-022-02160-z

Swift, B., Jain, L., White, C., Chandrasekaran, V., Bhandari, A., Hughes, D. A., & Jadhav, P. R. (2018). Innovation at the Intersection of Clinical Trials and Real-World Data Science to Advance Patient Care. *Clinical and Translational Science*, 11(5), 450–460. https://doi.org/10.1111/cts.12559

Testa, U., Castelli, G., & Pelosi, E. (2018). Lung Cancers: Molecular Characterization, Clonal Heterogeneity and Evolution, and Cancer Stem Cells. *Cancers*, 10(8), 248. https://doi.org/10.3390/cancers10080248

Thomas, A. (2024). Trial of EP0057, a Nanoparticle Camptothecin with Olaparib in People with Relapsed/Refractory Small Cell Lung Cancer. In ClinicalTrials.gov (No. NCT02769962). *National Library of Medicine*. https://clinicaltrials.gov/study/NCT02769962

Tripepi, G., Chesnaye, N. C., Dekker, F. W., Zoccali, C., & Jager, K. J. (2020). Intention to Treat and Per Protocol Analysis in Clinical Trials. *Nephrology*, 25(7), 513–517. https://doi.org/10.1111/nep.13709

Uliński, R., Kwiecień, I., & Domagała-Kulawik, J. (2022). Lung Cancer in the Course of COPD-Emerging Problems Today. *Cancers*, 14(15), 3819–3819. https://doi.org/10.3390/cancers14153819

Vikas, N., Sahu, H. K., Mehata, A. K., Viswanadh, M. K., Priya, V., & Muthu, M. S. (2022). Dual-Receptor-Targeted Nanomedicines: Emerging Trends and Advances in Lung Cancer Therapeutics. *Nanomedicine*, 17(19), 1375–1395. https://doi.org/10.2217/nnm-2021-0470

Wang, Q., Atluri, K., Tiwari, A. K., & Babu, R. J. (2023). Exploring the Application of Micellar Drug Delivery Systems in Cancer Nanomedicine. *Pharmaceuticals*, 16(3), 433. https://doi.org/10.3390/ph16030433

Wang, Y. (2025). Carbon Nanoparticle-Loaded iron [CNSI-FE(II)] in the treatment of advanced solid tumor (CNSI-FE(II)). In ClinicalTrials.gov (No. NCT06048367). *National Library of Medicine*. https://clinicaltrials.gov/study/NCT06048367

Wathoni, N., Puluhulawa, L. E., Joni, I. M., Muchtaridi, M., Mohammed, A. F. A., Elamin, K. M., Milanda, T., & Gozali, D. (2022). Monoclonal Antibody as a Targeting Mediator for Nanoparticle Targeted Delivery System for Lung Cancer. *Drug Delivery*, 29(1), 2959–2970. https://doi.org/10.1080/10717544.2022.2120566

Wright, B. (2017). Clinical Trial Phases. *A Comprehensive and Practical Guide to Clinical Trials*, 11–15. https://doi.org/10.1016/b978-0-12-804729-3.00002-x

Yan, F., Thall, P. F., Lu, K. H., Gilbert, M. R., & Yuan, Y. (2018). Phase I–II Clinical Trial Design: A State-of-the-Art Paradigm for Dose Finding. *Annals of Oncology*, 29(3), 694–699. https://doi.org/10.1093/annonc/mdx795

Yu, X.-Y., Jin, X., & Shou, Z.-X. (2021). Surface-Engineered Smart Nanocarrier-Based Inhalation Formulations for Targeted Lung Cancer Chemotherapy: A Review of Current Practices. *Drug Delivery*, 28(1), 1995–2010. https://doi.org/10.1080/10717544.2021.1981492

Zhang, Y., Luo, G., Etxeberria, J., & Hao, Y. (2021). Global Patterns and Trends in Lung Cancer Incidence: A Population-Based Study. *Journal of Thoracic Oncology*, 16(6), 933–944. https://doi.org/10.1016/j.jtho.2021.01.1626

Zhao, Q. (2013). Neoadjuvant Chemotherapy of Nanoparticle Albumin-bound Paclitaxel in Lung Cancer. In ClinicalTrials.gov (No. NCT02016209). *National Library of Medicine*. https://clinicaltrials.gov/study/NCT02016209

18 Future Directions in the Management of Lung Cancer

Mansi Verma[1,2], Niraj Kumar Singh[1], Aditi Singhal[3], Anurag Chaudhary[4], Silpi Chanda[5], and Ram Ji Gupta[6]*

[1] Division of Pharmacology, Institute of Pharmaceutical Research, GLA University, NH#19 Delhi Mathura Highway, Chaumuhan, Mathura 281406, Uttar Pradesh, India

[2] Sunder Deep Pharmacy College, SDGI Global University, NH-09, Delhi Hapur Highway, Ghaziabad 201015, NCR, Uttar Pradesh, India

[3] Department of Pharmaceutics, Meerut Institute of Engineering and Technology, N.H. 58, Delhi-Roorkee Highway, Baghpat Road Bypass Crossing, Meerut 250005, India

[4] Department of Pharmaceutical Chemistry, Meerut Institute of Engineering and Technology, N.H. 58, Delhi-Roorkee Highway, Baghpat Road Bypass Crossing, Meerut 250005, India

[5] Department of Pharmacognosy, Amity Institute of Pharmacy, Amity University, Lucknow Campus, U.P sector 125, Noida 201313, India

[6] Division of Pharmacology, R.V. Northland Institute, Gautam Budhh Nagar, Ghaziabad, India

*Corresponding author: Mansi Verma, mansi7184@gmail.com.

18.1 INTRODUCTION

Lung cancer accounts for approximately 13% of all cancer cases and is the foremost cause of cancer-related deaths worldwide, resulting in more fatalities than breast, brain, colorectal, and prostate cancers combined [1, 2]. As reported by the American Cancer Society, the United States recorded 236,740 cases of lung cancer and 130,180 related mortalities in 2022. Less than 5% of those with severe illness, such as those in stages IIIB or IV, have good prognosis [3].

Most patients (84%) have advanced disease present at the time of diagnosis, with only 15% receiving an early-stage diagnosis. Overall, individuals with non–small cell lung cancer (NSCLC) have dismal prognosis and low five-year OS of about (17.4%) [4–6].

NSCLC is the most common type of lung cancer, accounting for more than 85% of all cases. The three most common subtypes of NSCLC are adenocarcinoma, squamous cell carcinoma, and large cell carcinoma. [7–9]. Chemotherapy, radiation therapy, and surgery are common methods to treat cancer. Advanced and more effective approaches such as immunotherapy and target treatment have emerged as result of newly scientific techniques to treat illnesses, like infections, malignancies, autoimmune diseases, and other conditions [10–12].

In treating end-stage lung cancer, NSCLC, immunotherapy and targeted therapy among other therapies have shown encouraging outcomes within last ten years. An increase in overall survival was observed among patients treated with these therapies [13].

The primary reason is the presence of biomarkers that allow the chosen patients for immunotherapy-based and personalized treatment approaches. Or we can say, in an effort to enhance result

DOI: 10.1201/9781003595830-18

for patients having NSCLC, therapy is moving towards more recent targeted and small-molecule medicines. In fact, studies of the human genome have made it possible to more effectively identify gene changes that might serve as "targets" for treatment [14, 15].

18.2 STRATEGIES FOR TREATING NSCLC

The size, grade, and location of the tumour, the condition of the lymph nodes, and the patient's general health and lung function are the primary determinants of treatment options for NSCLC. For NSCLC the approved therapeutic options involve surgery, therapy of radiation, chemotherapy, and targeted therapy [16].

Surgery is only treatment option for stage 0 NSCLC; neither chemotherapy nor radiation therapy are required. Additionally, surgery may be replaced by photodynamic treatment, brachytherapy, or laser therapy. Surgery remains the main treatment approach for stage I NSCLC, and the addition of adjuvant chemotherapy or radiotherapy post-surgery may help lower the risk of cancer recurrence. In stage II NSCLC, adjuvant chemotherapy, radiotherapy, and immunotherapy may be used after the cancer has been removed. Chemotherapy, radiation therapy, surgery, or immunotherapy may be used as part of first treatment for stage IIIA NSCLC. Adjuvant treatment with targeted medication osimertinib may also be a possibility in the future for patients those cancer cells contain certain EGFR mutations. Surgery cannot completely eradicate stage IIIB; however, immunotherapy and chemoradiotherapy can be beneficial. Stages IVA and IVB lung cancers, typically considered incurable, may benefit from symptom management using treatments like photodynamic therapy, surgery, laser therapy, radiation, chemotherapy, immunotherapy, and targeted approaches [17].

18.3 TARGETED DRUG THERAPY

Cancerous cells vary from healthy cells due to changes in their DNA or proteins. As a result, they may grow more quickly and occasionally spread. In order to target specific genes and proteins that contribute to the development and survival of cancer, targeted cancer therapy makes use of these variations. Thus, it limits the harm to healthy cells by halting the development and spread of cancer cells. Due to their distinct side effects, targeted medications are frequently utilised for advanced lung malignancies when traditional medications are ineffective. Numerous medications that target these pathways have been created, and the FDA has authorised a large number of targeted compounds [5].

Most of the epidermal growth factor receptor (EGFR), Ras-mitogen-activated protein kinase (RAS-MAPK), phosphatidylinositol 3-kinase/protein kinase B/mammalian target of rapamycin (PI3K/AKT/mTOR), rearranged during transfection (RET), B-Raf proto-oncogene (BRAF), mesenchymal-epithelial transition (MET), and neurotrophic tyrosine receptor kinase/proto-oncogene 1, receptor tyrosine kinase (NTRK/ROS1) inhibitors that are now feasible for NSCLC are now standard of therapy for a subset of patients.

Even with these novel therapies options, patients having advanced NSCLC still face many obstacles because the majority of these individuals acquire resistance and their illness progresses [18].

18.4 VASCULAR ENDOTHELIAL GROWTH FACTOR

VEGF is also known as the growth-factor-promoting blood vessel formation and its receptors small molecule inhibitors targeting VEGF/VEGFR have demonstrated effectiveness in treating various cancers including NSCLC, across multiple clinical trials. Two FDA-approved anti-VEGF routes for advanced NSCLC are Bevacizumab (Avastin/anti-VEGF antibody) and Ramucirumab (Cyramza/anti-VEGFR antibody), which can be administered alone or alongside chemotherapy. Numerous studies have demonstrated the crucial function that the VEGF pathway plays in angiogenesis [19].

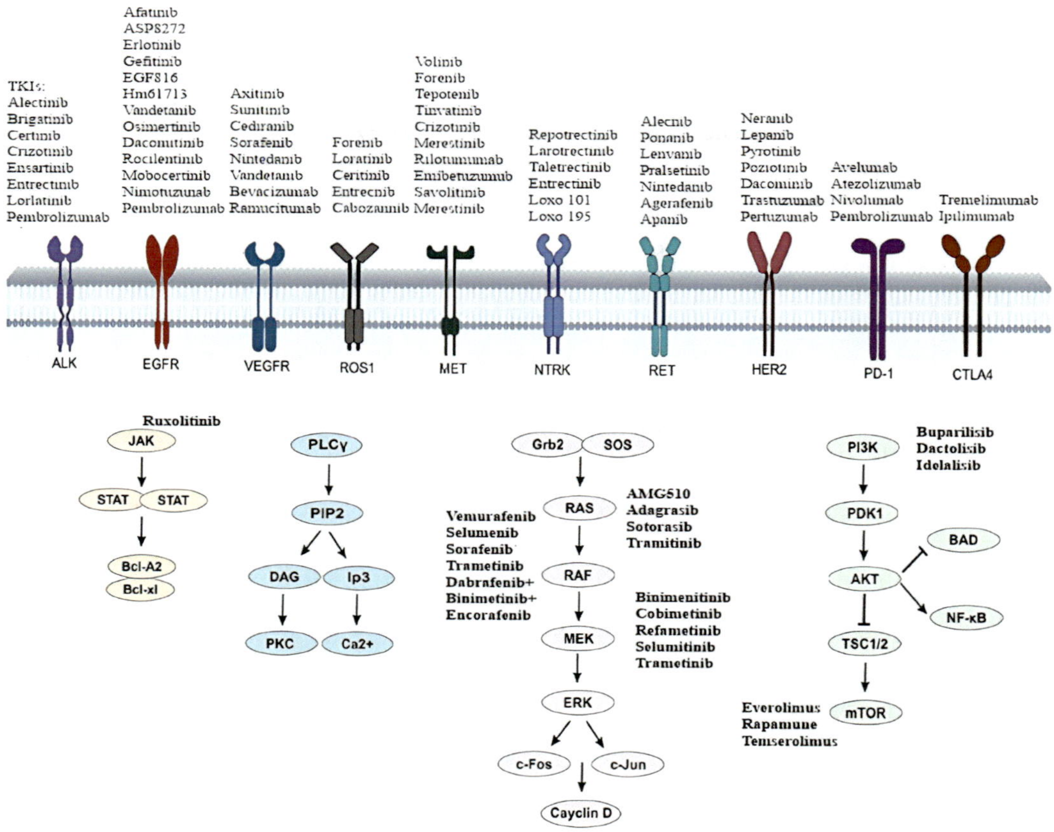

FIGURE 18.1 Target genes and drugs in treatment of NSCLC: a summary of targeted therapy medications that target important associated genes and their signalling pathways

Interestingly, new research suggests that VEGF may contribute to the pathophysiology of cancer in a broader sense than previously believed since it induces tumour-induced immunosuppression [20].

VEGFs have the ability to inhibit the development of haematopoietic lineages, including dendritic cells (DCs), which in turn inhibits the process of antigen presentation. It may also enhance the infiltration of tumour-associated macrophages (M2 macrophages), regulatory T (Treg) cells and myeloid derived suppressor cells (MDSC) into NSCLC niche [21]. Blocking VEGF signalling reverses these immunosuppression pathways increasing effector T cell infiltration and slowing the infiltration of suppressor cells into the tumour [22].

According to research, anti-VEGF chemotherapy or combination treatment with another anti-angiogenic inhibitor has a mainly synergistic anti-tumour effect [23]. In one research, VEGF and EGF-like ligands in human NSCLC were targeted by new bispecific decoy receptor called VEGFR EGFR/Fc. The dual-specific antibody that was created prevented angiogenesis, tumour invasion, migration, and proliferation [24]. Bevacizumab combined gemcitabine/cisplatin treatment demonstrated therapeutic effectiveness and extended patients' long-term survival in advanced NSCLC [25]. Additionally, In patients with NSCLC, studies have shown that combining bevacizumab with erlotinib leads to a higher occurrence of adverse events but results in a longer progression-free survival (PFS) compared to using either drug alone [26].

18.5 KRAS

KRAS-gene-derived protein impacts cellular signalling pathways that control cell development, division, and maturation. It is the most often mutated oncogene in human malignancies, which prevents the creation of potent medications that target KRAS [27, 28].

KRAS was therefore historically thought to be undruggable, despite enormous efforts that led to the creation of medications aimed at blocking KRAS or its signalling pathways. However, a number of recent studies with promising findings are ushering in an exciting new age [29].

Shokat and Ostrem identified the switch II pocket present in both the active and the inactive conformations of the KRAS protein and deepened our insight into the complex interactions within the RAS signalling protein family [30, 31]. Based on this body of research, the U.S. Food and Drug Administration (USFDA) acknowledged Sotorasib and Adagrasib as two direct KRAS G12C (OFF) inhibitors as significant advancements in therapy of patients having KRAS G12C mutant metastatic lung cancer. These inhibitors have demonstrated encouraging outcomes in early-phase clinical studies (phases I and II) (NCT04933695, NCT04303780, NCT04625647, NCT03785249) [32, 33].

Amgen's Sotorasib (AMG510) is the first FDA-approved drug which selectively and irreversibly inhibits KRAS G12C. It binds to surface groove of protein receptor near histidine 95, adjacent Targeting the switch II pocket at cysteine 12, stabilizing KRAS in inactive GDP bound state [34–37].

Adagrasib (MRTX849) and other KRAS G12C (OFF) inhibitor works similarly to sotorasib in that they secure the switch II pocket and bind KRAS G12C preferentially and irreversibly in its inactive form bound to GDP [38–40].

18.6 EGFR: A CELL SURFACE RECEPTOR FOR EPIDERMAL GROWTH FACTOR

The epidermal growth factor receptor (EGFR) also called as HER1 or ErbB1, is part of the receptor tyrosine kinase family and can activate multiple signalling pathways that support cell differentiation, proliferation, growth, and survival [41, 42].

Following ligand binding, EGFR tyrosine kinase undergoes homo or heterodimerization and auto-phosphorylates tyrosine rich cytoplasmic domain to activate the receptor. This starts two important subsequent intermediate pathways: PIK3CA/AKT1/mTOR pathway and RAS/RAF1/MAP2K1/MAPK1 kinases6 [43].

Among the emerging driving mutations, EGFR gene represents one of the most significant genetic alterations found in individuals having NSCLCs. Increased ligand generation by diseased cells activating mutations of EGFR mutations within cancerous cells and increased expression of EGFR on cancerous cells (account for 40–80% of NSCLCs) are three primary factors which activate EGFR [44, 45].

Gefitinib and Erlotinib, the first drugs developed to selectively inhibit the tyrosine kinase region of EGFR. In both phase I trials and later phase II studies, these medications show remarkable effectiveness in NSCLC patients previously treated with chemotherapy. Consequently, they received approval for use in treating advanced NSCLC [46].

Another EGFR-TKI that has shown promise in treating patients having advanced NSCLC who have activating mutations of EGFR is afatinib [47, 48].

18.7 ALK: A PROTEIN KINASE IMPLICATED IN TUMOUR DEVELOPMENT

Anaplastic lymphoma kinase (*ALK*), a receptor tyrosine kinase, has been widely investigated as a key target for tyrosine kinase inhibitor (TKI) therapies. This transmembrane tyrosine kinase receptor, which has 1620 amino acids, is produced by proto oncogene *ALK* which is primarily found in nervous system. When ligands attach to extracellular domain of *ALK* the intracellular kinase domain dimerises and is subsequently auto-phosphorylated. This protein helps control how cells grow. Neuroblastoma, anaplastic large cell lymphoma, and NSCLC are among the cancers in various alterations that can take place in the *ALK* gene. Alterations in the *ALK* gene can lead to uncontrolled growth and proliferation of cancer cells upon activation. In tumour cells, *ALK* signalling is typically driven by chromosomal translocations, gene fusions, gene amplification/dysregulation, and activating point mutations. [49–52].

ALK rearrangements were first discovered in approximately 60% of anaplastic large cell lymphomas, a rare subtype of non-Hodgkin lymphoma, where they were identified as a fusion involving a portion of the nucleophosmin (*NPM*) gene [53]. Among various fusion partners, echinoderm microtubule-associated protein-like 4 (EML4) is the most common, with breakpoints predominantly occurring within the EML4 gene. Another unusual fusion partners have also been found: TFG and KIF5B are two examples [54, 55]. It has been demonstrated that oncogenic fusion genes such as EML4 and *ALK* are present in around 2–7% of stage III or IV NSCLC patients; more commonly young people who smoke lightly or never are more prone, causing metastatic brain tumours [56, 57].

Treatment for *ALK*-positive people, especially those having NSCLC, is greatly aided by certain inhibitors such as alectinib, crizotinib, and ceritinib [58].

The medication crizotinib strongly and precisely inhibits *ALK* and c-MET. Using the inhibitor of cocrystal structure and c-MET to increase active site binding PHA-66752, the first-generation c-MET inhibitor was modified into a compound belonging to the 3-benzyloxy-2-aminopyridine class. Crizotinib has shown improved progression-free survival (PFS) compared to chemotherapy in both first-line and second-line treatment settings [59, 60]. Ceritinib, a second-generation *ALK* inhibitor, has been shown to be effective in treating advanced or metastatic *ALK*-positive NSCLC, including in patients previously treated with crizotinib [61, 62]. Preclinical studies revealed that ceritinib exhibits several times greater potency against ALK compared to crizotinib. Furthermore, it has been shown that Ceritinib works well for those who have developed a resistance to crizotinib, with brain metastases showing impressive tumour responses [63, 64].

Evidence from the following five clinical trials supports the effectiveness of ceritinib compared to Crizotinib in patients having advanced or metastatic NSCLC who have not previously been treated with Crizotinib: NCT01283516, NCT01685138, NCT00585195, NCT00932451, and NCT00932893 [65, 66].

18.8 C-ROS ONCOGENE 1 (ROS1)

ROS proto-oncogene 1 receptor tyrosine kinase (ROS1) rearrangement was initially identified in NSCLC cell lines in 2007. ROS1-rearranged NSCLC accounts for approximately 1–2% of all NSCLC cases, and these are exclusively classified as adenocarcinomas [67].

The overall rate of ROS1 positive, and NSCLC showed no discernible difference from ROS1-negative NSCLC, although ROS1-positive frequency was much greater in young patients who had never smoked [68]. ROS1 tyrosine kinase inhibitors (TKIs), including entrectinib and crizotinib, have FDA approval. In fact a number of other specific TKIs are under development/evaluation in clinical studies.

18.9 CRIZOTINIB (1ST GENERATION)

The amino acid composition of kinase domains of ALK and ROS1 is 77% similar [69].

The structural similarities between ALK and ROS1, along with crizotinib's proven efficacy in ALK-positive NSCLCs, suggested that ROS1 could serve as a promising therapeutic target for crizotinib. In 2016, the U.S. FDA and the European Medicines Agency (EMA) approved Crizotinib (Xalkori, PF-02341066) for the treatment of NSCLC with ROS1 rearrangements [70].

The preferred first-line therapeutic option for ROS1-rearranged NSCLC is Crizotinib. Despite crizotinib's notable effectiveness, resistance to crizotinib causes tumour recurrence in NSCLCs that have been rearranged by ROS1. The ROS1 G2032R mutation is primarily responsible for acquired secondary point mutations, resulting in the development of crizotinib resistance [71, 72].

18.10 ENTRECTINIB (2ND GENERATION)

Entrectinib, a powerful oral multitarget tyrosine kinase inhibitor (TKI) that acts against ALK, TRK and ROS1 kinases were initially introduced in 2016 and later received FDA approval for treating ROS1 positive NSCLC [73]. A research revealed entrectinib's in vivo and in vitro actions in NSCLC models [74]. In ROS1-positive animals, entrectinib had significant action. Furthermore, entrectinib has been shown to promote intracranial activity and well permeate the blood–brain barrier (BBB), which suggests that it may be effective in brain metastases [74]. E STARTRK-1, STARTRK-2, and ALKA-372–001 are three phase I/II cohort studies that have proven entrectinib's effective and controllable safety in patients having ROS1 + NSCLC. The median response time was 24.6 months and total objective response rate was 77%; 11% patients experienced serious adverse events and 34% reported high-grade adverse events (grades 3 or 4) [75]. The results of two FDA approved ROS1 inhibitors like Entrectinib and Crizotinib were compared in a recent research. The variables included in this study were PFS, OS and time to treatment discontinuation (TTD). According to these data, entrectinib (12.9 months) results in longer TTD than Crizotinib (8.8 months) [76].

18.11 LORLATINIB (3RD GENERATION)

Lorlatinib (PF-06463922) is a third-generation tyrosine kinase inhibitor developed for precision-targeted therapy. Anaplastic lymphoma kinase (ALK), a well-studied receptor tyrosine kinase, serves as a prominent target for TKI-based treatments. Lorlatinib demonstrated strong effectiveness in NSCLC cases harbouring the ROS1 G2032R mutation [77]. NSCLCs with G2032R alteration in the ROS1 gene that developed crizotinib resistance showed high Lorlatinib potency [71, 72]. In contrast to crizotinib, alectinib, and ceritinib, oral lorlatinib demonstrated a noticeably greater ROS1 inhibition, suggesting an intriguing substitute therapy for crizotinib resistant ROS1 positive NSCLCs. Furthermore, Lorlatinib increased CNS penetration and peaked in cerebrospinal fluid (CSF), suggesting it may be a viable option for brain metastases [77]. Between 2017 and 2019 phase I and II clinical studies examined the effectiveness of lorlatinib in patients with (ROS1 and ALK) rearranged NSCLC through the evaluation of objective response rate (ORR) and median progression-free survival (PFS) [61, 58]. In patients having advanced ROS1 positive NSCLC, the clinical trials demonstrated significant activity. Lorlatinib showed a promising role in CNS metastases or crizotinib resistance. Furthermore, lorlatinib was well tolerated and there was no discernible harm [61, 58].

18.12 SINGLE TREATMENT

In 2011, the US authorised Vemurafenib (PLX4032), the first medication to treat BRAF-mutant cancer [78]. Vemurafenib, an oral selective inhibitor targeting the oncogenic B-Raf kinase in melanoma patients with the BRAF V600E mutation, was commercially introduced in 2008 [79]. Vemurafenib suppresses the activity of the BRAF V600E mutation by blocking the RAF/MEK/ERK signalling cascade [80]. Patients having BRAF positive NSCLC who received Vemurafenib in 2015 had a 7.3-month PFS and 42% ORR [81]. Vemurafenib increased the PFS in those patients having V600 BRAF driven NSCLC, who had never been treated before, per 2017 open-label 2 phase study. Administration of vemurafenib in these individuals has been reported as safe in terms of toxicity [82].

Dabrafenib is an inhibitor that competes with adenosine triphosphate (ATP) for binding sites that specifically inhibits BRAF kinase [83]. An open-label 2 phase clinical trial investigation examined Dabrafenib impact in NSCLC patients having V600E BRAF positivity in 2016. According to reports, the majority of patients responded quickly to therapy; 58% of patients had their adenocarcinoma under control, whereas 33% of patients responded overall. This research was the first of three enrolled cohorts (cohort A) in the series [84, 85]. The trametinib and dabrafenib combination was studied by Cohorts B and C, this will be mentioned next.

18.13 ADJUVANT TREATMENT/CO-ADMINISTRATION (SUPPORTIVE TREATMENT)

18.13.1 Dabrafenib and Trametinib Combination Therapy Targeting the BRAF/MEK Pathway

A number of clinical studies have looked into how BRAF/MEK inhibition can work together to provide a synergistic effect. Planchard et al. examined toxicity and therapeutic results of dabrafenib and trametinib combination in the cohort B study [82].

Patients having metastatic NSCLC who had previously taken platinum-based chemotherapy and had BRAF V600E mutation were subjects of this combination approach. The reported median progression free survival exceeded nine months and overall response rate surpassed 50% indicating that this combination is a viable treatment option. Notably, more than half of the patients (56%), with melanoma, experienced severe side effects such pyrexia, anaemia, disorientation, and reduced appetite, despite the tolerable safety profile and the absence of treatment-related mortality [82].

Subsequently, in 2017, Planchard et al. carried out cohort C study to evaluate the previously mentioned combination therapy of Trametinib and Dabrafenib in patients having newly diagnosed BRAF V600E-positive advanced NSCLC [86]. As a result, the toxicity and response rate for individuals who have had prior treatment are similar to those who have not. The FDA authorised trametinib and dabrafenib combination therapy as a therapeutic option for patients having BRAF positive V600E metastatic NSCLC in 2017 due to the medication's encouraging outcomes [86].

18.13.2 Trametinib + Vemurafenib

A 2015 research examined the results of individuals receiving inhibitor of MEK trametinib with or without vemurafenib. V600E-BRAF was impacted by a single dose of vemurafenib, whereas both non-V600E-BRAF and V600E-BRAF were impacted by trametinib. When vemurafenib and trametinib were administered together, it has been shown to increase apoptosis and BIM (proapoptotic protein) more than when trametinib was administered alone. Therefore, the efficacy of combining inhibitor of MEK with BRAF inhibitor seemed encouraging. Notably, the study also examined the combination of erlotinib and vemurafenib in BRAF V600E cells; and no synergistic effects were discovered [87].

18.13.3 Selective RET Inhibitors

Selective tyrosines TKI and pralsetinib (BLU-667) effectively target and inhibits RET fusions.

In an open-label, multi-cohort, phase 1/2 study (ARROW) (NCT03037385), the safety and effectiveness of pralsetinib were investigated in patients having RET altered metastatic NSCLC who had previously taken treatment as well as those who had not [88]. A phase 1 trial was used to identify the maximum tolerable dosage which was 400 mg OD [79].

The ORR served as the basis for reporting the response rate; 6% patients who had previously had platinum-based chemotherapy demonstrated full response with ORR 61%. In contrast, the complete response rate was 11% and the ORR for individuals who had never had therapy was 70%. These findings indicated that treatment-naive individuals had a higher ORR: however, those who had received previously therapy had lingering median duration of response at 6 and 12 months [88].

A number of side effects, including as anaemia, hypertension, and neutropenia, were noted although no treatment-related fatalities were documented. Overall, these results underscore oral pralsetinib as a promising and tolerable therapeutic option for patients with RET-altered metastatic NSCLC [88].

Measuring changes/removal of the amount of RET circulating tumour DNA (ctDNA) in samples of blood was another way to estimate the response of pralsetinib for those patients having RET fusion positive NSCLC. Interestingly, practically all patients saw a fast drop in RET ctDNA over the whole range of pralsetinib dosages; 81% of NSCLC patients had no detectable RET ctDNA after eight weeks, indicating a quick response [89].

Pralsetinib and traditional first-line therapies are compared in randomised, recruited, open-label III phase study based on PFS of patients having RET altered NSCLC. Pralsetinib will be administered at random to the experimental arm. For the control group, patients having non-squamous cell lung cancer got platinum plus pemetrexed ± pembrolizumab, whereas those having squamous cell lung cancer received platinum + gemcitabine. The aim of this study was to identify good choice for first-line therapy [90].

18.13.4 LOXO-292

Selpercatinib is selective TKI inhibitor that is taken orally. It is a small molecule RET inhibitor that is ATP competitive and was created to block activating or resistant RET mutations. Selpercatinib has demonstrated exceptional RET-inhibition efficacy and significantly lower toxicity in vitro and in vivo when compared to multi-kinase inhibitors (MKIs) [79]. Selpercatinib demonstrated strong clinical efficacy in patient having lung cancer that was positive for KIF5B-RET fusion. Patients resistant to alectinib had previously undergone chemotherapy, immunotherapy, and whole-brain radiation. Selpercatinib has been shown to significantly lower neurological [79].

18.14 NON-SELECTIVE INHIBITORS OF RET

MKIs, which target several kinases, are non-selective RET inhibitors. They have demonstrated poor effectiveness and have caused off-target damage. The first MKIs to be tried in patients having RET altered NSCLC were non-selective inhibitors of RET such as vandetanib, cabozantinib, and lenvatinib. Cabozantinib was shown to have the highest median PFS (3.6 months) and complete response rate (37%) in a 2017 research.

Although no full response was seen, the response rates with vandetanib and sunitinib were reported to be 18% and 22%, respectively. Vandetanib and sunitinib were shown to have median PFSs of 2.9 and 2.2 months, respectively [91]. Selective RET inhibitors seem to be better options because multi-kinase inhibitors (MKIs) have limited effectiveness and may have considerable off-target effects. Resistance to both FDA-approved RET inhibitors is unavoidable and the most resistant RET mutation has been found to be G810C [67].

Therefore, more research is needed to create more selective RET inhibitors and combination treatments. The effectiveness, safety, and tolerability of TPX-0046 in drug resistant and naive RET-altered tumour models with Solvent Front Mutations (SMF) are the focus of phase I/II clinical research (NCT04161391). In vitro and in vivo, TPX-0046 has demonstrated strong anti-tumour effectiveness against a range of RET mutations, including SFM-mediated resistance. In order to examine the ORR, this trial's phase II will evaluate TPX-0046 in patients having stage IV NSCLC who are treatment naive or have already received therapy [75].

18.15 TRANSITION FROM MESENCHYMAL TO EPITHELIAL PHENOTYPE

A proto-oncogene member of tyrosine kinase receptor family mesenchymal-epithelial transition (MET) was triggered after attaching to its ligand and hepatocyte growth factor (HGF) [92], in addition to MET's physiological role in cell division [93].

This proto-oncogene experiences a number of changes, including overexpression, amplifications, and mutations, which may result in cancerous cells [94]. According to PM Comoglio, MET is found to be overexpressed in approximately 20% of NSCLC cases and amplified in about 1–5% of them [95, 96]. However, it has been demonstrated to be one of the key mechanisms of metastasis that is necessary to return mesenchymal phenotype to epithelial phenotype so that cancer cells can adhere to other tissues [97]. Accordingly, it has been demonstrated that a poor clinical prognosis for NSCLC is associated with this abnormal MET expression [98]. Therefore, MET targeting has been explored in clinical studies as a therapeutic strategy. Three main kinds of small molecule of inhibitors of MET are distinguished by the binding position on MET structure [99]. Types I and II prevent ATP binding receptor activation whereas type III binds allosteric sites rather than ATP binding site [100]. Furthermore, there are two subgroups of type I MET inhibitors: Ia (Crizotinib) and Ib (Tepotinib, Capmatinib, and Savolitinib). The Ib category is more selective to MET and has less off target effects [101]. Notably, crizotinib was initially authorised to treat ALK-rearranged and ROS1 rearranged NSCLC; however, it has now been shown that it may be used to attract NSCLC with MET amplification [102]. Following the FDA's approval of Crizotinib, Tepotinib and Capmatinib were authorised for treatment of metastatic NSCLC in patients with MET exon 14 (METex14) skipping mutation [103] as one of the primary mutations that independently causes NSCLC [104].

In addition to attaching to a hydrophobic pocket next to the ATP binding site of MET, type II inhibitors such as glesatinib, capozantinib, and merestinib can also block other kinases including RON, AXL, and VEGFR2. Although the drug is approved to treat advanced medullary thyroid cancer and advanced clear cell renal cell carcinoma, studies on its efficacy and safety in MET mutated NSCLC have produced positive results [105]. Additionally using type III inhibitor, Tivantinib in combination with RGFR inhibitor Erlotinib helped those with NSCLC [106].

18.16 NTRK: A RECEPTOR TYROSINE KINASE INVOLVED IN NEUROTROPHIC SIGNALLING

A set of three receptor tyrosine kinases embedded in the membrane TRKA, TRKB and TRKC are involved in neurotrophic tropomyosin receptor kinase (NTRK) and are in charge of the physiological maturation and development of central nervous system [1]. The PI3K/Akt/mTOR, Ras/Raf/MAPK, PLCc/PKC signalling pathway, all of which are essential for cell proliferation and plasticity, are mediated by these tyrosine kinase receptors [2, 3]. These receptors are naturally activated by three essential ligands: NGF, BDNF, and NT-3 [3, 4]. Otherwise, some genetic changes in NTRK genes, including deletions, splice variants, amplifications, and mutations, have been linked to cancers and may activate downstream signalling pathways without the need for ligands [5, 6]. Less than 1% of individuals having NSCLC have NTRK fusions common mutation found in many malignancies. A chimeric protein with carcinogenic potential is the outcome of this [7].

The high frequency of NSCLC worldwide made NTRK a tempting target for inhibitory drug development, even though NTRK fusions seem to be a relatively infrequent cause of NSCLC [8].

18.17 HER2: HUMAN EPIDERMAL GROWTH FACTOR RECEPTOR 2

The four members of human epidermal growth factor receptors (RTK) are HER1–4, which are encoded by the ErbB1–4 genes [107]. Since HER2 is transmembrane glycoprotein, no precise ligand for it has been identified [108]. Although HER2 activation triggers a number of proliferative or apoptotic signalling pathways such as, JAK/STAT, MAPK and PI3K/AKT, its carcinogenic and tumourigenic functions have been demonstrated in many malignancies [109].

In this context, it has been shown that three primary genetic changes—mutation, overexpression, and amplification—may result in NSCLC. Several studies have reported their prevalence to be around 2–4%, 11–32%, and 2–23%, respectively [110, 111]. However, there is still no consensus on the precise definition of HER2 positive cancer of lung.

Many inhibitors, including, lapatinib, poziotinib, afatinib, dacomitinib and neratinib, have been investigated as a treatment choice for individuals having NSCLC but regrettably none of them have been authorised as of yet [112–115].

Although many tyrosine kinase inhibitors (TKIs) have shown limited effectiveness in treating NSCLC, several clinical trials have reported encouraging outcomes. There have been reports of ORR of 53.3% and median PFS of 6.4 months with no serious side effects when pyrotinib, an oral pan-HER TKI, is used to treat NSCLC patients with HER2 exon 20 mutations [116].

Additionally mobocertinib, a novel small molecule TKI that specifically inhibits EGFR insertion and HER2 mutation, is being studied with generally positive outcomes. [117] Additionally, research on mobocertinib, a novel small molecule TKI that specifically inhibits EGFR insertion and HER2 mutation, is progressing with promising first findings. Monoclonal antibodies that target HER2 have been studied for use in immunotherapy to treat NSCLC. Pertuzumab and trastuzumab, two medications licensed for HER2 + breast cancer, have been utilised in clinical trials in this respect: nevertheless, further research is required to demonstrate their therapeutic efficacy on NSCLC [118, 119]. Moreover, antibody drug conjugations have been created as successful HER2 targeting method. Two antibody drug conjugates, T-DMI and T-DXd, have trastuzumab linked to deruxtecan and emantisine. Clinical trials have linked both of these medications, and their encouraging outcomes led to the FDA approving the T-DXd medicine to treat gastric cancer and HER2-mutant NSCLC [120, 121].

18.18 PI3K/AKT/MTORC

The PI3K/Akt/mTOR signalling cascade is frequently implicated in molecular alterations associated with NSCLC. Oncogenic mutations in EGFR and KRAS, two major drivers in NSCLC, can activate this pathway, thereby promoting cancer cell survival, proliferation metabolic activity [122].

RTKs are also recognised as an entrance site for the activation of PI3K/AKT/mTOR signalling cascade. Ligand-independent signalling is the outcome of RTK mutation and amplification in NSCLC. Thus, these signals may be inhibited by targeting PI3K/Akt/mTOR signalling pathway [123]. Based on its structure and function, the PI3K family of lipid kinases is divided into three classes. Different roles in signal transduction were demonstrated by each class. Regulated cancerous transformation and sustained tumour growth. However, somatic mutations and PI3K class amplification have been found in patients having NSCLC [124]. Squamous and nonsquamous NSCLC have been found to have activation mutations in PIK3CA, which codes for the catalytic subunit PI3Kα, as well as changes in tumour suppressor phosphatase and tensin homolog. According to preclinical and clinical research, PIK3CA-mutant lung cancer is regressed when PI3K pathway is targeted. Through they simultaneously inhibit several PI3K pathway components, the proliferation of PI3K-dependent NSCLC cell lines might be stopped, slowing the progression of NSCLC [125, 126]. There are two types of PI3K inhibitors: selective and pan-PI3K [122]. A number of pan PI3K inhibitors have undergone testing in clinical trial stages involving Buparlisib (BKM120), PX-866, Pilaralisib (XL-147), and Pictilisib (GDC-0941). The data pertaining to partial responders showed no appreciable change in PFS or OS [122].

As a result, the panclass I PI3K inhibitor treatments have not demonstrated sufficient effectiveness. There are several clinical trials using class I PI3K inhibitors that are isoform specific. These include potent PI3K inhibitors Taselisib (GDC-0032) and Alpelisib (BYL719), which target the p110δ, p110γ, and p110δ isoforms, respectively. In a II phase study of individuals with advanced NSCLC, they have been evaluated [123, 124]. Clinical phase I trials have examined several selective PI3K inhibitors such as GSK2636771, INK1117, AZD8186, and SAR260301 [122].

However, in individuals with NSCLC, overexpression of the Akt pathway is considerable and linked to elevated mTOR [125].

Preclinical research using NSCLC cell lines showed that Akt activation results in HER2 amplification, PTEN, EGFR, or PIK3CA mutations [126]. Consequently, as numerous studies have described, the wide range of Akt pathway inhibitors under development focus inhibition of Akt is effective therapeutic step.

Perifosine, an alkyl phospholipid (APL) which prevents Akt from translocating to membrane and preventing Akt phosphorylation and activation, is the most well-tolerated therapy for NSCLC. [127].

Data demonstrated that perifosine inhibits GSK3β's translocation to cell membrane and kinases phosphorylation of Akt and ERK.

In addition to blocking rapamycin, perifosine therapy significantly reduces the levels of GSK3β and p-GSK3β in A549-RR cells. These results suggest that perifosine may be able to inhibit GSK3β and other survival pathways that are activated by the inhibition of mTORC1. Recent research, however, has shown that rapamycin and perifosine together can be more effective than either drug alone since each one inhibits cell growth via controlling GSK3β activity [128]. Another strong and specific AKT inhibitor with anti-proliferative properties is MK-2206. Combined effects of MK-2206 having RTK inhibitors like erlotinib and lapatinib can considerably exhibit tumour inhibitory actions, according to in vitro and in vivo data [129]. Additionally, phase II research reported that among patients who had already received erlotinib, combination of MK2206 with erlotinib achieved its aim [130]. In the meantime, a sizable fraction of NSCLC showed that the mTOR pathway was upregulated. The mTOR complexes (mTORC1 and mTORC2) are crucial for preserving homeostasis and cellular growth [131].

Two important mTORC1 substrates Elongation Initiation Factor (EIF) 4E binding protein 1 (4E-BP1) and phosphorylation of p70-ribosomal protein S6 kinase 1 (p70-S6K1) control a variety of functions in NSCLC patients, including proliferation, cell growth, migration, and invasion [132], Protein Kinase Cα (PKCα) and Akt and are directly phosphorylated and activated by mTORC2 [133].

Consequently, mTOR inhibitors may be a desirable tactic to stop the spread of advanced non-NSCLC. Preclinical and clinical research has been used to generate and examine ongoing clinical trials for Temsirolimus, Sirolimus and Ridaforolimus, and Everolimus. Everolimus specifically targets mTORC1 as a mTOR inhibitor. According to conservative assessments, the majority of individuals receiving Everolimus experienced moderate side effects [134]. However, Everolimus and EGFR inhibitor combination therapy has demonstrated limited anticancer activity in patients having NSCLC with mutation in PI3K AKT-mTOR pathway cascade [135]. The combination of Sirolimus and Pemetrexed showed promise in the phase I/II research and didn't seem to pose any serious safety risks [136].

Temsirolimus's effectiveness endpoint was not met when used as a single targeted medication [102]. By inhibiting both PI3K and mTOR, dual targeting has been created.

Two significant downstream targets of PI3K/Akt signalling cascade are the transcription factors Cyclin D and FOXO [102]. Through transcriptional repression and proteasome-mediated degradation BEZ235 reduced the expression of cyclin D1 and cyclin D3 in NSCLC, ultimately stopping cell cycle at G1 phase and preventing PI3K/Akt activation [136]. Additionally, BEZ235 suppressed mTOR signalling by decreased production of cyclin D. [136]. Additionally, a phase Ib trial testing XL765 and GDC-0980 showed a satisfactory safety profile in individuals having NSCLC. [117, 137].

18.19 TARGETING PROGRAMMED DEATH RECEPTOR 1 (PD-1)/PD-L1 PATHWAY

The PD-1/PD-L1 signalling pathway plays a crucial role in regulating immune responses and maintaining tissue homeostasis and self-tolerance [138].

Activated B cells, T cells, activated monocytes, natural killer T cells, and dendritic cells (DCs) are the main cells that express PD-1, negative costimulatory receptor [139]. Garon EB, PD-1 links itself to its ligands, PD-L1 and PD-L2, which are members of B7 family and expressed by immune, normal, tumour cells [140].

However, efficient anti-tumour immune responses were inhibited by T cells that expressed PD-1. T cell transcription factors such as T-bet and GATA-3 are inhibited by PD-1/PD-L1 interaction. It causes tumour-infiltrating T cells to undergo apoptosis and suppresses the survival, activity, and proliferation of CD8+ cytotoxic T lymphocytes. Furthermore, PD-L1 expression can be mediated by the development of regulatory T cells (Tregs) and maintenance of their suppressive function.

Thus, inhibiting the PD-1/PD-L1 pathways can stimulate anti-tumour action mediated by both effector T cell activation and Treg suppression [141].

A tumour microenvironment gradually becomes immune suppressive, leading to development of tumour promoting cells such T regs, M2 macrophages, and myeloid-derived suppressor cells even if immune cell migration to the tumour exhibits anti-tumour action. To get around the immune system's defences, cancer cells employ immuno-inhibitory processes and avoid immune identification. Immune editing is a dynamic process that can create different immune evasion mechanisms by modifying the immunological milieu and enhancing immune cell activity [142–144].

In order to eradicate NSCLC, immune responses may be regulated by targeting PD-1/PD-L1 checkpoint. PD-1/PD-L1 pathway modulation presently is being researched as a possible immunotherapy for NSCLC patients. For the first- and second-line contexts, a number of PD-1/PD-L1 antibodies have been licensed because of their superior efficacy, longer duration of response, and acceptable safety profiles when compared to other conventional treatments [145].

A completely human monoclonal antibody against PD-1 called nivolumab was authorized for use in therapy of patients of NSCLC. Patients treated with nivolumab as first-line monotherapy do not have higher PFS than those treated with platinum based chemotherapy in large population with PD-L1 expression level of 5% or above. Nivolumab demonstrated a more favourable safety profile compared to chemotherapy [144].

Additionally, a phase III study showed that nivolumab significantly improved survival and response rates in advanced NSCLC when compared to docetaxel [145].

As first-line treatment for metastatic NSCLC having PD-L1 tumour percentage score at least 50%, MK-3475, pembrolizumab, humanized monoclonal IgG4 anti-PD-1 antibody is another anti-PD1 that has strong anti-tumour activity and considerably better OS and PFS than chemotherapy. Compared to platinum-based chemotherapy, there was reduced frequency of therapy-administration-related adverse events and long-term follow-up revealed no signs of increased toxicity [146].

In patients having recurrent/advanced NSCLC, the studies showed that pembrolizumab had a superior ORR than nivolumab, but there was no discernible difference in PFS between the two drugs [134]. The data showed that immunotherapy is ineffective for patients of NSCLC with EGFR or HER2-mutations and ALK rearrangements. Additionally, NSCLC with MET and BRAF mutations should respond as well to anti-PD1/PD-L1 immunotherapy [147].

Even while anti-PD1 monotherapy, such as pembrolizumab and nivolumab, was similarly successful in treating older and younger patients having NSCLC individuals with poor performance status had a significantly worse survival rate. Nonetheless, NSCLC patients with low PS and positive PD-L1 expression may benefit from an immune checkpoint inhibitor [148]. Programmed death receptor ligand 1 (PD-L1) and other immunological checkpoints were considered potential therapeutic targets.

A humanized IgG1 antagonist antibody to PD-L1, atezolizumab, prevents PD-L1 and PD-1 receptor activation complex from interacting. Its goal is to stop T cell inhibitory signals, which make tumours visible to cytotoxic T cells. Following atezolizumab treatment, patients having NSCLC who expressed PD-L1 had improved OS and ORR [149].

Compared to chemotherapy, cemiplimab monotherapy dramatically improved OS and PFS in patients having advanced NSCLC whose PD-L1 was at least 50% [151]. Durvalumab, a human IgG1 monoclonal antibody with high affinity and selectivity, also prevents PD-L1 from interacting with PD-1 enabling T cells to identify and eliminate tumour cells. A real-world prospective study found that durvalumab therapy improved OS and progression-free survival in patients having stage III NSCLC while also exhibiting a safety profile [152, 153].

In comparison to chemotherapy, anti-PD-1 and anti-PD-L1 therapy demonstrated lower ORR, a high rate of grades III and IV Aes, and high drug related death, as well as a significantly longer overall survival, according to meta-analysis data [154].

When compared to docetaxel, Avelumab does not substantially improve OS in patients having NSCLC with PD-L1 + tumours, according to a phase III trial [155].

18.20 CYTOTOXIC T-LYMPHOCYTE-ASSOCIATED ANTIGEN-4

CTLA-4 (cytotoxic T-lymphocyte-associated antigen-4, CD152) is one type of immunological checkpoint mechanism which controls negative regulation of T cell activation and maintenance of self-tolerance. However, in patients having NSCLC, CTLA-4 expression may be predictive and prognostic biomarker. There is evidence that the expression of CTLA-4 in metastatic NSCLC lymph nodes differs from that in original tumours in terms of prognosis. This discrepancy is caused by phenotypic variations between the lymph node and primary tumour microenvironments [156]. In the initial phase of naive T-cell activation, which often takes place in lymph nodes, CTAL-4, one of the most well-known immune checkpoint inhibitors, may decrease autoreactive T cells [157].

CTLA-4 was able to exert its inhibitory impact through a variety of methods, like competing with CD28-positive costimulation for binding to their shared B7 ligands (CD80/CD86) [158]. Targeting CTLA4, Ipilimumab is a completely humanized IgG1κmAb. In patients with stage IV NSCLC, ipilimumab treatment increased mWHO-PFS and irPFS as well as tolerability and safety and without having a significant negative effect on toxicities [157]. Tremelimumab is an extra monoclonal immunoglobulin G2 antibody that targets CTLA–4. It prolongs T-cell activation and stops natural T cell downregulation, which enhances immunological function [158].

Targeting CTLA4, Ipilimumab is a completely humanized IgG1κmAb. In patients with stage IV NSCLC, ipilimumab treatment increased irPFS and mWHO-PFS as well as tolerability and safety without having a significant negative effect on toxicities [159].

Tremelimumab is an additional monoclonal immunoglobulin G2 antibody that targets CTLA-4 and improves immune function by extending T-cell activation and prevents normal T cell downregulation [160].

In patients with advanced disease, such as NSCLC, Tremelimumab demonstrated safety profile and enhanced the durability of objective responses compared to standard treatment [161]. According to the findings of clinical trials, anti-CTLA4 monoclonal antibodies cause acute immune activation which raises the number of CD8+ CTL infiltrates in tumour and may result in temporary autoimmune symptoms.

To avoid side effects, corticosteroid treatment with CTLA-4 blocking has been planned. Additionally, GM-CSF and CTLA-4 expressing tumour cell vaccination combination may be therapeutically beneficial as it alters the immune response and targets important immune system regulatory mechanisms [158].

18.21 FDA-AUTHORIZED CLINICAL STUDIES FOCUSING ON TARGETED THERAPIES FOR NSCLC

Both clinical and preclinical studies have confirmed the safety and effectiveness of various therapies for NSCLC patients harbouring distinct genetic alterations, such as EGFR exon 19 deletions, exon 21 (L858R) mutations, RET fusion-positive metastases, ALK gene rearrangements, and EGFR exon 20 insertions.

Nonetheless, the U.S. FDA has approved a number of targeted medicines for NSCLC based on findings of clinical trials (Table 18.1). Erlotinib is one of the receptor tyrosine kinase inhibitors [162] gefitinib [163], osimertinib [164], crizotinib [163] lorlatinib [165], tepotinib [166, 167] first obtained approval to treat patients with METex14 skipping variations, TKI resistance, ALK-positive, acquired EGFR and EGFR mutation, respectively. The FDA has granted approval to immune checkpoint inhibitors, including atezolizumab [168]. Additionally, the FDA granted initial approval to amivantamab, an injectable bispecific antibody directed at EGFR and c-MET for NSCLC patients harbouring EGFR exon 20 insertion mutations [165].

Amivantamab had strong and long-lasting responses with acceptable safety, according to results from the CHRYSALIS clinical study (NCT02609776) with median OS of 22.8 months, median PFS of 8.3 months, and ORR rate of 40% [173].

Unfortunately, individuals have not responded to treatment and the condition has progressed, according to recent data [174]. Mobocertinib's recent clinical development for EGFRex20ins-mutated NSCLC was supported by earlier preclinical research [174].

18.22 COMBINATION THERAPY

Target therapy may be more successful when used in combination with other cancer treatment modalities in order to overcome its drawbacks and boost its effectiveness. Target therapy is frequently used in conjunction with other standard treatments including chemotherapy for patients having NSCLC.

Several studies have been carried out in this area that can aid in improving target therapy's effectiveness and overcoming its drawbacks, including drug resistance [174].

Various studies have compared targeted therapy medications with chemotherapy medicines, such as the use of cetuximab, an EGFR-targeting monoclonal antibody, alongside cisplatin and vinorelbine [175]: Figitumumab mAb that targets insulin like growth factor type 1 receptor (IGF-1R), bevacizumab, mAb that targets VEGF, and carboplatin and paclitaxel [175] in conjunction with a gemcitabine/cisplatinum or paclitaxel and carboplatin regimen [173]. Overall, all of the research findings were comparable.

The group that received these therapies had greater indicators of efficacy including median progression-free survival, overall survival, and response rates, and safety of combination therapy. The combination of a c-MET inhibitor and the EGFR inhibitor erlotinib showed enhanced antitumour activity compared to using either agent alone in an NSCLC mouse xenograft model [65].

EGF-R, VEGF-R, and RET tyrosine kinase activity are the specific targets of the multi-target TKI vannetanib. Vandetanib with docetaxel, like the previously described combination treatments, showed encouraging results with higher progression-free survival.

Additionally, it demonstrated that Vandetanib outlasted chemotherapy alone in controlling lung cancer symptoms [176, 177].

Promising outcomes have also been reported from trials that have been done and are still underway regarding the effects of sunitinib in conjunction with erlotinib- and platinum-based chemotherapy. Sunitinib is a multitargeted TKI which inhibits, PDGFR, VEGFR, KIT, RET, and Flt-3 and has antiangiogenic and anticancer properties [178]. Similar studies have demonstrated promising results when everolimus, mammalian target of rapamycin (mTOR) inhibitor, is used in conjunction with chemotherapy/chemotherapy plus gefitinib regimen. These results include better disease

TABLE 18.1

Approved Targeted Therapies for Non-Small Cell Lung Cancer (NSCLC): Mechanisms, Dosage, and Clinical Outcomes

Year	Therapeutic Agent	Mode of Inhibition	Healthcare Recipient	Prescribed Dosage	Organisation	Therapeutic Results	References
2011	Crizotinib	Small-sized TKI compound	NSCLC patients exhibiting ALK positivity	Twice-daily oral dose of 250 mg	Pfizer	An overall response rate (ORR) of 61%, with a 95% CI	162
2013	Erlotinib (tarceva)	EGFR TKI	NSCLC with EGFR exon 19 Deletions/exon 21 {L858R} substitution mutations	Administered at 150 mg each day	Astellas Pharmaceuticals	Treatment showed 65% response rate, 10.4 months PFS, and 22.9 months OS	158
2015	Osimertinib (Tagrisso™ AZD9291)	Advanced-generation EGFR inhibitor	NSCLC patients who have received EGFR TKI therapy	80 or 160 mg daily for a median of 260 and 171 days	Astellas Pharmaceuticals	ORR: 70%	169
2015	Gefitinib	Selective small molecule	NSCLC patients having EGFR exon 19 deletions/exon 21 (L858R) substitution mutations	250 mg daily	Astellas Pharmaceuticals	ORR: 50% (95% CI 41–59), DoR: 6.0 months	170
2018	Lorlatinib (lorbrena)	ATP-binding site competitor targeting tyrosine kinase	ALK-rearranged non–small cell lung cancer	100 administered every day	Pfizer	100 mg once daily 36.2% (95% CI 22.7–51.5), intracranial ORR: 56.0% (95%CI 34.9–75.6), PFS: 9.9 (95%CI 5.5, 21.0)	163
Year	**Drug**	**Type of inhibition**	**Patient**	**Recommended Dose**	**Company**	**Outcomes**	**References**
2020	Ramucirumab + erlotinib	VEGFR2-blocking compound	NSCLC patients with EGFR exon 19 deletions or exon 21 (L858R) mutations	10 mg/kg given once every 2 weeks	Eli Lilly	ORR: 76%, PFS: 19.4 months	171
2020	Pralsetinib (Gavreto™)	RET inhibitor	Metastatic NSCLC patients characterised by RET gene fusions	The dose is usually 400 mg once a day	Blueprint Medicines Corporation		82
2020	Co-administration of Opdivo (nivolumab) and Yervoy (ipilimumab)	Anti-PD1, anti-CTLA-4	NSCLC patients lacking EGFR or ALK genomic alterations	Nivolumab given at 360 mg every 3 weeks alongside ipilimumab at 1 mg/kg every 6 weeks	Bristol Myers Squibb Corporation	ORR: 38% (95% CI 33, 43) PFS: 6.8 months (95% CI 5.6, 7.7) OS: 14.1 months (95% C 13.2, 16.2)	83

(*Continued*)

TABLE 18.1 (Continued)
Approved Targeted Therapies for Non-Small Cell Lung Cancer (NSCLC): Mechanisms, Dosage, and Clinical Outcomes

Year	Drug	Type of inhibition	Patient	Recommended Dose	Company	Outcomes	References
2020	(Alunbrig) brigatinib	Tyrosine kinase inhibitor (TKI)	ALK-rearranged metastatic non–small cell lung cancer	Start with 90 mg of brigatinib taken orally once daily for the initial 7 days, followed by an increased dose of 180 mg orally once daily with or without food	ARIAD	The objective response rate (ORR) was 74% (95% CI: 66% to 81%), and the progression-free survival (PFS) was 24 months (95% CI)	84
2020	(Tecentriq) atezolizumab	Anti PD-L1	PD-L1-high NSCLC patients lacking EGFR and ALK gene aberrations	An intravenous infusion over 60 minutes of 840 mg biweekly, 1200 mg every 3 weeks, or 1680 mg every 4 weeks	Genentech	An ORR of 38% was reported (95% CI: 29–48), along with a median PFS of 8.1 months (95% CI: 6.8–11.0) and a median OS of 20.2 months	172
2020	(Tepmetko) tepotinib	MET TKI	Progressive or recurrent NSCLC patients with METex14 skipping	500 mg taken once per day	Merck	PFS: 9.5–12.2 months	64
2020	Oral capmatinib (marketed as Tabrecta™)	MET-targeting agent	NSCLC patients with MET exon 14 skipping mutations	400 mg, orally twice daily	Novartis	An ORR of 68% was observed (95% CI: 48–84)	71
2020	Retevmo (selpercatinib)	Selective ATP-binding site inhibitor of RET kinase in small-molecule form	Metastatic RET fusion positive NSCLC patients	120 mg for patients less than 50 kg and 160 mg for those 50 kg or greater	Eli Lilly	An ORR of 85% was observed (95% CI: 70–94%)	72
2021	Lumakras™ (sotorasib)	Ras-related GTPase inhibitors	NSCLC patients with KRA mutations	A 960 mg oral dose administered once per day, with or without food	Amgen	ORR of 35% and a DCR of 91% were observed. The median duration of response reached 10.9 months, while progression-free survival was 6.3 months	173
2021	Libtayo® (cemiplimab-rwlc)	Anti-PD-L1	NSCLC patients with high PD-L1 expression (tumour proportion score [TPS] > 50%) and no EGFR, ALK or ROS1 aberrations	Every 3 weeks, 350 mg is delivered via a 30-minute IV infusion	Regeneron Pharma Inc	37% response rate, with a median progression-free survival of 6.2 months and overall survival of 22.1 months	166

symptoms, higher survival rates, and a positive response to treatment [179, 180]. Additionally, target therapy plus radiation or chemotherapy plus radiation has shown encouraging outcomes in terms of increased response to therapy and survival. Gefitinib and erlotinib combined with chemoradiation [171, 181] and Cetuximab, an inhibitor of EGFR in combination with irradiation and chemotherapy for late-stage lung cancer, [182] are two examples.

The CALGB 30407 study was a randomized phase II trial assessing pemetrexed, carboplatin, and thoracic radiotherapy with or without cetuximab in patients with locally advanced unrespectable NSCLC.

18.23 THERAPEUTIC LIMITATIONS IN TARGETING NSCLC-SPECIFIC MUTATIONS

A number of important obstacles stand in the way of the extensive application of targeted treatments for advanced NSCLC in hopes of curing the illness. These include toxicity, medication resistance, and the expensive cost of these drugs, which may prevent some NSCLC patients from accessing them.

18.24 HETEROGENEITY IN TUMOURS AND THE ABSENCE OF RELIABLE BIOMARKERS

The absence of approved biomarkers to forecast treatment response because of tumour heterogeneity is another drawback of targeted therapy in NSCLC [174].

Even while targeted therapy can be used to find people who might benefit from it, not every patient with a certain molecular defect will react to treatment. Furthermore, just a portion of the cancer cells in a tumour may have some molecular alterations which could result in resistance and insufficient targeting [183]. Determining which patients are suitable for targeted therapy versus those who may benefit more from alternative treatments can be a complex decision [184]. Furthermore, even in cases when molecular alterations are detected, in some cases available data are not enough to determine the optimum course of treatment, including which targeted therapy to employ and how best to sequence medicines.

18.25 DRUG RESISTANCE

Selective pressure from targeted medicines can result in anomalies that trigger the driving oncogene. Selective inhibitors initially show good results in malignant tumours with particular defects. But resistance grows from time to time, frequently as result of target gene genetic abnormalities. For instance, following osimertinib therapy, mutations like EGFR C797S have emerged as secondary alterations, whereas EGFR T790M mutations may result in resistance to erlotinib/gefitinib [172].

Furthermore, it has been demonstrated that individuals with ALK and ROS-1 mutations (L2026M and L1196M) develop on-target resistance to crizotinib [172].

The therapeutic effectiveness of targeted treatments in NSCLC is severely limited by these genetic alterations and ensuing resistance. Acquired resistance to osimertinib or gefitinib in NSCLC cells has demonstrated EMT traits, including an increase in and stemness and vimentin, a decrease in E-cadherin, without any subsequent EGFR gene alteration [185]. AXL is linked to pathophysiology of EMT, and mesenchymal EGFR mutant cells may become more sensitive to erlotinib when AXL inhibitors are administered [186]. Although these inhibitors have demonstrated considerable effectiveness in certain NSCLC patients, resistance may emerge through various mechanisms such as reduced antigen display to immune cells, activation of alternative immune checkpoint pathways, and changes within the cancer microenvironment.

18.26 TOXICITY

While short duration use of targeted medications is usually tolerable, long-term chronic usage of these medications frequently has serious harmful adverse effects. Because targeted medicines selectively target cancer cells, they were once thought to be less hazardous.

However, toxicity may result from these treatments' "on-target" as well as "off-target" consequences. Additionally, some targeted therapies can lead to serious adverse effects such blood clots, liver or renal damage, and gastrointestinal issues [174].

These adverse effects may restrict the therapy's dosage and duration, which would lessen its efficacy [187]. On-target toxicity occurs when a drug designed to inhibit cancer-specific targets also affects the same proteins in normal cells, disrupting signalling pathways and causing cellular damage [188]. This can lead to various adverse effects, including skin rashes from EGFR inhibition, high blood pressure from VEGF inhibition, and elevated blood sugar levels from PI3K inhibition [189].

When a drug inhibits a protein it was not intended to target, it causes off-target toxicity, which can result in negative side effects. Each drug may have its own distinct off-target toxic effects and did not show a class effect in contrast to on target toxicity. For instance, gefitinib is not linked to cardiac toxicity; however, osimertinib is, and this includes impaired left ventricular function, heart failure, conduction problems, and cardiac infarction [188].

18.27 CONCLUSIONS AND FUTURE PERSPECTIVES

The development of molecular pathology over the past ten years has increased our knowledge of the underlying pathological processes and heterogeneity of NSCLC, leading to a notable advancement in patient treatment using more sophisticated and efficient techniques. Due to their positive results, immunotherapy and target therapy have created new and exciting avenues for lung cancer treatment. Multiple signalling pathways and specific oncogenic driver mutations have been identified as contributors to malignant transformation. Moreover, epidermal growth factor receptor new targets are continuously being identified, including VEGF, MET, HER2, microRNAs, ALK, immune checkpoint inhibitors (PD-1 and CTLA4), and signalling pathways like PI3K/AKT/mTOR, RAS/MAPK, and NTRK/ROS1 pathways. This leads to development of new treatments. A significant number of medications that have shown clinical success have targeted these variables and pathways: PI3K/AKT/mTOR inhibitors (everolimus), EGFR inhibitors (gefitinib and erlotinib), and NTRK/ROS1 (entrectinib) inhibitors are among the medications now utilised as first-line treatments in place of chemotherapy. It is interesting to note that checkpoint inhibitors, including the anti-PD1 antibodies pembrolizumab and nivolumab, could help many NSCLC patients. Furthermore, several studies have shown that immunotherapies in conjunction with specific targeted medicines can effectively treat NSCLC. Although target therapy has managed NSCLC, therapeutic resistance in cancer cells is unavoidable. To enhance the effectiveness of treatment, it is vital to understand resistance pathways and design combination therapies. Several clinical trials assessing drugs used in immunotherapy and targeted therapy have been carried out up to this point with promising results. The results of trials will help define the role of combined immunotherapies, immune based monotherapies, and combinations targeted therapy immunotherapy in treatment of NSCLC. For patients with limited therapeutic options, targeted therapy may eventually transform the way lung cancer is handled. To find the predicted characteristics reacting to targeted medicines, clinical research is still ongoing. Better understanding of molecular biomarkers and newly created combination treatments may offer the most potent remedy. In the future, theranostics and genuinely tailored oncological treatment will depend on appropriate patient selection utilizing predictive biomarkers to reduce vulnerability and optimize the few resources available.

Improved patient outcomes are anticipated if resistance is addressed, suitable inhibitors or their combinations are used, and adverse effects are kept to a minimum.

REFERENCES

1. Bray F, Ferlay J, Soerjomataram I, Siegel RL, Torre LA, Jemal A. Global cancer statistics 2018: GLOBOCAN estimates of incidence and mortality worldwide for 36 cancers in 185 countries. *CA*. 2018;68(6):394–424.
2. Siegel RL, Miller KD, Jemal A. Cancer statistics, 2020. *CA*. 2020;70(1):7–30.
3. Blumenschein G Jr, Kabbinavar F, Menon H, Mok T, Stephenson J, Beck J, et al. A phase II, multicenter, open-label randomized study of motesanib or bevacizumab in combination with paclitaxel and carboplatin for advanced nonsquamous non-small-cell lung cancer. *Ann Oncol*. 2011;22(9):2057–67.
4. García-Campelo R, Bernabé R, Cobo M, Corral J, Coves J, Dómine M, et al. SEOM clinical guidelines for the treatment of non-small cell lung cancer (NSCLC) 2015. *Clin Transl Oncol*. 2015;17(12):1020–9.
5. Majeed U, Manochakian R, Zhao Y, Lou Y. Targeted therapy in advanced non-small cell lung cancer: current advances and future trends. *J Hematol Oncol*. 2021;14(1):1–20.
6. Polverino A, Coxon A, Starnes C, Diaz Z, DeMelfi T, Wang L, et al. AMG 706, an oral, multikinase inhibitor that selectively targets vascular endothelial growth factor, platelet-derived growth factor, and kit receptors, potently inhibits angiogenesis and induces regression in tumor xenografts. *Can Res*. 2006;66(17):8715–21.
7. Lu T, Yang X, Huang Y, Zhao M, Li M, Ma K, et al. Trends in the incidence, treatment, and survival of patients with lung cancer in the last four decades. *Cancer Manag Res*. 2019;11:943.
8. Travis WD, Brambilla E, Nicholson AG, Yatabe Y, Austin JH, Beasley MB, et al. The 2015 World Health Organization classification of lung tumors: impact of genetic, clinical and radiologic advances since the 2004 classification. *J Thorac Oncol*. 2015;10(9):1243–60.
9. Clark SB, Alsubait S. *Non small cell lung cancer*. Petersburg: StatPearls Publishing Treasure Island; 2021.
10. Al-Haideri M, Tondok SB, Safa SH, Maleki AH, Rostami S, Jalil AT, et al. CAR-T cell combination therapy: the next revolution in cancer treatment. *Cancer Cell Int*. 2022;22(1):365.
11. Esmaeilzadeh A, Jafari D, Tahmasebi S, Elahi R, Khosh E. Immune-based therapy for COVID-19. In: Rezaei N, editor. *Coronavirus disease—COVID-19*. Cham: Springer; 2021.
12. Sadeghi S, Tehrani FR, Tahmasebi S, Shafiee A, Hashemi SM. Exosome engineering in cell therapy and drug delivery. *Inflammopharmacology*. 2023;31(1):145–69.
13. Daniels MG, Bowman RV, Yang IA, Govindan R, Fong KM. An emerging place for lung cancer genomics in 2013. *J Thorac Dis*. 2013;5(Suppl 5):S491.
14. Dearden S, Stevens J, Wu Y-L, Blowers D. Mutation incidence and coincidence in non small-cell lung cancer: meta-analyses by ethnicity and histology (mutMap). *Ann Oncol*. 2013;24(9):2371–6.
15. Pao W, Girard N. New driver mutations in non-small-cell lung cancer. *Lancet Oncol*. 2011;12(2):175–80.
16. Upadhya A, Yadav KS, Misra A. Targeted drug therapy in non-small cell lung cancer: clinical significance and possible solutions—part I. *Expert Opin Drug Deliv*. 2021;18(1):73–102.
17. Alexander M, Kim SY, Cheng H. Update 2020: management of nonsmall cell lung cancer. *Lung*. 2020;198(6):897–907.
18. Lin JJ, Shaw AT. Resisting resistance: targeted therapies in lung cancer. *Trends Cancer*. 2016;2(7):350–64.
19. Tian W, Cao C, Shu L, Wu F. Anti-angiogenic therapy in the treatment of non-small cell lung cancer. *Onco Targets Ther*. 2020;13:12113.
20. Zhao Y, Guo S, Deng J, Shen J, Du F, Wu X, et al. VEGF/VEGFR-targeted therapy and immunotherapy in non-small cell lung cancer: targeting the tumor microenvironment. *Int J Biol Sci*. 2022;18(9):3845–58.
21. Gabrilovich D, Ishida T, Oyama T, Ran S, Kravtsov V, Nadaf S, et al. Vascular endothelial growth factor inhibits the development of dendritic cells and dramatically affects the differentiation of multiple hematopoietic lineages in vivo: presented in part at the keystone symposium "cellular and molecular biology of dendritic cells", Santa Fe, NM, March 3–9, 1998, and at the annual meeting of the American association for cancer research, March 28–April 1, 1998. *Blood, J Am Soc Hematol*. 1998;92(11):4150–66.
22. Gabrilovich DI, Chen HL, Girgis KR, Cunningham HT, Meny GM, Nadaf S, et al. Production of vascular endothelial growth factor by human tumors inhibits the functional maturation of dendritic cells. *Nat Med*. 1996;2(10):1096–103.
23. Voron T, Colussi O, Marcheteau E, Pernot S, Nizard M, Pointet A-L, et al. VEGF-A modulates expression of inhibitory checkpoints on CD8+ T cells in tumors. *J Exp Med*. 2015;212(2):139–48.
24. Guo X-F, Zhang Y-Y, Kang J, Dou Q-H, Zhu X-F. A bispecific decoy receptor VEGFR-EGFR/Fc binding EGF-like ligands and VEGF shows potent antitumor efficacy. *J Drug Target*. 2022;30(3):302–12.
25. Duan J, Yang Z, Liu D, Shi Y. Clinical efficacy of bevacizumab combined with gemcitabine and cisplatin combination chemotherapy in the treatment of advanced non-small cell lung cancer. *J Buon*. 2018;23(5):1402–6.

26. Zhou K, Zhao S, Guo W, Ding L. Efficacy and safety of erlotinib combined with bevacizumab in the treatment of non-small cell lung cancer: a systematic review and meta-analysis. *Medicine.* 2020;99(3):e18771.
27. Mustachio LM, Chelariu-Raicu A, Szekvolgyi L, Roszik J. Targeting KRAS in cancer: promising therapeutic strategies. *Cancers.* 2021;13(6):1204.
28. Simanshu DK, Nissley DV, McCormick F. RAS proteins and their regulators in human disease. *Cell.* 2017;170(1):17–33.
29. Moore AR, Rosenberg SC, McCormick F, Malek S. RAS-targeted therapies: is the undruggable drugged? *Nat Rev Drug Discovery.* 2020;19(8):533–52.
30. Ostrem JM, Peters U, Sos ML, Wells JA, Shokat KM. K-Ras(G12C) inhibitors allosterically control GTP affinity and effector interactions. *Nature.* 2013;503(7477):548–51.
31. Ostrem JM, Shokat KM. Direct small-molecule inhibitors of KRAS: from structural insights to mechanism-based design. *Nat Rev Drug Discov.* 2016;15(11):771–85.
32. Jacobs F, Cani M, Malapelle U, Novello S, Napoli VM, Bironzo P. Targeting KRAS in NSCLC: old failures and new options for "non-G12c" patients. *Cancers.* 2021;13(24):6332.
33. Koga T, Suda K, Fujino T, Ohara S, Hamada A, Nishino M, et al. KRAS secondary mutations that confer acquired resistance to KRAS G12C inhibitors, sotorasib and adagrasib, and overcoming strategies: insights from in vitro experiments. *J Thorac Oncol.* 2021;16(8):1321–32.
34. Hong D, Bang Y, Barlesi F, Durm G, Falchook G, Govindan R, et al. 1257O Durability of clinical benefit and biomarkers in patients (pts) with advanced non-small cell lung cancer (NSCLC) treated with AMG 510 (sotorasib). *Ann Oncol.* 2020;31:S812.
35. Hong DS, Fakih MG, Strickler JH, Desai J, Durm GA, Shapiro GI, et al. KRAS(G12C) Inhibition with sotorasib in advanced solid tumors. *N Engl J Med.* 2020;383(13):1207–17.
36. Reck M, Spira A, Besse B, Wolf J, Skoulidis F, Borghaei H, et al. 1416TiP CodeBreak 200: A phase III multicenter study of sotorasib (AMG 510), a KRAS (G12C) inhibitor, versus docetaxel in patients with previously treated advanced non-small cell lung cancer (NSCLC) harboring KRAS p. G12C mutation. *Ann Oncol.* 2020;31:S894–5.
37. Skoulidis F, Li BT, Dy GK, Price TJ, Falchook GS, Wolf J, et al. Sotorasib for lung cancers with KRAS pG12C mutation. *New Engl J Med.* 2021;384(25):2371–81.
38. Fell JB, Fischer JP, Baer BR, Blake JF, Bouhana K, Briere DM, et al. Identification of the clinical development candidate MRTX849, a covalent KRASG12C inhibitor for the treatment of cancer. *J Med Chem.* 2020;63(13):6679–93.
39. Hallin J, Engstrom LD, Hargis L, Calinisan A, Aranda R, Briere DM, et al. The KRASG12C inhibitor MRTX849 provides insight toward therapeutic susceptibility of KRAS-mutant cancers in mouse models and patients therapeutic insight from the KRASG12C inhibitor MRTX849. *Cancer Discov.* 2020;10(1):54–71.
40. Janne P, Rybkin II, Spira A, Riely GJ, Papadopoulos KP, Sabari J, editors. *KRYSTAL-1: updated safety and efficacy data with adagrasib (MRTX849) in NSCLC with KRASG12C mutation from a phase 1/2 study.* Presented at the 32nd EORTC-NCI-AACR symposium; 2020.
41. Arteaga CL. The epidermal growth factor receptor: from mutant oncogene in nonhuman cancers to therapeutic target in human neoplasia. *J Clin Oncol.* 2001;19(18 Suppl):32s–40s.
42. Yarden Y, Sliwkowski MX. Untangling the ErbB signalling network. *Nat Rev Mol Cell Biol.* 2001;2(2):127–37.
43. Hynes NE, Lane HA. ERBB receptors and cancer: the complexity of targeted inhibitors. *Nat Rev Cancer.* 2005;5(5):341–54.
44. Jackman DM, Yeap BY, Sequist LV, Lindeman N, Holmes AJ, Joshi VA, et al. Exon 19 deletion mutations of epidermal growth factor receptor are associated with prolonged survival in non-small cell lung cancer patients treated with gefitinib or erlotinib. *Clin Cancer Res.* 2006;12(13):3908–14.
45. Rosell R, Moran T, Queralt C, Porta R, Cardenal F, Camps C, et al. Screening for epidermal growth factor receptor mutations in lung cancer. *N Engl J Med.* 2009;361(10):958–67.
46. Kris MG, Natale RB, Herbst RS, Lynch TJ Jr, Prager D, Belani CP, et al. Efficacy of gefitinib, an inhibitor of the epidermal growth factor receptor tyrosine kinase, in symptomatic patients with non-small cell lung cancer: a randomized trial. *JAMA.* 2003;290(16):2149–58.
47. Sequist LV, Yang JC, Yamamoto N, O'Byrne K, Hirsh V, Mok T, et al. Phase III study of afatinib or cisplatin plus pemetrexed in patients with metastatic lung adenocarcinoma with EGFR mutations. *J Clin Oncol.* 2013;31(27):3327–34.
48. Wu YL, Zhou C, Hu CP, Feng J, Lu S, Huang Y, et al. Afatinib versus cisplatin plus gemcitabine for first-line treatment of Asian patients with advanced non-small-cell lung cancer harbouring EGFR mutations (LUX-Lung 6): an open-label, randomised phase 3 trial. *Lancet Oncol.* 2014;15(2):213–22.

49. Iwahara T, Fujimoto J, Wen D, Cupples R, Bucay N, Arakawa T, et al. Molecular characterization of ALK, a receptor tyrosine kinase expressed specifically in the nervous system. *Oncogene*. 1997;14(4):439–49.
50. Lemmon MA, Schlessinger J. Cell signaling by receptor tyrosine kinases. *Cell*. 2010;141(7):1117–34.
51. Palmer RH, Vernersson E, Grabbe C, Hallberg B. Anaplastic lymphoma kinase: signalling in development and disease. *Biochem J*. 2009;420(3):345–61.
52. Tan DS, Araújo A, Zhang J, Signorovitch J, Zhou ZY, Cai X, et al. Comparative efficacy of ceritinib and crizotinib as initial ALK-targeted therapies in previously treated ADVANCED NSCLC: an adjusted comparison with external controls. *J Thorac Oncol*. 2016;11(9):1550–7.
53. Gandhi L, Jänne PA. Crizotinib for ALK-rearranged non-small cell lung cancer: a new targeted therapy for a new target. *Clin Cancer Res*. 2012;18(14):3737–42.
54. Choi YL, Soda M, Yamashita Y, Ueno T, Takashima J, Nakajima T, et al. EML4-ALK mutations in lung cancer that confer resistance to ALK inhibitors. *N Engl J Med*. 2010;363(18):1734–9.
55. Kang J, Zhang XC, Chen HJ, Zhong WZ, Xu Y, Su J, et al. Complex ALK fusions are associated with better prognosis in advanced non-small cell lung cancer. *Front Oncol*. 2020;10:596937.
56. Koivunen JP, Mermel C, Zejnullahu K, Murphy C, Lifshits E, Holmes AJ, et al. EML4-ALK fusion gene and efficacy of an ALK kinase inhibitor in lung cancer. *Clin Cancer Res Off J Am Assoc Cancer Res*. 2008;14(13):4275–83.
57. Rodig SJ, Mino-Kenudson M, Dacic S, Yeap BY, Shaw A, Barletta JA, et al. Unique clinicopathologic features characterize ALK-rearranged lung adenocarcinoma in the western population. *Clin Cancer Res*. 2009;15(16):5216–23.
58. Shaw AT, Felip E, Bauer TM, Besse B, Navarro A, Postel-Vinay S, et al. Lorlatinib in non-small-cell lung cancer with ALK or ROS1 rearrangement: an international, multicentre, open-label, single-arm first-in-man phase 1 trial. *Lancet Oncol*. 2017;18(12):1590–9.
59. Cui P, Li R, Huang Z, Wu Z, Tao H, Zhang S, et al. Comparative effectiveness of pembrolizumab vs nivolumab in patients with recurrent or advanced NSCLC. *Sci Rep*. 2020;10(1):13160.
60. Huang S, van Duijnhoven SMJ, Sijts A, van Elsas A. Bispecific antibodies targeting dual tumor-associated antigens in cancer therapy. *J Cancer Res Clin Oncol*. 2020;146(12):3111–22.
61. Felip E, Kim D, Mehra R, Tan DS, Chow L, Camidge DR, et al. Efficacy and safety of ceritinib in patients (pts) with advanced anaplastic lymphoma kinase (ALK)-rearranged (ALK+) non-small cell lung cancer (NSCLC): an update of ASCEND-1. *Ann Oncol*. 2014;25:iv456.
62. Solomon BJ, Mok T, Kim DW, Wu YL, Nakagawa K, Mekhail T, et al. Firstline crizotinib versus chemotherapy in ALK-positive lung cancer. *N Engl J Med*. 2014;371(23):2167–77.
63. Gainor JF, Tan DS, De Pas T, Solomon BJ, Ahmad A, Lazzari C, et al. Progression-free and overall survival in ALK-positive NSCLC patients treated with sequential crizotinib and ceritinib. *Clin Cancer Res*. 2015;21(12):2745–52.
64. Thomas RK. Overcoming drug resistance in ALK-rearranged lung cancer. *N Engl J Med*. 2014;370(13):1250–1.
65. Kim ES, Salgia R. MET pathway as a therapeutic target. *J Thorac Oncol*. 2009;4(4):444–7.
66. Shaw AT, Solomon BJ, Chiari R, Riely GJ, Besse B, Soo RA, et al. Lorlatinib in advanced ROS1-positive non-small-cell lung cancer: a multicentre, open-label, single-arm, phase 1–2 trial. *Lancet Oncol*. 2019;20(12):1691–701.
67. Rikova K, Guo A, Zeng Q, Possemato A, Yu J, Haack H, et al. Global survey of phosphotyrosine signaling identifies oncogenic kinases in lung cancer. *Cell*. 2007;131(6):1190–203.
68. Bergethon K, Shaw AT, Ou SH, Katayama R, Lovly CM, McDonald NT, et al. ROS1 rearrangements define a unique molecular class of lung cancers. *J Clin Oncol*. 2012;30(8):863–70.
69. Huber KV, Salah E, Radic B, Gridling M, Elkins JM, Stukalov A, et al. Stereospecific targeting of MTH1 by (S)-crizotinib as an anticancer strategy. *Nature*. 2014;508(7495):222–7.
70. Shaw AT, Ou SH, Bang YJ, Camidge DR, Solomon BJ, Salgia R, et al. Crizotinib in ROS1-rearranged non-small-cell lung cancer. *N Engl J Med*. 2014;371(21):1963–71.
71. Awad MM, Katayama R, McTigue M, Liu W, Deng YL, Brooun A, et al. Acquired resistance to crizotinib from a mutation in CD74-ROS1. *N Engl J Med*. 2013;368(25):2395–401.
72. Gainor JF, Tseng D, Yoda S, Dagogo-Jack I, Friboulet L, Lin JJ, et al. Patterns of metastatic spread and mechanisms of resistance to crizotinib in ROS1-positive non-small-cell lung cancer. *JCO Precis Oncol*. 2017. https://doi.org/10.1200/PO.17.00063.
73. Menichincheri M, Ardini E, Magnaghi P, Avanzi N, Banfi P, Bossi R, et al. Discovery of entrectinib: a new 3-aminoindazole as a potent anaplastic lymphoma kinase (ALK), c-ros oncogene 1 kinase (ROS1), and pan-tropomyosin receptor kinases (Pan-TRKs) inhibitor. *J Med Chem*. 2016;59(7):3392–408.
74. Ardini E, Menichincheri M, Banfi P, Bosotti R, De Ponti C, Pulci R, et al. Entrectinib, a Pan-TRK, ROS1, and ALK inhibitor with activity in multiple molecularly defined cancer indications. *Mol Cancer Ther*. 2016;15(4):628–39.

75. Drilon A, Siena S, Dziadziuszko R, Barlesi F, Krebs MG, Shaw AT, et al. Entrectinib in ROS1 fusion-positive non-small-cell lung cancer: integrated analysis of three phase 1–2 trials. *Lancet Oncol.* 2020;21(2):261–70.
76. Doebele RC, Perez L, Trinh H, Martinec M, Martina R, Riehl T, et al. Comparative effectiveness analysis between entrectinib clinical trial and crizotinib real-world data in ROS1+ NSCLC. *J Comp Eff Res.* 2021;10(17):1271–82.
77. Zou HY, Li Q, Engstrom LD, West M, Appleman V, Wong KA, et al. PF-06463922 is a potent and selective next-generation ROS1/ALK inhibitor capable of blocking crizotinib-resistant ROS1 mutations. *Proc Natl Acad Sci USA.* 2015;112(11):3493–8.
78. Bollag G, Tsai J, Zhang J, Zhang C, Ibrahim P, Nolop K, et al. Vemurafenib: the first drug approved for BRAF-mutant cancer. *Nat Rev Drug Discov.* 2012;11(11):873–86.
79. Subbiah V, Taylor M, Lin J, Hu M, Ou S-HI, Brose MS, et al. Abstract CT043: highly potent and selective RET inhibitor, BLU-667, achieves proof of concept in a phase I study of advanced RET-altered solid tumors. *Cancer Res.* 2018;78(13_Supplement):CT043-CT.
80. Tsai J, Lee JT, Wang W, Zhang J, Cho H, Mamo S, et al. Discovery of a selective inhibitor of oncogenic B-Raf kinase with potent antimelanoma activity. *Proc Natl Acad Sci.* 2008;105(8):3041–6.
81. Hyman DM, Puzanov I, Subbiah V, Faris JE, Chau I, Blay J-Y, et al. Vemurafenib in multiple nonmelanoma cancers with BRAF V600 mutations. *N Engl J Med.* 2015;373(8):726–36.
82. Planchard D, Kim TM, Mazieres J, Quoix E, Riely G, Barlesi F, et al. Dabrafenib in patients with BRAFV600E-positive advanced non-small-cell lung cancer: a single-arm, multicentre, open-label, phase 2 trial. *Lancet Oncol.* 2016;17(5):642–50.
83. King AJ, Arnone MR, Bleam MR, Moss KG, Yang J, Fedorowicz KE, et al. Dabrafenib; preclinical characterization, increased efficacy when combined with trametinib, while BRAF/MEK tool combination reduced skin lesions. *PLOS One.* 2013;8(7):e67583.
84. Planchard D, Besse B, Groen HJ, Souquet P-J, Quoix E, Baik CS, et al. Dabrafenib plus trametinib in patients with previously treated BRAFV600Emutant metastatic non-small cell lung cancer: an open-label, multicentre phase 2 trial. *Lancet Oncol.* 2016;17(7):984–93.
85. Planchard D, Smit EF, Groen HJ, Mazieres J, Besse B, Helland Å, et al. Dabrafenib plus trametinib in patients with previously untreated BRAFV600E-mutant metastatic non-small-cell lung cancer: an openlabel, phase 2 trial. *Lancet Oncol.* 2017;18(10):1307–16.
86. Food Drug Administration. *FDA grants regular approval to dabrafenib and trametinib combination for metastatic NSCLC with BRAF V600E mutation.* FDA; 2017.
87. Joshi M, Rice SJ, Liu X, Miller B, Belani CP. Trametinib with or without vemurafenib in BRAF mutated non-small cell lung cancer. *PLOS One.* 2015;10(2):e0118210.
88. Gainor JF, Curigliano G, Kim D-W, Lee DH, Besse B, Baik CS, et al. Pralsetinib for RET fusion-positive non-small-cell lung cancer (ARROW): a multi-cohort, open-label, phase 1/2 study. *Lancet Oncol.* 2021;22(7):959–69.
89. Curigliano G, Subbiah V, Gainor J, Lee D, Taylor M, Zhu V, et al. Treatment with BLU-667, a potent and selective RET inhibitor, provides rapid clearance of ctDNA in patients with RET-altered non-small cell lung cancer (NSCLC) and thyroid cancer. *Ann Oncol.* 2019;30:v790.
90. Griesinger F, Curigliano G, Thomas M, Subbiah V, Baik CS, Tan DSW, et al. Safety and efficacy of pralsetinib in RET fusion-positive non-small cell lung cancer including as first-line therapy: update from the ARROW trial. *Ann Oncol.* 2022. https://doi.org/10.1016/j.annonc.2022.08.002.
91. Gautschi O, Milia J, Filleron T, Wolf J, Carbone DP, Owen D, et al. Targeting RET in patients with RET-rearranged lung cancers: results from the global, multicenter RET registry. *J Clin Oncol.* 2017;35(13):1403–10.
92. Organ SL, Tsao MS. An overview of the c-MET signaling pathway. *Ther Adv Med Oncol.* 2011;3(1 Suppl):S7–19.
93. Uchikawa E, Chen Z, Xiao GY, Zhang X, Bai XC. Structural basis of the activation of c-MET receptor. *Nat Commun.* 2021;12(1):4074.
94. Comoglio PM, Trusolino L, Boccaccio C. Known and novel roles of the MET oncogene in cancer: a coherent approach to targeted therapy. *Nat Rev Cancer.* 2018;18(6):341–58.
95. Bubendorf L, Dafni U, Schöbel M, Finn SP, Tischler V, Sejda A, et al. Prevalence and clinical association of MET gene overexpression and amplification in patients with NSCLC: results from the European thoracic oncology platform (ETOP) lungscape project. *Lung Cancer.* 2017;111:143–9.
96. Drilon A, Cappuzzo F, Ou SI, Camidge DR. Targeting MET in lung cancer: will expectations finally be met? *J Thorac Oncol.* 2017;12(1):15–26.

97. Hamilton G, Rath B. Mesenchymal-epithelial transition and circulating tumor cells in small cell lung cancer. *Adv Exp Med Biol.* 2017;994:229–45.
98. Soltermann A, Tischler V, Arbogast S, Braun J, Probst-Hensch N, Weder W, et al. Prognostic significance of epithelial-mesenchymal and mesenchymal-epithelial transition protein expression in non-small cell lung cancer. *Clin Cancer Res.* 2008;14(22):7430–7.
99. Hardy-Werbin M, Del Rey-Vergara R, Galindo-Campos MA, Moliner L, Arriola E. MET inhibitors in small cell lung cancer: from the bench to the bedside. *Cancers.* 2019;11(10):1404.
100. Backes A, Zech B, Felber B, Klebl B, Müller G. Small-molecule inhibitors binding to protein kinase. Part II: the novel pharmacophore approach of type II and type III inhibition. *Expert Opin Drug Discov.* 2008;3(12):1427–49.
101. Gherardi E, Birchmeier W, Birchmeier C, Vande WG. Targeting MET in cancer: rationale and progress. *Nat Rev Cancer.* 2012;12(2):89–103.
102. Reungwetwattana T, Liang Y, Zhu V, Ou SI. The race to target MET exon 14 skipping alterations in non-small cell lung cancer: the why, the how, the who, the unknown, and the inevitable. *Lung Cancer.* 2017;103:27–37.
103. Drusbosky LM, Dawar R, Rodriguez E, Ikpeazu CV. Therapeutic strategies in METex14 skipping mutated non-small cell lung cancer. *J Hematol Oncol.* 2021;14(1):129.
104. Awad MM, Oxnard GR, Jackman DM, Savukoski DO, Hall D, Shivdasani P, et al. MET Exon 14 mutations in non-small-cell lung cancer are associated with advanced age and stage-dependent MET genomic amplification and c-Met overexpression. *J Clin Oncol.* 2016;34(7):721–30.
105. Garcia-Robledo JE, Rosell R, Ruíz-Patiño A, Sotelo C, Arrieta O, Zatarain-Barrón L, et al. KRAS and MET in non-small-cell lung cancer: two of the new kids on the "drivers" block. *Ther Adv Respir Dis.* 2022;16:17534666211066064.
106. Agwa ES, Ma PC. Targeting the MET receptor tyrosine kinase in nonsmall cell lung cancer: emerging role of tivantinib. *Cancer Manag Res.* 2014;6:397–404.
107. Comprehensive molecular profiling of lung adenocarcinoma. *Nature.* 2014;511(7511):543–50.
108. Wei XW, Gao X, Zhang XC, Yang JJ, Chen ZH, Wu YL, et al. Mutational landscape and characteristics of ERBB2 in non-small cell lung cancer. *Thoracic Cancer.* 2020;11(6):1512–21.
109. Lv Q, Meng Z, Yu Y, Jiang F, Guan D, Liang C, et al. Molecular mechanisms and translational therapies for human epidermal receptor 2 POSITIVE breast cancer. *Int J Molr Sci.* 2016;17(12):2095.
110. Pellegrini C, Falleni M, Marchetti A, Cassani B, Miozzo M, Buttitta F, et al. HER-2/Neu alterations in non-small cell lung cancer: a comprehensive evaluation by real time reverse transcription-PCR, fluorescence in situ hybridization, and immunohistochemistry. *Clin Cancer Res.* 2003;9(10 Pt 1):3645–52.
111. Tan D, Deeb G, Wang J, Slocum HK, Winston J, Wiseman S, et al. HER-2/neu protein expression and gene alteration in stage I-IIIA non-small-cell lung cancer: a study of 140 cases using a combination of high throughput tissue microarray, immunohistochemistry, and fluorescent in situ hybridization. *Diagn Mol Pathol.* 2003;12(4):201–11.
112. Dziadziuszko R, Smit EF, Dafni U, Wolf J, Wasąg B, Biernat W, et al. Afatinib in NSCLC With HER2 Mutations: results of the prospective, open-label phase II NICHE trial of European thoracic oncology platform (ETOP). *J Thorac Oncol.* 2019;14(6):1086–94.
113. Ekman S. HER2: defining a Neu target in non-small-cell lung cancer. *Ann Oncol.* 2019;30(3):353–5.
114. Prelaj A, Bottiglieri A, Proto C, Russo GL, Signorelli D, Ferrara R, et al. 1388P Poziotinib in advanced NSCLC with EGFR or HER2 exon 20 insertion mutation: Initial results from a single site expanded access program. *Ann Oncol.* 2020;31:S882.
115. Ross HJ, Blumenschein GR Jr, Aisner J, Damjanov N, Dowlati A, Garst J, et al. Randomized phase II multicenter trial of two schedules of lapatinib as first- or second-line monotherapy in patients with advanced or metastatic non-small cell lung cancer. *Clin Cancer Res.* 2010;16(6):1938–49.
116. Wang Y, Jiang T, Qin Z, Jiang J, Wang Q, Yang S, et al. HER2 exon 20 insertions in non-small-cell lung cancer are sensitive to the irreversible pan-HER receptor tyrosine kinase inhibitor pyrotinib. *Ann Oncol.* 2019;30(3):447–55.
117. Janne PA, Neal JW, Camidge DR, Spira A, Piotrowska Z, Horn L, et al. Antitumor activity of TAK-788 in NSCLC with EGFR exon 20 insertions. *Ann Oncol.* 2019;30:108.
118. Barthélémy P, Leblanc J, Goldbarg V, Wendling F, Kurtz JE. Pertuzumab: development beyond breast cancer. *Anticancer Res.* 2014;34(4):1483–91.
119. Mazières J, Barlesi F, Filleron T, Besse B, Monnet I, Beau-Faller M, et al. Lung cancer patients with HER2 mutations treated with chemotherapy and HER2-targeted drugs: results from the European EUHER2 cohort. *Ann Oncol.* 2016;27(2):281–6.

120. Network CGAR. Comprehensive molecular profiling of lung adenocarcinoma. *Nature*. 2014; 511(7511):543.
121. Peters S, Stahel R, Bubendorf L, Bonomi P, Villegas A, Kowalski DM, et al. Trastuzumab emtansine (T-DM1) in patients with previously treated HER2-overexpressing metastatic non-small cell lung cancer: efficacy, safety, and biomarkers. *Clin Cancer Res*. 2019;25(1):64–72.
122. Tan AC. Targeting the PI3K/Akt/mTOR pathway in non-small cell lung cancer (NSCLC). *Thorac Cancer*. 2020;11(3):511–18.
123. Papadimitrakopoulou V. Development of PI3K/AKT/mTOR pathway inhibitors and their application in personalized therapy for non-small cell lung cancer. *J Thorac Oncol*. 2012;7(8):1315–26.
124. Scheffler M, Bos M, Gardizi M, König K, Michels S, Fassunke J, et al. PIK3CA mutations in non-small cell lung cancer (NSCLC): genetic heterogeneity, prognostic impact and incidence of prior malignancies. *Oncotarget*. 2015;6(2):1315–26.
125. Engelman JA, Chen L, Tan X, Crosby K, Guimaraes AR, Upadhyay R, et al. Effective use of PI3K and MEK inhibitors to treat mutant Kras G12D and PIK3CA H1047R murine lung cancers. *Nat Med*. 2008;14(12):1351–6.
126. Guo Y, Du J, Kwiatkowski DJ. Molecular dissection of AKT activation in lung cancer cell lines. *Mol Cancer Res*. 2013;11(3):282–93.
127. Le Grand M, Berges R, Pasquier E, Montero M-P, Borge L, Carrier A, et al. Akt targeting as a strategy to boost chemotherapy efficacy in non-small cell lung cancer through metabolism suppression. *Sci Rep*. 2017;7(1):45136.
128. Ma Z, Zhu L, Luo X, Zhai S, Li P, Wang X. Perifosine enhances mTORC1- targeted cancer therapy by activation of GSK3β in NSCLC cells. *Cancer Biol Ther*. 2012;13(11):1009–17.
129. Hirai H, Sootome H, Nakatsuru Y, Miyama K, Taguchi S, Tsujioka K, et al. MK-2206, an Allosteric Akt inhibitor, enhances antitumor efficacy by standard chemotherapeutic agents or molecular targeted drugs in vitro and in vivoMK-2206 sensitizes tumors to chemotherapy. *Mol Cancer Ther*. 2010;9(7):1956–67.
130. Lara PN Jr, Longmate J, Mack PC, Kelly K, Socinski MA, Salgia R, et al. Phase II study of the AKT inhibitor MK-2206 plus erlotinib in patients with advanced non-small cell lung cancer who previously progressed on erlotinib. *Clin Cancer Res*. 2015;21(19):4321–6.
131. Ekman S, Wynes MW, Hirsch FR. The mTOR Pathway in lung cancer and implications for therapy and biomarker analysis. *J Thorac Oncol*. 2012;7(6):947–53.
132. Kim J-O, Kim K-H, Song IS, Cheon K-S, Kim O-H, Lee SC, et al. Potentiation of the anticancer effects of everolimus using a dual mTORC1/2 inhibitor in hepatocellular carcinoma cells. *Oncotarget*. 2017;8(2):2936–48.
133. Sparks CA, Guertin DA. Targeting mTOR: prospects for mTOR complex 2 inhibitors in cancer therapy. *Oncogene*. 2010;29(26):3733–44.
134. Cui JJ, Tran-Dubé M, Shen H, Nambu M, Kung PP, Pairish M, et al. Structure based drug design of crizotinib (PF-02341066), a potent and selective dual inhibitor of mesenchymal-epithelial transition factor (c-MET) kinase and anaplastic lymphoma kinase (ALK). *J Med Chem*. 2011;54(18):6342–63.
135. Fang W, Huang Y, Gu W, Gan J, Wang W, Zhang S, et al. PI3K-AKT-mTOR pathway alterations in advanced NSCLC patients after progression on EGFR-TKI and clinical response to EGFR-TKI plus everolimus combination therapy. *Transl Lung Cancer Res*. 2020;9(4):1258–67.
136. Komiya T, Memmott RM, Blumenthal GM, Bernstein W, Ballas MS, De Chowdhury R, et al. A phase I/II study of pemetrexed with sirolimus in advanced, previously treated non-small cell lung cancer. *Transl Lung Cancer Res*. 2019;8(3):247–57.
137. Dolly SO, Wagner AJ, Bendell JC, Kindler HL, Krug LM, Seiwert TY, et al. Phase I Study of apitolisib (GDC-0980), dual phosphatidylinositol-3-kinase and mammalian target of rapamycin kinase inhibitor, in patients with advanced solid tumors. *Clin Cancer Res*. 2016;22(12):2874–84.
138. Lastwika KJ, Wilson W 3rd, Li QK, Norris J, Xu H, Ghazarian SR, et al. Control of PD-L1 expression by oncogenic activation of the AKT-mTOR pathway in non-small cell lung cancer. *Can Res*. 2016;76(2):227–38.
139. Garon EB, Rizvi NA, Hui R, Leighl N, Balmanoukian AS, Eder JP, et al. Pembrolizumab for the treatment of non-small-cell lung cancer. *N Engl J Med*. 2015;372(21):2018–28.
140. Keir ME, Butte MJ, Freeman GJ, Sharpe AH. PD-1 and its ligands in tolerance and immunity. *Annu Rev Immunol*. 2008;26(1):677–704.
141. Ji M, Liu Y, Li Q, Li X-D, Zhao W-Q, Zhang H, et al. PD-1/PD-L1 pathway in non-small-cell lung cancer and its relation with EGFR mutation. *J Transl Med*. 2015;13:5.

142. Spranger S, Spaapen RM, Zha Y, Williams J, Meng Y, Ha TT, et al. Up-regulation of PD-L1, IDO, and T(regs) in the melanoma tumor microenvironment is driven by CD8(+) T cells. *Sci Transl Med*. 2013;5(200):200ra116.
143. Beatty GL, Gladney WL. Immune escape mechanisms as a guide for cancer immunotherapy. *Clin Cancer Res*. 2015;21(4):687–92.
144. Hudson K, Cross N, Jordan-Mahy N, Leyland R. The extrinsic and intrinsic roles of PD-L1 and Its receptor PD-1: implications for immunotherapy treatment. *Front Immunol*. 2020. https://doi.org/10.3389/fimmu.2020.568931.
145. Moya-Horno I, Viteri S, Karachaliou N, Rosell R. Combination of immunotherapy with targeted therapies in advanced non-small cell lung cancer (NSCLC). *Ther Adv Med Oncol*. 2018;10:1758834017745012.
146. Carbone DP, Reck M, Paz-Ares L, Creelan B, Horn L, Steins M, et al. Firstline nivolumab in stage IV or recurrent non-small-cell lung cancer. *N Engl J Med*. 2017;376(25):2415–26.
147. Reck M, Rodríguez-Abreu D, Robinson AG, Hui R, Csőszi T, Fülöp A, et al. Five-Year outcomes with pembrolizumab versus chemotherapy for metastatic non-small-cell lung cancer with PD-L1 tumor proportion score ≥ 50. *J Clin Oncol*. 2021;39(21):2339–49.
148. Matsubara T, Seto T, Takamori S, Fujishita T, Toyozawa R, Ito K, et al. Anti-PD-1 monotherapy for advanced NSCLC patients with older age or those with poor performance status. *Onco Targets Ther*. 2021;14:1961–8.
149. Ilie M, Hofman P. Atezolizumab in advanced non-small cell lung cancer. *J Thorac Dis*. 2017;9(10):3603–6.
150. Compared to chemotherapy, cemiplimab monotherapy dramatically improved OS and PFS in patients with advanced non-small-cell lung cancer whose PD-L1 was at least 50%.224.
151. Sezer A, Kilickap S, Gümüş M, Bondarenko I, Özgüroğlu M, Gogishvili M, et al. Cemiplimab monotherapy for first-line treatment of advanced non-small-cell lung cancer with PD-L1 of at least 50%: a multicentre, open-label, global, phase 3, randomised, controlled trial. *Lancet*. 2021;397(10274):592–604.
152. Yamamoto T, Tsukita Y, Katagiri Y, Matsushita H, Umezawa R, Ishikawa Y, et al. Durvalumab after chemoradiotherapy for locally advanced non-small cell lung cancer prolonged distant metastasis-free survival, progression-free survival and overall survival in clinical practice. *BMC Cancer*. 2022;22(1):364.
153. Antonia SJ, Villegas A, Daniel D, Vicente D, Murakami S, Hui R, et al. Overall survival with durvalumab after chemoradiotherapy in stage III NSCLC. *New Engl J Med*. 2018;379(24):2342–50.
154. Huang G, Sun X, Liu D, Zhang Y, Zhang B, Xiao G, et al. The efficacy and safety of anti-PD-1/PD-L1 antibody therapy versus docetaxel for pretreated advanced NSCLC: a meta-analysis. *Oncotarget*. 2018;9(3):4239–48.
155. Park K, Özgüroğlu M, Vansteenkiste J, Spigel D, Yang JCH, Ishii H, et al. Avelumab versus docetaxel in patients with platinum-treated advanced NSCLC: 2-year follow-up from the JAVELIN lung 200 phase 3 trial. *J Thorac Oncol*. 2021;16(8):1369–78.
156. Paulsen EE, Kilvaer TK, Rakaee M, Richardsen E, Hald SM, Andersen S, et al. CTLA-4 expression in the non-small cell lung cancer patient tumor microenvironment: diverging prognostic impact in primary tumors and lymph node metastases. *Cancer Immunol Immunother*. 2017;66(11):1449–61.
157. Buchbinder EI, Desai A. CTLA-4 and PD-1 pathways: similarities, differences, and implications of their inhibition. *Am J Clin Oncol*. 2016;39(1):98–106.
158. Tarhini AA, Kirkwood JM. Tremelimumab (CP-675,206): a fully human anticytotoxic T lymphocyte-associated antigen 4 monoclonal antibody for treatment of patients with advanced cancers. *Expert Opin Biol Ther*. 2008;8(10):1583–93.
159. Lynch TJ, Bondarenko I, Luft A, Serwatowski P, Barlesi F, Chacko R, et al. Ipilimumab in combination with paclitaxel and carboplatin as first-line treatment in stage IIIB/IV non-small-cell lung cancer: results from a randomized, double-blind, multicenter phase II study. *J Clin Oncol*. 2012;30(17):2046–54.
160. Zhao Y, Yang W, Huang Y, Cui R, Li X, Li B. Evolving roles for targeting CTLA-4 in cancer immunotherapy. *Cell Physiol Biochem*. 2018;47(2):721–34.
161. Khozin S, Blumenthal GM, Jiang X, He K, Boyd K, Murgo A, et al. U.S. food and drug administration approval summary: erlotinib for the first-line treatment of metastatic non-small cell lung cancer with epidermal growth factor receptor exon 19 deletions or exon 21 (L858R) substitution mutations. *Oncologist*. 2014;19(7):774–9.
162. Kazandjian D, Blumenthal GM, Yuan W, He K, Keegan P, Pazdur R. FDA approval of gefitinib for the treatment of patients with metastatic EGFR mutation-positive non-small cell lung cancer. *Clin Cancer Res*. 2016;22(6):1307–12.
163. Greig SL. Osimertinib: first global approval. *Drugs*. 2016;76(2):263–73.

164. Kazandjian D, Blumenthal GM, Chen HY, He K, Patel M, Justice R, et al. FDA approval summary: crizotinib for the treatment of metastatic non-small cell lung cancer with anaplastic lymphoma kinase rearrangements. *Oncologist*. 2014;19(10):e5–11.
165. Syed YY. Lorlatinib: first global approval. *Drugs*. 2019;79(1):93–8.
166. Markham A. Tepotinib: first approval. *Drugs*. 2020;80(8):829–33.
167. Sakai H, Morise M, Kato T, Matsumoto S, Sakamoto T, Kumagai T, et al. Tepotinib in patients with NSCLC harbouring MET exon 14 skipping: Japanese subset analysis from the phase II VISION study. *Jpn J Clin Oncol*. 2021;51(8):1261–8.
168. FDA approves atezolizumab for first-line treatment of metastatic NSCLC with high PD-L1 expression U.S. food and drug administration. 2020. https://www.fda.gov/drugs/resources-information-approved-drugs/fda-approves-atezolizumab-first-line-treatment-metastatic-nsclc-highpd-l1-expression,cemiplimab.
169. Rizvi NA, Cho BC, Reinmuth N, Lee KH, Luft A, Ahn MJ, et al. Durvalumab with or without tremelimumab vs standard chemotherapy in first-line treatment of metastatic non-small cell lung cancer: the MYSTIC phase 3 randomized clinical trial. *JAMA Oncol*. 2020;6(5):661–74.
170. Hellmann MD, Ciuleanu TE, Pluzanski A, Lee JS, Otterson GA, Audigier-Valette C, et al. Nivolumab plus ipilimumab in lung cancer with a high tumor mutational burden. N *Engl J Med*. 2018;378(22):2093–104.
171. Khuri F, Harvey R, Saba N, Owonikoko T, Kauh J, Shin D, et al. Everolimus, an mTOR inhibitor, in combination with docetaxel for recurrent/refractory NSCLC: a phase I study. *J Clin Oncol*. 2009;27(15_suppl):8060.
172. Gerlinger M, Rowan AJ, Horswell S, Math M, Larkin J, Endesfelder D, et al. Intratumor heterogeneity and branched evolution revealed by multiregion sequencing. *N Engl J Med*. 2012;366(10):883–92.
173. Pacheco JM. Mobocertinib: a potential treatment for NSCLC with EGFR exon 20 insertions. *Cancer Discov*. 2021;11(7):1617–9.
174. Rivera-Concepcion J, Uprety D, Adjei AA. Challenges in the use of targeted therapies in non-small cell lung cancer. *Cancer Res Treat*. 2022;54(2):315–29.
175. Johnson DH, Fehrenbacher L, Novotny WF, Herbst RS, Nemunaitis JJ, Jablons DM, et al. Randomized phase II trial comparing bevacizumab plus carboplatin and paclitaxel with carboplatin and paclitaxel alone in previously untreated locally advanced or metastatic non-small-cell lung cancer. *J Clin Oncol*. 2004;22(11):2184–91.
176. Gonzalvez F, Vincent S, Baker TE, Gould AE, Li S, Wardwell SD, et al. Mobocertinib (TAK-788): a targeted inhibitor of EGFR exon 20 insertion mutants in non-small cell lung cancer. *Cancer Discov*. 2021;11(7):1672–87.
177. Pirker R, Pereira JR, Szczesna A, von Pawel J, Krzakowski M, Ramlau R, et al. Cetuximab plus chemotherapy in patients with advanced nonsmall- cell lung cancer (FLEX): an open-label randomised phase III trial. *Lancet*. 2009;373(9674):1525–31.
178. Haluska P, Shaw HM, Batzel GN, Yin D, Molina JR, Molife LR, et al. Phase I dose escalation study of the anti insulin-like growth factor-I receptor monoclonal antibody CP-751,871 in patients. *Clin Cancer Res*. 2007;13(19):5834–40.
179. Heymach JV, Johnson BE, Prager D, Csada E, Roubec J, Pesek M, et al. Randomized, placebo-controlled phase II study of vandetanib plus docetaxel in previously treated non small-cell lung cancer. *J Clin Oncol*. 2007;25(27):4270–7.
180. Socinski MA, Novello S, Brahmer JR, Rosell R, Sanchez JM, Belani CP, et al. Multicenter, phase II trial of sunitinib in previously treated, advanced non-small-cell lung cancer. *J Clin Oncol*. 2008;26(4):650–6.
181. Soria JC, Shepherd FA, Douillard JY, Wolf J, Giaccone G, Crino L, et al. Efficsacy of everolimus (RAD001) in patients with advanced NSCLC previously treated with chemotherapy alone or with chemotherapy and EGFR inhibitors. *Ann Oncol*. 2009;20(10):1674–81.
182. Blumenschein GR Jr, Paulus R, Curran WJ, Robert F, Fossella F, Werner- Wasik M, et al. Phase II study of cetuximab in combination with chemoradiation in patients with stage IIIA/B non-small-cell lung cancer: RTOG 0324. *J Clin Oncol*. 2011;29(17):2312–8.
183. Govindan R, Bogart J, Stinchcombe T, Wang X, Hodgson L, Kratzke R, et al. Randomized phase II study of pemetrexed, carboplatin, and thoracic radiation with or without cetuximab in patients with locally advanced unresectable non-small-cell lung cancer: cancer and leukemia group B trial 30407. *J Clin Oncol*. 2011;29(23):3120–5.
184. Gatzemeier U, Pluzanska A, Szczesna A, Kaukel E, Roubec J, De Rosa F, et al. Phase III study of erlotinib in combination with cisplatin and gemcitabine in advanced non-small-cell lung cancer: the tarceva lung cancer investigation trial. *J Clin Oncol*. 2007;25(12):1545–52.

185. Lovly CM, Salama AK, Salgia R. *Tumor heterogeneity and therapeutic resistance*. American Society of Clinical Oncology Educational Book American Society of Clinical Oncology Annual Meeting; 2016;35:e585–93.
186. Pan Y, Deng C, Qiu Z, Cao C, Wu F. The resistance mechanisms and treatment strategies for ALK-rearranged non-small cell lung cancer. *Front Oncol*. 2021;11:713530.
187. Weng CH, Chen LY, Lin YC, Shih JY, Lin YC, Tseng RY, et al. Epithelial—mesenchymal transition (EMT) beyond EGFR mutations per se is a common mechanism for acquired resistance to EGFR TKI. *Oncogene*. 2019;38(4):455–68.
188. Zhu C, Wei Y, Wei X. AXL receptor tyrosine kinase as a promising anti-cancer approach: functions, molecular mechanisms and clinical applications. *Mol Cancer*. 2019;18(1):153.
189. Perez-Soler R. Rash as a surrogate marker for efficacy of epidermal growth factor receptor inhibitors in lung cancer. *Clin Lung Cancer*. 2006;8(Suppl 1):S7–14.
190. Robinson ES, Khankin EV, Karumanchi SA, Humphreys BD. Hypertension induced by vascular endothelial growth factor signaling pathway inhibition: mechanisms and potential use as a biomarker. *Semin Nephrol*. 2010;30(6):591–601.
191. Chu CY, Choi J, Eaby-Sandy B, Langer CJ, Lacouture ME. Osimertinib: a novel dermatologic adverse event profile in patients with lung cancer. *Oncologist*. 2018;23(8):891–9.
192. Herbst RS, Sun Y, Eberhardt WE, Germonpré P, Saijo N, Zhou C, et al. Vandetanib plus docetaxel versus docetaxel as second-line treatment for patients with advanced non-small-cell lung cancer (ZODIAC): a doubleblind, randomised, phase 3 trial. *Lancet Oncol*. 2010;11(7):619–26.

Index

Note: Page numbers in *italics* indicate a figure and page numbers in **bold** indicate a table on the corresponding page.

Q

R

S

T

U

V

W

X

Y

Z

For Product Safety Concerns and Information please contact our
EU representative GPSR@taylorandfrancis.com Taylor & Francis
Verlag GmbH, Kaufingerstraße 24, 80331 München, Germany